Soil Chemistry

IUPAC Periodic Table of the Elements

Key:

atomic number
Symbol
name
conventional atomic weight
standard atomic weight

1	2	3	4	5	6	7	8	9	10	11	12	13	14	15	16	17	18
1 **H** hydrogen 1.008 [1.0078, 1.0082]																	2 **He** helium 4.0026
3 **Li** lithium 6.94 [6.938, 6.997]	4 **Be** beryllium 9.0122											5 **B** boron 10.81 [10.806, 10.821]	6 **C** carbon 12.011 [12.009, 12.012]	7 **N** nitrogen 14.007 [14.006, 14.008]	8 **O** oxygen 15.999 [15.999, 16.000]	9 **F** fluorine 18.998	10 **Ne** neon 20.180
11 **Na** sodium 22.990	12 **Mg** magnesium 24.305 [24.304, 24.307]											13 **Al** aluminium 26.982	14 **Si** silicon 28.085 [28.084, 28.086]	15 **P** phosphorus 30.974	16 **S** sulfur 32.06 [32.059, 32.076]	17 **Cl** chlorine 35.45 [35.446, 35.457]	18 **Ar** argon 39.95 [39.792, 39.963]
19 **K** potassium 39.098	20 **Ca** calcium 40.078(4)	21 **Sc** scandium 44.956	22 **Ti** titanium 47.867	23 **V** vanadium 50.942	24 **Cr** chromium 51.996	25 **Mn** manganese 54.938	26 **Fe** iron 55.845(2)	27 **Co** cobalt 58.933	28 **Ni** nickel 58.693	29 **Cu** copper 63.546(3)	30 **Zn** zinc 65.38(2)	31 **Ga** gallium 69.723	32 **Ge** germanium 72.630(8)	33 **As** arsenic 74.922	34 **Se** selenium 78.971(8)	35 **Br** bromine 79.904 [79.901, 79.907]	36 **Kr** krypton 83.798(2)
37 **Rb** rubidium 85.468	38 **Sr** strontium 87.62	39 **Y** yttrium 88.906	40 **Zr** zirconium 91.224(2)	41 **Nb** niobium 92.906	42 **Mo** molybdenum 95.95	43 **Tc** technetium	44 **Ru** ruthenium 101.07(2)	45 **Rh** rhodium 102.91	46 **Pd** palladium 106.42	47 **Ag** silver 107.87	48 **Cd** cadmium 112.41	49 **In** indium 114.82	50 **Sn** tin 118.71	51 **Sb** antimony 121.76	52 **Te** tellurium 127.60(3)	53 **I** iodine 126.90	54 **Xe** xenon 131.29
55 **Cs** caesium 132.91	56 **Ba** barium 137.33	57-71 lanthanoids	72 **Hf** hafnium 178.49(2)	73 **Ta** tantalum 180.95	74 **W** tungsten 183.84	75 **Re** rhenium 186.21	76 **Os** osmium 190.23(3)	77 **Ir** iridium 192.22	78 **Pt** platinum 195.08	79 **Au** gold 196.97	80 **Hg** mercury 200.59	81 **Tl** thallium 204.38 [204.38, 204.39]	82 **Pb** lead 207.2	83 **Bi** bismuth 208.98	84 **Po** polonium	85 **At** astatine	86 **Rn** radon
87 **Fr** francium	88 **Ra** radium	89-103 actinoids	104 **Rf** rutherfordium	105 **Db** dubnium	106 **Sg** seaborgium	107 **Bh** bohrium	108 **Hs** hassium	109 **Mt** meitnerium	110 **Ds** darmstadtium	111 **Rg** roentgenium	112 **Cn** copernicium	113 **Nh** nihonium	114 **Fl** flerovium	115 **Mc** moscovium	116 **Lv** livermorium	117 **Ts** tennessine	118 **Og** oganesson

57 **La** lanthanum 138.91	58 **Ce** cerium 140.12	59 **Pr** praseodymium 140.91	60 **Nd** neodymium 144.24	61 **Pm** promethium	62 **Sm** samarium 150.36(2)	63 **Eu** europium 151.96	64 **Gd** gadolinium 157.25(3)	65 **Tb** terbium 158.93	66 **Dy** dysprosium 162.50	67 **Ho** holmium 164.93	68 **Er** erbium 167.26	69 **Tm** thulium 168.93	70 **Yb** ytterbium 173.05	71 **Lu** lutetium 174.97
89 **Ac** actinium	90 **Th** thorium 232.04	91 **Pa** protactinium 231.04	92 **U** uranium 238.03	93 **Np** neptunium	94 **Pu** plutonium	95 **Am** americium	96 **Cm** curium	97 **Bk** berkelium	98 **Cf** californium	99 **Es** einsteinium	100 **Fm** fermium	101 **Md** mendelevium	102 **No** nobelium	103 **Lr** lawrencium

For notes and updates to this table, see www.iupac.org. This version is dated 1 December 2018.
Copyright © 2018 IUPAC, the International Union of Pure and Applied Chemistry.

Soil Chemistry

5th Edition

Daniel G. Strawn
Hinrich L. Bohn
George A. O'Connor

WILEY Blackwell

This edition first published 2020
© 2020 John Wiley & Sons Ltd

Edition History
Wiley (1e, 1979); Wiley-Interscience (2e, 1985); Wiley (3e, 2001); Wiley-Blackwell (4e, 2015)

Registered Office(s)
John Wiley & Sons, Inc., 111 River Street, Hoboken, NJ 07030, USA
John Wiley & Sons Ltd, The Atrium, Southern Gate, Chichester, West Sussex, PO19 8SQ, UK

Editorial Office
9600 Garsington Road, Oxford, OX4 2DQ, UK

For details of our global editorial offices, customer services, and more information about Wiley products visit us at www.wiley.com.

Wiley also publishes its books in a variety of electronic formats and by print-on-demand. Some content that appears in standard print versions of this book may not be available in other formats.

Library of Congress Cataloging-in-Publication Data
Names: Strawn, Daniel, author. | Bohn, Hinrich L., 1934– author. |
    O'Connor, George A., 1944– author.
Title: Soil chemistry.
Description: Fifth edition / Daniel G. Strawn (University of Idaho),
    Hinrich L Bohn, George A O'Connor. | Hoboken, NJ : John Wiley & Sons,
    [2020] | Includes index.
Identifiers: LCCN 2019035987 (print) | LCCN 2019035988 (ebook) | ISBN
    9781119515180 (hardback) | ISBN 9781119515159 (adobe pdf) | ISBN
    9781119515258 (epub)
Subjects: LCSH: Soil chemistry.
Classification: LCC S592.5 .B63 2020 (print) | LCC S592.5 (ebook) | DDC
    631.4/1–dc23
LC record available at https://lccn.loc.gov/2019035987
LC ebook record available at https://lccn.loc.gov/2019035988

Cover image: Periodic table: © ALFRED PASIEKA/SCIENCE PHOTO LIBRARY/Getty Images
Soil image: Lithochrome_color by The John Kelly Collection / Soil Science downloaded via Flickr is licensed under CC BY
Cover design by Wiley

Set in 10/13pt Palatino by SPi Global, Pondicherry, India

10 9 8 7 6 5 4 3 2 1

CONTENTS

PREFACE TO FIFTH EDITION

This new edition of *Soil Chemistry* contains more examples, more illustrations, more details of calculations, and reorganized material within the chapters, including nearly 100 new equations and 51 new figures. Our goal remains to provide an introductory text for senior-level soil chemistry students. This requires compromise on depth of explanation of the topics so that the main points are not lost on students, while providing sufficient information for explanation. We strive to achieve this balance throughout. Students wanting more details can review the more than 200 references provided in figure and table captions and at the end of the chapters. Additional details can also be found in textbooks in chemistry, geology, pedology, geochemistry, colloid science, soil chemistry, and soil fertility.

The textbook's focus is on species and reaction processes of chemicals in soils, with applications to environmental and agricultural issues. Topics in the 13 chapters range from discussion of fundamental chemical processes to review of properties and reactions of chemicals in the environment.

In producing this new edition, we have corrected a few errors from the previous edition. However, with the addition of new material, introduction of new errors is inevitable. Please send notifications of errors to lead author Dan Strawn at the University of Idaho. An erratum will be made available on Dr. Strawn's website.

PREFACE TO FOURTH EDITION

The goal of the First Edition of *Soil Chemistry* published in 1976 was to provide a textbook for soil science students to learn about chemical processes occurring in soils. The First Edition, and subsequent Second Edition (1985) and Third Edition (2003), focused on explaining the principles of chemical reactions in soils and the nature of soil solids. Intricacies and advanced details of theories were omitted for clarity. In the Fourth Edition, Dr. Dan Strawn, a professor at University of Idaho for 15 years, has led a revision of the classic text, working closely with original authors Dr. Hank Bohn (Professor Emeritus, University of Arizona) and Dr. George O'Connor (Professor, University of Florida). The collaboration has resulted in a new version for this classic text.

The Fourth Edition of *Soil Chemistry* is a major revision, including updated figures, tables, examples, and explanations; but it maintains the goal of the early editions in that it is written at a level to teach chemical properties and processes to undergraduate students. Graduate students and professionals, however, will also find the textbook useful as a resource to understand and review concepts, as well a good source to look up soil chemical properties listed in the many tables. To improve readability for undergraduates, citations have been omitted from the discussions.

Many of the ideas covered, however, are emphasized in graphs and tables from the literature, for which citations are included.

The topics covered in *Soil Chemistry,* Fourth Edition, are presented in the same order as previous versions. We have added a chapter on surface charge properties (Chapter 9), and the adsorption chapters of the previous editions have been completely reorganized—explanation of cation, anion, and organic chemical adsorption processes are presented in Chapter 10, and quantitative modeling of adsorption processes is presented in Chapter 11. New to this version are special topics boxes that provide highlights of topics, historical information, and examples. Enhanced discussion of carbon cycling, new theories of SOM formation and structure, details of soil redox properties, and information on chemicals of emerging concern have been added. In each chapter, key words are bolded; students should use key words as study aides to ensure they understand main concepts.

We have made every effort to minimize errors. It is said that *with each edit, only half the errors are found*, and thus, 100% accuracy is fleeting. Please forward errors or questions to Dr. Dan Strawn at the University of Idaho. We will make an erratum available.

ACKNOWLEDGMENTS

We are like dwarves perched on the shoulders of giants, and thus we are able to see more and farther than the latter. And this is not at all because of the acuteness of our sight or the stature of our body, but because we are carried aloft and elevated by the magnitude of the giants.

John of Salisbury (1159 AD)
after Bernard of Chartres, circa 1115 AD.

As this well-known quote elegantly illustrates, the ideas presented in this book are not my own, but are compiled from the years of research and teachings of soil scientists, chemists, and physicists. I drew from many resources to write this textbook, including other textbooks, review articles, and research articles. I acknowledge the *giants* who have benefited my understanding.

Every effort was made to present the work of others in a careful and meaningful way to illustrate and explain the discipline of soil chemistry. In several cases, authors provided clarification, as well as encouragement to use their data and figures. I am grateful for their generosity.

Hank Bohn was unavailable to assist in the Fifth Edition. He was instrumental in the Fourth Edition overhaul, and his spirit and contributions live on in the Fifth Edition.

I appreciate George O'Connor's collaboration in preparing the Fifth Edition. The content greatly benefited from his keen eye for relevance and attention to detail.

Writing a textbook is a labor of love. I gratefully acknowledge those who inspired my passion for soil science at University of California, Davis, where I was an undergraduate student, and Dr. Don Sparks at the University of Delaware, where I completed my PhD.

I am indebted to the students at the University of Idaho who inspired me to work hard to teach better; this text is for them. I am also grateful to my colleagues in the Department of Soil and Water Systems at the University of Idaho for providing me a home to practice my profession.

The thrill of writing a textbook is soon overshadowed by the seemingly infinite time sink. I am grateful to Kelly, Isabell, and Serena for their patience and support as I completed this project.

Dan Strawn, 2019
University of Idaho

Soil Chemistry

1 INTRODUCTION TO SOIL CHEMISTRY

No one regards what is at his feet; we all gaze at the stars. Quintus Ennius (239–169 BCE)

Heaven is beneath our feet as well as above our heads. Henry David Thoreau (1817–1862)

The earth was made so various that the mind of desultory man, studious of change and pleased with novelty, might be indulged. William Cowper (The Task, 1780)

The Nation that destroys its soil destroys itself. Franklin Delano Roosevelt (1937)

1.1 The soil chemistry discipline

The above quotations illustrate how differently humans see the soil that gives them life and sustenance. In recent decades, great strides in understanding the importance of soils for healthy ecosystems and food production have been made, but the need for preservation and improved utilization of soil resources remains one of society's greatest challenges. Success requires a better understanding of soil processes.

Soil is a complex mixture of inorganic and organic solids, air, water, solutes, microorganisms, plant roots, and other types of biota that influence each other, making soil processes complex and dynamic (**Figure 1.1**). For example, air and water weather rocks to form soil minerals and release ions; microorganisms catalyze many soil weathering reactions; and plant roots absorb and exude inorganic and organic chemicals that change the distribution and solubility of ions. Although it is difficult to separate soil processes, soil scientists have organized themselves into subdisciplines that study physical, biological, and chemical processes, soil formation and distribution, and specialists that study applied soil science topics such as soil fertility.

The discipline of soil chemistry has traditionally focused on abiotic transformations of soil constituents, such as changes in oxidation state of elements and

Soil Chemistry, Fifth Edition. Daniel G. Strawn, Hinrich L. Bohn, and George A. O'Connor.
© 2020 John Wiley & Sons Ltd. Published 2020 by John Wiley & Sons Ltd.

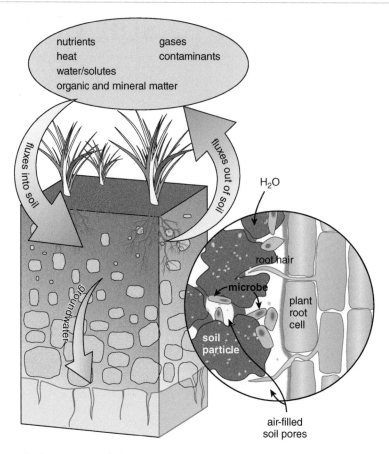

Figure 1.1 Soils are composed of air, water, solids, ions, organic compounds, and biota. The soil in the microscopic view shows soil particles (e.g., aggregates of minerals and organic matter), air and water in pore spaces, microbes, and a plant root. Fluxes of material or energy into and out of the soil drive biogeochemical reactions, making soils dynamic. Fluxes can be to the atmosphere, eroded or leached offsite into surface water, or percolated to groundwater.

association of ions with surfaces. Chemical reactions in soils often lead to changes between solid, liquid, and gas states that dramatically influence the availability of chemicals for plant uptake and losses from soil that in turn are important aspects of fate and transport of nutrients and contaminants in the environment. With the ever-increasing pressures to produce more food and extract resources such as timber, oil, and water from the environment, pressures on soil resources are increasing. Addressing these pressures and challenges requires detailed knowledge and understanding of soil processes. Modern soil chemistry strives to understand interactions occurring within soils, such as interactions between soil microbes and soil minerals.

The focus of soil chemistry is chemical reactions and processes occurring in soils. A chemical reaction defines the transformation of reactants to products. For example, potassium availability for plant uptake in

soils is often controlled by cation exchange reactions on clay minerals, such as:

$$K^+ + Na\text{-clay} = Na^+ + K\text{-clay} \qquad (1.1)$$

where reactants are aqueous K^+ and Na^+ adsorbed on a clay mineral (Na-clay), and products are aqueous Na^+ and K^+ adsorbed on a clay mineral (K-clay). The adsorption reaction exchanges ions between aqueous solution in the soil pore and the soil solids (clay mineral in this case) and is thus a solid–solution interface reaction. Cation exchange reactions are a hallmark of soil chemistry.

A goal of soil chemistry is predicting whether a reaction will proceed, which can be done using thermodynamic calculations. Soils are complex, however, and predicting the fate of chemicals in the environment requires including multiple competing reaction pathways occurring simultaneously. In addition, many soil reactions are slow and fail to reach equilibrium before the system undergoes a perturbation, making prediction of chemical species a moving target. The complexity and dynamic aspect of soils make understanding chemical reactions in nature a challenging problem, but, over the past 150 years, great advances have been made. The goal of this book is to present the current state of knowledge about soil chemical processes so that students can use them to understand the environmental fate of chemicals.

1.2 Historical background

About 2500 years ago, the senate of ancient Athens debated soil productivity and voiced the same worries about sustaining and increasing soil productivity heard today: *Can this productivity continue, or is soil productivity being exhausted?*

In 1790, Malthus noticed that the human population was increasing exponentially, whereas food production was increasing arithmetically. He predicted that by 1850 food demands would overtake food production, and people would be starving and fighting like rats for morsels of food. Although such predictions have not come to fruition, there are real challenges to feeding the world's increasing population, especially considering predicted changes in climate that will have significant impacts to food production systems and regional populations.

It is encouraging that food productivity has increased faster than Malthus predicted. Earth now feeds the largest human population ever, and a larger fraction of that population is better fed than ever before. Whether this can continue, and at what price to the environment, is an open question. One part of the answer lies in wisely managing soil resources so that food production can continue to increase and ecosystem functions can be maintained. Sustainable management requires careful use of soil and knowledge of soil processes. Soil chemistry is an important subdiscipline required for understanding soil processes.

Agricultural practices that increase crop growth, such as planting legumes, application of animal manure and forest litter, crop rotation, and liming were known to the Chinese 3000 years ago. These practices were also learned by the Greeks and Romans, and appeared in the writings of Varro, Cato, Columella, and Pliny, but were unexplained. Little progress on technology to increase and maintain soil productivity was made thereafter for almost 1500 years because of lack of understanding of plant–soil processes, and because of undue dependence on deductive reasoning. Deduction is applying preconceived ideas, broad generalities, and accepted truths to problems without testing if the preconceived ideas and accepted truths are valid. One truth accepted for many centuries and derived from the Greeks was that all matter was composed of earth, air, fire, and water; a weak basis, as we later learned, on which to increase knowledge.

In the early fifteenth century, Sir Francis Bacon promoted the idea that the scientific method is the best approach to gaining new knowledge: observe, hypothesize, test and measure, derive ideas from data, test these ideas again, and report findings. The scientific method brought progress in understanding our world, but the progress in understanding soil's role in plant productivity was minimal in the ensuing three centuries.

Palissy (1563) proposed that plant ash came from the soil, and when added back to the soil could be reabsorbed by plants. Plat (1590) proposed that salts from decomposing organic matter dissolved in water and were absorbed by plants to facilitate growth. Glauber (1650) thought that saltpeter (Na, K nitrates) was the key to plant nutrition by the soil. Kuelbel (ca. 1700) believed that humus was the principle of vegetation. Boerhoeve (ca. 1700) believed that plants absorbed the "juices of the earth." While these early theorists proposed reasonable relationships between plants and soils, accurate experimental design and proof was lacking, and their proposals were incomplete and inaccurate.

Van Helmont, a sixteenth-century scientist, tried to test the ideas of plant–soil nutrient relationships. He planted a willow shoot in a pail of soil and covered the pail so that dust could not enter. He carefully measured the amount of water added. After five years, the tree had gained 75.4 kilograms. The weight of soil in the pot was still the same as the starting weight, less about two ounces (56 g). Van Helmont disregarded the 56 grams as what we would today call experimental error. He concluded that the soil contributed nothing to the nutrition of the plant because there was no loss of mass, and that plants needed only water for their sustenance. Although he followed the scientific method as best he could, he came to a wrong conclusion. Many experiments in nature still go afoul because of incomplete experimental design and inadequate measurement of all essential experimental variables.

John Woodruff's (1699) experimental design was much better than Van Helmont's. He grew plants using rainwater, river water, and sewage water for irrigation, and added garden *mould* to the soils. The more solutes and solids in the growth medium – the "dirtier" the water – the better the plants grew, implying that something in soil improved plant growth. The idea developed that the organic fraction of the soil supplied the plant's needs.

In 1840 Justus von Liebig persuasively advanced the idea that inorganic chemicals were key to plant nutrition and that an input-output chemical budget should be maintained in the soil. Liebig's theory was most probably based on Carl Sprengel's work in 1820–1830 that showed that mineral salts, rather than humus or soil organic matter, were the source of plant growth.

Liebig's influence was so strong that subsequent findings by Boussingault (1865) showing that more nitrogen existed in plants than was applied to the soil, implying nitrogen fixation, was disregarded for many years. Microbial nitrogen fixation did not fit into the Sprengel-Liebig model.

Soil chemistry was first recognized as distinct from soil fertility in 1850 when J.T. Way, at Rothamsted, England, reported on the ability of soils to exchange cations (**Figure 1.2**). Their work suggested that soils could be studied apart from plants to discover important aspects for soil fertility. Van Bemmelen followed with studies on the nature of clay minerals in soils and popularized the theory of adsorption (published in 1863). These founding fathers of soil chemistry stimulated the beginning of much scientific inquiry into the nature and properties of soils that continues to this day.

Despite the significant advances in understanding soils and environmental processes, environmental complexity is too great for a single discipline to fully understand. Scientific training necessarily tends to

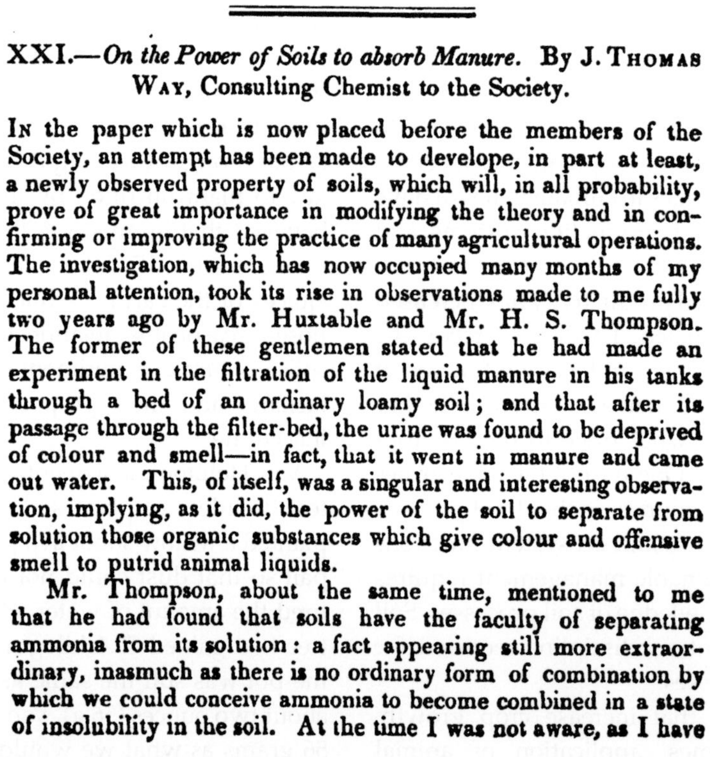

XXI.—*On the Power of Soils to absorb Manure.* By J. THOMAS WAY, Consulting Chemist to the Society.

IN the paper which is now placed before the members of the Society, an attempt has been made to develope, in part at least, a newly observed property of soils, which will, in all probability, prove of great importance in modifying the theory and in confirming or improving the practice of many agricultural operations. The investigation, which has now occupied many months of my personal attention, took its rise in observations made to me fully two years ago by Mr. Huxtable and Mr. H. S. Thompson. The former of these gentlemen stated that he had made an experiment in the filtration of the liquid manure in his tanks through a bed of an ordinary loamy soil; and that after its passage through the filter-bed, the urine was found to be deprived of colour and smell—in fact, that it went in manure and came out water. This, of itself, was a singular and interesting observation, implying, as it did, the power of the soil to separate from solution those organic substances which give colour and offensive smell to putrid animal liquids.

Mr. Thompson, about the same time, mentioned to me that he had found that soils have the faculty of separating ammonia from its solution : a fact appearing still more extraordinary, inasmuch as there is no ordinary form of combination by which we could conceive ammonia to become combined in a state of insolubility in the soil. At the time I was not aware, as I have

Figure 1.2 Snippet of paper authored by J. Thomas Way in the 1850 *Journal of the Royal Agriculture Society of England* describing the discovery of the ability of soil to _absorb_ ammonia from manure. It is now known that ammonium exchanges for other cations on the soil clay particles, which is an _adsorption_ reaction. Source: Way (1850).

specialize, learning more and more about less and less. Nature, however, is complex, and scientists of various disciplines apply their background to the whole environment with mixed results. Eighteenth-century naturalist Alexander von Humboldt popularized the concept that natural systems are interconnected, and proposed a link between soils, flora, and fauna in many essays and books published from 1800 to 1825. von Humboldt also proposed that human activity could have devastating effects on ecosystem functions – a radical idea for his time.

Specialists often try to compartmentalize natural systems, and bring along biases, one of which is that their area of study is the most important. Atmospheric scientists, for example, naturally believe that the atmosphere is the most important part of the environment. The authors of this book are no different. We argue, without apology, that the soil plays the central and dominant role in the environment. However, the truth is more in line with von Humboldt's ideas proposed over two centuries ago: soils are an intricate part of the web of nature, and a soil's characteristics are intimately tied to the plants, microbes, atmosphere, geology, climate, and landscape surrounding it. The linkages and influences go both ways. The unique relationship between soils and plants and microbe communities associated with them is referred to as an *edaphic* quality.

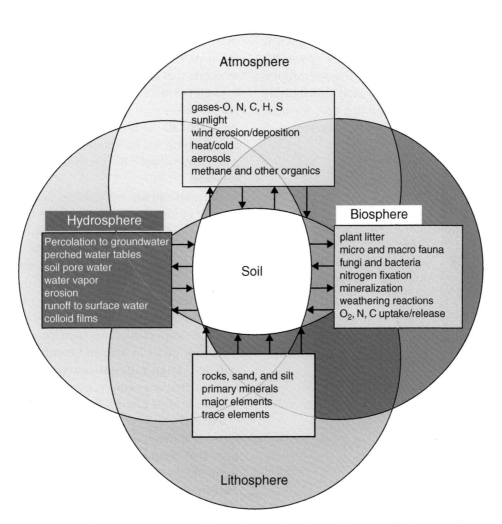

Figure 1.3 Soils interface with Earth's other spheres. Biogeochemical cycling within the soil influences flows of chemicals and energy into the hydrosphere, biosphere, and atmosphere. Arrows between the different spheres and the soil indicate important transformations.

1.3   The soil environment

Soils are the *skin of the earth*, and interface with the atmosphere, hydrosphere, lithosphere, and biosphere (**Figure 1.3**). The interaction of Earth's spheres within soil results in a mixture of solid, liquid, gas, and biota, called the **pedosphere**. A fifth Earth component is the anthrosphere, which describes human's interaction and influence on the environment. The **critical zone** is a concept that encompasses all life-supporting parts of the earth, including soils, groundwater, and vegetation. Regardless of how the environment is compartmentalized for study, chemical processes occurring in the soil are important aspects affecting healthy and sustainable environments.

In this section, we discuss the relationships between soil chemicals, the biosphere, soil solid phases, the hydrosphere, and the soil atmosphere. A typical soil is composed of ~50% solid, and ~50% pore space; the exact amount varies as a function of the soil properties, such as aggregation, particle size distribution, and so on (**Figure 1.4**). Throughout this text, the term *soil chemical* is used as a general term that refers to all the different types of chemicals occurring in soil, including ions, liquids, gases, minerals, soil organic matter, and salts.

1.3.1   Soil chemical and biological interfaces

A basic tenet of biology is that life evolves and changes to adapt to the environment, driven by reproductive success. Because soils have a significant impact on environmental conditions, there is a direct link between evolutionary processes and soils. Some even theorize that the first forms of life evolved from interactions of carbon and nitrogen with clay minerals of the type commonly observed in soils; where clays are hypothesized to have catalyzed the first organic prebiotic polymers. While such a theory is controversial, one cannot deny the role of soils in maintaining life and the environment. Even marine life is affected by chemicals and minerals that are transferred from the land to the sea by water flow or airborne dust particles. Thus, chemical processes in soils are critical for maintenance and growth of all life forms, and soils are locked in a partnership with the biosphere, hydrosphere, lithosphere, and atmosphere in providing critical **ecosystem services**.

The atmosphere, biosphere, and hydrosphere are weakly buffered against change in chemical composition and fluctuate when perturbed. Soils, in contrast, better resist chemical changes and are a steadying influence on the other three environmental compartments. Detrimental changes in the hydrosphere, atmosphere, and biosphere due to human activities often occur because the soil is bypassed, causing imbalances in important ecosystem processes that would otherwise occur in soil, and would therefore be better buffered. High nutrient concentrations entering surface waters, for example, bypass soil's nutrient cycles, and cause algal blooms that deteriorate water quality.

Ion exchange on mineral and organic matter surfaces, and mineral dissolution and precipitation reactions that occur in soils are soil reactions that regulate elemental availability to organisms (**Figure 1.5**), mediated by the root–soil interface called the **rhizosphere** (**Figure 1.6**). Soils act as sources and sinks of most of the essential nutrients required by organisms. Plants, for example, derive almost all their essential nutrients from the soil; with the exception of carbon, hydrogen,

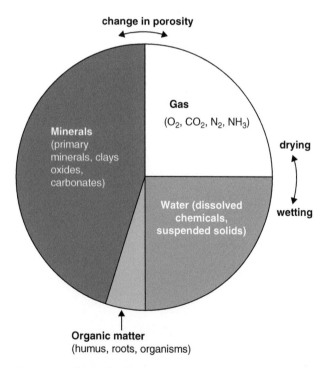

Figure 1.4 Typical volumetric composition of soils. Soil gases and solution fill the pore spaces at different ratios, depending on soil moisture content. The ratio of solid to pore spaces is controlled by the soil porosity.

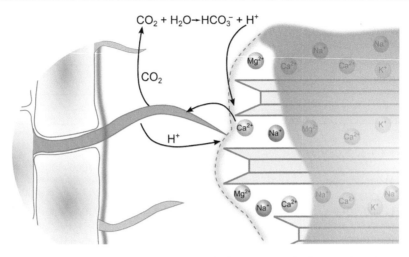

Figure 1.5 Plant root interaction with a soil aggregate composed of clay minerals. Carbon dioxide released by root respiration acidifies the rhizosphere. Cation exchange of protons and Ca^{2+} is depicted, where the Ca^{2+} is taken up by the plant root. Plant roots release protons to maintain charge balance when absorbing cationic nutrients. The protons released by the plant root and from CO_2 respiration cause weathering reactions that release additional cations from the soil minerals.

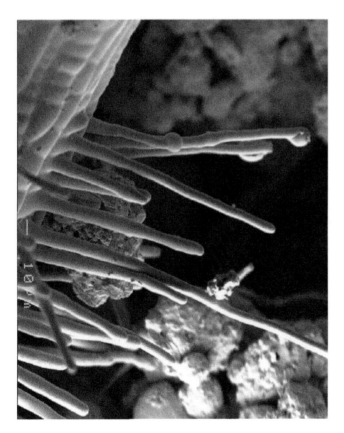

Figure 1.6 Cryo-SEM image of a plant root showing the root hairs interacting with the soil aggregates. ~10-micron diameter droplets of root exudate are seen on the tips of some of the root hairs. SEM image courtesy of Margret McCully, Australian National University.

oxygen, and minor amounts of nitrogen and sulfur gases (NO_x, NH_3, and SO_2) directly absorbed from the atmosphere by leaves.

Plants depend on soil's ability to buffer essential nutrients, and otherwise retard nutrient losses, such as leaching out of the plant–root zone. Thus, the entire terrestrial ecosystem depends on the **biogeochemical cycling** of potassium, calcium, magnesium, phosphorus, nitrogen, and micronutrients within soils. Under natural conditions, the major factors affecting ion availability to plants are: (1) ion concentration in the soil solution; (2) the degree of ion interaction with, and rate of release from, soil solid phases; (3) the activity of soil microorganisms; and (4) discrimination by the plant root during ion uptake. This book is concerned primarily with the first two factors: the soil solution and ion interaction with the solid phase.

Soil is an oxygen-silicon-aluminum-iron matrix containing relatively small amounts of the essential elements. But the small amounts of ions held by that matrix are vital to plant growth. **Table 1.1** shows representative contents of important elements in soils and plants. Concentrations of elements in plants depend on plant species, and soil concentrations. Concentrations of elements in soils depend on the **soil formation factors**, especially parent material, weathering processes, and biologically driven fluxes of elements. As a result of the many factors that influence

Table 1.1 Concentration ranges (mg kg^{-1} dry weight) of elements in plants and soils. *Reference Plant* (column 3) is a concept proposed to represent a *typical* nonaccumulator plant species; soil median represents the median for *typical* soil samples. Data compiled from: Bowen (1979); Markert (1992); Marschner (1995).

Element	Plant Range	Reference Plant	Soil Range	Soil Median
		————(mg kg^{-1})————		
		Major Elements		
Calcium	1000–50000	10000	700–500000	15000
Carbon	–	445000	7000–500000	20000
Hydrogen	41000–72000	65000	–	
Magnesium	1000–9000	2000	400–9000	5000
Nitrogen	12000–75000	25000	200–5000	2000
Oxygen	400000–440000	425000	–	490000
Phosphorus	120–30000	2000	35–53000	800
Potassium	5000–34000	19000	80–37000	14000
Sodium	35–1000	150	150–25000	5000
Sulfur	600–10000	3000	30–1600	700
		Micronutrients		
Boron	30–75	40	2–270	20
Chlorine	2000–20000	20000	8–1800	100
Cobalt	0.02–0.5	0.2	0.05–65	8
Copper	2–20	10	2–250	30
Iron	5–200	150	2000–550000	40000
Manganese	1–700	200	20–10000	1000
Molybdenum	0.03–5	0.5	0.1–40	1.2
Nickel	0.4–4	1.5	2–750	50
Silicon	200–8000	10000	250000–410000	330000
Zinc	15–150	50	1–900	90
		Other		
Aluminum	90–530	80	10000–300000	71000
Arsenic	0.01–1.5	0.10	0.1–40	6
Barium	10–100	40	100–3000	500
Cadmium	0.03–0.5	0.05	0.01–2	0.40
Chromium	0.2–1	1.5	5–1500	70
Iodine	0.07–10	3	0.1–25	5
Lead	0.1–5	1	2–300	35
Mercury	0.005–0.2	0.1	0.01–0.5	0.06
Selenium	0.01–2	0.02	0.1–2	0.4
Strontium	3–400	5	4–2000	250
Vanadium	0.001–10	0.5	3–500	90

plant uptake and soil concentration, average values in **Table 1.1** are instructive only; actual concentrations vary widely as shown by the ranges in **Table 1.1**.

Figure 1.7 shows typical elemental concentrations in plants as a function of typical soil concentrations. The data points represent *approximate* indices of the relative availability of elements in soil compared to plant uptake. Thus, values should be interpreted with caution because individual plant species and soil properties greatly influence element bioavailability. The

general trend in **Figure 1.7** is that uptake of the elements by plants is directly proportional to the concentration of the element in soils. Elements that deviate from this relationship suggest that a process in the soil is affecting availability of the element for plant uptake. Nitrogen, sulfur, chlorine, and boron concentrations in plants are above the 1:1 trend line, suggesting these elements are readily available for uptake by plants. The plant concentrations of potassium, calcium, magnesium, molybdenum, zinc, and copper occur on the

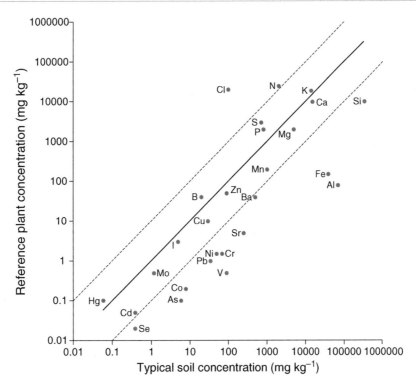

Figure 1.7 Log-log plot of elemental concentration of a typical plant as a function of total elemental concentration of a typical soil (See **Table 1.1**). Solid line is the 1:1 relationship indicating that plant concentrations are directly proportional to soil concentrations. Dashed lines represent one tenth and ten times the 1:1 relationship. Elements above the 1:1 line are more readily absorbed by plants than elements below the line.

1:1 line, suggesting uptake of these ions is correlated to soil concentration. Concentrations of iron, nickel, and cobalt in plants are below the 1:1 line, suggesting relatively low availability of these elements for plant uptake. Mercury is a contaminant and appears to be readily taken up in plants compared to typical soil concentrations. Other contaminants, such as arsenic, lead, cadmium, and aluminum, appear to have relatively low concentrations in plants compared to soil concentrations. One goal of soil chemistry is to explain why soil chemicals have such variability for plant availability.

In natural systems, the biosphere is at equilibrium with respect to nutrient cycling in soils, and in properly managed agricultural systems, soils are integral to supplying adequate nutrients for sustainable crop production. The Exhaustion Plot at Rothamsted Experiment Station in England, for example, has operated continuously since 1845 and revealed that even extreme management practices such as removing all plant material at harvest from the soil each year with no fertilization did not stop plant growth, although crop yields decreased. While such fertility management demonstrates soil's natural resilience, the decreased yields are clear indicators that this type of management of agronomic systems is not desirable for food production.

Immediate sources of the elements required for plant growth are solutes and electrolytes in the soil solution. Nearly all plant-essential nutrients exist in soils as ions. Some elements can have several different **oxidation states** and **valences**. Oxidation state is the difference in electrons and protons in an atom, while valence is the net charge on an ion or molecule. For example, nitrogen has several oxidation states that occur as different types of molecules in soils; such as nitrate (NO_3^-), ammonium (NH_4^+), nitrite (NO_2^-), amino N (R-NH_2), and nitrogen gas (N_2). The valence of the different nitrogen forms ranges from negative, zero, to positive, which has an important effect on the reactivity and availability of the various nitrogen species. Many elements in soils have only a single oxidation state, such as magnesium (Mg^{2+}) and calcium (Ca^{2+}), but their valence depends on the molecular species in the soil solution. For example, some of the aqueous Ca^{2+} in soil solutions may be complexed to SO_4^{2-}, and present as the $CaSO_4^0$ aqueous complex, thus making it an uncharged species (denoted by the superscript zero).

Ion availability in solution is renewed by soil reactions that add ions back to the soil solution after they are depleted. Main sources of ions to soil solution are: (1) mineral weathering, (2) organic matter decay, (3) rain, (4) irrigation waters, (5) fertilization, and (6) release of ions adsorbed by clays and organic matter in soils. With respect to ion availability for plant uptake, soils play an integral role in the delicate balance between preventing losses by leaching and supplying nutrients to plants and microorganisms (see **Figure 1.5**). In other words, ion retention by soils does not completely prevent leaching losses but is sufficient to maintain ions within the soil so that they can recycle between soils and plants before they are finally lost to groundwater, rivers, and the sea. Details of the different soil reactions that control ion availability for plant uptake are discussed later in this and other chapters.

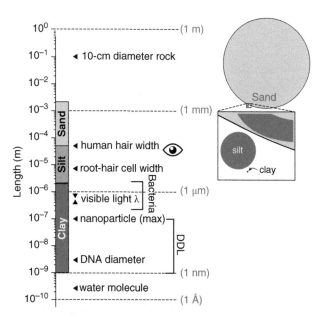

Figure 1.8 Diameter of sand, silt, and clay particles compared to other common objects. Drawing shows relative sizes of sand, silt, and clay (in blown up section the lower curve from the *S* from the *Sand* is shown to emphasize relative size). DDL is the size of a typical diffuse double layer thickness. Eye symbol indicates the approximate resolution of the human eye.

1.3.2 Soil solids

Soils contain both organic and inorganic solids. Inorganic solids in soils comprise mixtures of various types of minerals existing as rocks, sand, silt, and clays (**Figure 1.8**). Organic solids, such as soil organic matter, are equally important solid phases in soils, particularly in the O and A horizons. Soil chemical processes are greatly influenced by interactions with soil solids, thus, the study of solid-phase properties (soil mineralogy) is of great importance for soil chemistry.

The interface between solids and solution is called the **solid–solution interface**. The greater the **surface area** of the solid in a given volume or mass, the greater the area of the solid–solution interface, which means more reactive surfaces. Smaller particles have greater solid–solution interface (specific surface) and are more reactive than larger particles. For example, a 1-mm spherical sand particle has a surface area/mass ratio of about $0.002\ m^2\ g^{-1}$; a 1-μm clay particle, $2\ m^2\ g^{-1}$; and a 1-nm particle, $2000\ m^2\ g^{-1}$. **Figure 1.9** shows the change in surface area if a hypothetical rock of volume $1\ m^3$ were crushed into different particle sizes.

In soils, the smallest particles are termed **colloids**, which are particles that are 1 to 1000 nm in size. Colloids form unique mixtures when suspended in air or water. The components in colloidal mixtures tend to lose their individual identities so that the mixtures can be considered unique substances. Other examples of colloidal mixtures are fog, smoke, smog, aerosol, foam, emulsion, and gel. All are small particles suspended in a liquid or gaseous fluid. Colloidal particles interact strongly with fluid, but the individual solid–phase particles are still separate from the liquid, so they cannot be said to be homogenously dissolved. A colloidal mixture behaves so distinctively because of the large surface area of interaction between the particles and water or air. Thus, colloids in soils are important solid phases that greatly influence soil reactions.

The important properties that colloidal clays impart to soils include adsorption and exchange of ions, and absorption of water and gasses; processes that are essential to life. The surface properties of colloids in soils are dictated by the type of colloid, which are typically either degraded plant materials, termed humus, or clay-sized mineral particles (soil clays) such as phyllosilicates and iron and aluminum oxides.

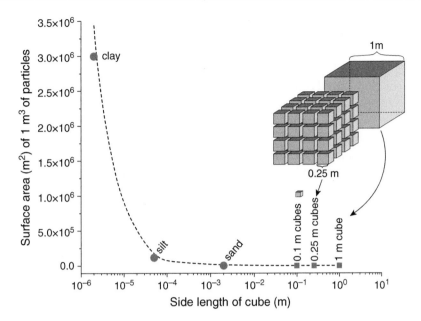

Figure 1.9 The effect of particle size on total surface area (SA) of 1 m³ stacked particles (no space between sub-cubes). The edges of the different cubes are six different lengths: 1 m (total SA = 6 m²), 0.25 m (total SA = 24 m²), 0.1 m (total SA = 60 m²), 0.002 m (sand size, total SA = 3 000 m²), 0.00005 m (silt size, total SA = 120 000 m²), 2 × 10⁻⁶ m (clay size, total SA = 3 000 000 m²). The graph illustrates that the surface area of 1 m³ of clay-sized cube particles is 10 000 times greater than the surface area of the same volume of sand-sized cubic particles.

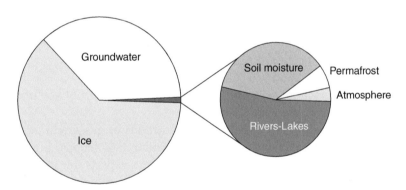

Figure 1.10 Estimate of distribution of non-ocean water on Earth. Ocean water volume (not shown) is 96.5% total water volume. Soil moisture (122 000 km³) is only a small fraction (9×10⁻⁵ %) of the total water volume on Earth. Data from Trenberth et al. (2007).

1.3.3 Soil interaction with the hydrosphere

The amount of water in soils is only a tiny fraction (0.001% to 0.0005%) of Earth's total water supply (**Figure 1.10**). Such a small amount of soil water seems perilously small to supply all terrestrial life. Yet, this miniscule portion is the water that supports plants, weathers rocks, forms soil, and is the medium in which most soil chemical reactions occur. The periodic droughts and resulting devastation around the world

emphasize the importance of soil water for agriculture and ecosystem health.

Surface and ground waters receive most of their dissolved solutes from the soil because rain first reaches Earth's surface, causing the most intense weathering in soils. As water percolates to greater depths, the composition changes less dramatically because the water already contains the salts from the above soil. However, the composition and concentration of dissolved solutes can change at depths if the water contacts soluble minerals in the subsurface (e.g., $CaCO_3$), or if it is stored for long periods in underground basins.

Stream water comes from soil and groundwater drainage, plus surface runoff. Natural drainage waters contain relatively low concentrations of the essential ions. The steady input of ions from drainage waters into surface water supports what is generally regarded as a natural and desirable aquatic population in streams and lakes. However, when concentrations of chemicals in runoff water increase above typical levels, such as in agriculturally modified watersheds, surface waters become contaminated.

Proper management can minimize runoff from agricultural lands, but some changes in water composition due to agriculture may be inevitable. Urbanization also increases runoff by creating less permeable surfaces such as rooftops and streets. The velocity of runoff increases as the overall permeability of the land surface decreases. In arid regions, surface runoff is a considerable fraction of stream flow when intense storms, sparse plant cover, and relatively low soil permeability create intermittent streams. The solute concentration in such waters is high and is important to the downstream ecology.

The **soil solution** is a component of the hydrosphere. Soils exist at an interface between the hydrosphere, atmosphere, biosphere, and lithosphere, and soil solutions are greatly influenced by interactions with these other spheres. The soil solution is the direct source of mineral nutrients for all terrestrial plants and is the most important transfer medium for chemical elements essential to life.

Soil solution includes the aqueous solution in soil pores and the film of water associated with mineral surfaces. Many soil clay particles have a net negative charge that extends into the soil solution, where the charge is balanced by an excess of solution cations. Thus, soil solution (including the film of water on negatively charged colloids) differs from other aqueous solutions in that it is not electrically neutral and usually contains more cations than anions. Old and heavily weathered soils in regions lying within the tropic and subtropic latitudes, or soils of volcanic origin, as in Japan, New Zealand, and the Pacific Northwest, USA, may have a net positive charge. In this case, the soil solution has an excess of anions.

At field capacity water content, most soil solution exists in small contacts and pores (<10 μm) between sand and silt particles (e.g., water-filled spaces illustrated in **Figure 1.1**). Clay particles and microbes congregate at the contacts, so the soil solution interacts closely with the reactive bodies in the contact zones. Because soil–particle charge extends into the film of water on the particle's surface, the boundary between soil solids and the soil solution is diffuse. The water and ions at the interface are a distinct aqueous phase associated with the soil solids. This solid–solution water layer makes up a **diffuse double layer, which describes** the charge of the particle and the charge from the ions in the surface film of water. In soils containing considerable clay, a large part of the soil solution occurs in the diffuse double layer.

Soil solution contains a wide variety of solutes, including most natural occurring elements; albeit some at very low concentrations. The reactivity of soils greatly influences the composition of the soil solution. Aqueous reactions that require days and years in air, and hours in water, may require only seconds and minutes in soils. Because of the direct linkage between soil solution and surface water and groundwater, studying soil properties is an important aspect of water quality in rivers, lakes and aquifers.

1.3.4 Interaction of soil and the atmosphere

Interactions of gases with soils are less obvious than soil–water interactions, but soil–gas interactions are important aspects of carbon, nitrogen, and sulfur cycling. Soils absorb sulfur dioxide, hydrogen sulfide, hydrocarbons, carbon monoxide, nitrogen oxide, and ozone gases from the air. Soils release gases, such as H_2O, CO_2 and CH_4 that result from organic decay, and N_2 and N_2O from natural soil nitrogen compounds and fertilizers. Global cycles of carbon, nitrogen, and sulfur have prominent soil chemical processes that include soil–gas reactions (see Special Topic Box 1.1 for carbon cycle discussion).

The nitrogen cycle is an important example of soil–gas interactions. Nitrate (NO_3^-) and ammonium (NH_4^+) ions in rainwater interact with soils, plant roots, and microorganisms, and plants and microbes convert nitrogen to amino acids, or to N_2 and N_2O gases that diffuse back to the atmosphere. Ammonia gas (NH_3) is

Special Topic Box 1.1   Biogeochemical cycling of carbon in soils

Climate change and global biogeochemical cycling of carbon is a topic of great interest and research. Important aspects of the global carbon biogeochemical cycle are the magnitude of carbon pools and fluxes of carbon in and out of soil, which are critical parts of Earth's active carbon system where fluxes are faster than changes in geologic carbon (carbon stored in deep rock or the deep sea). Understanding soil carbon biogeochemical cycling requires knowledge of soil chemical processes.

Active carbon in Earth is divided into five reservoirs: soil, biomass, ocean surface, atmospheric, and extracted fossil fuels (**Table 1.2**). The amount of carbon in the soil far exceeds the other pools, and the soil's role in the carbon cycle is very large. Soil carbon consists of organic and inorganic fractions (**Table 1.3**). Wetlands, peatlands, and permafrost soils are amongst the largest reservoirs of organic carbon. The size of the soil carbon pool, although reported to some certainty in (**Table 1.2**), is a subject of current research and debate because it depends on modeling assumptions, such as soil bulk density and depth of

measurement; errors from spatial modeling extrapolation errors are also problematic. Despite the uncertainty, however, it is known that the soil organic carbon pool is the largest *active* global carbon reservoir.

Inorganic carbon that exist as carbonates in soils is also an important carbon reservoir. In shrublands, grasslands, and arid lands, pedogenic inorganic carbon (caliche or calcium carbonates) may exceed organic carbon in the soil. The turnover of carbon from the inorganic soil carbon pool is much slower than the soil organic reservoir. However, because it is a large pool, even small changes from management and changing climate can create significant fluxes between the soil and the atmosphere.

Decay of organic matter in soils and vegetative respiration are the largest fluxes of CO_2 to the atmosphere. Thus, changing land management can significantly impact the global carbon cycle. For example, clearing and cultivation of virgin soils from 1850 to 1997 is estimated to have released 140 Pg (1 Pg = 10^{15} g) of carbon to the atmosphere, approximately 50% of the total carbon emitted by fossil fuel use in the same period (see Lal, 2008).

Factors that affect soil biogeochemical cycling of carbon include temperature, microbial reactions, mineral composition, carbon inputs, soil permeability, and moisture content. Understanding details of the soil biogeochemical carbon cycle requires knowledge of the different carbon species, reactions, and how they respond to fluxes. **Table 1.3** lists common carbon species and examples of their reactions in soils. Soil carbon reactions and species depend on various soil processes, such as microbial respiration and association of carbon with soil minerals, and the type of SOM molecules present. Because of the natural variability in processes and SOM inputs, developing reactions to quantify carbon release from SOM is challenging, and a topic of much research.

Organic carbon cycling reactions and inorganic cycling are inextricably linked by the production and consumption of CO_2. For example, aerobic microbial activity degrades organic carbon compounds and increases the partial pressure of CO_2 gas within the soil's pore space. The increased partial pressure increases carbon dioxide dissolution in soil water, which affects soil pH and precipitation of carbonate minerals. Increased accumulation of carbonate minerals in subsoil leads to a Bk soil horizon development (**Figure 1.11**).

Table 1.2 Estimates of amount of carbon in the active reservoirs on earth (one petagram (Pg) = 1 gigaton (GT) = 10^{15} g).

Reservoir	Carbon (Pg)
Surface of ocean[1]	700–1000
Atmosphere (2005 level) [1]	805
Fossil fuel (recoverable) [1]	5000–10 000
Vegetation[2]	560
Soil organic carbon[2*]	1550 (1 m)
	2344 (2m)
Inorganic soil carbon[2]	720 (1 m)
	950 (2m)
Northern circumpolar permafrost region[3**] (to 3m)	1672

[*] Estimates of soil carbon depend on modeling inputs and assumptions (e.g., soil bulk density). Since 1950, numerous papers have published soil carbon reservoirs sizes, with the range from 504–3000 Pg carbon, and a median of 1460.5 Pg carbon (Scharlemann et al., 2014).
[**] Total soil carbon mass in the 0–300 cm depth, and deep carbon in the permafrost (yedoma) and deltaic deposits
[1] (Houghton, 2007).
[2] (Lal, 2014).
[3] (Tarnocai et al., 2009).

Table 1.3 Examples of carbon species and reaction in soils.

Types of Soil Carbon	Example Species	Example Reactions
Organic		
Biomolecules	Organic carbon in living or dead organisms and their metabolites (e.g., proteins, carbohydrates, methane)	Microbial oxidation of biomolecule to form CO_2 and SOM:
Soil organic matter (SOM)	Humic substances, humic, and fulvic acids	$CH_2O - R + H_2O = CO_2 + SOM + 4H^+ + 4e^-$
		$CH_2O\text{-}R$ represents a carbohydrate moiety in a biomolecule; SOM is a partially degraded biomolecule
Anthropogenic chemicals	Pesticides and industrial chemicals (e.g., trichloroethylene, DDT)	Adsorption of pesticides on soil organic matter and soil minerals
Incomplete organic matter combustion products	Fire-derived soil carbon, black carbon, biochar, soot, charcoal, graphite, tar	$Biomass + heat = C_xH_y + C_xH_yO_z + H_2O + CO_2 + CO + H_2 + etc.$ Heating of organic materials at high temperatures and limited oxygen presence produces gasification products C_xH_y and charcoal $C_xH_yO_z$, among other compounds
Inorganic		
Solid: carbonates	Calcite ($CaCO_3$), dolomite ($Ca_xMg_{1-x}(CO_3)$, siderite ($FeCO_3$)	Precipitation: $Ca^{2+}(aq) + CO_3^{2-}(aq) = CaCO_3(s)$
Gas: carbon dioxide	CO_2	Gas dissolution: $H_2O + CO_2(g) = H_2CO_3(aq)$
Liquid: carbonic acid	H_2CO_3, HCO_3^-, CO_3^{2-}, ion-carbonate complexes, e.g., $FeHCO_3^{2+}$ (aq)	Deprotonation: $H_2CO_3(aq) = HCO_3^-(aq) + H^+(aq)$ $HCO_3^-(aq) = CO_3^{2-}(aq) + H^+(aq)$ Complexation: $Ca^{2+}(aq) + CO_3^{2-}(aq) = CaCO_3(aq)$

A Bk horizon significantly impacts the flow of water and gases, the ability to plow, and chemical properties of soil.

Because soil carbon can exist in solid, liquid, and gas phases, and involves both organic and inorganic chemical processes, understanding the complete soil biogeochemical carbon cycle requires all disciplines of soil science. Soil chemistry predicts carbon reactions in the soil, and therefore is integral to the study of carbon cycling within ecosystems and between the carbon pools (soil, ocean, atmosphere, and biosphere). Important aspects of the soil–carbon cycle studied in soil chemistry include dissolution of carbon dioxide into soil water, precipitation of carbon into carbonates, adsorption of dissolved carbon species such as bicarbonate onto mineral surfaces, formation and properties of soil organic matter, and oxidation of soil organic matter to CO_2. Soil organic matter is one of the most reactive components of soils and imparts many of its important physical and chemical properties.

also emitted and absorbed in soils. Under natural conditions, gaseous nitrogen loss to the atmosphere is approximately balanced by N_2 uptake and conversion to amino acids by symbiotic and free-living soil microorganisms.

Direct soil absorption of gases is perhaps most obvious in the case of the rapid disappearance of atmospheric sulfur dioxide in arid regions. The basicity of arid soils represents an active sink for acidic compounds from the atmosphere. The amount of direct soil absorption of atmospheric gases, inappropriately termed dry-fallout by atmospheric scientists, is less in humid regions where plant absorption and rain washout of the gases are substantial.

Gas fluxes in soil are controlled by diffusion gradients and the permeability of the soil to the overlying atmosphere. Low permeability can result in gas concentrations within the soil that are several orders of magnitude greater or less than the overlying atmosphere. For example, root and soil microbial respiration

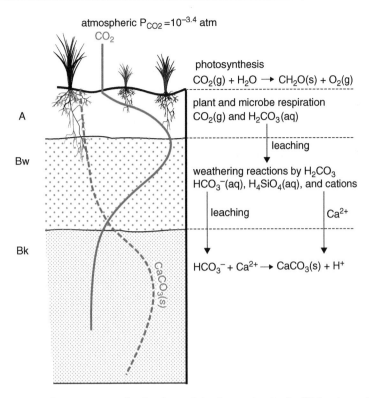

atmospheric $P_{CO2} = 10^{-3.4}$ atm

CO_2

photosynthesis
$CO_2(g) + H_2O \rightarrow CH_2O(s) + O_2(g)$

plant and microbe respiration
$CO_2(g)$ and $H_2CO_3(aq)$

leaching

weathering reactions by H_2CO_3
$HCO_3^-(aq)$, $H_4SiO_4(aq)$, and cations

leaching Ca^{2+}

$HCO_3^- + Ca^{2+} \rightarrow CaCO_3(s) + H^+$

A
Bw
Bk

$CaCO_3(s)$

Figure 1.11 Soil profile showing carbon reactions that lead to calcite formation in the Bk horizon. Carbon dioxide concentrations (solid line in profile) are elevated above atmospheric concentrations in the A horizon where plants and microbes respire carbohydrates (CH_2O). Elevated carbon dioxide increases carbonic acid (H_2CO_3), which weathers rocks. Leaching of cations and carbonic acid into the lower profile creates local saturation of calcium carbonate minerals (dashed line in profile), and development of a Bk horizon, where the k is a subordinate descriptor indicating carbonate accumulation. The concentration of carbon dioxide then decreases in the profile as a result of precipitation of carbonates in Bk horizon. Adapted from Chadwick and Graham (2000).

consume oxygen and nitrogen, and produce CO_2, leading to increased carbonic acid concentrations in soil water. Increased carbonic acid promotes mineral weathering and increased concentrations of cations such as calcium in the soil solution. In arid regions, leaching of bicarbonate and calcium to lower horizons, as opposed to out of the soil profile such as occurs in soils in humid climates, leads to development of a cemented horizon called a Bk horizon (**Figure 1.11**) (B is the horizon designation and k designates the horizon as calcite cemented).

If a soil has low gas permeability due to small soil pores, high pore tortuosity, high clay content, or high water content, fluxes of gas to the atmosphere are hindered, and soil pore concentrations of gas are depleted or increased relative to the overlying atmosphere. Some wetland plants can transfer oxygen gas internally to the root zone where the fluxes of oxygen

alter the availability of nutrients and contaminants, and the formation of iron and manganese solids.

1.4 Chemical reactions in soils

Chemistry is the study of chemical reactions and species, and the factors that influence them. To study reactions, a *chemical system* is defined in which inputs and outputs can be controlled and monitored. For example, NaCl dissolved in a beaker of water is a chemical system. Compared to a beaker used in laboratory chemistry, *soil systems* are much more complicated. Studying soil chemical processes requires that system boundaries be defined so that fluxes of energy and matter can be accounted for and monitored (**Figure 1.1**). For example, a field plot has fluxes of heat, moisture, chemicals, organisms, and gases that impact soil chemical processes and are continuously changing. The multiple fluxes affecting field experiments make measuring soil chemical processes challenging.

An alternative to monitoring soil processes at the field-scale is to impose constraints on the soil system. In the field, adding constraints on a system is difficult, but has been done by isolating a section of the soil using barriers as separations. Such **mesocosms** typically range in size from 10 cm to 2 m. An extreme case of attempting to isolate a system is the biosphere experiments conducted at the University of Arizona (*Biosphere 2*), where an entire ecosystem with humans living in it was sequestered in a controlled enclosure. A follow-up to Biosphere 2 is a large ecosystem mesocosm consisting of three 11-m by 30-m replicate hillslopes made from 1-m deep ground basalt rock and enclosed in a controlled temperature and moisture regime greenhouse. The experiment is called the Landscape Evolution Observatory (LEO) (**Figure 1.12**). Experiments in this observatory will characterize

physical and chemical processes in a well-controlled system, offering unique opportunities for understanding natural processes under conditions that are less variable than occur in nature because composition and fluxes can be controlled, the experiments can be replicated, and because it is easier to instrument and monitor.

Given the difficulty of constraining natural systems for study of soil chemical reactions, an alternative is studying system properties ex-situ. There are two types of ex-situ experimental system approaches:

1. Characterize the chemical properties on a soil sample taken from a field setting using various laboratory analyses, thus rendering a snapshot in time of the soil's conditions when sampled. For example, measuring water extractable concentration of an element in a soil sample, or measuring the pH of a soil sample.
2. Conduct experiments that monitor reactions and species in a **microcosm** that mimics field conditions.

Figure 1.12 The Landscape Evolution Observatory (LEO) in Oracle, Arizona is part of the Biosphere 2 complex. LEO contains over 1800 sensors to monitor soil, water and air properties to track weathering processes of basalt in a highly controlled environment. Photo courtesy of Till Volkman, The University of Arizona.

For example, a greenhouse experiment, or an anaerobic soil incubation in the laboratory.

Regardless of whether the system to be studied is in-situ in the field, or ex-situ in a laboratory, defining system boundaries and parameters is a requirement for conducting experiments and understanding chemical reaction processes occurring in soils.

1.4.1 Flow of chemical energy in soils

In soil, as in all the universe, energy flows towards the minimum (**Figure 1.13**). Science has put a theoretical framework on this axiom – the **laws of thermodynamics** – that allow predicting the state of a system and direction in which it will move. Thermodynamic treatment of relatively simple chemical reactions within a system is a well-developed area of study. However, as systems become more complex, as in soils and nature,

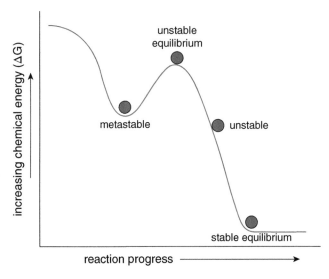

Figure 1.13 Chemical energy and reaction progress are analogous to a ball affected by gravity. Metastable and unstable positions are not the lowest energy and will therefore react. The lowest energy position is equilibrium, where no net reaction occurs. Metastable and unstable equilibrium energy positions have some stability, and species may exist in these states for a fraction of seconds to very long periods; but, given time, the species will convert to the lower energy species. Typically, activation energy is needed for metastable species to overcome the higher energy (small bump). In nature, microbes, mineral catalysts, enzymes, moisture, sunlight, temperature, or pressure provide activation energy.

implementing thermodynamics to predict reactions is more challenging. The difficulty in predicting thermodynamic processes of natural systems, however, does not mean that the principles are not binding.

In natural systems, complexity of the reactants and products are not well defined, and fluxes into and out of the system are difficult to constrain and measure. In addition, many natural systems rarely reach their minimum energy before fluxes cause changes in the reactants, and thus the system equilibrium state changes. Examples of such fluxes in soils are changes in moisture, temperature, and vegetation. Fluxes change over time scales of minutes to years. Any input into a system, whether it is energy (heat), matter (chemicals), or pressure, causes the system to move toward a new equilibrium with a different distribution of chemical species. This requirement is referred to as **Le Chatelier's principle**. To account for the fact that a system is undergoing change, but is not at equilibrium, scientists define system reactions as time-dependent or kinetically controlled.

The dynamic nature of natural systems causes some systems to never achieve a stable equilibrium. However, even in the absence of equilibrium, the total energy, whether it is energy in forms of chemical, pressure, or heat, will flow towards a minimum. In chemistry, theoretical energy of a system is quantified as a function of two thermodynamic factors: **enthalpy** (H) and **entropy** (S). Enthalpy is a measure of the heat of the system. Entropy is a measure of the tendency for energy to spontaneously go towards disorder. These factors embody the laws of thermodynamics. Enthalpy and entropy can be combined into a factor called **Gibbs free energy** (G):

$$\Delta G = \Delta H - T\Delta S \tag{1.2}$$

where ΔG is the change in free energy of the system and T is the temperature of the system. The delta symbols indicate a relative change from a known specified standard state.

Products and reactants have associated free energies, and their difference is the change in free energy of reaction (ΔG_{rxn}). Using the change in free energy to predict reaction status is useful for understanding how much of a chemical species should exist in a system. For example, in a soil containing solid phase calcium carbonate, thermodynamic equations and system properties such as partial pressure of carbon dioxide can be used to predict the soil pH (see Chapter 4).

A more detailed treatment of thermodynamic equations in soil chemistry will be given later. For now, the following points are emphasized:

1 Soil systems will move toward the lowest energy state.
2 A system at the lowest energy state is at equilibrium.
3 At Earth's surface, solar and geothermal (e.g., volcanic) energy continuously create new, unstable states of matter.
4 How fast chemicals go toward the lowest energy state is termed **reaction kinetics**.
5 Fluxes into and out of natural systems make them dynamic, changing the energy state of the systems; thus, the equilibrium state of soil chemical systems is constantly changing.

An example of the last point is the metastable iron oxide mineral ferrihydrite ($Fe_5HO_8 \cdot 4H_2O$) that exists in many soils (Color plate **Figure 1.14**). If the soil were static, that is, no input of energy or matter over time, the ferrihydrite would convert to more stable iron oxide minerals such as lepidocrocite (FeOOH), hematite (Fe_2O_3), or goethite (FeOOH with a distinct crystallography compared to lepidocrocite), and ferrihydrite would cease to exist. Yet, ferrihydrite is found in many soils because it is continuously being created, at least as fast as it is disappearing. A system where a product is created as fast as it is disappearing is at **steady-state**. In soils, steady-state may rarely occur, but the concept nevertheless illustrates the important point that chemicals in soils are continuously undergoing change.

1.4.2 Soil chemical speciation

Once a system of study has been defined, the next step in evaluating soil chemical processes is determining the species of the chemicals in the system.

Speciation infers the phase (gas, liquid, solid), oxidation, or valence state, isotopic state, bonding environment, and structure of an element or molecule.

For example, in soil, iron exists in two oxidation states (Fe^{2+} (ferrous) and Fe^{3+} (ferric)). Ferric iron has low

Figure 1.14 Top section of soil showing Fe-enriched (red) and Fe-depleted (gray) zones in a wetland soil. Iron in this soil exists predominantly as the metastable mineral ferrihydrite, with lesser amounts of the thermodynamically stable minerals lepidocrocite and goethite. The dynamic redox conditions in wetland soils are responsible for the persistence of metastable ferrihydrite in the soil. (See **Figure 1.14** in color plate section for color version.)

solubility and commonly exists as solid iron oxide minerals such as goethite (FeOOH). Ferrous iron is more soluble and commonly exists in solution phase as aqueous ions, such as Fe^{2+} or $Fe(OH)^+$. Ferrous iron is a reduced phase that is common in soils that are wet and have limited oxygen, such as wetland soils.

A basic understanding of speciation for many chemicals in soils exists. However, there remain many unknowns about chemical species because soil environments are highly varied, with numerous different chemical, physical and biological processes occurring–thus, no two are alike. Speciation of a chemical in soil can be determined by either direct measurement or prediction. Prediction can be accurate but should be supported with measurements. Measurements of soil chemical species are often difficult or time consuming. Advances in technology, however, are allowing for better speciation determinations, and better ability to predict reactions and fate of chemicals in the environment.

1.4.3   Chemical reaction types in soils

Change in chemical speciation can be expressed as a **chemical reaction**. For example, reduction of ferric iron in the mineral goethite (FeOOH) to ferrous iron can be written as:

$$FeOOH(s) + 3H^+(aq) + e^- = Fe^{2+}(aq) + 2H_2O(l) \quad (1.3)$$

where e⁻ is an electron. The letters (s), (aq), and (l) indicate *solid, aqueous,* and *liquid* states, respectively; the letter (g) in a written reaction indicates *gas* state. The symbols for chemical solid, liquid, or gas are added to reactions for clarity, but are not always required, especially when the state is implied in the associated text or in the reaction (e.g., for free ions the aqueous phase is implied). In the reaction in **Eq. 1.3**, oxidized ferric iron in goethite is a reactant that accepts an electron from an electron donor (biotic or abiotic, not shown in the reaction), and reduces to aqueous ferrous iron.

Writing soil chemical reactions is a straightforward exercise, but assigning the correct reaction to a complex soil system is more difficult. Thus, making species predictions using reactions and thermodynamic or kinetic constants is challenging. Fortunately, by constraining the system parameters and making measurements of chemical species in the soil, relevant reaction processes can be identified, and predictions of soil chemical processes are possible.

A chemical reaction may go forward or backward. Reactions that go in both directions are called *reversible*. Some reactions are irreversible or unidirectional, depending on the system's properties and state of the chemicals. For example, weathering of the primary mineral muscovite in a soil to form the secondary clay mineral vermiculite is irreversible; meaning vermiculite will not revert to muscovite under soil conditions because heat and pressure are insufficient. Whether a reaction is irreversible is species and system dependent. Although some specific reactions may be irreversible in practice, reverse reactions can be written for all reaction types.

The six basic reversible reaction processes that occur in soils are:

1 Sorption/desorption
2 Precipitation/dissolution
3 Immobilization/mineralization
4 Oxidation/reduction
5 Complexation/dissociation
6 Gas dissolution/volatilization

A chemical undergoing one of the above six reactions changes its state of matter, its oxidation state, or its molecular composition or structure. An overview and examples of these reactions are discussed below. Additional details are the topics of chapters in the textbook.

1.4.3.1   Sorption and desorption

Sorption and desorption reactions describe association and release of a chemical from a particle surface, where the particles are minerals, soil organic matter (SOM), or perhaps a biological cell. Often, sorption reactions are termed **adsorption**, which implies that the chemical resides on the solid surface, but is distinct from the solid or bulk solution, and is not forming a three-dimensional network of atoms on the surface (called surface precipitation). There are many different types of sorption, which will be covered in more detail in Chapter 10. One example is the adsorption of a sodium ion onto a clay mineral surface (**Eq. 1.1**). The solid in this reaction, which, could be the clay species montmorillonite, maintains its compositional integrity because the adsorbed ion is only associated with the surface of the mineral. The release of the potassium from the clay in **Eq. 1.1** is a desorption reaction. Together, the adsorption and desorption reactions depicted in **Eq. 1.1** are an example of a **cation exchange** reaction. Another example of a sorption/desorption reaction is adsorption of phosphate on an iron oxide mineral surface:

$$\equiv FeOH^{-0.5} + H_2PO_4^- == \equiv FeOPO_3H_2^{-0.5} + OH^- \quad (1.4)$$

where the $\equiv FeOH^{-0.5}$ indicates a functional group on the surface of an iron oxide mineral that is created by an incompletely coordinated bond on the terminus of the mineral, and thus has a charge associated with it. The triple bar symbol ($\equiv$) attached to the $FeOH^{0.5-}$ in **Eq. 1.4** indicates that the iron occurs on the *surface* of a mineral; i.e., it is an iron hydroxide functional group on the edge of the mineral interacting with solution. The phosphate adsorption reaction is a **ligand-exchange reaction** because it is displacing a OH⁻ functional group on the surface of the iron oxide with oxygen ligands from the phosphate. The forward reaction is adsorption of phosphate on the iron oxide surface, and the reverse is desorption of phosphate from the iron oxide surface into the soil solution.

1.4.3.2 Precipitation and dissolution

These reactions describe the change in a chemical from solution to the solid state, where a new solid is formed from solution constituents. Dissolution is the reverse of precipitation, meaning ions from the solid are released to the solution. An example of a precipitation–dissolution reaction is the formation of the calcium carbonate mineral calcite in soils (see e.g., **Figure 1.11**):

$$CaCO_3(s) + H^+ = Ca^{2+} + HCO_3^- \tag{1.5}$$

where the forward reaction is a dissolution reaction, and the reverse is a precipitation reaction. Note in this reaction the *(s)* is placed on the calcite to indicate that it is a solid and not an aqueous complex; the phases for the other ions are left out because they are obvious.

1.4.3.3 Immobilization and mineralization

Immobilization and mineralization reactions are generally biologically mediated. Immobilization refers to the uptake of chemical into the cellular structure of an organism, such as a microbe, fungi, or plant. The chemical within the organism is considered a biologically formed molecule (biomolecule). An example is the uptake of nitrate from soil solution into a plant, where it is utilized as a cellular metabolite to produce amino acids, such as glutamate ($C_5O_4H_6NH_3$), which are components of proteins:

$$NO_3^- + \text{N-reductase} + C_5O_1H_9 = C_5O_4H_6NH_3 \tag{1.6}$$

In this reaction, the $C_5O_1H_9$ is simply an *element placeholder* for cellular compounds to provide the reactants needed for the stoichiometry to balance and does not represent a molecular species in the cell. In plants, this reaction occurs in plant chloroplast, where the enzyme nitrogen reductase (N-reductase) reduces nitrate using photosynthetic energy and fixes or immobilizes it in the glutamate molecule.

Mineralization implies degradation, release, or conversion of a chemical to a form that is no longer a biomolecule. Products of mineralization reactions are inorganic chemicals and degraded organic chemicals. Degradation of organic nitrogen in glutamate to ammonium is an example of a mineralization reaction:

$$C_5O_4H_6NH_3 + 4H_2O + O_2 = NH_4^+ + 5CO_2 + 13H^+ \tag{1.7}$$

This is a summary reaction describing complete degradation of the glutamate biomolecule produced in the reaction in **Eq. 1.6** to produce ammonium, carbon dioxide, and protons. The reaction is carried out by microbes in soils.

Immobilization and mineralization reactions are important soil processes that determine the fate of chemicals in soils. Because immobilization and mineralization reactions require detailed discussion of microbiology and biochemistry, which is beyond the scope of fundamental soil chemistry, they are only broadly covered in this text.

1.4.3.4 Oxidation and reduction

The gain and loss of electrons from an element cause a change in oxidation state. Often, redox reactions result in changes in the physical state or molecular structure, and thus may be combined with other reaction types. For example, the redox reaction shown for iron in **Eq. 1.3** describes both reduction and dissolution of iron and is thus termed a reductive-dissolution reaction. Similarly, many mineralization and immobilization reactions are also redox reactions; for example, reactions in **Eqs. 1.6** and **1.7**.

1.4.3.5 Complexation and dissociation

These reactions describe interactions of two or more chemicals or aqueous ions. Protonation and deprotonation (gain and loss of H^+ ions) are specific types of complexation and dissociation involving acceptance and loss of a proton by an acidic ion or molecule. Hydrolysis is a dissociation reaction in which H^+ is released from a water molecule. Carboxylic acid, a common functional group on soil organic matter, is a weak acid that deprotonates between approximately pH 3 and 6:

$$R\text{-}COOH = R\text{-}COO^- + H^+ \tag{1.8}$$

The *R* indicates the rest of the organic compound that the carboxylic acid functional group is attached. The acidity of the carboxylic acid functional group depends on the composition of the rest of the organic molecule.

Chelation is a type of complexation involving formation of chemical bonds between a molecule that has two or more bonding sites and a cation. Chelation increases nutrient availability for plants and microbes. EDTA, an organic molecule that simulates many natural chelates, complexes metal cations by forming up to six ligand bonds with cations:

$$EDTA^{4-} + Pb^{2+} = PbEDTA^{2-} \qquad (1.9)$$

The chelation of the Pb^{2+} cation by EDTA would increase the mobility of lead and promote dissolution of solid lead phases if present because it is a soluble *sink* for the Pb^{2+} cation. Solid phase lead, if present in soil, will be released to soil solution by either desorption or dissolution to maintain the Pb^{2+} activity, thus causing total dissolved lead concentration ($Pb_T = Pb^{2+}$ + $PbEDTA^{2-}$ + other aqueous lead species) to increase.

Complexation reactions change the valence and molecular properties of chemicals in soil solutions, thereby changing the chemical's solubility, plant availability, and transport through the soil. Aqueous complexation of ions occurs in soil solution and changes the concentrations of free ions (noncomplexed ions). For example, the inorganic ligand bicarbonate readily forms aqueous complexes in solution with dissolved metal cations, such as Zn^{2+}:

$$Zn^{2+} + HCO_3^- = ZnHCO_3^+ \qquad (1.10)$$

In this reaction, the $ZnHCO_3^+$ aqueous complex would occur in the soil solution instead of the free hydrated Zn^{2+} ion.

1.4.3.6   Dissolution and volatilization
Dissolution and volatilization of gases in soils refers to reactions occurring between the soil atmosphere and the soil solution – specifically, transfer of gaseous chemicals into the aqueous phase, and the reverse. Since this reaction involves movement of gas into and out of liquid water, it is different from condensation and vaporization of a pure liquid to gas, and vice versa. Henry's gas law is used to predict equilibrium partitioning of gases into water (see Chapter 4). An example of gas dissolution in soil solution is the reaction of carbon dioxide (gas) with water to form carbonic acid (aqueous), as discussed in Special Topic Box 1.1. Oxygen, nitrogen, and sulfur also have important gas dissolution and volatilization reactions. Ammonium is a common ion in soil solution; however, when it deprotonates, it forms ammonia that will volatilize:

$$NH_4^+ (aq) = NH_3 (g) + H^+ (aq) \qquad (1.11)$$

The reverse reaction is used to dissolve ammonia gas (anhydrous ammonia) into soil solution to produce ammonium ions for soil fertilization. Anhydrous ammonia is liquid ammonia under high pressure that is injected into the soils. The reaction in **Eq. 1.11** predicts that adding anhydrous ammonia to soils would cause the pH of the soil solution to increase because the ammonia would protonate to ammonium, thereby consuming protons; or in other words, hydrolyzing water and releasing hydroxide ions.

In soils, most reactions involve the soil solution. **Figure 1.15** is a comprehensive view of the six

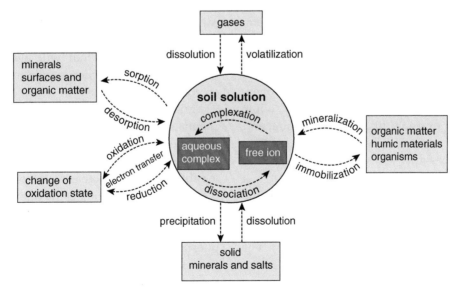

Figure 1.15 Chemical reactions in soil. Soil solution is in the center because most reactions occur between soil solution and either organisms, soil air, or solid phases.

reactions described above, as they occur in a soil system. Chemicals in soils exist in one of the states shown in the boxes, or soil solution. If the system is perturbed, the soil chemical may change via one of the six reactions to reestablish equilibrium. A detailed version of this diagram can be developed for any soil chemical. For example, the nitrogen cycle is usually illustrated with a figure detailing biogeochemical cycling of nitrogen between solid, gas, organic, and solution phases, showing the different nitrogen oxidation states and molecular species (see Figure 3.20).

1.5 Soil biogeochemical cycling

Because of the importance of interacting processes in soils, many soil chemists study soil **biogeochemical cycles**. Biogeochemistry implies that geochemical processes are coupled with biological processes and includes cycling of energy and matter within a system. Two basic concepts of biogeochemical cycling are: (1) magnitude of pools of chemicals; and (2) transfer of chemicals between the pools via fluxes of chemicals and energy. A biogeochemical pool, or reservoir, is a conceptual unit of the earth considered to represent a distinct part of its systems; e.g., the plant biosphere, soils, groundwater, ocean, or atmosphere.

Biogeochemical cycles are studied at numerous scales – from the global scale to the soil-pore scale – depending on how one defines the system. Regardless of the scale of the system, fluxes must be considered, and are used to understand the mass balance of the system pools. Elements readily involved in metabolic processes, such as carbon, oxygen, nitrogen, and sulfur, have active biogeochemical cycles, while other elements, for example, titanium, aluminum, and cesium, have less active biogeochemical cycles (less active does not necessarily imply simplicity).

An important example of the application of soil biogeochemistry is the carbon cycle. Soil carbon is the largest *active* carbon pool on Earth, and thus soil reactions are an important component of global carbon cycling (**Figure 1.11** and Special Topic Box 1.1). For example, in the ~10 000 years since the last ice age, massive amounts of carbon have moved from the

equatorial regions of Earth to the newly unglaciated polar regions. As a result, soil carbon stored in the soils and deep deposits in the northern circumpolar permafrost region represents a large deposit of Earth's current terrestrial carbon pool. Ten thousand years is a very short geologic time, and active transfer of carbon between eco-regions illustrates how dynamic Earth's biogeochemical cycles are. Present-day climate change may alter the rates of decomposition of the organic carbon in the northern circumpolar permafrost region. Thus, a massive flux of carbon could potentially *be on the move* again and have significant impacts on the global climate and ecosystem processes.

1.6 Soil chemical influences on food production

Soil is the main source of human nutrition. The oceans supplement our food supply, but their productivity is limited. Terrestrial plants remain the cheapest and most efficient means of converting solar energy into life support for this planet. The growth of plants is a large fraction of the world's economy and is fundamental to a nation's well-being.

Understanding soil processes and developing best management practices is critical for sustainably growing plants needed for food and fiber. Early researchers designed experiments to better understand soil chemical processes so that crop growth could be maximized. Modern soil chemistry researchers continue to strive for new discoveries that will allow for more and healthier food production, while minimizing impacts on ecosystems.

Agriculturalists can influence and modify soil chemistry to a considerable extent. The amounts of essential elements needed by plants over a season are small enough that supplementing the soil supply is feasible. However, increasing the efficiency of that fertilization is a continuing challenge because producing and applying fertilizers is expensive. For example, nitrogen fertilizer production is energy intensive, and phosphorus fertilizer sources are limited and therefore expensive to mine. Understanding soil chemical processes of amendments and fertilizers is important because they can change the availability of the applied nutrients.

Application of pesticides or herbicides also increases the health of crops or grazing pasture. Soil chemical properties and processes affect the efficacies of such chemicals towards the pests, and control unintended side effects, such as damage to plants or other beneficial organisms, or leaching of pesticides to surface and ground waters. Thus, understanding chemical reaction processes of fertilizers and pesticides in soils is a topic of great focus for increasing food production and maintaining healthy soils and ecosystems worldwide.

1.7 Soils and environmental health

In earlier times, when the population was less dense and industries were few and small, wastes were distributed widely on soil. Negative impacts were usually minor, and soils could readily assimilate insults,

such as contaminants, with minimal impacts to their natural biogeochemical cycles. Concentrating wastes in urban areas, industrial facilities, landfills, feedlots, and sewer treatment plants is causing contamination of the environment and suggests that humanity has exceeded the rate at which these materials can be assimilated by the soil and return to their natural biogeochemical cycles. The need to deal with polluted environments is an important application of soil chemistry (**Figure 1.16**).

The elements that humans release as wastes are derived from the soil and the earth. Chemical contamination is the diversion of chemical elements from their *natural* biogeochemical cycles. For example, nitrogen and phosphate from wastewater treatment plants and agricultural operations that flow into streams and lakes are removed from the soil–plant cycle. Water bodies have a lower chemical buffering capacity than soils, and readily suffer nutrient overload effects that degrade water quality. If nitrogen and phosphorus were instead put back into the soils at levels that do not

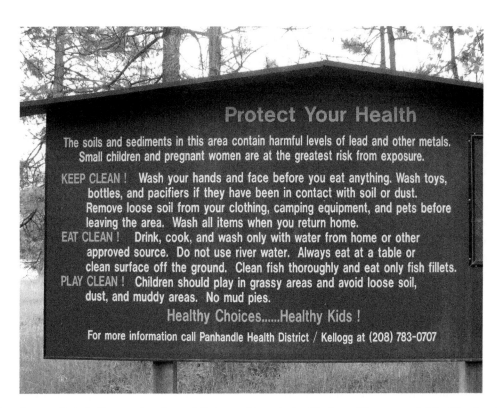

Figure 1.16 Sign from polluted site warning the public to avoid contact with the soil because of risk of lead poisoning.

drastically perturb natural processes, surface water quality would be much less degraded.

Despite soil's natural ability to buffer or attenuate soil chemicals, soil degradation and environmental pollution are tremendous challenges for civilization. There are three general types of soil degradation:

1 Decrease in physical, chemical, or biological properties such that the soil is less productive for plant growth. For example: depletion of nutrients or organic matter, increase in soil temperatures, compaction or surface crusting.
2 Reduction in soil depth by erosion.
3 Accumulation of chemicals to levels that detrimentally affect plant growth or ecosystem health. For example: salt accumulation, hydrogen ion accumulation (decreases pH), chemical contamination, excess nutrient buildup.

The most common soil degradation is desertification and salinization, either caused by overutilization of the soil for crop growth without regard to salt buildup, or overgrazing. Other issues that stem from environmental pollution occur when too much of a chemical exists in a soil, creating a potential toxicity issue for soil organisms, plants, or other organisms exposed to the chemical. Additionally, soil pollution creates off-site risks by leaching of the pollutant to ground or surface water.

Chemicals of concern can be either organic or inorganic. Most organic chemicals, such as pesticides and industrial chemicals, degrade over time, and contaminated soil can eventually return to a nonpolluted state, although some chemicals take a long time to degrade. Other contaminants such as inorganic chemicals or recalcitrant organic chemicals do not degrade, but persist in soil until they are leached by water, are volatilized and outgassed, or are sequestered by a plant. Common examples of inorganic pollutants are metals from industrial, mining, or agricultural sources. Even though inorganic chemicals do not degrade, their availability for uptake or leaching, and their hazard potential, is variable. Some chemicals are partially removed from the soil by natural leaching (e.g., zinc), while others are less soluble and more recalcitrant and remain in the soil solid phase (e.g., lead). Many potentially toxic metals are sorbed so tightly onto soils that they are immobile and unavailable for plant uptake or leaching.

Some inorganic chemicals are nutrients, but at elevated levels are contaminants. This includes macronutrients for plants, such as nitrate, and micronutrients such as boron, zinc, and copper. The amount and speciation of chemicals are the most important factors for determining whether soils are contaminated. Chapter 3 discusses the occurrence and speciation of many common chemicals that occur in soils and the environment.

1.7.1 Soil chemistry and environmental toxicology

Environmental toxicology is a discipline that specializes in the study of chemical risks or hazards in the environment. Because of the critical role of soil chemistry in chemical processes in the environment, there is an overlap with environmental toxicology. Thus, concepts and terms from environmental toxicology are frequently used when discussing soil chemistry topics. Definitions of toxicological terms commonly used in soil chemistry are:

- *Contaminant.* A chemical of concern that is present at elevated concentrations. Does not necessarily imply an organism is at risk.
- *Pollutant.* A contaminant judged to represent a hazard.
- *Toxicant or toxic chemical.* A chemical present in an amount and form that may cause damage to an organism (animal, plant, or microbe). Does not imply exposure to an organism.
- *Toxin.* A toxic chemical of *biological* origin, such as venoms and the active agents in poisonous plants. This term is often misused; soil chemistry does not typically deal with toxins.
- *Toxicity.* The relative degree to which a chemical is poisonous.
- *Poison.* A chemical that causes an adverse effect when an organism is exposed.
- *Bioavailability.* The *measured* availability of a chemical for uptake into an organism as measured by talking samples from an organism. May consider only uptake into organism, or uptake into targeted organ (e.g., blood stream, or leaf tissue).
- *Bioaccessibility.* The *potential* availability of a chemical to be taken up by an organism measured using an in-vitro method.

Often in soil chemistry, toxicity, bioavailability, or bioaccessibility are referred to in a more general sense using the term *availability* or *fate and transport*. Although not accurate for describing risks of chemicals, these broad terms are useful for generalizations or non-specific references to a chemical's environmental behavior.

To determine a chemical's potential toxicity, a risk-assessment screening model that categorizes risk based on four criteria is used:

1 *Sources*. Identify the chemical of concern and the environment in which it exists.

   Example: A mine-contaminated soil has elevated lead concentration.

2 *Pathways*. Identify the routes of exposure and factors that affect exposure from the chemical of concern to the *receptor*.

   Example: Children have lead-laden soil particles stuck to their hands and put their hands in their mouth. The species of the lead in the soil causes bioavailability to be variable.

3 *Receptors*. The organism(s) at risk from exposure to the chemical of concern.

   Example: Humans (children) ingest soil with elevated lead.

4 *Controls*. Natural and engineered solutions to reduce risks. Site management decisions can be made to mitigate risk; or provide a management scenario that minimizes exposure of the risk to the receptors.

   Example: Lead-contaminated soil remediation options may entail removing the soil, amending the soil with a product that will decrease lead bioavailability, or isolating the site so that children will not come in contact with contaminated soil.

Remediation of contaminated soil often involves in-situ treatment and monitoring. Ex-situ treatment is costly and greatly disturbs the soil, plants, and nearby lakes and streams, and is thus only done when contamination cannot be remediated in-situ, or if risks merit a costly remediation. **Figure 1.17** summarizes key factors for consideration of in-situ soil remediation.

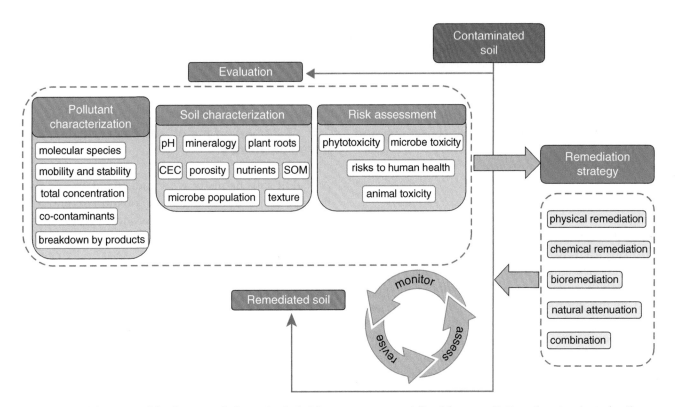

Figure 1.17 Schematic of the framework for ecological risk assessment specialized for remediation of contaminated soils. Adapted from US Environmental Protection Agency (1992).

An important aspect of risk assessment is prediction of the toxicity risks to a receptor. This is typically done using models with inputs from experiments. Toxicity risk experiments are either in-vivo or in-vitro. In-vivo models measure contaminant effects on an organism, while in-vitro models measure contaminant effects in a simulated, nonliving system. Often, contaminated soil studies use plants to assess bioavailability, or feed the soil to an animal, such as a pig, bird, or earthworm, to study its bioavailability and toxicity effects. From the bioavailability study results, researchers evaluate the toxicity of the soil and determine important soil properties that control toxicity. Plant uptake, seedling germination, or microbial respiration are also useful methods for evaluating soil toxicity.

In-vitro tests are done using simulated conditions to evaluate soil toxicity. For example, to measure metal availability from soils, extraction of the soil with a solution is often done. In some cases, the extractions are specifically designed to simulate bioavailability, but they are only estimates of bioavailability; thus, results from such extraction tests are referred to as measures of bioaccessibility. Prediction of toxicity risks typically have inputs of soil contaminant concentration and some soil properties, such as pH or organic matter concentration, to account for speciation effects on contaminant availability. While bioaccessibility assessments are useful, knowing the real contaminant speciation is the best way to predict bioavailability, however, studies on speciation and toxicity in organisms are time consuming and expensive.

1.8   Units in soil chemistry

The International Union of Pure and Applied Chemistry (IUPAC) has developed recommendations of a common language to describe chemical nomenclature, symbols, and units used for describing quantities. Most of the information is published in "The Gold Book," which is available on the Internet. For the most part, soil chemistry adheres to IUPAC's convention of using SI units (Système International d'Unités). However, as with many disciplines, expression of some soil chemical properties deviates from SI units because of historical or convenience reasons.

Units of measurement are either extensive or intensive. **Extensive properties** measure the amount of a substance or energy. For example, mass of an object is an extensive unit because it describes *how much* something weighs, and changes depending on how much of the object is present. **Intensive properties** describe a property of an object or energy that does not change when the total amount increases or decreases. Density is an intensive unit, for example, because it describes the mass of a substance per volume – no matter how much of the substance is present, its density does not change. Concentration is another example of an intensive unit. **Table 1.4** provides a listing of common units used in soil chemistry. Other units used in soil chemistry not listed in **Table 1.4** follow SI convention.

1.8.1   Converting units

To describe chemical processes and quantities of chemicals, it is often necessary, or desirable, to convert units. **Dimensional analysis** is a mathematical treatment used to change units of intensive or extensive properties. Even practiced scientists should always do careful dimensional analysis, as this is often a source of error, sometimes leading to disastrous and costly effects. Below are examples of dimensional analysis problems commonly encountered in soil chemistry.

1.9   Summary of important concepts in soil chemistry

A goal of soil chemistry is to understand and predict the fate, availability, and mobility of nutrients and contaminants, including both organic and inorganic chemicals, in the environment. To do this, soil chemistry studies chemical processes in soils; specifically, chemical reactions, species, and transformations within and between solid, gas, and liquid phases.

Understanding speciation of solids and chemicals in soils is key to predicting soil properties and how they will interact with plants, microbes, and animals. Soil solids consist of minerals and organic matter that can have high specific surface areas, creating high solid–solution interfaces and reactivity that facilitates surface reactions in the soils.

Table 1.4 Some specific units commonly used in soil chemistry.

Quantity	Unit	Symbol	Definition	Alternate unit
Land area	Hectare	Ha	10^4 m^2	1 Ha = 2.471 acres
Mass of 1 Ha of soil to 15 cm	Hectare furrow slice	HFS	2200 Mg Ha^{-1} 15 cm^{-1}	assuming soil bulk density of 1300 kg m^{-3}
Volume	Cubic meter	m^3		
	Liter	L	10^{-3} m^3	
Conductance	Siemens per meter	S m^{-1}	ohm^{-1}	1 S m^{-1} = 10 mho cm^{-1}; 1dS m^{-1} = 1 mho cm^{-1} 1 mho = 1 ohm^{-1}
Amount of ion charge	Moles charge	mol(+) or mol(−) or	mol ion times ion charge	equivalent = 1 mol charge; meq = 10^{-3} mol charge
Concentration	Moles per unit volume	M	mol L^{-1}	1M = 10^3 mol m^{-3} 1 mol m^{-3} = 1 mmol L^{-1}
	Moles charge per unit volume (normality)	N	mol charge L^{-1}	
	Millimolar	mM	10^{-3} mol L^{-1}	
	Micromolar	µM	10^{-6} mol L^{-1}	
Cation exchange capacity	Millimoles charge per kg solid	CEC	mmol(+) kg^{-1}	10 mmol(+) kg^{-1} =1 cmol kg^{-1} = 1 milliequivalent 100 g^{-1}
Specific surface area	Square meters per kilogram	SA	m^2 kg^{-1}	
Interatomic spacing	Nanometer	nm	10^{-9} m	1 nm = 10 angstroms (Å)

Example: Percent to Parts per Million

One percent is equal to a part per hundred. Units of soil constituents that are present less than 1% are often listed as parts per million (ppm). Parts per million are not SI units; rather, they are ratios of the amount of a part of one substance over one million parts of another substance. A solution 1 ppm in Ca^{2+} ions contains 1 g of Ca^{2+} ions in 1 million grams of solution. Care is needed when using the terms percent or ppm, or the similar unit part per billion (ppb), because they can be used to represent parts per whole on a volume, mass, or mixed basis. For example, the amount of chemical in soil can be represented as mg kg^{-1}, or a soil extraction can be represented by mg L^{-1}– both use the notation *ppm*. So, stating a soil has 100 ppm Zn, for example, without appropriate context is ambiguous. Because of the ambiguity, scientists would do well to move away from using the abbreviation ppm altogether, and instead use the actual units, such as mg kg^{-1}, and so on.

One percent is equivalent to 10 000 mg kg^{-1}. For example, if a soil contains 1.8% organic carbon, then it is 18 000 mg kg^{-1} (ppm) of organic carbon. The dimensional analysis is:

$$1.8\% \times \frac{10\,000\,mg\,kg^{-1}}{\%} = 18\,000\,mg\,kg^{-1} \tag{1.12}$$

Note that the percent units cancel.

Example: mg L⁻¹ to mmol L⁻¹

To convert a unit from mass basis to mole basis, such as mg L⁻¹ to mmol L⁻¹ requires using the atomic or molecular weight, which is the mass of a substance in 1 mole (6.022×10^{23} atoms or molecules) of substance. Units of atomic and molecular weights are g mol⁻¹. For convenience, because the unit mg L⁻¹ is often used to describe concentrations in soil chemistry, the atomic mass unit of mg mmol⁻¹ is often used instead of g mol⁻¹ (note both units are numerically equivalent), which saves from having to convert the milligram units to gram units. An example of conversion of mg L⁻¹ to mmol L⁻¹ follows. One g of soil is extracted with 10 mL of deionized (DI) water and measured to have a lead concentration of 8.06 mg L⁻¹, its molar concentration is 0.0389 mmol L⁻¹ (0.0000389 mol L⁻¹). A mol L⁻¹ is molar concentration (M). The dimensional analysis for this example is:

$$8.06 \frac{\text{mg Pb}}{\text{L}} \times \frac{1 \text{ mmol Pb}}{207.2 \text{ mg Pb}} = 0.0389 \text{ mmol Pb L}^{-1} \quad (1.13)$$

$$0.0389 \frac{\text{mmol Pb}}{\text{L}} \times \frac{1 \text{ mol Pb}}{1000 \text{ mmol Pb}} = 0.0000389 \text{ mol Pb L}^{-1} \quad (1.14)$$

IUPAC recommends not including the chemical identity in the units, for example, 0.0000389 mol L⁻¹ instead of 0.0000389 mol Pb L⁻¹, however in dimensional analysis it is convenient to include this for clarity. Lines through the units are used to illustrate the cancelation of the units in the numerator and denominator. Note that the measurement with the *least* number of significant figures dictates the number of significant figures in the answer.

Example: mg L⁻¹ in Soil Solution to mg kg⁻¹

In soil science, chemical concentrations in the soil are often represented by amount of chemical (either mass or number of moles) per mass of soil. Often, soils are extracted or digested in solutions and chemical concentrations are measured in the extracting solution as mg L⁻¹. To convert to mass of soil basis (mg kg⁻¹), the solution concentration is multiplied by the solution to solid ratio, sometimes referred to as solution:solid. For example, to convert the lead extracted from the problem in the above example to mg kg⁻¹, the following calculation is done:

$$8.06 \frac{\text{mg}}{\text{L}} \times \frac{10 \text{ mL solution}}{1 \text{ g soil}} \times \frac{1000 \text{ g}}{1 \text{ kg}} \times \frac{1 \text{ L}}{1000 \text{ mL}} = 80.6 \text{ mg kg}^{-1} \quad (1.15)$$

where the solution:solid ratio of 10 mL:1 g was used in the unit conversion. Thus, the result from the soil extraction is 80.6 mg kg⁻¹ of Pb solubilized by DI water. This can be assumed to be the water-soluble lead fraction from the soil.

An underlying principle of soil chemistry is that soils continually undergo fluxes of matter and energy that drive chemical reactions. The reactions that occur in soils are:

1 Adsorption and desorption
2 Precipitation and dissolution
3 Immobilization and mineralization
4 Oxidation and reduction
5 Complexation and disassociation
6 Dissolution and volatilization

Understanding how and what soil properties control reactions in soils allows for prediction of the behavior of nutrients and contaminants in soils. Soil properties such as soil pH, microbial activity, soil pore space, mineral composition, organic matter composition, chemical concentration, and soil texture are important soil properties that influence the reactivity and speciation of chemicals in soils. Topics covered in this textbook provide students with the knowledge to understand and predict the chemical processes in soils based on these properties.

Example: Moles of Charge on Soil Particles

An important measurement in soil science is the total charge on surfaces of the soil particles. It is typically measured in units of moles of charge per mass of soil. For example, for a negatively charged soil particle this would be mol(–) kg^{-1}; or in units of millimole of charge, mmol(–) kg^{-1}. The total negative charge on soil is typically expressed as total moles of charge of cations associated with the surface, or **cation exchange capacity (CEC)**. For example, consider 10 g soil that all of the cations on the surfaces were exchanged off the soil particles into 100 mL of extracting solution. Upon measuring the solution for all the exchanged cations, the total positive charge in the solution was calculated to be 32 mmol(+) L^{-1}. The cation exchange capacity is calculated as follows:

$$3.2 \frac{mmol(+)}{L} \times \frac{100 \text{ mL}}{10 \text{ g}} \times \frac{1000 \text{ g}}{1 \text{ kg}} \times \frac{1 \text{ L}}{1000 \text{ mL}}$$

$$= 320 \, mmol(-)kg^{-1} \tag{1.16}$$

In much of the soil science literature, CEC is expressed in units of centimoles of charge per kg (cmol kg^{-1}). This unit is not used in this textbook. The unit conversion to change between mmol(+) kg^{-1} to cmol(+) kg^{-1} follows.

$$320 \frac{mmol(+)}{L} \times \frac{1 \text{ cmol}(+)}{10 \text{ mmol}(+)} = 32 \text{ cmol}(+)kg^{-1} \tag{1.17}$$

Another unit used for CEC measurement is mmol per 100 g (mmol/100 g), which has the same numeric value as cmol kg^{-1}.

Questions

1. Starting with a cube 1 m on a side, calculate the change in surface area if sand, silt, and clay-size particles (assume perfectly stacking cubes) are making up the cube. How many particles would be in each size group?

2. What is the difference between chemical species, an ion, and an element?

3. Discuss how soil formation factors are affected by LeChatelier's principle.

4. Provide examples of all possible reactions that can occur in soils.

5. Which of the following are not considered a soil chemical: Pb^{2+}, Fe^{3+}, plant root, ferrihydrite, soil organic matter, soil microbe, $NaHCO_3$ (s), $AlOH_2^+$, CO_2 (g)?

6. How is free energy used to understand biogeochemical cycling of chemicals in soils?

7. If phosphorus availability to a plant is measured using an extraction of the soil, does the measured value represent bioaccessibility or bioavailability?

8. What factors cause elements in soils to deviate from the 1:1 line in **Figure 1.7**?

9. A soil is analyzed for total iron content by dissolving 0.1 g of soil in 10 mL of acid and digesting in a pressure bomb. The digest is filtered, diluted to a final volume of 100 mL, and analyzed for Fe. The total Fe in solution is 40 mg L^{-1}. What is the total iron concentration in the soil in mg kg^{-1} (ppm) and mmol kg^{-1}? Assuming all the iron occurs as the mineral ferrihydrite with a hypothetical formula of $Fe(OH)_3$, what mass percentage of the soil is iron oxide? Note: 1% = 10 000 ppm.

10. Why are total soil concentrations poor indicators of the amounts of ions that may enter the food chain?

11. By what mechanisms are ions held by soils?

12. How is soil chemistry knowledge useful to agriculturalists, environmentalists, toxicologists, public health professionals, and concerned citizens?

Bibliography

Bowen, H.J.M. 1979. Environmental Chemistry of the Elements Academic Press, London; New York.

Chadwick, O., and R. Graham. 2000. Pedogenic Processes, In M. E. Sumner, (ed.) Handbook of Soil Science. CRC Press, Boca Raton, FL.

Houghton, R.A. 2007. Balancing the global carbon budget. Annual Review of Earth and Planetary Sciences 35:313–347.

Lal, R. 2008. Sequestration of atmospheric CO_2 in global carbon pools. Energy and Environmental Science 1:86–100.

Lal, R. 2014. World soils and the carbon cycle in relation to climate change and food security, p. 32–66, In J. Weigelt, et al., (eds.) Soils in the Nexus. Oekom Verlag, München, Germany.

Markert, B. 1992. Presence and Significance of Naturally-Occurring Chemical-Elements of the Periodic System in the Plant Organism and Consequences for Future Investigations on Inorganic Environmental Chemistry in Ecosystems. Vegetatio 103:1–30.

Marschner, H. 1995. Mineral nutrition of higher plants. 2nd ed. Academic Press, London.

Scharlemann, J.P.W., E.V.J. Tanner, R. Hiederer, and V. Kapos. 2014. Global soil carbon: understanding and managing the largest terrestrial carbon pool. Carbon Management 5:81–91.

Tarnocai, C., J.G. Canadell, E.A.G. Schuur, P. Kuhry, G. Mazhitova, and S. Zimov. 2009. Soil organic carbon pools in the northern circumpolar permafrost region. Global Biogeochemical Cycles 23.

Trenberth, K.E., L. Smith, T.T. Qian, A. Dai, and J. Fasullo. 2007. Estimates of the global water budget and its annual cycle using observational and model data. Journal of Hydrometeorology 8:758–769.

US Environmental Protection Agency. 1992. Framework for Ecological Risk Assessment. Washington DC.

Way, J.T. 1850. On the power of soils to absorb manure. Journal of the Royal Agriculture Society of England Eleventh Part 1 No. XXV:313.

2 PROPERTIES OF ELEMENTS AND MOLECULES

2.1 Introduction

More than 150 years ago, Dmitri Mendeleev arranged the elements according to their properties, and introduced the concept of the periodic table (**Figure 2.1**). The elements are grouped as columns in the periodic table based on occupancy of their outer electron shells. They are further categorized into groups based on whether they are metallic, non-metallic, gas, or inert. Another categorization defines the elements as alkali elements, alkaline earth elements, transition metals, post-transition metals, metalloids, nonmetals, halogens, noble gases, rare earth elements (REE), actinides, and lanthanides. Rare earth elements are lanthanide elements plus scandium and yttrium. Despite their name, REE are quite common in soils and rocks in low concentrations; they are so named because they do not commonly occur in easily recovered ore-rich bodies. Elements that are within a group (column) often have similar properties, such as oxidation state (ion form) and reactivity. For example, Group 1 elements, the alkali metals lithium to francium, are all monovalent cations. In addition, element atomic size and mass increases from top row down: Li < Na < K < Rb < Cs < Fr.

The fundamental properties of elements and importance of these properties for understanding and predicting chemical processes are listed in **Table 2.1**. Some of the most important chemical properties in soils that control reactivity, availability, and speciation of chemicals are **oxidation state**, **ionic radius**, and type of **bonds** that occur within chemicals and in soil solids.

This chapter presents fundamental properties of atoms and molecules. The ionic and bonding properties discussed are key concepts used to understand how chemicals react in the environment. For example, the fundamental properties of elements explain ion exchange, activity of ions in soil solution, bonding of ions within minerals and organic matter, soil mineralogy, and organic chemical reactivity. The fundamental chemical concepts presented in this chapter

Soil Chemistry, Fifth Edition. Daniel G. Strawn, Hinrich L. Bohn, and George A. O'Connor.
© 2020 John Wiley & Sons Ltd. Published 2020 by John Wiley & Sons Ltd.

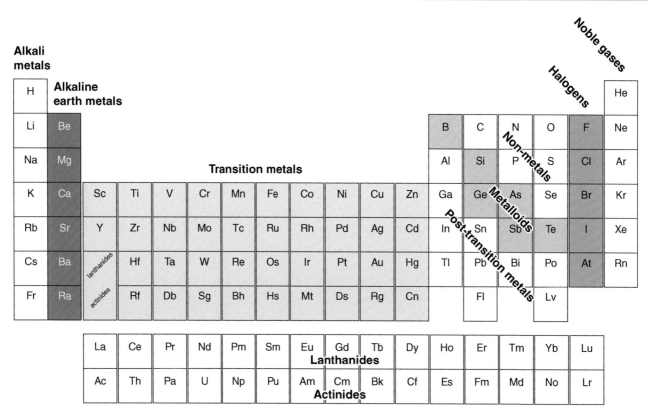

Figure 2.1 Periodic table of elements showing major categories.

Table 2.1 Fundamental properties of atoms, ions, and molecules and their relevance to reactivity and chemical speciation.

Element Property	Explanation
Atomic or molecular weight	Atom mass (g mol^{-1}) is a basic property used to calculate concentration.
Ionic radius	Ion size determines hydration properties and molecular and mineral structure.
Oxidation state	Ion charge determines electrostatic interactions.
Valence	The net charge on an ion or molecule, e.g., valence of $H_2PO_4^-$ is minus one.
Radioactivity	Emission of radioactive beta, gamma, alpha particles.
Electron affinity	The electron attractive power of an atom; useful for predicting ion–ion interactions and bonding.
Ionization energy	The energy required to remove an electron from an atom is important for oxidation state of atoms.
Electronegativity	Relative quantity that uses electron affinity and ionization energy to determine covalent and ionic character in molecular bonds.
Electron orbital geometry	Electrons reside in regions (electron orbitals), the shape of which is important for atomic and molecular bonding and reactivity, e.g., water molecules have lone pairs of electrons that account for many of the properties of H_2O.
Polarization of electron orbitals	Charge distribution in electron shells is shifted due to interaction of electron shells with neighboring elements
Molecular bond	Ions and molecules interact with each other and surfaces via either covalent, ionic, or van der Waal bonds.
Ionic potential	Charge density on an ion influences electrostatic interactions of ions with each other and soil particles.
Molecular geometry	Shape of molecules influences reactivity, e.g., chelates.

are used throughout this text to understand reactivity of chemicals in soils.

2.2   Ionization and ionic charge

In nature, most elements occur as ions (**Figure 2.2**). Notable exceptions are O_2, N_2, the inert gases, and the precious metals Au, Pt, and Ag. In reduced soils, such as occur in wetlands, some elements (e.g., sulfur and selenium) also occur in elemental forms. In the elemental state, atoms bind only to each other. As ions, elements are active and react with other ions and mineral or organic surfaces. Knowing the ionic species present in nature is a good starting point for determining chemical properties because ionization creates charge on cations and anions, which is a principal factor for determining elemental interactions, such as bonding and hydration. Special Topic Box 2.1 discusses the oxidation states of sulfur to illustrate how ionization affects chemical speciation.

Although oxidation state is a good place to start when predicting chemical behavior, many ions exist as molecules that have different valences than the individual element. Phosphorus, for example, is oxidation state positive five (P^{5+}), but exists in nature as the oxyanion phosphate (PO_4^{3-}), with a valence of negative three. In soils, phosphate exists as protonated acids (e.g., $H_2PO_4^-$ and HPO_4^{2-}), cation complexes, solids, organic phosphate compounds, or adsorbed onto mineral surfaces. Many other elements with small ionic radius and high positive charge exist as oxyanions, for example, sulfate (SO_4^{2-}), nitrate, (NO_3^-), carbonate (CO_3^{2-}), silicate (H_4SiO_4), selenate (SeO_4^{2-}), and arsenate (AsO_4^{3-}).

2.3   Ionic radius

Ionic radius is a function of the size of the nucleus, electron shell configuration, and interactions of the electrons and protons. Ionic radius is greater or less than atomic radius (**Table 2.3** and **Figure 2.3**). In soil chemistry, ionic radius is used to predict and understand mineral structure, ion exchange, and hydration of ions. For example, cation exchange preference depends partially on ionic radius of the cations as well as cation charge.

Ionic potential (IP) describes the charge density (magnitude of charge per area) of ions (**Figure 2.4**). It is

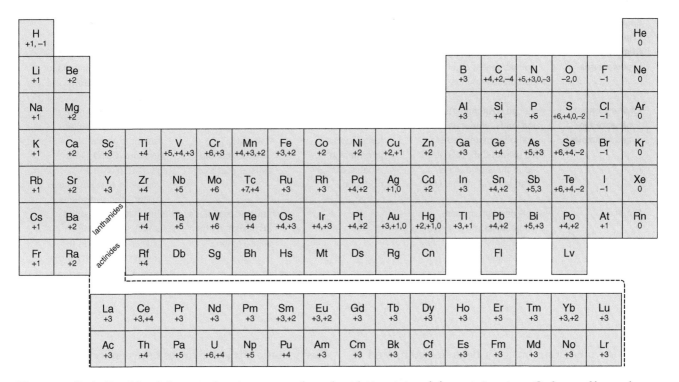

Figure 2.2 Periodic table of elements showing *common* formal oxidation states of elements in nature. Carbon, sulfur, and nitrogen have oxidation states in addition to those listed.

Special Topic Box 2.1   Oxidation states of sulfur in soil

In nature, sulfur has many different oxidation states, and exists as numerous different species including S^{2-} (sulfide), S^0 (elemental sulfur), S^{4+} (sulfur dioxide or sulfite, SO_2, SO_3^{2-}), and S^{6+} (sulfate, SO_4^{2-}) (**Table 2.2**). In some compounds, the formal oxidation state of sulfur is ambiguous because of the extremely covalent nature of sulfur bonds; thus, the sulfur atoms take on intermediate or average oxidation states between two bonded atoms.

In soils, the oxidation state of sulfur changes with soil redox state. In reducing environments, sulfide and elemental sulfur occur. Elemental sulfur is observed either as an intermediate from the oxidation of sulfide, or production of elemental sulfur in anaerobic respiration. Sulfide (S^{2-}) exists as metal sulfides, hydrogen sulfide gas, or thiol functional groups (R-SH) in organic compounds. The rotten egg smell coming from flooded soils is escaping hydrogen sulfide gas.

In oxidizing environments, sulfate is the stable sulfur species. It is a soluble anion that has intermediate adsorption on oxide mineral surfaces compared to chloride (weakly sorbed), and phosphate (strongly sorbed). In arid regions, sulfate forms salts such as gypsum or anhydrite.

Organic sulfur compounds are common in soils, existing as both reduced and oxidized sulfur functional groups. An important class of sulfur-containing organic functional groups are thiols, which consist of a sulfide bonded to a carbon compound (R-S-H; where R is the carbon-based compound). Another interesting sulfur functional group class occurring in soils is thiocyanate, which consist of reduced sulfur bonded between a carbon and cyanide molecule ($R - S - C \equiv N$). Thiocyanate compounds are produced by cruciferous plants (e.g., mustard, cabbage, and broccoli) and have natural pesticide properties.

Table 2.2 Sulfur oxidation states and species, and their common compounds and occurrences.

	S^{2-}	S^0	S^{4+}	S^{6+}
Species observed in nature	Sulfide anion	Elemental sulfur	Sulfite anion SO_3^{2-} sulfur dioxide SO_2	Sulfate anion SO_4^{2-}
Common compounds	Metal sulfides, hydrogen sulfide gases, organic sulfide compounds	Elemental sulfur	Sulfite is conjugate base in salts	Sulfuric acid, sulfate salts, metal sulfate minerals, or aqueous complexes, thiosulfate ($S_2O_3^{2-}$)
Environments observed in and examples	Reduced environments: wetlands, estuaries, flooded soils. Examples include FeS_2 (pyrite) CuS, H_2S, R-SH as in the amino acids cysteine and methionine.	Shale rocks, lake or marine sediments, fossil fuels, intermediate product of H_2S oxidation	SO_2 is emitted into atmosphere by fossil-fuel combustion and causes acid rain.	Oxidized soils and natural waters contain sulfate anion; examples: gypsum ($CaSO_4$) and jarosite ($FeK(SO_4)_2$) occur in arid soils that have high sulfate concentrations.
Notes	Highly covalent in many compounds; thus, formal oxidation state of S^{2-} may not be applicable.		In aqueous solutions, SO_2 exists as the sulfite anion.	Thiosulfate has a sulfide anion substituting for an oxygen position in the sulfate anion ($S(O_3S)^{2-}$).

useful for predicting an ion's behavior compared to other ions. Ionic potential combines ionic charge (z) and ionic radius (r):

$$IP = \frac{z}{r} \tag{2.1}$$

Electrostatic attractive forces between ions can be predicted by ionic potential, which are important factors controlling bonding between cations and anions. Ionic potential is also useful for understanding the amount of water surrounding an ion (**hydration sphere**), interactions of ions with surfaces (e.g., cation exchange),

Table 2.3 Ionic radius and hydration properties of metal ions.

Ion	Ionic radius (nm)[1]	Inner shell H_2O coordination number[2]	Hydrated radius (nm)	
			Diffusion measurement[3]	Molecular modeling or spectroscopic[4]
Alkali ions				
Na^+	0.116	6	0.358	0.441–0.48
K^+	0.165	6–8	0.331	0.460–0.530
Cs^+	0.195	8–12	0.329	
Alkaline earth ions				
Mg^{2+}	0.086	6	0.428	0.410–0.447
Ca^{2+}	0.126	6–8	0.412	0.448–0.460
Sr^{2+}	0.140	8	0.412	0.492–0.494
Metals				
Mn^{2+}	0.081	6	0.438	0.417–0.443
Fe^{2+}	0.075	6	0.428	0.430–0.451
Fe^{3+}	0.069	6	0.457	0.409–0.480
Co^{2+}	0.079	6	0.423	0.393–0.428
Ni^{2+}	0.083		0.404	0.399–0.433
Cu^{2+}	0.087	6	0.419	0.395–0.420
Zn^{2+}	0.088	6	0.430	0.395–0.426
Al^{3+}	0.068	6	0.475	0.399–0.415

[1] Shannon (1976).
[2] First shell water coordination estimates from Persson (2010).
[3] Nightingale (1959).
[4] Ohtaki and Radnai (1993).

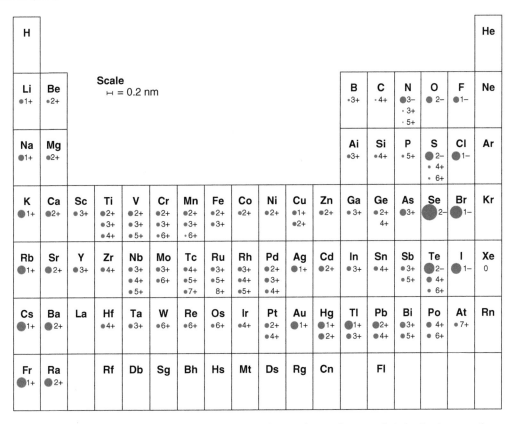

Figure 2.3 Periodic table showing relative sizes of some ions. Valence of ions changes their ionic size; see, for example, sulfur. Data are from Shannon (1976).

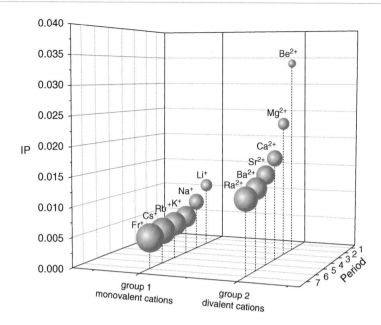

Figure 2.4 Ionic potential of alkali metals and alkaline earth metals. Ionic potential units are electron charge per picometer. The spheres represent relative sizes of the cations. For a given charge, the IP of the smallest cations are greater than the larger cations because the charge is distributed over a smaller total surface area. Ionic potential of divalent cations is greater than monovalent cations of similar size. For example, Ca^{2+} and Na^+ are similar in size, but Ca^{2+} has twice the ionic potential. As a result, Ca^{2+} cations have greater electrostatic interaction forces than Na^+ cations.

and ion distribution in the *swarm* of ions occurring on mineral surfaces (diffuse double layer; covered in Chapter 10).

2.4 Molecular bonds

Reactions of chemicals within soils usually involves making and breaking of bonds. Bonds are attractive inter-molecular forces, and are categorized as either **covalent, ionic,** or **van der Waal bonds.** The van der Waal bonds are intermolecular electrostatic forces that include **hydrogen bonds,** dipole–dipole bonds, and London dispersion forces (**Figure 2.5**). In soil systems, covalent bonds are the strongest, while London dispersion forces are the weakest (**Figure 2.5**). For example, the Si-O bond in quartz (mineral formula SiO_2) is a strong covalent bond, thus, quartz resists weathering and typically exists in soils as unweathered sand-size particles. Similarly, the strong covalent bonds between lead and oxygen-functional groups on organic matter and oxide minerals cause Pb^{2+} to be adsorbed strongly in soil, and thus not readily desorbed into soil solution and minimally available for plant uptake.

Linus Pauling, two-time Nobel Laureate, quantified the relative bond strength between atoms using a quantity called **electronegativity**, which describes how strongly an ion attracts an electron to itself. Two bonding atoms that have similar electronegativity share electrons. When one atom in a molecule is much more electronegative than another, the bond is ionic. There is a range of electronegativity differences in molecular bonds, and bonds often occur as a hybrid between covalent and ionic. The strongest molecular bonds with respect to weathering are those with the greatest degree of covalent bonding. For example, the mica mineral muscovite $(KAl_2(Si_3Al)O_{10}(OH)_2)$ has Si-O bonds, K-O bonds, and Al-O bonds. Si-O bonds are the most covalent, Al-O bonds are next in covalent bond character, and K-O bonds are more ionic in character than covalent. The K-O bonds in mica are the most susceptible to weathering because water molecules *break* ionic bonds more easily than the more covalent Al-O and Si-O bonds. Thus, K^+ dissolution is the first step in weathering of mica and is an important process for release of K^+ for use as a plant nutrient. The concept of electronegativity only partially predicts the nature of molecular bonds;

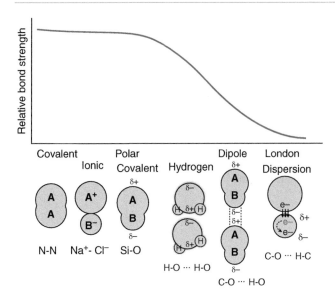

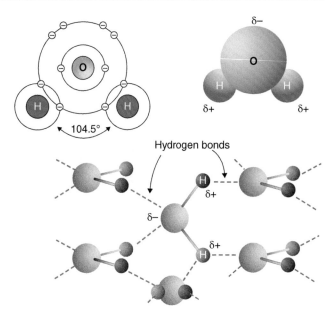

Figure 2.5 Six basic types of bonds that occur in soil chemicals and minerals are illustrated, along with their relative bond strength, and examples of element bond types. *A* and *B* denote different ions. δ+ and δ− represent partial positive and negative charges within atoms or molecules. Ionic and polar covalent bonds are depicted as having similar strength; however, bond strength depends on whether the stress is physical or chemical, as well as the ion properties within the molecule. For example, in the presence of water, polar-covalent bonded ions in a solid may be stronger than many ionic-bonded solids. In contrast, on the interior of the solid (away from the solution), ionic bonds are relatively stronger than covalent bonds to physical forces.

Figure 2.6 Electron orbital dot structure of a water molecule showing the two lone-pairs of electrons on oxygen and the electron poor region on the hydrogen atoms. The electron shell structure creates a polar molecule with two δ− and two δ+ regions. The charged regions of the molecules interact with each other via dipole-dipole interactions, which in water are called hydrogen bonds. In the bottom figure, a network of water-bonded molecules is joined by hydrogen bonds. The network is dynamic, breaking and making new bonds in fractions of a second.

however, determining percent ionic or covalent character of bonds is useful for basic understanding of molecular bonding and reactivity in minerals, organic matter, and solutions.

2.5   Nature of water and hydration of ions

The geometric shape of water molecules and their electronic properties dictates the behavior of ions in aqueous solution. Water molecules interact with each other through **hydrogen bonds** (**Figure 2.6**). The interaction is responsible for water's relatively high boiling point and specific heat (the amount of energy needed to raise its temperature). The H^+ ions in H_2O are 105° apart, creating a nonlinear, polar molecule (**Figure 2.6**). The water molecule exists as a **dipole** with a positive end (the hydrogen side) and a negative end. The positive

end of one water molecule attracts the negative end of another water molecule. Water molecules cluster together in groups, ranging from dimers to more than 20 water molecules per cluster. Because of thermal motion, the groups continually break apart and reform. In ice, the structure is a rigid hexagonal packing of water molecules. In liquid water, clusters of water molecules are also structured, although not as rigidly. The clusters can slide closer together, so water is denser than ice. On a molecular scale, water is like a flexible and dynamic ice slurry.

Ions and charged surfaces can break down the *ice slurry* structure of water. The electric charges are stronger than hydrogen bonds and tend to pull water molecules away from each other by attracting the positive or negative ends of the water dipoles (**Figure 2.7**). Ions and water molecules are constantly in motion, but they remain in the vicinity of each other for various periods. Water molecules that remain near an ion comprise a sphere of water molecules around the ion

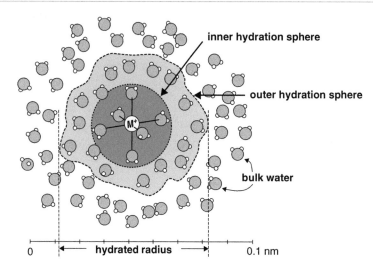

Figure 2.7 Hydration sphere of water on a cation M⁺. Water in the inner hydration sphere is electrostatically attracted to the ions, orienting the partially negative side of the water molecule toward the charged cation. The water molecule orientations are reversed in anions. The outer hydration sphere holds water more loosely, meaning the water molecules are much more mobile. Water molecules in the bulk solution have little to no interaction with the cation, however are exchanged with water molecules in the outer and inner hydration sphere.

(a **solvation sphere** or sheath) (**Figure 2.7**). Four to 12 water molecules exist in the closest solvation sphere surrounding an ion. The exact primary hydration number depends on the central ion radius (**Table 2.3**). Smaller ions fit fewer water molecules in their primary hydration sphere than larger ions – consider the analogy of the maximum number of cows that can surround a small water trough compared to a larger water trough.

Outside of the primary solvation sphere is a second sphere of water molecules also affected by the ion's charge (**Figure 2.7**). These water molecules have been torn away from the bulk water to some extent but are not as closely associated with the ion as those in the primary solvation sphere. The orientation of water molecules in the primary solvation sphere, and the more random orientation of water molecules in the secondary sphere dissipate the ion's charge within 0.1 to 0.3 nm of the central ion. This charge dissipation causes the ions to interact less electrostatically with each other. There are typically 10–30 water molecules in the second hydration sphere. The distance between the center of the ion and the outside of the second hydration sphere is the **hydrated radius (Figure 2.7)**. Measuring hydrated radius is difficult because water molecules in the second hydration sphere are not *fixed* – meaning they exchange with the bulk water and inner-sphere water on time frames of microseconds to

picoseconds. Estimates of hydrated radii for ions using spectroscopic measurements and molecular modeling are reported in **Table 2.3**. The relationship between ionic size and hydrated radius for alkali and alkaline earth metals is shown in **Figure 2.8**. The relationship holds for many ions in solution, but for some ions, atomic and molecular forces can increase or decrease the attraction of water molecules affecting the hydrated radius. The force of attraction of water molecules increases as the charge density (ionic potential (IP)) increases. Thus, for example, lithium is the smallest alkali metal and has a greater ionic potential and hydrated radius than the larger alkali metals below it on the periodic table. Similarly, due to the increase in charge of the alkaline earth metals (divalent cations), their IP is much greater than the monovalent alkali metals, and the force between the alkaline earth cations and water molecules is greater, making their hydrated radius greater.

The force (F) between an ion and a water dipole charge is calculated using **Coulomb's law**

$$F = k\frac{q_1 q_2}{r^2} \tag{2.2}$$

where q_i is the magnitude of charge, which can be either the same or different in sign; r is the separation

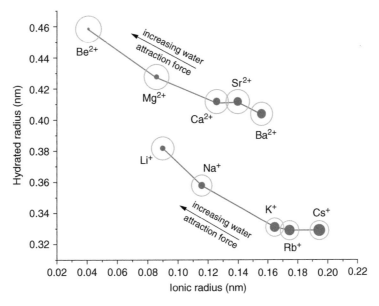

Figure 2.8 Hydrated radii of alkali and alkaline earth cation plotted as a function of their ionic radii. The hydrated radius is shown as an open circle and drawn to relative size of the ionic radius shown as a filled circle. Data are listed in **Table 2.3**.

distance between charges; and k is a constant related to the dielectric properties of water. Coulomb's law states:

1 The closer the charges (smaller r), the greater the attractive or repulsive force.
2 Greater magnitude charges will generate greater electrostatic forces.

A good analogy for Coulomb's law is holding two magnets near each other and feeling the force of attraction or repulsion as a function of separation distance or size of the magnet. Extending this concept to ions in water, the separation distance (r) between small ions and water dipoles is less than larger ions, and thus smaller ions have greater attractive forces for water than larger ions. Also, higher charged ions have more attractive force for water than lower charged ions.

Although measurement of hydrated radius is ambiguous, **heat of hydration** for ions is a reliably measured parameter correlated to the size of the hydration sphere. Heat of hydration is the *energy* released when an ion is hydrated, which is primarily enthalpy $\left(\Delta H^{\circ}_{hyd}\right)$ because energy from hydration entropy is small. Heat of hydration is proportional to the amount of work required to remove the water molecule from the charge of the ion. As force of attraction between ion and water dipole increases, heat of hydration increases (becomes more negative, where more negative values are a greater release of energy).

The work in moving an electrostatic charge, such as a water dipole, through a charged field, such as an ion, is described by integrating (summing up over small changes in ion separation distance) Columbic force (**Eq. 2.2**), resulting in Coulomb's energy formula

$$E = k\frac{q_1 q_2}{r} \tag{2.3}$$

where E is the total energy to move a charge outside of the electric field, q_i is the magnitude of charges, and r is the separation distance of charges. For a hydrated ion, Eq. 2.3, shows that the total energy (E), which is equivalent to the hydration energy of an ion $\sim \left(\Delta H^{\circ}_{hyd}\right)$, is a linear function of the magnitude of the charges on the ion and water molecule divided by their separation distance ($q_1 q_2/r$). The relationship between ionic radius and charge magnitude to $\left(\Delta H^{\circ}_{hyd}\right)$ is shown in **Figure 2.9**. The linearity of the data in **Figure 2.9** verifies that the strength of water hydration is controlled by ion charge magnitude and radius, as predicted by Eq. 2.3.

The heat of hydration for an ion can be used to quantify the nature of the hydration sphere, and is useful for comparing ion behavior in solution, cation and

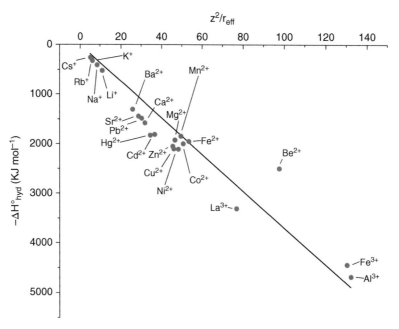

Figure 2.9 Negative heat of hydration $\left(\Delta H^{\circ}_{\text{hyd}}\right)$ for cations as a function of Coulomb-energy term (z^2/r_{eff}), where r_{eff} is the effective radius of the ion and its first shell of water molecules. The more negative the value of $\left(\Delta H^{\circ}_{\text{hyd}}\right)$, the stronger the interaction between the ion and water molecules in hydration spheres. Hydration energy is greater (increases negatively) with increasing ion charge and with decreasing ion size. Modified from Richens (1997), with permission.

anion interactions in aqueous solutions, and ion interaction with charged surfaces such as cation exchange.

A closely related concept to hydration energy is water exchange rate; the speed at which water molecules exchange from the hydration sphere of an ion to the bulk solution. Water exchange rates are experimentally measured, and for the same reasons discussed above, are proportional to ionic radius and charge. Water exchange rate is often used in a similar manner as hydration energy to compare aqueous ion properties (see, e.g., Special Topic Box 2.2).

2.6   Ligands and metal bonds

In nature, ligands are ions or molecules with lone pairs of electrons available for ionic or covalent bonding. Common ligands in nature include many anions, such as HCO_3^-, OH^-, SO_4^{2-}, HPO_4^{2-}, Cl^-, F^-, NO_3^-, $B(OH)_3$, R-COO$^-$, and HS$^-$, and some nonionic molecules like NH_3. The concentrations of ligands in soil solutions and natural waters vary, depending on the source and properties of waters. Carbonate, chloride, and sulfate are the predominant anion ligands in most natural systems, with concentrations ranging from 10^{-2} to 10^{-5} M. Phosphate,

fluoride, nitrate, and organic ligands are the second most common anion ligand types in natural waters. Arsenate (AsO_4^{3-}), selenate (SeO_4^{2-}), and borate ($B(OH)_3$) are less common ligand anions in the environment.

Ligands commonly form bonds with metal cations that have unoccupied electron orbitals, for example, an iron hydroxide aqueous complex such as $Fe(OH)^{2+}$ is an important metal hydroxide-ligand species in soil solution. Many transition metals have ligand coordination spheres of six, meaning six ligands are coordinated to the metal. For example, the aqueous $FeOH^{2+}$ molecule is actually a $FeOH(H_2O)_5^{2+}$ molecule, with one hydroxide and five water molecule ligands in the first shell directly surrounding the iron atom (**Figure 2.11**).

Some molecules have multiple ligand functional groups on them and may form more than one bond with a metal; such coordination complexes are called **chelates**. The organic molecules, oxalate, citrate, and ethylenediaminetetracetate (EDTA) are examples of common chelates in soil and water systems (see Figure 4.17).

The importance of ligand-metal bonding is that it changes the speciation of the ions and molecules involved in the complex. For example, in the $Fe(OH)^{2+}$ complex, iron with an oxidation state of three occurs as a *divalent* iron-hydroxide ligand species. In many

Special Topic Box 2.2   Using ion hydration to predict bioavailability

Figure 2.10 shows an example of the relationship between water exchange rates of cations and the bioaccessibility of ions in soil. Bioaccessibility is distinct from bioavailability because it measures relative availability of a chemical in a *simulated* organism. Such tests are useful for making assessments of potential bioavailability.

The data in **Figure 2.10** show that bioaccessibility of ions from contaminated soils is a function of water exchange rate, implying that bioaccessibility is controlled by *kinetics* (e.g., desorption and dissolution kinetics) as opposed to thermodynamic *equilibrium*.

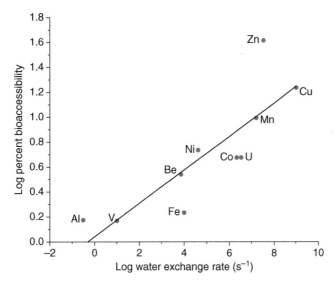

Figure 2.10 Effect of water exchange rate on the bioaccessibility of metals from soils collected from a contaminated site in Iqaluit, Nunavut, Canada. Soil physicochemical properties also affect metal bioaccessibility, so the regressed slope and intercept of this relationship are not necessarily applicable to other soils. Adapted from Laird et al. (2011). Reproduced with permission of American Chemical Society.

cases, this changes the solubility of metals in solutions (see Section 4.8). Plants use this phenomenon to increase availability of iron in their rhizosphere by releasing organic ligands called **siderophores** that have a high specificity for iron, thus increasing the availability of Fe(III)[1] in the soil solution for uptake by the plant root.

The bond strength between a ligand and metal varies, depending on the nature of the bonding electrons in the ligand and the metal. For example, ligands such as the oxygen in a carboxylic acid functional group (R-COOH) form strong covalent bonds with many metals, while other ligands, such as nitrate, form relatively weaker electrostatic bonds with most metals. The strength of metal-ligand bonds is related to the ionization energy, electron orbital geometry, and ionic radius of the metal and ligand. The varying metal-ligand bond strengths creates a selectivity or affinity series of the metal for different ligands. This affinity is quantified using thermodynamics to calculate the stability constant, which is represented using the letter K. For example, for the displacement of one of the six water molecules surrounding an Fe^{3+} ion by a hydroxide, the following reaction occurs:

$$Fe(H_2O)_6^{3+} + OH^- = FeOH(H_2O)_5^{2+} + H_2O \qquad (2.4)$$

The stability constant for this reaction, $K_{FeOH^{2+}}$, can be used to predict which iron species will be stable under

---

[1] The roman numeral in *Fe(III)* is used to indicate the oxidation state of the iron atom regardless of whether it occurs as a free ion or an ion complexed to ligands or molecules. The valence of the different ferric iron species can vary depending on the molecular species. The notation is used to refer to all species in the solution (or solid) that have iron with oxidation state three in it.

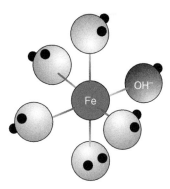

Figure 2.11 Aqueous Fe(III) ion showing a hydrolysis complex ($FeOH^{2+}$), where one of the six ligand positions is occupied by a hydroxide ligand.

a specified set of solution conditions. This topic is dealt with in more detail in Chapter 4.

2.7 Summary of important concepts of elemental and molecular properties

Diameter and electron configuration are characteristics of elements that affect how they interact with each other, and their speciation. These elemental properties can be used to predict and understand formation of ions, molecular bonds, and hydration. Properties such as electron affinity, valence, ionic potential, ionic radius, hydration energy, and bond type describe how the elements interact, react, and change species.

The fundamental concepts presented in this chapter are used to understand chemical reactions and are drawn upon in all of the chapters in this book. For example, cation adsorption selectivity on a clay mineral is predicted by ion valence and ion size; and desorption behavior of phosphate from iron oxide mineral surfaces is slow because phosphate forms strong ligand bonds with the Fe-OH surface functional groups.

Application of fundamental chemistry to nature is often difficult because the systems are complex. However, even in complex systems, the fundamentals of physics and chemistry are preserved and can be used to advance understanding and prediction of the fate and behavior of chemicals in the environment.

Questions

1 What is the oxidation state of the cationic element and valence of molecule in the following chemical species: CO_2, HCO_3^-, NO_3^-, Pb^{2+}, $Fe(OH)_2^+$, SO_4^{2-}, and SO_3^{2-}?

2 Arrange the following in order of ionic radius, hydrated radii, ionic potentials, and water exchange rates: Na^+, Ca^{2+}, Al^{3+}, Zn^{2+}, Pb^{2+}.

3 Silicon has an electronegativity of 1.9 and Al^{3+} has an electronegativity of 1.6. Which will have a more covalent bond with oxygen (electronegativity = 3.44)? How might this affect dissolution of a mineral containing these elements?

4 What is the relationship between hydration energy, hydrated radius, and water exchange rate? What is a principle force that explains water–ion interactions?

5 Describe the difference between ionic, covalent, polar-covalent, hydrogen, and van der Waals bonding.

6 Which bonds in the mineral biotite ($K(Si_3Al)$ $(Mg,Fe)_2O_{10}(OH)_2$) would be the least likely to be broken in an aqueous solution? (Hint: Each of the cations is coordinated by oxygen or hydroxide ligands; see clay structures in Chapter 7.)

Bibliography

Laird, B.D., D. Peak, and S.D. Siciliano. 2011. Bioaccessibility of metal cations in soil is linearly related to its water exchange rate constant. Environmental Science & Technology 45:4139–4144.

Nightingale, E.R. 1959. Phenomenological theory of ion solvation – effective radii of hydrated ions. Journal of Physical Chemistry 63:1381–1387.

Ohtaki, H., and T. Radnai. 1993. Structure and dynamics of hydrated ions. Chemical Reviews 93:1157–1204.

Persson, I. 2010. Hydrated metal ions in aqueous solution: How regular are their structures? Pure and Applied Chemistry 82:1901–1917.

Richens, D.T. 1997. The Chemistry of Aqua Ions. John Wiley and Sons, New York.

Shannon, R.D. 1976. Revised effective ionic-radii and systematic studies of interatomic distances in halides and chalcogenides. Acta Crystallographica Section A 32:751–767.

3 CHARACTERISTICS OF CHEMICALS IN SOILS

3.1 Introduction

Many questions about chemicals in soils relate to how much is there and what is the mobility and bioavailability of the chemical. Predicting bioavailability requires knowing something about the chemical's solubility, speciation, and the target organism. **Figure 3.1** shows a classic diagram often used to illustrate how soil pH affects plant nutrient availability. The graph is a useful picture of the relationship between soil properties, chemical properties, and plant nutrition, but is not accurate enough to base best management practices for chemicals in the environment. Furthermore, interactions of soil microbes with soil chemicals has significant effects on chemicals, and thus viewing soil physicochemical properties alone to predict chemical processes is inherently an incomplete picture.

This chapter discusses some of the characteristics of chemicals in the environment that are important for determining their fate, transport, and bioavailability.

Classically, soil chemistry has focused more on inorganic chemicals than organic chemicals. This chapter continues the focus on inorganic chemicals but provides an overview of common organic chemical types that occur in the environment, including pesticides, industrial chemicals, pharmaceuticals, and chemicals of emerging concern.

3.2 Occurrence of elements in soils

As Earth formed and cooled during its origin, the lighter elements tended to come to the surface. Earth's center is thought to be iron-rich and has a density of >6000 kg m^{-3}. The density of the outer crust, or mantle, is about 2800 kg m^{-3}. The density of the rock minerals at the Earth's surface is about 2650 kg m^{-3}. The elements in rock minerals at Earth's surface are the starting materials for soils and contain the essential elements from which soil and life evolved. Some of the

Soil Chemistry, Fifth Edition. Daniel G. Strawn, Hinrich L. Bohn, and George A. O'Connor.
© 2020 John Wiley & Sons Ltd. Published 2020 by John Wiley & Sons Ltd.

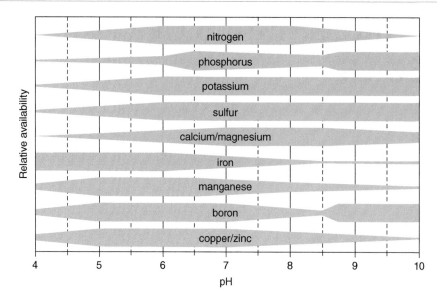

Figure 3.1 General trends in nutrient availability as a function of soil pH. The wider portions of the bars imply greater availability of the ions for plant uptake. Disclaimer in Truog's original paper: *"The writer wishes to emphasize that 'the chart' represents a generalized presentation, which is, tentative in certain respects"* … *"in the reaction range of pH 5.5 to 8.5, it is believed that the chart presents a fairly reliable picture."* Source: Adapted from Truog (1946). Reproduced with permission of ACSESS.

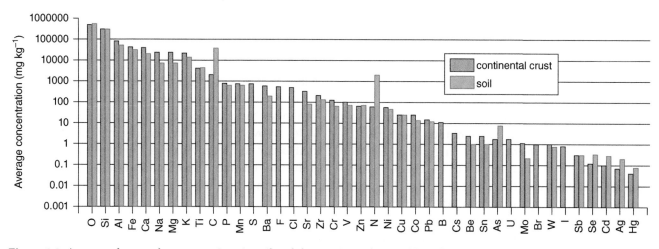

Figure 3.2 Average elemental concentrations in soil and the continental crust. Note the *y-axis* is logarithmic. Soil data are from the mean of more than 1000 non-anthropogenic-influenced soils collected from regions across the United States (Wilson et al., 2008). Soils were sampled to represent the major soil types found in most of the regions within the United States. Continental crust concentrations are from Wedepohl (1995).

heavier elements are also essential for life, and some are detrimental.

Only a few elements make up more than 99% of the mass of Earth's crust: oxygen, silicon, aluminum, iron, calcium, sodium, magnesium, and potassium (**Figure 3.2**). Other elements are present in small, but important amounts. By volume, most of the minerals in Earth's crust are made up of oxygen atoms. Soils

contain all the natural chemical elements in the periodic table. Even the rarest natural element, the alkali metal francium (^{87}Fr), which has never been isolated from a natural sample, occasionally appears as faint lines in the emission spectrum of arid soil extracts where alkali metal salts accumulate.

Like the crustal rocks, most of the mass of soil is comprised of oxygen, followed by silicon, aluminum, carbon, and iron. (**Figure 3.3**). Elemental contents of soils are highly variable, especially for nutrients such as nitrogen, sulfur, phosphorus, etc., which have standard deviations that are greater than the means reported in **Figure 3.3**. Soils are also distinct from Earth's crust because carbon may comprise a significant percentage of the mass, ranging from less than 1% to more than 50% in Histosols. Because of the variability in element concentrations in soils, averages derived from broad geographic areas should not be used to infer element composition at a particular site, nor on which to base management decisions.

For most purposes, the important elements in soil chemistry are those that are essential or toxic to living organisms. **Figure 3.4** shows the **essential elements** for plants and animals, and commonly occurring toxic elements. Essential elements, commonly referred to as essential nutrients, are the elements from which plants, and therefore animals, evolved. Essential nutrients are required for an organism to complete its life cycle. The amounts of the elements required vary with species of organism. Most organisms have a lower and upper limit of required elements. Organisms suffer deficiency when the bioavailable concentrations are below the lower limit required. Increasing bioavailable amounts above the upper boundary does not increase health of the organism; and if the concentration is well above the beneficial range, elements can be toxic. For some essential nutrients, the range between deficiency and toxicity is very narrow, making over-dosage by way of fertilization or inadvertent contamination common;

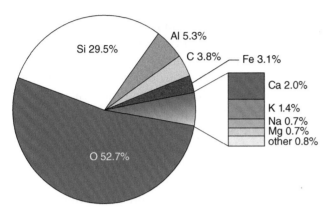

Figure 3.3 Percent mass composition of elements in soil. Total carbon includes inorganic and organic carbon. Nine elements make up >99% of the mass of soil, and oxygen predominates. Soil data are from **Figure 3.2**.

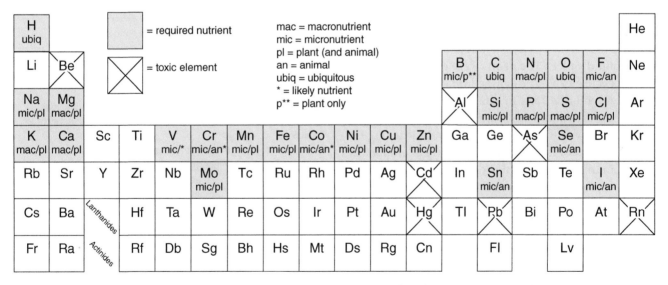

Figure 3.4 Periodic table of elements, not including lanthanides and actinides, showing nutrients and common contaminants. Nutrients are differentiated as being micro or macro nutrients for plants and animals. Most plant nutrients are also nutrients for animals. However, animals require some micronutrients that plants do not. Plant requirements for Cr and Co are uncertain. Requirement of V by animals is uncertain. Sodium is a micronutrient in C4 plants, and a macronutrient for animals. Silicon is a micronutrient for some plants; for example, grasses require significant amounts of Si. Most of the elements have no known nutrient status and are present in amounts too small to be of concern as potential contaminants. In some environments, concentrations of uncommon elements are elevated and present contamination problems. For example, radionuclide elements, such as Cs and U, are present at elevated levels at nuclear testing and waste disposal sites.

these elements are micronutrients. Boron is an example of a plant micronutrient in soils that has a narrow range between deficiency and toxicity. Selenium is an example of an animal micronutrient with a narrow range between deficiency and toxicity.

Water, plants, and dust contain many ions, so the exposure of organisms to most of the elements in the periodic table is continuous, but in very small amounts. Many of the elements in **Figure 3.4** are shown as neither essential nor toxic because our knowledge is incomplete. Some elements not highlighted are likely essential, and many elements not highlighted are certainly toxic if enough is taken up by an organism.

Figure 3.4 omits the lanthanides and actinides. The lanthanides, though widespread at low concentrations in soils, have not been investigated for their essentiality or toxicity to organisms. The actinides, particularly plutonium, are generally toxic to animals, even when present at low concentrations in soils. Contamination of soils with actinides has occurred from radionuclide research and use in energy production and weapons.

Proving essentiality of chemicals is often difficult. For example, in the 1950s experiments were done to determine if chlorine was an essential element for plants. Experiments used annual plants grown in greenhouses with carefully filtered air to remove dust and water droplets. The plants were grown hydroponically because even acid-washed quartz sand contains sufficient chloride ions to satisfy the plant's requirements. Additionally, the water was doubly distilled and not allowed to contact glass to avoid leaching Cl^- out of the glass. Plants had to be second generationally grown in this environment because normal seeds contained enough Cl^- to satisfy the plant's life-cycle requirement. These experiments provided new insight into plant micronutrient requirements for chloride. Because chloride input to most regions comes from rain sourced from the oceans, and because soils weakly retain chloride, continental regions far from the oceans can have low concentrations in the soil–plant–water system, causing chloride deficiencies that subtly affect plant and animal health.

Humans rely on ecosystem services, whether it is agriculture, watershed processes, or parks and open spaces, thus, managing the environment is a topic of great concern. Best management practices include adding chemicals to soils or onto plants (e.g., pesticides and fertilizers), understanding the fate and bioavailability of chemicals in soils, and in many cases, remediating environments that have levels of chemicals in them that are detrimental to the ecosystem. In addition, chemical cycles in natural systems are important to understand because they affect water quality and ecosystem health. The most common human sources of chemicals in the environment are industry, agriculture, municipal, and residential (**Table 3.1**).

Plants take up nutrients and contaminants from soil solution, which are supplied from soil processes (e.g., desorption, mineralization, and mineral dissolution). **Figure 3.5** shows processes of soil chemical uptake by an organism, such as a plant root. Soil chemicals in the *inert fraction*, such as insoluble minerals, are only slowly available to the soil solution and absorption by an organism. The *reactive phases* of chemicals in soil readily transfer between the soil solid phase and the solution phase (adsorption and desorption), which are important mechanisms for buffering of nutrients in soil solutions. For example, a goal of in-situ

Table 3.1 Sources and types of chemicals input into the environment from human activities.

Source	Chemicals introduced into environment
Agriculture	
Agriculture fields	Fertilizers, pesticides, manure, biosolids, CO_2, sediments, dust, gases (nitrogen)
Livestock	Veterinary pharmaceuticals, manure, liquid waste, gases (methane, CO_2 NH_3), odors
Industry	
Stack emissions	Particulates, gases (CO_2), Hg, As, Se, polyaromatic hydrocarbons
Mining	Metal and metalloid-laden mine-waste, smelting, and refining waste
Industrial discharges	Contaminated solids, gases and liquids, organic and inorganic hazardous waste
Municipal	
Water treatment plants	Phosphorus and nitrogen, biosolids, cosmetics and pharmaceuticals, nanoparticles, perfluorinated chemicals (e.g., PFOS, PFOA)
Waste disposal facilities	Gases (methane), metals and metalloids, organic liquids, and gases
Residential	
Homeowners	Garden fertilizers and pesticides, septic and leach fields, automobile fluids and exhaust, runoff from rooftops and yards
Individuals	Automobile emissions, litter

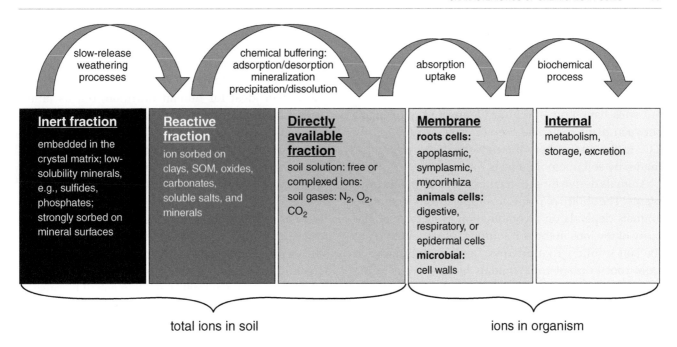

Figure 3.5 Soil and biological processes that affect uptake of nutrients and contaminants in soils. Adapted from Rocha et al. (2011). Reproduced with permission of Elsevier.

remediation of contaminated soils is to move chemicals from the reactive fraction to the inert fraction. Over time, many contaminants transform into more inert phases, and soil amendments can promote further transformation of the chemicals to nonreactive phases. Examples of amendments that can immobilize metals in soils are lime, organic matter, and phosphate.

3.3 Essential elements

All organisms need the essential macronutrients C, H, O, P, K, N, S, Ca, Fe, and Mg, in relatively large amounts to complete their life cycle (**Figure 3.4**). Iron is sometimes categorized as a micronutrient. A mnemonic phrase to remember the macronutrients is, "<u>see Hopkins cafe</u>, <u>m</u>ighty <u>g</u>ood." Animals additionally require Na and Cl as macronutrients, but since global animal biomass is only 1/10 000 that of plant biomass, animal requirements are insignificant in global-soil biogeochemical cycling of nutrients. Animal nutrient requirements, however, can affect local biogeochemical processes in a field or watershed, such as a confined animal feeding operations (CAFO) or wastewater treatment plant. The nutrient requirements and functions for plants are summarized in **Table 3.2**.

Table 3.2 Categorization of known essential elements according to their physiological function in plants and animals. Silicon and sodium are not required by all plants. Cobalt is required by nitrogen-fixing rhizobia. Adapted from Markert (1992a). Reproduced with permission of Springer.

Structural elements	C, H, O, N, P, S, (Si), Ca
Electrolytic elements	K, (Na), Ca, Cl, Mg
Enzymatic elements	
Plants	B, (Co), Cu, Fe, Mg, Mo, Mn, Ni, Zn
Animals	Co, Cr, Cu, Fe, I, Mg, Mo, Mn, Ni, Sb, Se, V, Zn

The micronutrients B, Si, F, Cl, V, Mn, Fe, Cr, Co, Ni, Cu, Zn, Mo, Se, Sn, Na, and I are required in smaller amounts than the macronutrients but are just as essential. Some micronutrients are only needed for animals, and not plants (F, V, Cr, Co, Sn, and Se). Boron is not known to be required by animals. Nickel may be required by animals in very small amounts. Whether all the micronutrients are required by all organisms is unclear. Several have been shown to be essential for only one species. Because proving essentiality is tedious and expensive, an element shown essential for one or a few species is often assumed to be essential for organisms of the same family or genus. The list of essential micronutrients will probably grow as experimental techniques become more refined.

Most essential elements are present as ions in the soil solution, and flow into the plant as it absorbs water. Plants obtain hydrogen, carbon, and oxygen from air, but soils have pore space for O_2 and CO_2 movement between plant roots and the atmosphere, and supply CO_2 to the atmosphere through the decay of organic matter by soil microorganisms.

Animals derive most of their essential elements from plants. The ability of plants to supply these elements to animals depends on a combination of factors: availability of the ions in the soil solution, plant selectivity at the soil solution–root interface, and ion translocation from root to plant top. Animals have evolved in the presence of soils and plants, and thus do not typically suffer microelement deficiencies. However, essential elements in natural systems are occasionally too high or too low for animals because plants can tolerate a much wider range of elemental concentrations than animals. Plant contents of iron, for example, tend to be lower than ideal for human nutrition; but as omnivores, humans obtain needed iron from meats or iron-rich plant parts. The supply of essential elements by plants to animals is generally adequate; cases of too little or too much are noteworthy because of their rarity. Grazing animals that are restricted to their feeding domain sometimes suffer from toxicities or deficiencies. For example, in a few semi-arid parts of North America, grazing animals suffer from high selenium, whereas in other regions, grazing animals suffer from selenium deficiency and require diet supplements. In Australia, grazing animals have low cobalt because of unsatisfactory amounts in soils.

Special Topic Box 3.1   Rhizosphere

The soil–root interface, called the **rhizosphere,** has distinct physical and chemical properties as compared to the bulk soil not in direct contact with roots. **Figure 3.6** shows a micrograph of plant root cross section that includes the attached rhizosphere. There are many ways to sample and study rhizosphere soil, such as placing small probes in-situ to measure a soil property like pH, EC, or Eh, or removing the soil from the plant root to conduct ex-situ measurements. A popular method for removing the rhizosphere soil is to gently pull the plant from the soil, with the goal to keep the root intact, and then shake off the soil attached to the roots into a sample container. The soil shaken from the roots is operationally defined as the rhizosphere soil.

Plant roots exude organic chelates, ions, and gases, and actively exchange protons with the soil solution in the rhizosphere to maintain charge balance within the plant root cells. The close association of the rhizosphere soil with plant roots creates unique soil properties, such as pH, mineralogy, microbial population, carbon content, and nutrient availability. For example, Figure 1.6 shows a microscopic image of a root sampled from a frozen soil section where a small drop of root exudate is frozen in time.

Due to the increased carbon in root exudates and dead root cells, the population of microbes increases in the rhizosphere. Rhizosphere bacteria, such as nitrogen-fixing rhizobia on legumes, are sometimes symbiotic with the plant roots. The increased microbial activity can be a major driver of the unique characteristics of the rhizosphere soil.

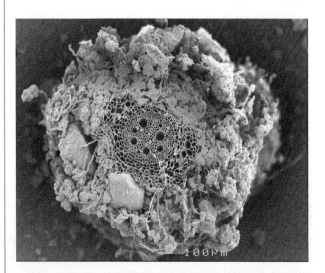

Figure 3.6 Cryo-SEM micrograph of cross section of a wheat root showing the attached soil. Image courtesy of Michelle Watt, CSIRO.

Figure 3.7 Mean percent of extractable selenium from a soil collected from rhizosphere of Aster plants grown compared to the soil collected next to the plants (bulk soil). The soil was collected from a reclaimed mine soil from Idaho, USA. The three extractants, AB-DTPA, phosphate, and hot-water extracts assess the availability of selenium for plant uptake, where the AB-DTPA is the most aggressive, and the hot water extract is the least aggressive. Error bars represent the standard error. The data show that in the rhizosphere, biotic processes are occurring that are increasing selenium availability. Data are from Oram et al. (2011).

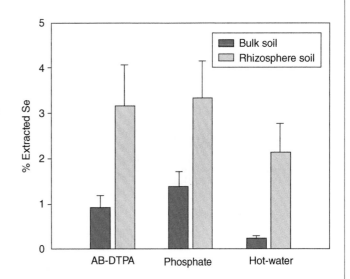

For example, it can create microsites of reduced redox potential because of the high rates of respiration from the microbes near the root–carbon mycorrhizae.

Figure 3.7 shows the availability of selenium from the rhizosphere of an aster plant grown in a reclaimed soil. In the rhizosphere soil, the availability of selenium is much greater than in the surrounding bulk soil, indicating that rhizosphere processes alter the availability of the selenium. The rhizosphere processes can cause direct changes to the speciation of the selenium, such as oxidation of the reduced selenium species selenite to selenate, which is a more soluble species, thus enhancing its availability for uptake into the plant root.

3.3.1   Plant deficiency

Nutrient deficiencies in plants are usually evident by reduced yield or productivity, abnormal coloration, and plant and fruit deformities. The nitrogen content of most soils is low enough that plants benefit from added nitrogen. In many soils, plants (especially food crops) also grow better with added phosphorus, potassium, and sulfur.

Table 1.1 lists the total amounts of the essential ions in soils, which includes ions unavailable for plant uptake because they are trapped within the crystal lattices of clay, silt, sand, and gravel particles. Total concentrations are poor indicators of plant availability. A more useful value would be the amount of each ion available for plant uptake during the plant's growing season. Measuring this availability has been, and remains, the subject of much research. Assessing availability is difficult because ion uptake by plants varies with plant variety and growth conditions, as well as with soil properties. A major factor influencing an ion's availability is its soil chemistry (**Figure 3.5**), but differences in plant biology are also important. Modern methods of assessing plant nutrient status based on hand-held detectors that can measure plant *health* in-situ, or remote detection via satellite or unmanned aerial drones are promising tools for more accurate nutrient management.

Nitrogen and phosphorus are the most common macronutrients applied to soils to promote crop growth, however, excess fertilization causes problems in many agricultural regions because the nutrients can leach or runoff into groundwater or surface water. Nitrogen and phosphate from fertilizers, animal manures and wastewater from human wastes are the most common chemicals that degrade water quality.

3.4   Inorganic contaminants in the environment

The boundaries between toxicity, sufficiency, and deficiency are vague. **Figure 3.8** shows the occurrence and relative toxicity of several inorganic chemicals at

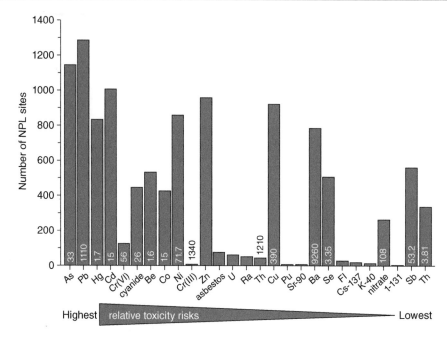

Figure 3.8 Number of sites in the USA as of 2017 that have selected inorganic contaminants present, as listed on the US Environmental Protection Agency (EPA) National Priority List (NPL). Numbers at base of columns are the median (mg kg⁻¹) concentrations of the contaminants in the soils at the sites. The NPL is a list of sites in the United States that have contaminated water, air, or soil to a degree requiring further investigation and possible remediation to prevent adverse impacts to humans or the environment. The US Agency for Toxic Substances and Disease Registry (ATSDR) ranks the hazardous substances at sites on the NPL based on *total priority risks points*. Points are calculated from the number of sites in which the contaminant is present, the toxicity of the contaminants, and risks of exposure to humans at the site. The triangle below the *x-axis* qualitatively indicates toxicity risks based on the ASTDR total priority risks points calculation. Arsenic is the number one risk among the 1313 NPL sites, thorium is the lowest risks amongst contaminants shown in this graph. Data are reported in Agency for Toxic Substances and Disease Registry (2017).

sites in the United States. Amongst the contaminants at the US sites, arsenic poses the greatest risks to human health. Other countries have similar contaminants of concern as those shown in **Figure 3.8**. In Southeast Asia, for example, arsenic is widespread in groundwater and floodplains, and threatens contamination of water and food supply, particularly rice. In northern Nigeria, soil contaminated with lead from gold mining is causing illness and death within some communities.

Toxic elements are those that are more likely to be found at excessive concentrations, or elements that are toxic at low concentrations. The most prevalent and widespread *natural* soil toxicity problem is aluminum toxicity to plants in acid soils. Excessive concentrations of contaminants are typically due to inputs from human activity. However, in some environments, *natural* concentrations of elements are excessively high, and organisms, including plants, microbes, and

invertebrates have evolved special adaptation mechanisms to tolerate the relatively high concentrations of elements. For example, soils developed from serpentine rocks have elevated concentrations of Mg, Fe, Ni, Cr, and Co. Plant communities that thrive in this environment withstand high concentrations of these elements. Other examples of environments with naturally elevated concentrations of potentially toxic elements are hot springs, saline lakes, and thermal vents on the seafloor. Organisms that have adapted to live in these environments are called extremophiles. Extremophiles are exceptions to biological systems on earth and are interesting for study because they provide insights into biological evolution, physiology, and biochemistry.

The toxic elements shown in **Figure 3.4** are distinctively prevalent in many soil environments, do not benefit living organisms at any concentration (as far as yet known), and are toxic at low concentrations in the food/water/air web. Arsenic, listed as toxic in

Figure **3.4**, has been recently reported by some researchers as being beneficial for animals. However, there are no instances where arsenic deficiency has been reported, and there are many environments where arsenic contamination is reported, thus it is not an element of *concern* with respect to nutritional requirements. Another instance of a rare use of a primarily toxic element is the use of cadmium in place of zinc by a species of phytoplankton that live in zinc deficient ocean environments.

In many environments, soils contain synthetic chemicals, radioactive isotopes, and elements that have been concentrated by human activities. **Figure 3.8** shows the number of sites on the US National Priority List (NPL) that report the presence of inorganic hazardous chemicals. The listing of chemicals provides a good relative reference to the prevalence and relative hazards of inorganic chemicals that threaten human and environmental health. **Figure 3.8** also shows the median concentration of the elements at contaminated sites. Most governments have agencies to regulate contaminants in the environment. Agencies establish regulatory limits based on toxicology and risk evaluation for many chemicals to protect and gauge soil, water, and environmental quality.

In the absence of bioavailability data, regulatory agencies have little recourse but to use total soil concentration as the determining criterion for soil contamination level. For elements strongly retained by soil solids, such as lead and beryllium, risks are overestimated if based on total concentration alone. To properly assess bioavailability, pathways, and receptors must be considered. For example, for children playing in Pb-contaminated soils, hand-to-mouth behavior creates a direct ingestion pathway and a very high exposure risk.

The challenges of excess chemicals in soils are not only a concern for contaminated sites (e.g., the NPL list). Agricultural fields and feedlots, or other nonpoint sources, introduce many surface and groundwater contaminants. Contaminants from agriculture include veterinary drugs, pesticides, phosphorus, and nitrogen.

The listing of chemicals at contaminated sites shown in **Figure 3.8** is provided in order of their relative toxicity and threat to human health. The data show that arsenic, lead, mercury, cadmium, and chromium are amongst the most serious environmental inorganic contaminants in the United States. These elements present toxicity risks to humans and most other organisms when they are present in the soil and water at elevated concentrations. Except for micronutrient requirements for chromium, human exposure to these elements is detrimental. At many sites, the presence of multiple contaminants, including organic and inorganic chemicals, exacerbate the risks for human and ecosystem health. A worldwide survey of toxic contaminants in 2016 by Pure Earth and Green Cross Switzerland is shown in **Table 3.3**.

3.4.1 Assessing contamination status of soils

Due to agriculture, mining, industrial, and other human activities, contaminated soils are widespread, and the number of contaminated sites is greater than can be feasibly remediated. Thus, regulatory agencies are forced to prioritize which soils to clean up to get the most remediation for the money spent. Although knowing how much of a chemical exists in soil is a good first step to understand contaminant risks at a site and can be suggestive of how to remediate the site, total contaminant concentration alone is insufficient.

Table 3.3 Top ten sources of pollution based on number of humans put at risk. In total, pollution sources listed in this table puts over 32 million people at risk. Adapted from Pure Earth (2016).

Rank	Industry	Principal contaminants
1	Lead acid battery recycling	Pb
2	Mining and ore processing	Pb, Cr, As, Cd, Hg, cyanide
3	Lead smelting	Pb, Cd, Hg
4	Tanneries	Cr
5	Artisanal gold mining	Hg, Pb
6	Industrial dumpsites	Pb, Cr, Cd, As, pesticides, volatile organic compounds
7	Industrial facilities	Pb, Cr
8	Chemical manufacturing	Pesticide, volatile organic compounds, As, Cd, cyanide, Hg, Cr, Pb
9	Product manufacturing	Hg, Cr, Pb, As, Cd, volatile organic compounds, dioxins, cyanide, sulfur dioxide
10	Dye industry	Cr, Pb, Hg, As, Cd, Ni, Co, nitrates, chlorine compounds

The mere presence of a chemical in soils is rather insignificant with respect to how and if the site needs remediation. What matters is the availability of a substance to plants and animals, or to the soil solution, which is a function of the contaminant and soil properties. In some cases, a soil labeled as contaminated because of its total element concentration poses little risk because the bioavailability or mobility of the chemical of concern is minimal. For example, plants and groundwater at sites with copper, lead, and zinc ore deposits are not necessarily contaminated with these elements even though the total amounts of the metals *in the soils* can be very high.

Relative availability is a function of the species distribution. Thus, for example, lead bioavailability is less in a soil with lead present as mostly mineral phases that have low solubility (e.g., lead sulfide minerals) as compared to a soil that has lead present adsorbed onto mineral and organic matter particles (**Figure 3.9**); even if the soil with the lead minerals in it has a greater *total* lead concentration in it than the soil with lead adsorbed onto soil particle surfaces.

Of course, if one is concerned where risks are likely to exist, measuring the total amount of the chemical of concern is a good screening tool. However, to understand and predict bioavailability and solubility of contaminants requires more information about the forms of the chemicals in the soil than simply total concentration.

Instead of using total concentration at a site to determine contamination status, some contaminated sites can be managed based on relative availability of the contaminant for leaching, plant uptake, or bioavailability. This may be assessed by an extraction (in vitro) test. For example, to assess the risks of lead to humans, soil extraction tests have been developed that simulate the conditions in the digestive system of a target organism (the extraction tests are referred to as physiologically based extraction tests). The in vitro tests are typically calibrated by measuring the lead levels in blood or tissue of a target organism. For tests to predict risks for humans, mammals such as pigs or mice have been used as test organisms. Once validated, the in-vitro test can be used to assess the risks of poisoning from the contaminated soil samples. Although this is sometimes more efficient than using total soil concentration for deciding where contaminated soil risks are greatest, inaccuracies due to the possible influence of untested soil properties are problematic, especially when risks to human health are high. Thus, regulatory agencies are, justifiably, cautiously reviewing the use of in-vitro tests.

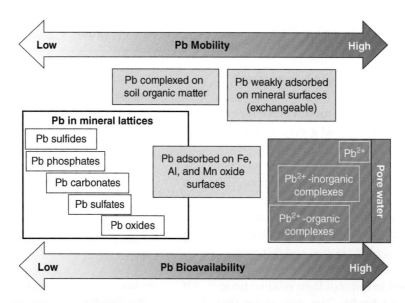

Figure 3.9 Relative mobility and bioavailability of different species of lead in a soil. Mobility is determined by the distribution of the lead between the solution and solid phases, which is controlled by dissolution, desorption or disassociation reactions. Factors such as solubility, particle size, and speed of dissolution are included in the relative bioavailability assessments. The sum of all the lead phases in a soil is the total soil lead concentration.

Alternative approaches to empirically based extraction tests utilize measures of soil chemical and physical properties and make predictions based on chemical speciation and physical transport properties. This physicochemical approach is referred to as **mechanistically based modeling**. Characterizing chemical speciation and understanding transport processes is not trivial. As science and analytical methods become more powerful, using a more mechanistically based approach to predict chemical fate and availability will become more accurate and useful. More accurate measurements of chemical risks at contaminated sites will allow money to be spent on sites posing the greatest *real* risks, as opposed to simply going by total concentration of contaminants.

3.5 Anthropogenic organic chemicals in the soil environment

In addition to the inorganic chemicals found in soils, many artificially produced (*xenobiotic*) organic chemicals occur in the environment. Xenobiotics may be applied to soils for beneficial purposes, such as pesticides, or may be unintentionally applied to soils and have negative impacts on soil and environmental quality. **Table 3.4** lists common organic chemicals that occur in the environment, as well as man-made organic and inorganic nanoparticles (nanometer-sized colloids). The number and type of organic chemicals that occur in the environment are numerous. This section discusses some of the general types of organic chemicals

Table 3.4 Sources and examples of common organic chemicals and chemicals of emerging concern (C_hEM) that occur in the environment.

Type	Source	Classes	Examples
Pesticides	Agriculture, home owners, industry spills	Herbicides, insecticides, fungicides, bactericides	2, 4,-D (herbicide), atrazine (herbicide) glyphosate (herbicide), chlorpyrifosm metachlor (herbicide), metam sodium (fumigant pesticide), DDT (insecticide), carbofurna (insecticide), diazanon (pesticide)
Industrial chemicals	Manufacturing	Adhesives, preservatives, solvents, refrigerants, propellants	Surfactants: sodium stearate, 4-(5-dodecyl) benzenesulfonate Solvents: methyl ethyl ketone, benzene, carbon tetrachloride, chloroform, dioxins Preservatives: formaldehyde Fluoridated resins and lubricants: PFOA, PFOS
Manmade nanoparticles	Consumer products, waste water, industrial waste	Catalysts, drugs, antimicrobials, sunscreens	Nano-Ag, Zn, Fe, Ti, Si (including oxides), organic metal polymers, nano-plastics
Veterinary medicines and animal production chemicals	Agriculture runoff, animal waste	Antibiotics, hormones, pesticides, resistant pathogens, VOCs, endocrine disruptors	Hormones: 17alpha-estradiol, estrone, and estriol, estrogen, progesterone, testosterone, rBST; parasite pesticides: organophosphates, amides, pyrethroids
Pharmaceuticals and personal care products (PPCP)	Wastewater treatment, hazardous waste disposal	Steroids, medicines, preservatives, surfactants, insecticides	Coprostanol (steroid), N-N-diethyltoluamide (insect repellent), caffeine, triclosan (antimicrobial disinfectant), antibiotics (quinolones such as ciprofloxacin, sulphonamides, roxythromycin, dehydrated erythromycin)
Waste organics in landfills	Landfills for domestic and industrial waste	Paint, pesticides, pharmaceuticals, detergents, personal care products, fluorescent tubes, waste oil, treated wood, electrical equipment	Aromatic compounds, chlorinated aliphatics, phenols, phthalates, pesticides (e.g., DEET, and degradation products of glyphosate, atrazine, and simazine), CFC, perfluoroalkyls
Stack emissions	Coal burning and industrial	Volatile organic compounds, polyaromatic hydrocarbons (PAH)	Benzene, formaldehyde, toluene, inorganics (Hg, N_2O, CO_2, SO_2), ethylene dibromide, benzo(a)pyrene (BaP)
Petroleum extraction, processing, and distribution	Crude oil and gas spills	Hydrocarbons: gas, oil, tar, polyaromatic hydrocarbons	Benzene, xylene, toluene, benzo(a)pyrene (BaP)

that occur in the environment and their reactivity in soils. Additional details of organic chemical behavior in soils are presented in later chapters.

3.5.1   Pesticides in the environment

Pesticides are necessary to grow food to feed the world's population but are harmful if not properly managed. Agriculture accounts for more than 80% of world pesticide use; industry and homeowners account for approximately 10% each. Global trends have shown that a country's food production is directly correlated with pesticide use. However, at a certain point, increased use of pesticides does not facilitate additional food productivity. **Figure 3.10** shows the amounts of pesticide used in the world in the year 2012. Most of the pesticides produced are **herbicides**, including glyphosate and atrazine (**Figure 3.11**). **Insecticides** comprise 18% of the total world pesticide usage. **Fumigants** are applied as gas to sterilize soil for treatment of pathogenic nematodes, fungi, or microbes. Greater than 80% of the pesticides used are applied to the crops corn (39.5%), soybeans (21.7%), potatoes (10.2%), cotton (7.3%) and wheat (4.5%).

Important aspects of organic pesticide chemistry in the environment are volatility, solubility, adsorption to minerals and soil organic matter, degradation rate, and degradation products. For example, the most commonly used fumigant in the United States, metam sodium, is an amine with thiol functional groups ($C_2H_4NNaS_2$) (**Figure 3.12**). Within minutes to hours after application to soils, metam sodium degrades to methylisothiocyanate (MITC). MITC is the active form of the pesticide, and is volatile and soluble, and does not adsorb strongly to soil particles because it is uncharged. Eventually, MITC is degraded by microorganisms, is lost through volatilization, or leaches into ground and surface waters. Thus, including the speciation, solubility and degradation rates are part of the strategy for correct application of metam sodium, and are important for determining its eventual fate in the environment.

Because of their widespread use, many pesticides end up contaminating surface and groundwaters. This is a problem in both urban and agricultural watersheds. Atrazine ($C_8H_{14}ClN_5$), one of the most commonly used herbicides, is commonly present in surface waters downstream of agricultural fields and causes anomalies in aquatic vertebrates. It is suspected to be the cause of deformities in reproductive organs of wild frogs and fish. The main factors affecting atrazine fate in the environment are its solubility and degradation. Atrazine has low solubility, and readily adsorbs on soil organic matter. The half-life of atrazine in the environment is on the order of days to a year, implying that degradation can be rapid or slow, depending on soil physicochemical properties and microbial activity. Atrazine runoff from agricultural systems occurs as both dissolved and sediment bound forms. With improved application practices, such as timing of application to avoid rain events, dissolved atrazine runoff has been decreasing, but sediment-bound atrazine runoff remains problematic. For example, research on occurrence of pesticides in a stream in Nebraska, USA (**Figure 3.13**) showed that during storm events, sediment-bound pesticide concentration was high, while during low flow events, sediment-bound pesticides were negligible. A study of fish growth in the high-flow events suggested that pesticide concentrations were high enough to cause deformities. These data indicate that more effort to prevent loss of pesticides such as atrazine through both erosion and dissolved forms is needed.

3.5.2   Chemicals of emerging concern in the environment

Chemicals of emerging concern (C_hEC) are a recently defined category of chemicals in the environment that include pharmaceuticals and other consumer products

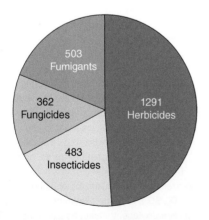

Figure 3.10 World pesticide use in 2012. Numbers are 10^6 kg of pesticide used. Data from Atwood and Paisley-Jones (2017).

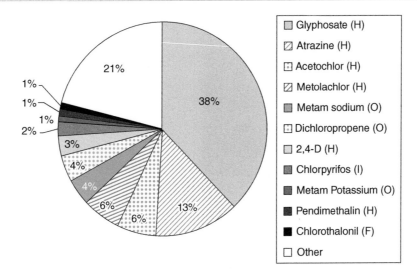

Figure 3.11 Percentage of pesticide usage on 21 major crops in US in 2008. Letters after pesticide name indicate herbicide (H), insecticide (I), fungicide (F), or other (O). Modified from JorgeFernandez-Cornejo et al. (2014).

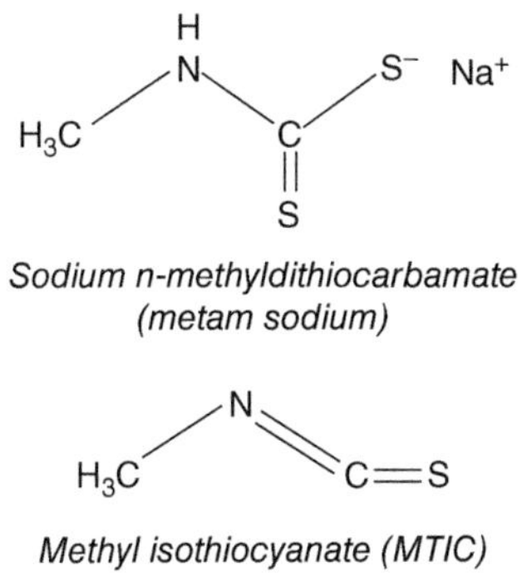

*Sodium n-methyldithiocarbamate
(metam sodium)*

Methyl isothiocyanate (MTIC)

Figure 3.12 Chemical structure of metam sodium and its transformation product methyl isothiocyanate (MITC).

common in modern life. Although it is a relatively new categorization of chemicals, the occurrence of many of these chemicals is not new. Awareness of them is increasing because of their increased frequency of occurrence in soil, air, and water. In addition, lower detection limits of analytical instruments have improved, allowing for detection of the chemicals at very low levels. Scientists are discovering that at even very low levels, some C_hEC create risks to human and ecosystem health. In this text, the common abbreviation for chemicals of emerging concern, *CEC*, is not used to avoid confusion with the abbreviation for cation exchange capacity.

Examples of C_hEC are compounds such as endocrine disruptors, antibiotics, and pharmaceuticals that end up in soils, sediments, and groundwater or surface waters.

Recent toxicology studies suggest such chemicals can negatively affect native microbial populations, aquatic vertebrates and invertebrates, birds, and pose risks to humans. **Figure 3.14** shows a survey of C_hEC occurrence in selected streams across the United States in 1999–2000, with a focus on streams that are directly below wastewater treatment plant discharges and watersheds that have intense urbanization or agriculture impacts. Steroids and nonprescription drugs were found in most of the streams surveyed. Although C_hEC are reported as being present in many waters, potential risks of impact depend on several factors, including chemical persistence, reactivity, exposure, and concentration.

Much of the focus of C_hEC studies are their occurrence in water, but soil is a key aspect of the watershed contributing solutes and eroded particles to surface and groundwater. Within the watershed, animal husbandry and waste landfills are common sources of C_hEC to soils. However, little is known about how C_hEC move from soil to water. As new C_hEC are used in manufacturing and agriculture, and usage becomes more widespread, risks of environmental impacts will continue to be a concern.

Polyaromatic hydrocarbons (PAH) are common C_hEC found in the environment. PAHs are toxic compounds that consist of at least two bonded benzene rings, hence the name *poly* (many) *aromatic* (six carbon ring compound). PAHs are introduced into the environment from many sources, including stack emissions, wastewater, wood burning (including forest

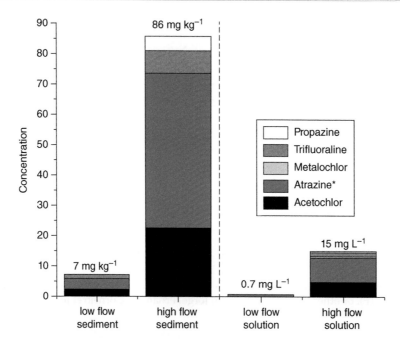

Figure 3.13 Pesticide concentrations in solution and on sediments in stream water in the Elkhorn River, Nebraska USA. The stream is located downstream of an agricultural watershed. Atrazine* is the sum of atrazine and the degradation product DEA. Data from Zhang et al. (2015).

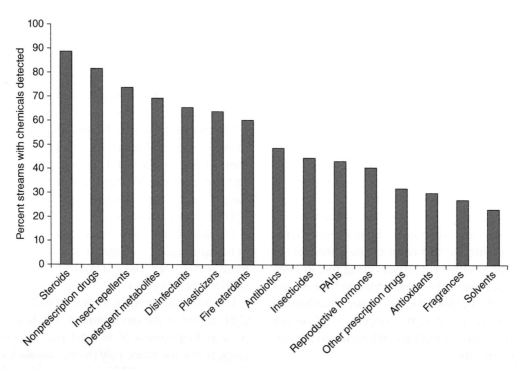

Figure 3.14 Frequency of occurrence of various classes of chemicals of emerging concern in 139 streams in 30 US states during 1999 and 2000. Streams were sampled downstream from intense urbanization or livestock production. Thus, the results are not representative of all streams. Adapted from Kolpin et al. (2002).

fires), and automobiles; and are broadly distributed in air, water, and soil. Oil extraction and refining facilities are common sources of PAH contamination to soils and sediments. PAH toxicity has been observed in aquatic vertebrates and invertebrates, and they are suspected carcinogens for humans. A common type of PAH in the environment is benzo(A)pyrene, which consists of six fused benzene rings, and is a combustion byproduct. PAHs typically have long lifetimes in the environment, and their breakdown products may be as toxic as the parent molecules. They generally have low solubility because they are hydrophobic and strongly adsorb to organic matter in soils and sediments, which fortunately limits their bioavailability.

3.5.3 Chemical factors affecting organic chemical reactions in soil

Fate, transport, and reactivity of organic chemicals in soils is a function of stability of the chemical to transformation and degradation, charge associated with the chemical in solution, hydrophobicity, molecular size, and types of functional groups on the organic chemical. Some organic chemicals readily sorb onto soil mineral or organic matter particles, while others remain in soil solution or in soil pores as gases. Distribution of chemicals between solid and solution is quantitatively described by a distribution (or partition) coefficient (K_d), which is the ratio of the amount of chemical on the solid to the amount in solution (further discussed in Chapters 10 and 11). Distribution of organic chemicals on solids and compound transformation rates are important factors for pesticides because the efficacy of the chemical is largely dictated by its availability to react with the targeted pest rather than the soil. The same principles used to understand pesticide fate and transport in soils apply to organic C_hEC.

Organic chemicals have varying degrees of polarity, which causes **hydrophobic** and **hydrophilic** characteristics. Halogen-containing organic chemicals typically have a high degree of hydrophobicity. In soil, hydrophobic chemicals resist dissolution in water, tending to associate with each other as **non-aqueous phases liquids** (NAPL) that reside in soil pores, or they partition to hydrophobic regions of soil organic matter.

Many organic chemicals have acidic or basic functional groups that govern the charge characteristics of the compound, and thus its solubility and sorption on soil particles. Examples of **acidic functional groups** are carboxyls, phenols, and thiols (**Figure 3.15**). Amino groups are the primary **basic functional groups** that can become positively charged.

In addition to creating charge on organic chemicals, functional groups can form covalent bonds with cations and mineral surfaces. Carboxylates are particularly effective at forming covalent bonds, and organic chemicals that have carboxylate groups can adsorb on positively charged soil particles, for example:

$$X^+ + R\text{-}COO^- = X \equiv OOC\text{-}R \qquad (3.1)$$

where X^+ is a positively charged soil surface that reacts with the negatively charged deprotonated oxygen of the carboxylate functional group to form a carboxylate-surface bond, indicated by the triple bar. Carboxylate functional groups can adsorb on negatively charged soil particles through a cation bridge, which consists of

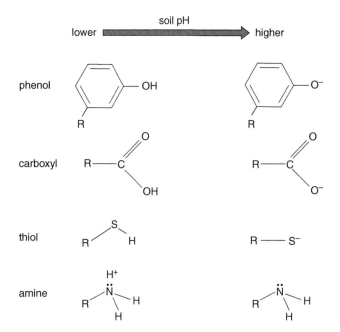

Figure 3.15 Examples of organic functional groups that occur on pesticides that have charge states of non-ionic, negative charge, or positive charge, depending on pH. The difference in charge causes the pesticides to behave differently. The pH at which the functional groups protonate or deprotonate can be predicted by the acidity constant (K_a). *R* represents any organic carbon-based molecule.

a cation between the negatively charged surface and the negatively charged oxygen in the carboxylate functional group; this complex is called a **ternary complex**.

Adsorption decreases the availability of pesticides to interact with organisms and decreases compound mobility. Such immobilization may be considered beneficial from the perspective of protecting ground or surface waters, however, if the chemical is unavailable for the targeted pest, adsorption can decrease the chemical's efficacy. Alternatively, adsorption causes the chemical to persist in the soil instead of leaching out of the soil profile and continue working as a pesticide long after it was applied. Thus, to best manage pesticide application, it is critical to include adsorption processes.

Persistence of organic chemicals in the environment is controlled by susceptibility of the chemical to biological or abiotic transformation, and eventual degradation. How fast organic chemicals such as pesticides degrade strongly affects their fate and efficacy. Biotic transformation happens by microbial processes or through biochemical degradation inside an organism. Abiotic transformations are influenced by environmental conditions, such as water, pH, temperature, sunlight exposure, and association of the chemical with mineral surfaces that act as a catalyst or protect the chemical from degradation.

Lifetimes of organic chemicals in the environment can range from hours to several decades. Short-lived organic chemicals pose less of a threat for environmental contamination. However, breakdown products may be equally or more problematic than the original organic chemical.

Environmental lifetimes of pesticides are important considerations in development of best management practices. Longer environmental lifetimes allow the chemical time to interact and react with the intended target, while shorter lifetimes may help protect the environment from unintended pesticide exposure. Lifetimes are typically characterized by the time required for ½ the concentration of the chemical to degrade; called the *half-life*. For example, the half-life of atrazine in an aerobic soil is predicted to be 146 days, and the half-life of the herbicide acetochlor is predicted to be 14 days.

3.6   Properties of the elements in soils

The elements of primary interest for soil chemistry exhibit a wide range of chemical behavior that causes highly varied mobility and bioavailability. As a useful perspective for soil chemistry, the elements of common interest are grouped according to general chemical properties and reactivity in soil:

1. Exchangeable cations: Ca^{2+}, Mg^{2+}, Na^+, K^+, NH_4^+, Al^{3+}
2. Soluble anions: NO_3^-, SO_4^{2-}, Cl^-, H_2CO_3, HCO_3^-, SeO_4^{2-}
3. Poorly soluble anions: $H_2PO_4^-$, HPO_4^{2-}, H_3BO_3, $H_4BO_4^-$, $Si(OH)_4$, MoO_4^{2-}, $HSeO_3^-$
4. Poorly soluble metal cations: Al^{3+}, Fe^{3+}, Mn^{4+}, Mn^{2+} Cu^{2+}, Zn^{2+}, Ni^{2+}, Co^{2+}
5. Toxic ions: Al^{3+}, H_3AsO_4, H_3AsO_3, Cd^{2+}, CrO_4^{2-}, Hg^{2+}, Pb^{2+}
6. Biogeochemical cycling elements: C, N, O, S

The *exchangeable cations* are retained by the negative charge of soil minerals and organic matter. *Soluble anions* neutralize cations and the positive charge of soil particles but are more soluble and sorbed less strongly than anions such as phosphate. *Poorly soluble anions* are present at low concentrations in soil solution (typically less than 10^{-5} M) and often occur as minerals, salts, or sorbed to mineral surfaces. Concentrations of *poorly soluble metal cations* (e.g., transition metals) in soil solutions are typically much less than alkali and alkaline earth cations. *Toxic ions* are commonly present in soils at concentrations that pose toxicity risks. Aluminum is included in this group because of its widespread plant toxicity in acid soils. The soil chemistry of the major *biogeochemical cycling elements* carbon, nitrogen, and sulfur are dominated by biological processes – principally primary production by plants, fixation by microbes, and microbial driven mineralization. **Table 3.5** shows the major categories of important soil chemicals, their most common oxidation states in soils, and likely ion species in soil solutions.

Although elements in the groups listed in **Table 3.5** may have similar behavior, the chemical form or species of each element causes them to have unique properties within soils. Investigators often focus on a particular element because it has an overwhelming influence on some property of interest. For example, poorly soluble metal cations and aluminum both produce acidity via hydrolysis, but because aluminum has a much greater abundance, Al^{3+} hydrolysis is a more common acid production reaction in soils. Some of the distinguishing chemical properties of chemicals important in soil chemistry are discussed below.

Table 3.5 Ions of major interest in soil chemistry and their most common species. The ions are grouped according to major characteristics.

Ion	Comments
Exchangeable cations	
Ca^{2+}	Occur as exchangeable cations in soils; these ions are relatively easily manipulated by
Mg^{2+}	liming, irrigation, or acidification; exchangeable Al^{3+} is characteristic of acid soils;
Na^+	productive agricultural soils are rich in exchangeable Ca^{2+}; NH_4^+, and K^+; K^+ can become
K^+	fixed in clay minerals.
NH_4^+	
Al^{3+}	
Soluble anions	
NO_3^-	Present in lower concentrations than the major cations in all but the most coarse-textured
SO_4^{2-}	and strongly saline soils; sulfate and nitrate are important nutrient sources for plants;
Cl^-	sulfate, chloride, and bicarbonate salts accumulate in saline and alkaline soils; selenate
H_2CO_3, HCO_3^-	(SeO_4^{2-}) anion is more soluble than selenite (SeO_3^{2-}).
Se^{6+} as SeO_4^{2-}	
Poorly soluble anions	
$H_2PO_4^-$, HPO_4^{2-}	Strongly retained by soils; retention or fixation by soils is pH dependent; borates are the
H_3BO_3, $H_4BO_4^-$ $(B(OH)_4^-)$	most soluble of the group; molybdate and silica are more soluble at high pH; phosphate
$Si(OH)_4$	is more soluble at neutral or slightly acid pH; selenite is more reduced and less mobile
MoO_4^{2-}	than selenate.
Se^{4+} as $HSeO_3^-$	
Poorly soluble metals	
Al^{3+}	As silica and other ions leach during weathering, insoluble hydroxides accumulate in soils;
$Fe(OH)_2^+$, Fe^{2+}	iron and manganese are more soluble in waterlogged or reduced soils; availability
Mn^{4+}, Mn^{3+} Mn^{2+}	increases with increasing soil acidity; metals are complexed by SOM and strongly
Cu^{2+}	adsorbed on mineral surfaces.
Zn^{2+}	
Ni^{2+}	
Co^{2+}	
Toxic ions	
Al^{3+}	Often not readily soluble; Al^{3+} is a hazard to plants, the others are of more concern to
As(V): $H_2AsO_4^-$, $HAsO_4^{2-}$	animals; Cd^{2+} is relatively available to plants; As and Cr oxanions increase in solubility
As(III): $H_2AsO_3^-$, $HAsO_3^{2-}$	with pH; As(III) is more soluble than As(V); cation contaminants are less available to
Cr(VI): CrO_4^{2-}	plants with increasing pH; many contaminants form poorly soluble sulfides in reduced
Cd^{2+}	soils.
Hg^{2+}	
Pb^{2+}	
Major biogeochemical cycling elements	
C^{4-} to C^{4+} $(CH_4$ to $CO_2)$	Soil biochemistry revolves around the oxidation state changes of soil carbon, nitrogen, and
N^{3-} to N^{5+} $(NH_4^+, N_2, NO_2, NO_3^-)$	sulfur compounds; nitrogen occurs in oxidation states from N(-III) to N(V); molecular
O^{2-}, O_2	oxygen is the main electron acceptor; nitrate, and sulfate are electron acceptors when
S^{2-} to S^{6+} $(H_2S$ to $SO_4^{2-})$	oxygen supply is low.

3.6.1 Alkali and alkaline earth cations

The major alkali and alkaline earth cations Na^+, K^+, Ca^{2+}, and Mg^{2+} are prominent in soils, even though large amounts of these ions are lost during weathering of rocks to soils. They are often referred to as **base cations** because of their association with hydroxide producing, or acid neutralizing reactions. The other alkali and alkaline earth metals Li^+, Cs^+, Rb^+, Fr^+, Sr^{2+}, Ba^{2+}, and Ra^{2+} are present in only trace amounts in rocks and soils. Although Be^{2+} is in the alkaline earth metal column, its large ionic potential causes it to chemically react more like Al^{3+} than the other alkaline earth cations.

Alkali and alkaline earth cations commonly occur as hydrated ions in soil solution and neutralize the negative surface charge on soil colloids. For example, exchangeable cations held on soil clay minerals are attracted to, and effectively neutralize, the negative surface charge of the clay minerals. Hydrated ions associated with soil surfaces can easily be exchanged for other cations. For example, Ca^{2+} can displace a Na^+ cation adsorbed on negatively charged surface:

$$Ca^{2+} + Na_2\text{-clay} = 2Na^+ + Ca\text{-clay} \qquad (3.2)$$

where the *-clay* represents negative electric potential on a clay mineral surface that neutralizes the positive charge from two Na^+ cations or one Ca^{2+} cation. The exchange reaction is conservative with respect to the exchange of charges.

In productive agricultural soils, cations are typically present in the order $Ca^{2+} > Mg^{2+} > K^+ > Na^+$. Deviations from this order create ion-imbalance problems for plants. High Mg^{2+}, for example, can occur in soils formed from serpentine rocks, and the excess Mg^{2+} inhibits Ca^{2+} uptake by plants. High Na^+ occurs in soils where water drainage is poor and evaporation rates exceed rainfall. High Na^+ creates problems of low water flow and availability in soils. In acid soils, Ca^{2+} concentrations are typically low, and deficiency may exacerbate the plant stress created from Al^{3+} toxicity. Of the alkali and alkaline earth cations, Na^+, K^+, Ca^{2+}, and Mg^{2+} are the most important, and commonly discussed in soil chemistry. Additional details of their soil chemical reactions are discussed below.

3.6.1.1   Calcium

Calcium is an essential element for plants and animals, and the amounts are rarely deficient in soils because other problems appear before the calcium is deficient. In most highly productive soils, Ca^{2+} dominates the adsorption sites on mineral surfaces. Weathering causes dissolution and leaching of calcium, as well as other alkali and alkaline earth elements from soils. Thus, soils from regions that have intense rainfall and high weathering rates have less calcium than less weathered soils. In the subsoils of arid and semi-arid regions, calcium commonly precipitates as calcite ($CaCO_3$) rather than being leached away (see Figure 1.11). In many arid soils, calcium carbonate minerals

create indurated layers (caliche). Precipitation of $CaCO_3$ in soils is affected by the rates of soil water movement, CO_2 production by roots and microbes, CO_2 diffusion to the atmosphere, and water loss by soil evaporation and plant transpiration. $CaCO_3$ layers are also derived from upward movement and evaporation of Ca-rich waters. Calcium carbonate accumulations can amount to as much as 90% of the mass of affected soil horizons. Gypsum ($CaSO_4 \cdot 2H_2O$), which is much more soluble than calcium carbonate minerals, precipitates in some arid soils.

3.6.1.2   Magnesium

Although it is the second most abundant exchangeable cation in soils, magnesium is the least-studied ion in this group. Most likely because excessive or deficient amounts of magnesium for plant growth are uncommon. Adsorption of Mg^{2+} as exchangeable cations is like Ca^{2+}; although Mg^{2+} is slightly less preferentially adsorbed than Ca^{2+} in most cases. Magnesium is an important constituent of many primary and secondary aluminosilicate minerals. Weathering of magnesium from mafic minerals (magnesium and iron basalts) often leads to the formation of chlorite and montmorillonite clay minerals in soils. Soils that are inherently low in magnesium, or soils growing crops having high magnesium requirements, benefit from dolomite ($CaMg(CO_3)_2$) or limestone-containing dolomite amendment.

3.6.1.3   Potassium

Potassium is the third-most commonly added fertilizer element, after nitrogen and phosphorus. Many humid and temperate region soils are unable to supply sufficient potassium for crops. Farmers in these areas long ago recognized the benefits of applying wood ash and other liming materials to acid soils. Both the alkalinity of the ash (to counter Al^{3+} toxicity) and its potassium and calcium content are beneficial.

Soils retain potassium more strongly than sodium because the *hydrated* K^+ cation is smaller than the hydrated Na^+ cation. Thus, K^+ selectively adsorbs on clay minerals due to the closer approach of the charged ion (i.e., greater ion attraction force, see Eq. 2.2). In soils containing vermiculite or illite, potassium can become fixed or entrapped in the interlayer of the clays. In soil solutions, potassium concentration is typically low, but is replenished by diffusion from clay minerals, and from slow weathering of potassium-containing feldspar minerals.

The ability of some soils to supply potassium for plants is remarkable. Tropical soils derived from easily weathered basaltic rocks have supplied up to 250 kg ha^{-1} yr^{-1} of potassium to banana plants for many years without noticeable soil depletion in potassium availability. In temperate regions, soils that supply adequate potassium for crop needs often contain considerable potassium-containing mica (illite) in their clay fractions that rapidly weathers releasing the potassium; the rapid weathering is facilitated by their small crystal size and the climate.

Continual potassium fertilization is necessary for deficient soils because much of the added potassium is fixed by clay minerals and therefore unavailable. Soils containing appreciable sand- and silt-sized vermiculite are particularly troublesome in this regard because the amount of potassium required to saturate potassium-fixation sites is uneconomical.

3.6.1.4 Sodium

Sodium is required by all animals. Some *higher* plants, and many halophytes (salt-loving plants) require sodium to regulate osmotic potential. Due to its ubiquity, sodium deficiency in plants is rare. More commonly, sodium is a concern when it occurs in excess of 5–15% of the exchangeable cations within the soil. Sodium can accumulate in these amounts in areas inundated by seawater, in arid areas where salts naturally accumulate from the evaporation of incoming surface or groundwater, and in irrigated soils because irrigation water often contains high sodium. Exchangeable Na$^+$ inhibits water movement into and through many soils.

Saline soils are a problem for plants because the high osmotic potential of the soil solution causes the plant to expend so much energy to take up water that little energy is left for growth and crop yield. Sodium is toxic to some plants at high concentrations, but for most plants this is a relatively minor problem compared with the restricted water uptake that normally precede sodium toxicity. Fruit and nut trees and berries are sensitive to sodium and may show toxicity symptoms before water deprivation. The high soil pH that accompanies sodium accumulation in arid soils is generally of secondary importance compared to the microelement deficiencies induced by the high pH, as well as the water uptake problems.

Most plant roots repel most of the sodium in the soil solution. Halophytes are distinct because they uptake

Na$^+$, which is a great advantage in high-salt soils and waters because it reduces the difference in osmotic potential that the plant must overcome to uptake water.

3.6.2 Major soluble anions in soils

In weakly weathered soils, such as those that predominate in North America and Europe, most soil colloids have a net negative charge, which is balanced by cations in the water surrounding the colloidal particles. In soils where weathering has been intensive, soil clays that are both positively and negatively charged are common. The surface charge on soil clays in variable charged soils is pH dependent. Positively charged surfaces retain anions in the same way as the negatively charged soils retain cations. For example, SO$_4^{2-}$ can displace Cl$^-$ anions adsorbed on a positively charged iron oxide mineral surface:

$$SO_4^{2-} + Cl_2\text{-oxide} = 2Cl^- + SO_4\text{-oxide} \qquad (3.3)$$

Where the *-oxide* represents positive electric potential on an iron oxide surface that neutralizes the negative potential from two Cl$^-$ anions.

The major soluble anions in the soil solution are Cl$^-$, HCO$_3^-$, SO$_4^{2-}$, and, to a lesser extent, NO$_3^-$. In humid region soils, the anion sum rarely exceeds 0.01 M in the soil solution. In arid regions, anion concentration can reach 0.1 M in agricultural soils, and > 1 M in extremely saline soils. Note the cation charge and anion charge in soil solution must be balanced. The relative amounts of anions vary with fertilizer and management practices, mineralogy, microbial and higher-plant activity, saltwater encroachment, irrigation water composition, and atmospheric fallout.

Typical concentrations of anions in soil solutions are Cl$^-$ > HCO$_3^-$ > SO$_4^{2-}$ > NO$_3^-$. At high pH (pH > 8.5) and high salinity, the distribution might be (HCO$_3^-$ + CO$_3^{2-}$) > Cl$^-$ > SO$_4^{2-}$ > NO$_3^-$.

The major soluble anions are retained weakly by most soils. Nitrate and Cl$^-$ anions move through soils at virtually the same rate as the water. Sulfate and HCO$_3^-$ lag slightly behind the wetting front because they interact with Ca^{2+}, Mg^{2+}, and Al^{3+}, and with positively charged sites on clay particles. This interaction is weak, however, compared to the strong retention of anions such as phosphate. In strongly weathered soils with abundant aluminum and iron oxide minerals, retention of sulfate increases greatly. Such soils can develop significant positive charge, particularly at low pH, and their anion adsorption (exchange) capacities can exceed their cation adsorption (exchange) capacities.

3.6.2.1 Nitrate

Plants absorb nitrogen as nitrate or ammonium from soil solutions. Soil solution nitrate concentration is a small fraction of total soil nitrogen. Soil solution nitrate concentration reflects a balance, or steady state, between nitrogen turnover in the soil and plant uptake of nitrogen from the soil. Fertilization can temporarily change this steady state until denitrification, leaching, and nitrogen uptake by plants and microbes restore the nitrogen balance.

The nitrate concentration in the soil solution is greatest near the soil surface, where the ion is produced by microbial decay of organic matter. Root and microbial uptake reduce the nitrate concentration in the root zone. Other than microbial reduction in anoxic soils, nitrate has little reactivity with the soil. If leached below the surface horizons and root zone, nitrate tends to move unhindered and unchanged through the vadose zone and groundwater. Unwise fertilization and organic waste disposal cause nitrate concentrations in groundwaters and drainage waters to increase. If downward water flow is very slow, and the groundwater table is deep, as in arid regions, and there is a source of organic nutrients for the microbes, microbial transformation of NO_3^- to N_2 and N_2O gases (denitrification) may cause nitrate to slowly disappear.

Agriculture greatly contributes to nitrogen concentrations in groundwater and surface waters. Land clearing and leaching of nitrogen from geologic strata in arid regions contributes to nitrate in the ground and surface waters. Organic nitrogen compounds in soils erode and oxidize (mineralize) to nitrate in aerated streams.

The major sources of soil NO_3^- in soils are fertilizers, animal feedlots, septic systems, and grazing animals. Rainfall contributes an additional small amount: The first raindrops in a storm front contain nitrate from natural and anthropogenic NO_x (NO plus NO_2/N_2O_4) washed out of the air, plus HNO_3 formed by lightning. In industrial regions, absorption of NO_x gases released from the industrial activities to the atmosphere may add additional NO_3^- to soils.

3.6.2.2 Sulfate

Sulfate concentrations in soil solutions are good indicators of sulfur availability to plants. Sulfur is increasingly recognized in short supply in humid region soils. The sulfate supply in arid soils is typically adequate for plant needs.

As redox conditions of soils change, sulfur readily changes oxidation state between sulfate (SO_4^{2-} with sulfur oxidation state S^{6+}), elemental S, and sulfide (S^{2-}) (see Special Topic Box 2.1). Several species of sulfur bacteria catalyze these changes.

In acid soils and in soils having considerable aluminum and iron oxides, sulfate anions are retained by adsorption on the positively charged mineral surfaces. The adsorption of sulfate depends on pH because positive charge on most minerals is pH dependent. The adsorption strength of sulfate, that is, the strength of bond between the sulfate anion and positively charged surfaces, is intermediate with respect to other anions such as weakly retained chloride and strongly adsorbed phosphate.

In arid soils, gypsum ($CaSO_4 \cdot 2H_2O$) often precipitates in sulfate-rich soils. Gypsum is a soil fertilizer/conditioner added to some soils to supply sulfur as a plant nutrient, and because of the beneficial effects of calcium cation addition to soil.

3.6.2.3 Halides

In nature, negative one is the only oxidation state of the halogens F^-, Cl^-, Br^-, and I^-. Recent research suggests that organohalides such as chlorinated organic molecules are an important phase of halogens in soils; they may be the predominant form in organic matter rich soils. Apparently, the organohalide compounds are produced by plants or microbes and released into the soil as either exudates or during cellular degradation. Whereas organochlorides have been reported to comprise most of the chlorine phases in some soils, chloride anions comprise the majority of the water-soluble species in soils, especially in arid soils that facilitate buildup of soluble salts.

Chloride is essential for plants in trace amounts, and for animals in greater amounts. As a group, halide ions are retained weakly and occur in low amounts in well-drained soils, although fluoride is held to some degree in soils. Plants are much more tolerant of high chloride concentrations than of high concentrations of other micronutrient ions. Except for the specific chloride sensitivity of fruit and nut trees and berries, the effect of excess chloride in soils is mainly to increase the osmotic pressure of soil water, and thereby reduce water availability to plants. Chloride absorbed by plants from the soil solution is a significant reservoir in terrestrial environments. The natural rate of chloride supply to animals from soils through plants may be less than optimal.

Chloride is only a minor constituent of igneous rocks, although apatite contains up to 0.01 mol fraction of Cl^- and micas can contain small amounts. Most of the chloride input to soils is from rain, marine aerosols, salts trapped in soil parent materials of marine origin, and volcanic emissions. Chloride fallout at seacoasts can be as much as $100\,kg\,ha^{-1}\,yr^{-1}$, but this rate decreases rapidly to 1 or $2\,kg\,ha^{-1}\,yr^{-1}$ with distance toward the continental interiors. Since natural chloride deficiencies in plants are not observed, the low rates input to inland soils is apparently adequate for plant nutrient requirements.

The soil chemistry of iodide and bromide resembles that of chloride, except that I^- and Br^- are retained more strongly, especially by acid soils. The major input of iodine to soils appears to be atmospheric. Endemic iodine deficiency (goiter in humans) occurs in mountainous and continental areas isolated from the sea. Fortunately, supplementing NaCl with small amounts of iodide supplies the necessary amount for animal (and human) nutrition. Iodide and bromide are both potentially toxic, but no natural cases have been reported. Bromide has been used as a tracer for the movement of water, nitrate, and soil solutions in soils.

Fluoride has a chemically distinct behavior compared to the other halides and is the most common halide in igneous rocks. The igneous minerals fluorspar (CaF_2) and apatite ($Ca_5(F,OH)(PO_4)_3$) are both insoluble in water. Fluoride can substitute for OH^- to some extent in soil minerals. This mechanism is probably also responsible for fluoride retention by aluminum and iron oxides in acid soils. Fluoride also associates strongly with H^+. HF is a weak acid, $pK_a = 3.45$, and fluoride adsorption to minerals decreases below this pH.

Fluoride concentrations in soil solutions, groundwaters, and surface waters of humid and temperate regions are generally low ($<1\,mg\,L^{-1}$). The amount of fluoride in human diets has increased due to fortification in domestic water and toothpaste for dental carie

prevention, and because of increased phosphate fertilization. The fluoride impurities in fertilizer are absorbed to some extent by plants. In arid regions, the fluoride concentrations in groundwater can be elevated (e.g., some groundwaters in SW United States have native fluoride concentrations exceeding $10\,mg\,L^{-1}$). Excessive fluoride ingestion can cause dental fluorosis and is proposed to lead to skeletal fluorosis, a concern in many countries.

3.6.2.4 Selenate

The 6+ oxidation state of selenium occurs as selenate oxyanions (SeO_4^{2-}). Selenium is a micronutrient for many animals and is often fortified in diets of grazing livestock. However, the range between toxicity and deficiency in animals is narrow. In arid regions that have high selenium in the soil parent materials (e.g., some shale rocks), excess selenium intake by animals can be a problem.

Like sulfur, selenium can have multiple oxidation states, including Se(-II), Se(0), Se(IV), and Se(VI). In well-drained soils, selenium exists primarily as oxyanions selenite (Se(IV)) or selenate (Se(VI)). The soil chemistry of selenate is very similar to sulfate: moderately soluble, sorbed to positively charged surfaces, and forms some salts in arid regions. Selenite is less soluble than selenate, and more strongly sorbed to positively charged soil particles.

3.6.3 *Poorly soluble anions*

The oxyanions borate (H_3BO_3), silicate (H_4SiO_4), phosphate (H_3PO_4), molybdate (H_2MoO_4), and selenite (H_2SeO_3), and their related deprotonated forms have low solubility in soils because of strong adsorption to soil clays and formation of mineral precipitates with low solubility. Arsenic and chromium also exist as oxyanions in soils, but are less common, and are discussed in the toxic ion categories. Ions that form oxyanions are small highly charged cations that form strong molecular bonds with oxygen or hydroxyl ions to create oxyanions.

The weakly soluble oxyanions are weak acids – in aqueous solutions they gain and lose H^+ as the acidity increases and decreases. The oxyanions tend to be weakly soluble in the soil solution because they associate strongly with Fe^{3+}, Al^{3+}, Ca^{2+}, and Mg^{2+} cations to form minerals that have low solubility, and because they adsorb onto mineral surfaces more strongly than the soluble anions. For example, phosphate exists in many soils as calcium phosphate minerals, such as hydroxyapatite, that are not very soluble. The dissolution reaction for hydroxyapatite is

$$Ca_5(PO_4)_3 OH(s) + 7H^+ = 3H_2PO_4^- + 5Ca^{2+} + H_2O \qquad (3.4)$$

Because hydroxyapatite has low solubility, it is not a major source of soluble phosphate and calcium ions in soil solutions.

3.6.3.1 Silicic acid

Silicic acid ($Si(OH)_4$ or H_4SiO_4), or silicate is ubiquitous in soils because silicon is the second most common element in soils. Although $Si(OH)_4$ is technically an acid, it is fully protonated until pH 9, and thus in most soil solutions it does not occur in its anion form. Silicates are the predominant mineral class in soils and have important roles in mineral weathering reactions. Weathering releases $Si(OH)_4$ to the soil solution, which forms secondary minerals, leaches to lower horizons, or out of the soil profile. In semi-arid regions, the subsoil can become enriched in silicon, causing opaline silicate minerals to precipitate and cement the horizon. Soil horizons cemented with silicate minerals are called duripans. The silicate in duripans occurs as amorphous silicon dioxide (SiO_2 (am)) called opal or silica glass.

Silicon is necessary for some plants and animals in trace amounts. Some scientists consider silicon as a *quasi-essential* element. Silica (SiO_2) is incorporated in plant cell walls and provides physical strength and resistance to insect and fungal attack. Silica may also aid some plants in overcoming soil salinity. These benefits are more evident in hydroponics because many growth solutions lack silicon. Silica in plants forms a cast of the cell wall's morphology. These can remain intact in soil as phytoliths *(plant stone)* after the plant decays.

Amorphous silica (opal) that forms when soluble silica concentrations are high, and aluminosilicate clay minerals kaolinite, smectite (montmorillonite), vermiculite, hydrous mica (illite), are important secondary silicate minerals. Kaolinite tends to form at low silicate concentrations typical in soils developed in humid regions, whereas smectite forms at higher silicate and

calcium concentrations typical in soils formed in arid and semi-arid regions. The clay fraction of soils usually contains a mixture of these clay minerals, plus considerable amorphous silicate material, including allophane and imogolite, which may not be identifiable by X-ray diffraction.

3.6.3.2 Borate

Boron exists in solution as boric acid (H_3BO_3 or $B(OH)_3$). At high pH (>8.5), boric acid takes on another hydroxide anion to form the borate anion ($B(OH)_4^-$):

$$B(OH)_3 + OH^- = B(OH)_4^- \qquad (3.5)$$

Boron is a trace nutrient that is toxic when bioavailable amounts are high. The concentration range between boron deficiency and toxicity to plants is very narrow, and dependent on the plant species. Many soils in agriculturally productive southern and western California have naturally high levels of boron. Evapotranspiration enriches soil boron concentration in soils, leading to decreased crop yields. Several soil management techniques have been used to reduce or eliminate toxic concentrations of boron. Unlike chloride, sulfate, and other soluble anions, the protonated borate molecule (H_3BO_3) is not as easily leached from the plant root zone. Leaching soil with high-quality irrigation water is one method used to remove boron and salt from soils. However, to reduce the boron concentration to acceptable levels requires about three times as much water as it does to reduce the concentration of chloride salts to acceptable levels. Boron desorption from soils is frequently observed to be slow, and many researchers claim that desorption is hysteretic from adsorption.

In highly weathered soils, boron plant deficiencies occur. Thus, soils in humid regions sometimes benefit from boron additions. The range between deficient and excess is narrow and spreading a few kg ha⁻¹ uniformly over the soil is important to avoid toxicity problems. In practice, borate salts are mixed in with other fertilizers or inert materials to aid in even application.

Boron is widely and rather uniformly distributed in rocks and sediments. Tourmaline is a boron-containing mineral. More commonly, boron exists in minerals as an impurity. Boron released to solution by weathering interacts primarily with iron and aluminum oxides, with maximum adsorption at pH 7 to 9. Aluminosilicates adsorb only a small amount of boron because they have mostly negative charge, but in some cases, the small pH-dependent charge on their edges may be enough to retard boron leaching through the soil profile.

3.6.3.3 Molybdate

Molybdenum occurs in soil solutions as the molybdate oxyanion (H_2MoO_4), with an oxidation state of 6+. Molybdate deprotonates starting around pH 4, and by pH 5 exists almost entirely as MoO_4^{2-}. Molybdenum has several stable oxidation states, but in nature Mo(VI) (molybdate), and to a lesser extent Mo(IV) are most common. Molybdate forms complex polymeric ions at high concentrations in water. In soil solutions, however, only the monomer exists.

Molybdate solubility and plant availability increase with increasing pH. Correspondingly, in acid soil solutions the molybdate concentration is low. Molybdenum is essential for the symbiotic nitrogen-fixing microorganisms growing on root nodules of legumes and some other plants. Molybdenum deficiency occurs in some acid soils because of low solubility due to strong adsorption on positively charged minerals. Liming corrects soil acidity, and thus ameliorates low molybdate availability. Direct molybdate fertilization is also effective, although molybdate salts, like boron fertilizer, must be diluted and spread carefully to prevent toxicity.

3.6.3.4 Selenite

Selenite is a Se(IV) oxyanion (H_2SeO_3) retained more strongly by soils than selenate, which is a Se(VI) oxyanion. Selenite is a stable oxidation state in many soil conditions, but selenate is the form most often reported; perhaps because it is more soluble and therefore leached and detected more readily. Selenate is stable under strongly oxidizing conditions, and elemental selenium is stable under reducing conditions. Some reports indicate that selenite may be lost from soils by reduction to H_2Se gas.

Selenium gained notoriety in California when drainage water from a large irrigation project evaporated in a reservoir that was also a waterfowl nesting area. Evaporation concentrated selenium, which caused excessive waterfowl exposure that led to deaths and malformations of embryos. Other instances of elevated selenium in soil have occurred, primarily in arid or semi-arid regions. Legumes growing in these soils often accumulate excessive selenium and grazing animals are at risk of selenosis from consuming these plants.

3.6.3.5 Phosphate

Next to nitrogen, phosphorus is often the most limiting of the major fertilizer elements (N, P, K, S) for plant growth. Thus, crops generally respond to phosphate fertilization. Plant availability belies the actual *total* concentration of phosphorus in the soils because phosphate in many soils is strongly retained and unavailable for plant uptake.

Phosphate chemistry in soils has been studied more intensively than most other elements, except perhaps nitrogen. In the year 2017, 294 research articles were published that included the terms *phosphorus* and *soil* in their title. Such prolific research illustrates the great interest and importance in understanding phosphorus soil chemistry for agriculture and the environment.

Phosphate concentrations of waters draining from soils unaffected by human activities are typically low. However, fertilization and application of animal waste to soils increases dissolved and particulate phosphate in soils and runoff waters. Worldwide, 70% of croplands have phosphorus application rates in excess of plant uptake (**Figure 3.16**). Although phosphate is strongly retained by soils, excess phosphorus may leach into surface waters stimulating algae blooms and water degradation because it is often the limiting nutrient for aquatic plant growth.

Phosphate added to soils is first adsorbed quickly and is later *fixed* into increasingly less soluble forms as time increases. Well-weathered soils have the lowest phosphorus availability for leaching and plant uptake (**Figure 3.17**). As soil acidity increases, phosphate is increasingly unavailable due to retention by aluminum and iron oxide minerals. Highly weathered soils also have the least amount of total phosphorus (**Figure 3.18**); likely because all the available phosphorus has been leached from the soils.

Phosphate is continually released from organic matter decay, desorption of adsorbed phosphate, and solubilization of soil minerals; however, these resupply processes are usually too slow to meet requirements of agricultural crop plants. Despite the intense research effort on soil phosphorus chemistry and fertility,

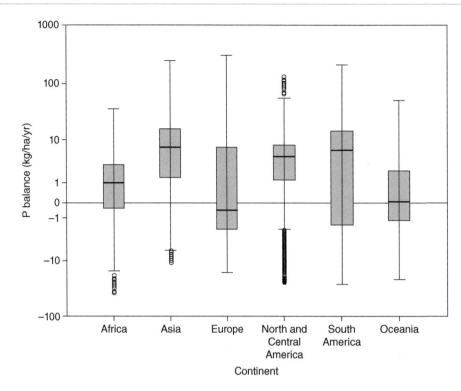

Figure 3.16 Plot of the balance of the phosphorus application rates and plant uptake rates for different continents (Europe and Russia are combined). Positive values indicate that more phosphorus is applied than the plants use, and vice versa (note log scale). The line in the box is median; upper- and lower-box borders are quartiles; line bars are second quartiles; and points are outliers. In most regions of the world, more phosphorus is applied to soils than is taken up by plants. Europe is an exception. North and Central America, South America, and Asia have some of the greatest excesses of P. Source: MacDonald et al. (2011). Reproduced with permission of PNAS.

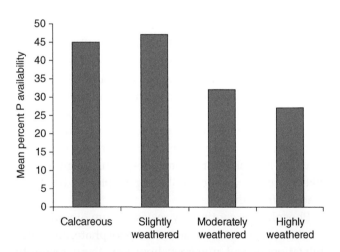

Figure 3.17 Mean percent of soil phosphorus available to plants categorized by amount of soil weathering. Data represent analysis of over 200 soils. Data from Sharpley (1995).

quantitative predictions of phosphate concentrations in soil solutions are poor, and no techniques have been devised to release the large amounts of unavailable phosphate in soils, nor to prevent fixation of fertilizer phosphate by soils. The uncertainties about soil phosphate chemistry and the difficulty of increasing phosphate availability are due to phosphate's strong interaction (adsorption bonds) with soil minerals. Recent research has isolated microbes that stimulate phosphate availability for plant uptake if applied to soils, however, more research and innovation is required to meet the crop demands for more phosphorus; especially considering the limited phosphorus supply in the world's mines.

In alkaline soils with high Ca concentrations, the phosphate mineral apatite is common Apatite has a mineral formula $Ca_5(OH,F)(PO_4)_3$, where either a

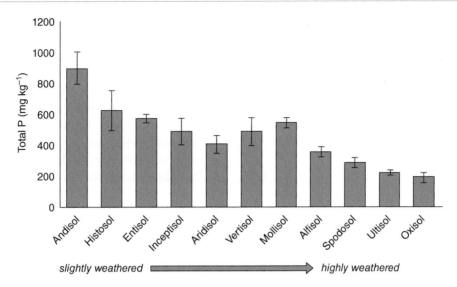

Figure 3.18 Mean total soil phosphorus concentration in 11 soil orders (error bars are one standard error). Data represent analysis of over 178 natural, unfertilized, and uncultivated soils. From Yang and Post (2011). Reproduced with permission.

hydroxide or fluoride is in the mineral structure. In acid soils, most solid-phase phosphate is associated with iron and aluminum oxides. Phosphate fixation is appreciable in all but very coarse-textured soils and is particularly strong in soils rich in amorphous iron and aluminum oxides or allophane-rich soils (i.e., soils with high concentrations of volcanic ash input). In paddy soils, release of phosphate occurs by reduction of Fe(III)-oxide minerals.

Organic phosphorus compounds, like phytic acid $(C_6H_{18}O_{24}P_6)$, are another important type of phosphorus that occurs in soils. Organic phosphorus can be the predominant phosphorus species in organic rich horizons.

For soil testing, many workers devised extraction procedures yielding phosphate values related to crop response of phosphate fertilizer. However, such indices are often site- and crop-specific. Mechanistic predictions of phosphorus availability use equilibrium solubility models to calculate the dissolution of various phosphorus mineral phases. While mineral solubility modeling provides insight into dissolved phosphorus concentrations in soils, phosphorus reactions in soils are complex and dynamic, and predictions based on mineral solubility alone are inaccurate. Future research to better understand phosphate chemistry may improve the accuracy of mechanistic approaches for predicting the availability of phosphate.

Human activities affect the global phosphorus cycle to an ever-increasing degree. The rate of phosphate mining equals or exceeds the rate of phosphate lost naturally from the continents. This suggests that terrestrial ecosystems are being loaded with phosphorus, causing pollution of freshwater sources. Specifically, phosphorus from human waste and runoff and erosion from landscapes with excess phosphorus loading is negatively affecting freshwater and near-shore marine ecology in most regions of the world.

The most feasible alternative to orthophosphate fertilization of soils is the centuries-old practice of increasing the pH of acid soils by liming. Phosphorus availability from organic manures seems to be relatively available for plant uptake, but total phosphorus concentration of manure per mass of manure is low, making the prospect of using manures as a fertilizer for wide-scale agronomic crops inefficient and expensive due to transport and spreading costs. Near animal production facilities, however, soils have become loaded with phosphorus from animal manures, causing degradation of surface water quality.

3.6.4 Poorly soluble metal cations

The metal cations of interest in soil chemistry that are poorly soluble are primarily Al^{3+}, Cu^{2+}, Fe^{3+}, Mn^{4+}, and Zn^{2+}. Many metal ions have low solubility and behave similarly; particularly the first-row transition metals, including Ni^{2+} and Co^{2+}. All tend to become more soluble with increasing soil acidity.

3.6.4.1 Aluminum

Soluble and exchangeable aluminum in soils is closely related to soil acidity. Soil acidity is due to reactions of water with exchangeable Al^{3+} that is on the surface of soil particles:

$$Al^{3+} + H_2O = Al(OH)^{2+} + H^+ \qquad (3.6)$$

This is a hydrolysis reaction that produces a proton and increases the solution acidity. Exchangeable Al^{3+} and its hydrolyzed and polymerized forms, $Al(OH)_x^{(3-x)+}$, produce acidity as low as pH 4.5. Stronger acidity (lower pH) means other H^+-yielding reactions are active, such as organic acids from soil organic matter decay, sulfur and sulfide oxidation, phosphate fertilizers, ammonia oxidation, acid rain, and Fe- and Mn-hydrolysis.

In the most highly weathered soils, Al-water reactions weather aluminosilicate minerals to gibbsite $(Al(OH)_3)$. While other major exchangeable cations are leached from soils during weathering, aluminum is retained in soils as solid-phase $Al(OH)_3$. Under moderate weathering, aluminum forms the intermediate aluminosilicates: smectite and kaolinite clay minerals. The large amounts of limestone required to neutralize acid soils is due to the Al^{3+} that converts to $Al(OH)_3$ releasing protons that are consumed by the carbonate anion released when the amended limestone dissolves.

3.6.4.2 Transition metals Fe, Mn, Cu, and Zn

Transition metals are distinct from the elements at either end of the periodic table in that electrons are added to and removed from *d*-orbitals. The chemistry of the transition metals changes more subtly from element to element than elements having electron gains or losses in the *s* and *p* orbitals. Many transition metals can have more than one oxidation state in soil.

Under oxidizing conditions, transition metals associate strongly with O^{2-} and OH^- ions, and tend to precipitate as insoluble oxides or hydroxides, or as minor components of insoluble aluminosilicates. The precipitation may be on the surface of minerals, but the effect on water solubility and plant availability is the same. Some of the metals present in lower concentrations in soils, such as Cd^{2+}, Hg^{2+}, Zn^{2+}, and Cu^{2+} also react extensively with organic matter and sulfides.

The first-row transition metals titanium, vanadium, chromium, manganese, iron, cobalt, nickel, copper, and zinc are important either because their amounts in soils are large (e.g., iron) and therefore important to soil development, and/or because they are essential elements to living organisms. A few transition metals below the first row, for example, molybdenum, cadmium, and mercury, are also of interest in soil chemistry because they are essential (Mo) or toxic (Cd, Hg) elements.

Transition metal ions precipitate from soil solutions as oxides, such as $FeOOH$, Fe_2O_3, and MnO_2. In solution, the cations form hydrolysis complexes with one to four hydroxides, and go from cationic to anionic form, depending on pH. For example, a summary reaction for goethite ($FeOOH$) dissolution and precipitation at the pH of normal soils is:

$$FeOOH(s) + H^+ = Fe(OH)_2^+(aq) \qquad (3.7)$$

In addition to hydrolysis and precipitation, soils apparently retain trace metal cations by other mechanisms, such as coprecipitation as minor constituents in Fe, Al, Ti, Mn oxides, or aluminosilicates; adsorption on soil surfaces; complexation with organic matter; and incorporation into plant tissues and decay products. These mechanisms often reduce metal concentrations in soil solutions to well below those predicted by their pure oxide solubility products.

The transition metals manganese, iron, copper, nickel, and zinc are all essential for plants and animals. Vanadium, chromium, and cobalt are also essential for animals. The soil solution concentrations and plant availabilities of these ions generally decrease with increasing pH. Compared to the major fertilizer elements nitrogen, phosphorus, potassium, and sulfur, plant deficiencies of micronutrients are infrequent. Plant variety and growth rate seem to be at least as important as soil factors in determining microelement deficiencies. Iron and zinc deficiencies are common in irrigated crops, especially fruit trees grown on alkaline soils. Such deficiencies are caused in part by high growth rates, and consequently high plant demands for these elements. Native plants without irrigation apparently have adapted to the natural rates of trace metal recycling.

Deficiencies of zinc and copper have been recognized in some acid soils. The deficiency is mostly attributable to the low concentrations remaining in highly weathered, acidic, soils. Weathering removes these ions from the soil profile faster than iron, aluminum, and manganese.

Metal cations form complexes with chelates and dissolved organic molecules, which controls their concentrations in the soil solution. The solubility of trace metals that are complexed with chelates in soils, especially in the rhizosphere, is much greater than the solubility of the hydrated cations. Natural chelates are important for metal availability for plant uptake, and some plants use energy to produce such chelates. For example, a siderophore is natural chelate that increases iron availability for plant uptake. Chelates added to soils can increase the solubility and plant availability of trace metals. Chelates from decomposing organic matter can dissolve Fe- and Al-containing minerals and move them downward, where they re-precipitate in the soil profile. This process causes the spodic horizons in Spodosols.

Attempts to estimate the amounts of micronutrients available to plants have been somewhat successful. Soils to be tested are usually extracted with solutions containing chelates, such as DTPA or EDTA. The amount of metal extracted is then correlated with plant response. However, like most soil nutrition test based on extractions, for some soil and plant systems accuracy of DTPA for plant available metals may be poor.

3.6.5 Common toxic elements in soils

The elements aluminum, arsenic, chromium, cadmium, mercury, and lead are commonly found in contaminated soils. At high soil concentrations, manganese, zinc, cobalt, and nickel may also be toxic. All these elements, except for arsenic, are metals. Arsenic is a metalloid. Historically these elements were referred to as heavy metals, however this term is losing favor. Elements are considered potentially toxic when their ions are available to an organism at concentrations that do harm or occur in soils or groundwater at concentrations higher than native conditions. **Table 3.6** lists common concentration ranges and averages of metals in soils, water, and plants. For comparison, the range of regulatory limits in residential soils is listed as reference to provide guidance of concentrations that are potentially harmful for humans. Toxicity to natural ecosystems may occur at

much lower or higher concentrations than the residential limit concentrations.

Naturally occurring high concentrations of toxic elements are rare in soils, except for widespread Al^{3+} phytotoxicity in acid soils. Soil contamination from smelting, metal plating, manufacturing, municipal and industrial wastes, and automobile traffic can increase soil concentrations of these ions to possibly toxic levels. Oxidation states of common contaminant elements are: Be^{2+}, F^-, Cr^{3+}, Cr^{6+}, Ni^{2+}, Zn^{2+}, As^{3+}, As^{6+}, Cd^{2+}, Hg^0, Hg^+, and Hg^{2+}, Pb^{2+}, and Pb^{4+}.

3.6.5.1 Metal contaminants
Figure 3.19 shows the distribution of Cd^{2+}, Cu^{2+}, Pb^{2+}, Ni^{2+}, and Zn^{2+} cations in soils between the solid and solution phases (called the distribution or partition coefficient) as a function of pH. Many data were compiled from the literature to derive the relationships. Since pH is a dominant factor controlling metal solubility, a relationship is expected; the regression line, however, represents only the best linear fit to the data, and the actual variability about the lines is great. The high variability indicates that the simple linear relationships between distribution coefficient and pH are not sufficient to accurately predict metal solubility at a given site, but the lines show the general trends observed for metals.

Removal of metals from solution and release from the solid phase back to the solution occurs from metal adsorption/desorption or precipitation/dissolution processes. Distribution coefficients represent an equilibrium state, thus if metal concentrations in soil solutions vs the solid phase exceed the distribution coefficient at a given pH, the metal will either precipitate or adsorb. Similarly, values below the distribution coefficient cause metals to be released by either dissolution of a solid phase, or desorption of the metal from a soil particle surface (e.g., cation exchange). Amongst the common contaminant metals, Pb^{2+} has the lowest solubility in soil pore water (**Figure 3.19**). Below pH 7, Cu^{2+} is the next lowest solubility, followed by Zn^{2+}, Cd^{2+}, and Ni^{2+}. At higher pH, Ni^{2+} and Zn^{2+} solubility exceed that of Cu^{2+}.

Cadmium is a rather soluble transition metal that behaves somewhat like Ca^{2+}, except that Cd^{2+} reacts more strongly with organic matter and sulfides. Thus, compared to calcium, cadmium movement and plant availability in soils is small. But, because it is an

Table 3.6 Representative concentrations of potential contaminants in soils, plants, and fresh water. Reference plant concentrations are for a plant grown in noncontaminated soil. Residential soil limits are minimum, and maximum of values gleaned from United Kingdom, United States of America, Netherlands, Australia, and World Health Organization. Residential soil limits are intended to provide general guidance on concentrations in residential soils used for screening purposes by regulatory agencies; soil concentrations above or below these values may pose risks of poisoning to humans (site-specific assessment of toxicity potential is required to determine actual toxicity risks).

Element	Soil range[a] (mg kg^{-1})	Soil median[a] (mg kg^{-1})	Contaminated soil mean[b] (mg kg^{-1})	Residential soil limits (mg kg^{-1})[e]	Fresh water median (mg L^{-1})[a]	Reference plant[c] (mg kg^{-1} (dw[f]))
Al	10^4–3×10^5	71 000			0.3	80
As	0.1–40	6	48	2–110	0.0005	0.10
Be	0.01–40	0.3	2.6	160	0.0003	0.001
Cd	0.01–2	0.40	19	1–70	0.0001	0.05
Cr	5–1500	70	46[d]	0.29[d]	0.001	1.5
Co	0.05–65	8	17	23[e]	0.0002	0.2
Cu	2–250	30	410	30–3100	0.003	10
Hg	0.01–0.5	0.06	2.6	0.5–23	0.0001	0.1
Mn	20–10 000	1000	1210	1800	0.008	200
Ni	2–750	50	89	40–600	0.0005	1.5
Pb	2–300	35	862	100–530	0.003	1
Sb	0.2–10	1	76	15–31	0.0002	0.1
Se	0.01–12	0.4	5.4	35–390	0.0002	0.02
Zn	1–900	90	1300	200–23 000	0.015	50

[a] Bowen (1979).
[b] Contaminated soil mean concentrations are from the National Priority List (NPL) sites established by the USEPA (see Figure 3.8 in this textbook).
[c] Markert (1992b).
[d] Cr(VI)
[e] Data compiled from: Ohandja et al. (2012); Rodríguez Eugenio et al. (2018); and USEPA (2013) (http://www.epa.gov/reg3hwmd/risk/human/rb-concentration_table/Generic_Tables/index.htm; checked March 2019)) (minimum of carcinogenic or of toxicity risks values is shown).
[f] dry weight

extreme poison, even low Cd^{2+} availability is a serious concern in soils where it is present in abnormally high concentrations.

3.6.5.2 Mercury

Divalent mercury, Hg^{2+}, is retained strongly by soil as either solid minerals or adsorbed on soil particle surfaces. Hg^{2+} can be reduced to Hg^+ or Hg^0. Hg^{2+} probably predominates in oxidized soils, but in wetlands or sediments, reduction occurs. Elemental mercury is slightly volatile and diffuses as a gas through soil pores. Because of the redox cycling (driven by microbes), mercury is relatively mobile compared to some other metals (e.g., Pb^{2+}). The toxic compound dimethyl mercury ($Hg(CH_3)_2$), formed in contaminated and highly reduced aquatic sediments seems to be rare in soils. A common pathway for mercury uptake into the food web is as suspended particles and sediments.

3.6.5.3 Arsenate and chromate

The availability of the oxyanions arsenate ($H_2AsO_4^-$) and chromate (CrO_4^{2-}) decreases as iron and aluminum oxide content increases, and increases with pH. Arsenic is a toxic element. Movement of arsenic through soils is minimal unless large quantities are concentrated in a small or unreactive soil volume, such as might be the case for industrial waste disposal on coarse-textured soils; or if the soil undergoes changes in oxidation state, in which case arsenite (H_3AsO_3) is produced, which is more soluble than arsenate.

Arsenic problems in soils are primarily the result of anthropogenic activities. Lead arsenate and copper acetate-arsenate (Paris green) were once common insecticides. Spraying apple trees with Paris green for many years led to arsenic concentrations high enough to harm the trees and hinder replanting. After several decades since Paris green was last applied, risks from lead and arsenic in the soils are somewhat diminished because they have either been dispersed and effectively diluted, or they have transformed to less soluble species. However, some historic orchards may still pose risks for lead and arsenic contaminated soils, especially if developed for housing.

In Southeast Asia, some soils and groundwaters have naturally elevated levels of arsenic. Because of agriculture and aquifer management practices, arsenate that is stable and attached to iron oxides, is reduced to more soluble arsenite (H_3AsO_3). As a result, arsenic poisoning to humans is occurring in the region from consumption of arsenic-contaminated drinking water and food (e.g., rice).

Chromium is a micronutrient for animals. Its oxidized form (Cr(VI)) occurs as the chromate anion (CrO_4^{2-}). Cases of chromium poisoning from chromate leaching into drinking water sources have occurred. Under reducing conditions, Cr(VI) reduces to Cr(III), which is insoluble.

3.6.6 Major biogeochemical elements: carbon, nitrogen, and sulfur

Carbon, nitrogen, and sulfur are grouped together because (1) their geochemical cycles are rapid and include the atmosphere; (2) they change oxidation states rapidly and cyclically in soils and the environment; (3) the changes in oxidation state provide biological energy; and (4) carbon, nitrogen, and sulfur processes in soils are interrelated, although not similar. Oxidation and reduction are principal chemical reactions that occur during carbon, nitrogen, and sulfur cycling. **Table 3.7** shows the different oxidation states and species these elements exist in soils.

Photosynthesis of carbon and its oxidation back to CO_2 provides the energy that drives most life. Nitrogen and sulfur oxidation-reduction also involves energy changes, but the amounts are small in comparison, and often require energy from carbon sources. The changes and turnover rates of carbon, nitrogen, and sulfur in some soil processes occur within hours and days. Turnover rate is the time required for a complete change of all the substances in the soil from one phase to another.

The oxidation-reduction reactions of carbon, nitrogen, and sulfur (and O) in soils are catalyzed by microbes. Redox reaction rates without such catalysis are very slow and are often irreversible. Due to the

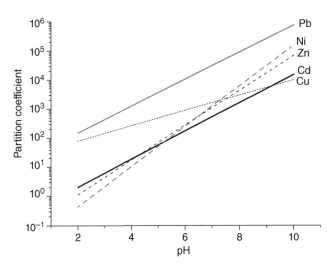

Figure 3.19 Best fit of partition coefficients of metals in soils as a function of pH. The lines are regression from 140 to 840 points from the literature. The partition coefficient is the ratio of the concentrations on the solid over the concentrations in solution, with units of L kg^{-1}. A high-partition coefficient indicates that the metal is associated with soil particles as opposed to dissolved into solution. Adapted from Sauve et al. (2000). Reproduced with permission of American Chemical Society.

Table 3.7 Oxidation state and examples of C, N, S in the soil.

Element	Common oxidation state in soil	Example species of oxidization states*	Common occurrences in soil
Carbon	Inorganic: (II), (IV); Organic: (–IV to III)	CH_4, glucose, CO, R-carboxyl, CO_2	Soil organic matter, biota, calcite
Nitrogen	(–III), (0), (I), (III), (V)	NH_4^+, N_2, N_2O, NO_2^-, NO_3^-	Adsorbed-NH_4, proteins, DNA, N_2, nitrate
Sulfur	(–II), (0), (IV), (VI)	S^{2-}, S, SO_3^{2-} or SO_2, SO_4^{2-}	Sulfate, H_2S gas, metal sulfides, thiols, gypsum ($CaSO_4$), sulfate in soil solution

*Listed as most reduced to most oxidized

continuous fluxes in soils, carbon and nitrogen are in a constant state of flux (turnover rates for nitrogen and carbon are months to a year). The other elements in the periodic table cycle annually in smaller concentrations between plants and soil, and on geologic time scales between soils and rocks.

In soils developed in temperate and humid regions, carbon, nitrogen, and sulfur exists predominantly in organic compounds, resulting in a C/N/S mass ratio of about 100/10/1. With increasing aridity, the ratio of changes as the amount of soil organic matter decreases and the amounts of carbonate, sulfate, and nitrate anions in the soil solution increases.

Change of oxidation state causes changes in the physical and chemical properties of elements. Some redox reactions produce volatile compounds (CO_2, CH_4, N_2O, N_2, NH_3, SO_2, H_2O, etc.), causing gasses to exchange between soils and the atmosphere. Because of the active movement between solid, solution and gas, and the multiple oxidation states, the behavior of carbon, nitrogen, and sulfur is more complex than that of many other elements.

The amounts of carbon in soil organic matter are almost as large as all other active carbon reservoirs on Earth combined. This is also true for sulfur, except for ocean water. The largest fraction of nitrogen, on the other hand, is in the atmosphere (**Table 3.8**). The environmental effects of human alterations of the natural carbon and nitrogen cycles are subjects of intense interest because of the effects of gaseous compounds of these elements on climate. Soil emission and absorption of CO_2, N_2O, and CH_4 to and from the atmosphere are important processes in assessing managed and natural ecosystem contributions to global climate change.

Concentration of carbon, nitrogen, and sulfur gases in the atmosphere are buffered by cycling from soil reservoirs. Soil organic matter is the largest *active* carbon reservoir in the environment, and its amount changes with the CO_2 concentration in the air, rainfall, and temperature. For example, since the retreat of the glaciers 12000 years ago, carbon has been transferred *through the atmosphere* from soils and vegetation in southern latitudes and deposited as carbon-rich peat and soil organic matter deposits in the northern latitudes (permafrost, bogs, and deltaic deposits). The full extent of soil's effectiveness at buffering atmospheric CO_2, and the time scales under which it operates, are poorly understood.

3.6.6.1 Carbon

There are two categories of carbon chemistry in soils: inorganic and organic (see Special Topic Box 1.1). Inorganic carbon in soils exists as carbonate minerals, carbonic acid anions in soil solution, and CO_2 in soil pore spaces. An important process in soils for inorganic carbon is dissolution of carbon dioxide gas into water to form carbonic acid:

$$CO_2(g) + H_2O = H_2CO_3(aq) \qquad (3.8)$$

Carbonic acid deprotonates and buffers soil solution pH, and forms complexes with dissolved cations.

Organic carbon compounds in soils exist in several oxidation states (**Table 3.7**). Partially degraded biomolecules comprise soil organic matter. Both inorganic and organic soil carbon chemistry are discussed in more detail in other chapters of this text.

3.6.6.2 Nitrogen

After water, nitrogen is the most important substance for optimizing plant yield. There are many different nitrogen species, and it has a complex biochemical

Table 3.8 Global estimates of carbon, nitrogen, and sulfur in Earth's active reservoirs. Bold values indicate the major active pool of each element on Earth. Data collected from: Reeburgh (1997); Lal (2008); Lavelle et al. (2005).

Global pool	Carbon	Nitrogen	Sulfur
Soil	**2600 Pg**	95 Pg	300 Pg
Plants	650 Pg	35 Pg	0.0085 Pg
Atmosphere	780 Pg	**4 × 10^6 Pg**	0.0048 Pg
Surface ocean water	900 Pg	2.2 × 10^4 Pg[a]	**1.3 × 10^6 Pg**
Freshwater (including sediments)	150 Pg		0.3 Pg

[a] Predominantly N_2; organic N = 2 × 10^2 Pg

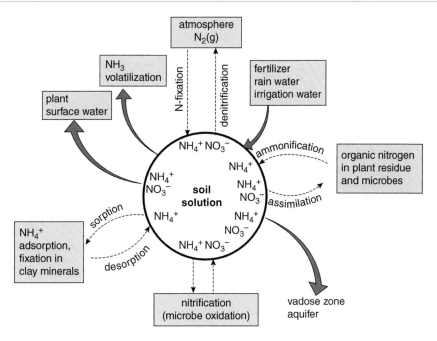

Figure 3.20 Nitrogen cycle in soils showing fluxes of N_2 gas, NH_4^+, and NO_3^-.

cycle in nature. As a result, nitrogen in soil has been studied for centuries, and is still the most-studied element in soil chemistry, soil microbiology, and soil fertility.

Many of the important nitrogen chemical reactions in soils require microbial catalysis, except for ammonium adsorption on clay minerals (fixation). Nitrate is stable under oxidizing conditions, and amino-N is stable under strongly reducing conditions. Nitrogen gas (N_2) is the predominant nitrogen species in the environment. The nitrogen cycle reactions in soil consist of five general processes:

1 Nitrogen gas incorporation into soil (N_2-fixation)
2 Ammonium conversion to nitrate (nitrification)
3 Nitrate conversion to N-gases (denitrification)
4 Plant and microbial absorption and conversion of inorganic nitrate or ammonium to organic nitrogen (assimilation or immobilization)
5 Conversion of organic nitrogen to inorganic nitrogen (ammonification or mineralization) by organisms

Other less common nitrogen transformation processes not commonly studied in soils include anammox (anaerobic ammonium oxidation to nitrogen gas), and dissimilatory nitrate reduction to ammonium (DNRA), both of which occur in anaerobic soils. Anammox has

been reported to be a major contributor of nitrogen loss from paddy soils and other flooded soils, and has been observed in nonflooded soils, although to a much smaller extent.

Figure 3.20 shows the major fluxes of nitrogen in the environment (see Chapter 5 for additional details of nitrogen redox transformations). Fluxes of nitrogen in and out of the soil include fertilization, rainwater, gas diffusion, and percolation to groundwater. A key aspect of the nitrogen cycle is the transfer of electrons to or from nitrogen species, largely driven by microbial reactions. Nitrogen has multiple stable intermediate oxidation states between the most reduced species (N^{3-}, (NH_3)), and the most oxidized species (N^{5+}, (NO_3^-)). Thus, there is a great diversity in the nitrogen species that occur in soils (**Table 3.7**).

Many nitrogen transformations depend on photosynthetic energy stored in carbon compounds. Alfred Redfield first proposed that C/N molar ratios of phytoplankton in the oceans were constant of 6.6. This concept has been used for decades to understand biogeochemical processes in oceans and freshwater bodies. Similarly, carbon and nitrogen cycling in soils from plant input and microbial activity leads to a C/N mole ratio in soils of approximately 14.3 (± 0.5) (Cleveland and Liptzin, 2007). Microorganisms maintain the soil C/N ratio at a steady state by oxidizing

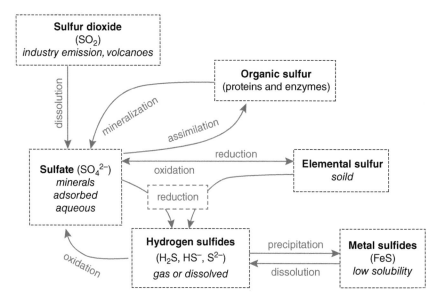

Figure 3.21 Pathways of biogeochemical processes for sulfur showing species, phases, sources, and reactions.

carbon, fixing nitrogen, or by denitrification. Thus, soils act as buffers to carbon and nitrogen concentrations. Some crop residues have C/N ratios up to 80:1 (e.g., grain straw), and thus when incorporated into soils, nitrogen is limited because of the excess microbial activity utilizing all the added carbon. But, over time, the soil will reestablish its natural C/N ratio and nitrogen will start to be mineralized and released to the soil solution.

Soil nitrogen contents fluctuate with soil and fertilizer management, including growth of leguminous versus nonleguminous crops. The production and use of nitrogen fertilizers is a significant part of the nitrogen cycle in agricultural areas. About half of the nitrogen applied as fertilizer is recovered by plants, the remainder is denitrified or leached from the soil.

3.6.6.3 Sulfur

Although sulfur is less volatile than carbon or nitrogen, organic decay in saturated wetland soils may produce organic sulfur compounds such as methyl mercaptan (CH_3SH), dimethyl sulfide (CH_3-S-CH_3), dimethyl disulfide (CH_3-S-S-CH_3), and inorganic hydrogen sulfide (H_2S). Within the soil, these gases oxidize, or the sulfide reacts strongly with metal ions and precipitates as FeS_2, MnS, and other metal sulfides rather than escaping to the atmosphere. If soils emit H_2S and organic sulfide gases, their strong odors are noticeable.

Sulfur has a very active biogeochemical cycle (**Figure 3.21**), with reactions mediated by microbes, abiotically, and plants and animals. Sulfate is the stable oxidation state in aerobic soils, and sulfide is stable in anaerobic soils (see Special Topic Box 2.1 for detailed list of sulfur species). Sulfur changes its oxidation state by microbial catalysis, and the changes are much more reversible than nitrogen and carbon reactions. Elemental sulfur rarely occurs naturally in soils but is sometimes added to soils as an amendment. Sulfides are common in wetlands, and in soils affected by mining wastes. When elemental sulfur and sulfides are exposed to oxygen, they oxidize to H_2SO_4. Soil acidities as low as pH 2 may persist until the sulfide or sulfur has all been oxidized and leached away.

Major sulfur inputs to soils include atmospheric SO_2 and its various oxidation products from coal combustion, petroleum processing, ore smelting, and sulfate from sea spray. Most of the atmosphere's sulfur falls near the areas where it is produced. Fallout occurs both as rain and as direct plant absorption. In arid regions, SO_2 and its oxidation products, H_2SO_3, and H_2SO_4, are absorbed directly and rapidly by the basic soils and their dust.

Atmospheric sulfur from smokestacks can be carried hundreds of kilometers over flat terrain. In regions of sulfur-deficient soils, atmospheric sulfur at low concentrations can benefit plants. Such benefits, however, must be weighed against the associated acidification of freshwater, phytotoxicity, health hazards, smog, and building deterioration at the higher concentrations near the sources.

In contrast to phosphate anions, sulfate anions are adsorbed less strongly by soils. Sulfate anions are absorbed readily by plants and incorporated into biomass. Hence, biomass and SOM constitute large sulfur reservoirs at Earth's surface (**Table 3.8**). The C/S mass ratio in soil organic matter is typically about 100/1. The sulfate content of soils increases with aridity and with salt accumulation.

Widespread sulfur deficiencies in many agricultural soils have become more obvious in recent decades as triple superphosphate (TSP) began to supplant superphosphate as a phosphate fertilizer. Superphosphate is made with H_2SO_4 and contains about 11–12% by weight sulfur. This advantageous sulfur fertilization ended when higher purity triple superphosphate, made with H_3PO_4, was substituted. Until the sulfur deficiency was recognized and corrected, the *improved* triple superphosphate sometimes produced lower crop yields than did superphosphate.

3.7 Summary of important concepts for chemicals in the soil environment

In soils, chemicals occur as ions, organic compounds, gases, minerals, and salts. Chemicals in soils are generally classified as either contaminants or nutrients. Most required nutrients become contaminants when their concentrations exceed the natural limits of their biogeochemical cycles.

Another classification of soil chemicals is inorganic and organic chemicals. Organic chemicals are often introduced into soils and the environment intentionally to control pests or weeds, and sometimes accidentally from spills or uncontrolled industrial processes. The fate of organic chemicals is a function of the charge characteristics of the organic chemical and its degradation rate and products, as well as the soil properties. Inorganic chemicals occur as dissolved ions, gasses, and solids. In this chapter, inorganic ions were categorized into six groups based on similarities in their chemical behavior.

Understanding chemical behavior in soils and the environment starts by knowing the species and possible reactions occurring in soils. Information in this chapter is a useful reference when studying specific reaction processes and soil chemical characteristics in the remaining text.

Questions

1 Why are total elemental concentrations in soils poor indicators of the amounts of ions that enter the food chain?

2 What are the major exchangeable cations in soils? How do the relative proportions change between acid and basic soils? What soil conditions are typical of deficiency, sufficiency, and toxicity (if any) for these ions?

3 How are Fe(III) and Al(III) reactions both different and similar in soils?

4 Which transition metals are essential to (a) Plants? (b) Animals? Which are only toxic, so far as is currently known? Which can be both essential and toxic?

5 What are the forms of the micronutrients in soil solutions? How does their availability change with pH?

6 Explain why trends for total concentrations of some elements in soils are very similar to elements in the lithosphere, while for other elements, concentrations are different.

7 Why is pH considered a master variable when evaluating chemical processes and bioavailability in soils?

8 Why are chemicals of emerging concern a distinct category of chemicals?

9 What factors are important for pesticide adsorption, mobility, and toxicity?

10 Compare oxidation states, valences, molecular species, and mobility of the oxyanions phosphate, sulfate, arsenate, arsenite, selenate, and selenite.

11 Discuss the chemical species that occur in the global active pools for carbon and nitrogen.

12 At a contaminated site, what factors should be used to determine the risks at the site?

13 Describe how soil fluxes might affect bioavailability of a chemical in the soil.

Bibliography

Agency for Toxic Substances and Disease Registry. 2017. Substance Priority List (SPL) Resource Page. Atlanta, GA.

Atwood, and Paisley-Jones. 2017. Pesticides industry sales and usage: 2008 and 2012 market estimates. U.S. Environmental Protection Agency, Washington, DC.

Bowen, H.J.M. 1979. Environmental Chemistry of the Elements Academic Press, London; New York.

Cleveland, C.C., and D. Liptzin. 2007. C : N : P stoichiometry in soil: is there a "Redfield ratio" for the microbial biomass? Biogeochemistry 85:235–252.

JorgeFernandez-Cornejo, R. Nehring, C. Osteen, S. Wechsler, A. Martin, and A. Vialou. 2014. Pesticide Use in U.S. Agriculture: 21 Selected Crops, 1960–2008. US Department of Agriculture, Economic Research Service, May 2014.

Kolpin, D.W., E.T. Furlong, M.T. Meyer, E.M. Thurman, S.D. Zaugg, L.B. Barber, and H.T. Buxton. 2002. Pharmaceuticals, Hormones, and Other Organic Wastewater Contaminants in U.S. Streams, 1999–2000: A National Reconnaissance. Environmental Science & Technology 36:1202–1211.

Lal, R. 2008. Sequestration of atmospheric CO_2 in global carbon pools. Energy and Environmental Science 1:86–100.

Lavelle, P., R. Dugdale, R. Scholes, A. A. Berhe, E. Carpenter, L. Codispoti, A.-M. Izac, J. Lemoalle, F. Luizao, M. Scholes, P. Tre´guer, and B. Ward. 2005. Chapter 12: Nutrient Cycling Ecosytems and Human Well-being: Millenium Ecosytem Assessmnet. Island Press, United States of America.

MacDonald, G.K., E.M. Bennett, P.A. Potter, and N. Ramankutty. 2011. Agronomic phosphorus imbalances across the world's croplands. Proceedings of the National Academy of Sciences of the United States of America 108:3086–3091.

Markert, B. 1992a. Presence and significance of naturally occurring chemical-elements of the periodic system in the plant organism and consequences for future investigations on inorganic environmental chemistry in ecosystems. Vegetation 103:1–30.

Markert, B. 1992b. Establishing of reference plant for inorganic characterization of different plant-species by chemical fingerprinting. Water Air and Soil Pollution 64:533–538.

Ohandja, D., S. Donovan, Castle, P., Voulvoulis, N. and Plant, J.A. 2012. Regulatory systems and guidelines for the management of risk, *In* J. A. Plant, Voulvoulis, N., and Ragnarsdottir, K.V. (eds.) Pollutants, Human Health, and the Environment: A Risk-Based Approach. Wiley-Blackwell, Chichester, West Sussex, U.K.

Oram, L.L., D.G. Strawn, and G. Moller. 2011. Chemical speciation and bioavailability of selenium in the rhizosphere of symphyotrichum eatonii from reclaimed mine soils. Environmental Science & Technology 45:870–875.

Pure Earth. 2016. The Toxins Beneath Our Feet: 2016 World's Worst Pollution Problems. http://www.worstpolluted.org/.

Reeburgh, W.S. 1997. Figures summarizing the global cycles of biogeochemically important elements. Bulletin of the Ecological Society of America 78:260–267.

Rocha, L., S.M. Rodrigues, I. Lopes, A.M.V.M. Soares, A.C. Duarte, and E. Pereira. 2011. The water-soluble fraction of potentially toxic elements in contaminated soils: Relationships between ecotoxicity, solubility and geochemical reactivity. Chemosphere 84:1495–1505.

Rodríguez Eugenio, N., M.J. McLaughlin, D.J. Pennock, Food, N. Agriculture Organization of the United, and P. Global Soil. 2018. Soil pollution: a hidden reality. Food and Agriculture Organization of the United Nations, Rome.

Sauve, S., W. Hendershot, and H.E. Allen. 2000. Solid-solution partitioning of metals in contaminated soils: Dependence on pH, total metal burden, and organic matter. Environmental Science & Technology 34:1125–1131.

Sharpley, A. 1995. Fate and Transport of Nutrients: Phosphorus. USDA Natural Resources Conservation Service:http://www.nrcs.usda.gov/wps/portal/nrcs/detail/national/technical/nra/rca/?&cid=nrcs143_014203.

Truog, E. 1946. Soil reaction influence on the availability of plant nutrients. Soil Science Scoeity of America Proceedings: 305–308.

USEPA. 2013. Residential Soil Screening Guidance (HQ=1). Available at http://www.epa.gov/reg3hwmd/risk/human/rb-concentration_table/Generic_Tables/index.htm (accessed April 2014).

Wedepohl, K.H. 1995. The Composition of the Continental-Crust. Geochimica Et Cosmochimica Acta 59:1217–1232.

Wilson, M.A., R. Burt, S.J. Indorante, A.B. Jenkins, J.V. Chiaretti, M.G. Ulmer, and J.M. Scheyer. 2008. Geochemistry in the modern soil survey program. Environmental Monitoring and Assessment 139:151–171.

Yang, X., and W.M. Post. 2011. Phosphorus transformations as a function of pedogenesis: A synthesis of soil phosphorus data using Hedley fractionation method. Biogeosciences 8:2907–2916.

Zhang, Y., R.G. Krysl, J.M. Ali, D.D. Snow, S.L. Bartelt-Hunt, and A.S. Kolok. 2015. Impact of sediment on agrichemical fate and bioavailability to adult female fathead minnows: A field study. Environmental Science & Technology 49:9037–9047.

4 SOIL WATER CHEMISTRY

4.1 Introduction

Soil water is the aqueous solution that resides in pore spaces of soil and reacts with the soil particles, gases, roots, and microbes (Figure 1.1). An aqueous **solution** is a mixture of dissolved chemicals (**solutes**) and water. In saturated, or near saturated soils, the water (called **bulk solution**) is readily available for plant uptake (**Figure 4.1**). As soils dry out, larger soil pores dewater, and solution resides only in small pores as hygroscopic-water and progressively thinner films on particle surfaces. The solution chemistry in a film of water on a particle is distinct compared to the bulk solution because the charge of soil particles extends into the water film, interacting with ions and polar water molecules. The properties of the **solid–solution interface** are discussed in Chapters 9, 10, and 11. In soils that are very dry, water only exists in mineral particles and organic matter molecules, called *molecular water*. In this chapter, chemical reactions that occur in bulk solution are discussed, including:

1. Protonation and deprotonation of acids
   example: $H_2CO_3 = H^+ + HCO_3^-$
2. Hydrolysis
   example: $Zn^{2+} + H_2O = ZnOH^{1+} + H^+$
3. Complexation
   example: $Ca^{2+} + SO_4^{2-} = CaSO_4^0(aq)$
4. Gas dissolution and volatilization
   example: $CO_2(g) + H_2O = H_2CO_3(aq)$
5. Solid precipitation and dissolution
   example: $Pb^{2+} + SO_4^{2-} = PbSO_4(s)$

The first three reaction types are aqueous reactions. Gas dissolution and volatilization involve liquid–gas interactions. Precipitation and dissolution reactions deal with movement of ions between aqueous and solid phases.

Soil Chemistry, Fifth Edition. Daniel G. Strawn, Hinrich L. Bohn, and George A. O'Connor.
© 2020 John Wiley & Sons Ltd. Published 2020 by John Wiley & Sons Ltd.

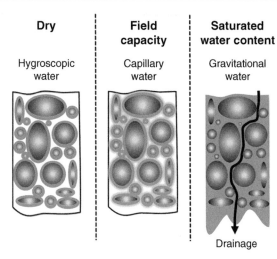

Dry	**Field capacity**	**Saturated water content**
Hygroscopic water	Capillary water	Gravitational water

Drainage

Figure 4.1 Schematic showing different soil water states. Under dry conditions, soil water is associated with particle surfaces such as clay interlayers, and total dissolved solids are high due to concentration by evapotranspiration. At field capacity, capillary water and gas are present in soil pores, and water and nutrients are available for plant uptake. Under saturated conditions, there is no pore space for gas, and water will drain out of soil carrying chemicals to groundwater or surface waters.

Most methods for studying soil solution properties rely on separating the solution from the soil using a lysimeter, by filtration, or centrifugation. Once the solution is extracted, chemical composition or properties can be measured. However, the chemical properties of the soil solution ex-situ may not be the same as the soil solution in-situ (*within* the soil). For example, measurement of pH, ion complexation, ion activity, and gas dissolution rate may be affected by the solids present in soil. Some soil solution properties, such as soil pH and electrical conductivity, can be measured without extracting the solution from the soil.

The focus of this chapter is equilibrium reactions. In natural systems, fluxes of solutions through the soil are often faster than reactions come to equilibrium, and the prediction of soil–solution interface reactions requires use of reaction kinetics instead of equilibrium models. Even in soil systems where reactions are kinetically controlled, predicting *expected* speciation at *equilibrium* is useful because thermodynamics describes the species in the system that are being produced, allowing for insight into the behavior of the chemicals in soil.

4.2 Thermodynamic approach to aqueous soil chemistry

To make equilibrium predictions, thermodynamic chemical relationships are used. Thermodynamics is a well-developed science for predicting chemical reactions. As discussed in Chapter 1, thermodynamic calculations are centered on the three laws of thermodynamics:

1 Energy is conserved.
2 Matter in a closed system not at equilibrium will change so that it has maximum disorder and lowest relative energy.
3 At absolute zero Kelvin, the entropy of a perfect crystal is zero.

The first law is the most familiar from physics and chemistry classes. The second law is more abstract but is often conceptualized as a system that spontaneously moves toward randomness; for example, a drop of dye in water diffuses throughout the water until equally distributed. The third law is best regarded as a rule needed to quantify the *relative energy* part of the second law.

For chemistry, the three laws are used to quantify energy, enthalpy (H), and entropy (S) of a chemical system relative to a reference state (**standard state**). This allows chemical speciation to be calculated since the chemical energy of a system will progress to the lowest energy state (Figure 1.13). Full treatment of thermodynamic application to aqueous chemistry is beyond the scope of this text. However, coverage of some of the basic thermodynamic relationships is provided because they are needed for quantifying equilibrium reactions in aqueous systems.

Table 4.1 summarizes four basic thermodynamic relationships used in predicting chemical equilibrium processes. *Equation I* in **Table 4.1** uses enthalpy ($\Delta H°$) and entropy ($\Delta S°$) values to calculate the **Gibbs free energy** of a system ($\Delta G°$). Enthalpy and entropy values at standard state temperature (298 K), pressure (1 atmosphere), and concentration (1 molar) are available in reference tables. Free energy values calculated for reactions at standard state are called **free energies of formation** ($\Delta_f G°$), which are also available in reference tables.

Equation II in **Table 4.1** allows the calculation of **free energy of reaction** ($\Delta_r G°$) using available $\Delta_f G°$ for

Table 4.1 Four principal equations used for thermodynamic predictions of reactions.

Equation	Terms	Explanation
Equation I $\Delta G^\circ = \Delta H^\circ - T\,\Delta S^\circ$	ΔG° = free energy (kJ mol^{-1}) ΔH° = enthalpy (kJ mol^{-1}) T = standard state temperature (K) ΔS° = entropy (kJ mol^{-1} K^{-1}) $\Delta_r G^\circ$ = free energy of reaction	Relates enthalpy and entropy to free energy. Values are in reference tables.
Equation II $\Delta_r G^\circ = \sum n\Delta_f G^\circ_{products} - \sum n\Delta_f G^\circ_{reactions}$	$\sum n\Delta_f G^\circ_{products}$ = sum of free energy of formation of products $\sum n\Delta_f G^\circ_{reactants}$ = sum of free energy of formation of reactants n = stoichiometric coefficients	Free energy of reaction is the difference between sum of products and sum of reactants.
Equation III For reaction bB + cC … = dD + eE … $\Delta_r G^\circ = -RT\ \ln\dfrac{a_D^d a_E^e\ldots}{a_B^b a_C^c\ldots}$	R = universal gas constant T = standard state temperature a_i = activity of reactants and products B,C… = reactants D,E… = products b,c,d,e… = stoichiometric coefficients	Relates free energy to activity of reactants and products.
Equation IV For reaction bB + cC … = dD + eE … $K = \dfrac{a_D^d a_E^e\ldots}{a_B^b a_C^c\ldots} = e^{\frac{-\Delta_r G^\circ}{RT}}$ or $K = \dfrac{(D)^d (E)^e\ldots}{(B)^b (C)^c\ldots}$	K = equilibrium constant a_D^b, a_E^e = (D)^d,(E)^e = activity of products to power of stoichiometric coefficient a_B^b, a_C^e = (B)^b,(C)^c = activity of reactants to power of stoichiometric coefficient	The ratio of activities of the product of products over the product of reactants is defined as the equilibrium constant. By substituting equation IV into equation III, the equilibrium constant (K) can be calculated from $\Delta_r G^\circ$.

products and reactants. Tables often provide $\Delta_r G^\circ$ for a listed reaction, thereby avoiding the necessity to calculate them, so long as the reaction listed is appropriate for the system of interest. Entropy and enthalpy of reaction are calculated from $\Delta_f S^\circ$ and $\Delta_f H^\circ$ using an equation analogous to *Equation II* (e.g., $\Delta_r H^\circ = \sum n\Delta_f H^\circ_{products} - \sum n\Delta_f H^\circ_{reactants}$).

Equation III in **Table 4.1** relates free energy of a system at equilibrium to the activities of products and reactants. For species in solution, **activity** is the *effective concentration of a species in solution and* can be calculated from solution species concentrations.

Equation IV in **Table 4.1** relates the activity terms in the *Equation III* to an **equilibrium constant** (K). Thus, at equilibrium the ratio of the activities of products over reactants is a function of the free energy of reaction, and if $\Delta_r G^\circ$ is calculated using *Equations I–III* or using tabulated $\Delta_r G^\circ$ values, the theoretical activity of individual species in a system can be calculated. Where does that get you? Since concentrations in solution can be measured, solution activities can be calculated and compared to *theoretical* values. This predicative ability

is used to model soil solution chemistry, including solubility of minerals, complexation of ions, acid and base reactions, and dissolution and volatilization of gas in aqueous solutions.

4.2.1 Example using thermodynamics to calculate gypsum solubility in soils

An example of the use of thermodynamics to predict system chemistry is the solubility of gypsum. Gypsum is a common soil mineral that occurs in arid and semi-arid soils and is also a soil amendment added to increase calcium and sulfate availability. The dissolution reaction for gypsum into its ion constituents is written

$$CaSO_4 \cdot 2H_2O(s) = Ca^{2+}(aq) + SO_4^{2-}(aq) + 2H_2O(l) \quad (4.1)$$

The parenthetical notations *(s)*, *(aq)* and *(l)* indicate *solid, aqueous,* and *liquid* phases; they are not always used in writing reactions, especially where the phase

Table 4.2 Thermodynamic data for gypsum (CaSO$_4$·2H$_2$O), Ca^{2+}, SO$_4^{2-}$, and H$_2$O at standard state (25 °C). From Drever (1997).

Species	$\Delta_f G°$ (kJ mol^{-1})	$\Delta_f H°$ (kJ mol^{-1})	$\Delta_f S°$ (kJ mol^{-1} K^{-1})
Ca^{2+}	−552.8	−543	−0.0562
SO$_4^{2-}$	−744	−909.34	0.0185
H$_2$O(l)	−237.14	−285.83	0.06995
CaSO$_4$·2H$_2$O	−1797.36	−2022.92	0.1939

of the species is obvious. Thermodynamic data for the products and reactants of gypsum dissolution are provided in **Table 4.2**.

Using data from **Table 4.2** and *Equation II* in **Table 4.1**, the free energy of reaction is

$$\Delta_r G° = \Sigma n\Delta_f G°_{products} - \Sigma n\Delta_f G°_{reactants}$$
$$= \left(\Delta_f G°_{Ca} + \Delta_f G°_{SO_4} + 2\Delta_f G°_{H_2O}\right) - \Delta_f G°_{CaSO_4·2H_2O} \quad (4.2)$$

$$\Delta_r G° = (-552.8) + (-744) + 2(-237.14)$$
$$- (-1797.36) = 26.28 \, kJ \, mol^{-1} \quad (4.3)$$

$\Delta_r G°$ can also be calculated from $\Delta_r H°$ and $\Delta_r S°$ values in **Table 4.2** using an equation analogous to *Equation II* in **Table 4.1** to calculate the enthalpy of reaction:

$$\Delta_r H° = \Sigma n\Delta_f H°_{products} - \Sigma n\Delta_f H°_{reactants}$$
$$= \left(\Delta_f H°_{Ca} + \Delta_f H°_{SO_4} + 2\Delta_f H°_{H_2O}\right) - \Delta_f H°_{CaSO_4·2H_2O}$$
$$= -1.08 \, kJ \, mol^{-1} \quad (4.4)$$

Using a similar approach, $\Delta_r S° = -0.0917$ kJ mol^{-1} K^{-1}.

Using the entropy and enthalpy of reaction values calculated in **Eq. 4.4** together with *Equation I* in **Table 4.1**,

$$\Delta_r G° = \Delta_r H° - T\Delta_r S°$$
$$= -1.08 \, kJ \, mol^{-1} - \left(298.15 K \times -0.0917 \, kJ \, mol^{-1} K^{-1}\right)$$
$$= 26.26 \, kJ \, mol^{-1} \quad (4.5)$$

The enthalpy- and entropy-derived value is very close to the value calculated using free energy of formation values in **Eq. 4.3**. Note the temperature variable in *Equation I* in **Table 4.1** must be the standard state temperature that the constants are reported (298.14 K), not the *system* temperature for which the problem is being calculated.

The equilibrium constant for gypsum dissolution is calculated from the free energy of reaction using *Equation IV* in **Table 4.1**:

$$K_s = e^{\left(\frac{-\Delta_r G°}{RT}\right)} = \exp\left(\frac{-26.26 \, kJ \, mol^{-1}}{8.314 \times 10^{-3} \, kJ \, mol^{-1} K^{-1} \times 298.15 K}\right)$$
$$= 2.51 \times 10^{-5} \quad (4.6)$$

Thus, the activities of species in the gypsum reaction are equal to the equilibrium constant by

$$K_s = \frac{a_{Ca} a_{SO_4} a_{H_2O}^2}{a_{CaSO_4·2H_2O}} = 2.51 \times 10^{-5} \quad (4.7)$$

where a_{Ca} is the activity of Ca^{2+} in solution, and so on. The activity of gypsum is a solid phase activity, which, if it is pure and crystalline is equal to one; otherwise the solid phase activity must be calculated, which, in practice, is not as accurate as the method for calculating aqueous species activities. The notation most used for activities is to place the species in parentheses, e.g., $a_{Ca} \equiv$ (Ca^{2+}) (sometimes braces or brackets are used instead of parentheses, but not in this textbook).

The activity of water in most natural systems is close to unity, and thus is not typically calculated. However, in solutions that have high concentrations of dissolved ions, water activity can deviate from unity sufficiently to require including the calculated water activity in the equilibrium constant evaluation.

The solubility equilibrium constant, K_s, for the gypsum dissolution reaction (**Eq. 4.1**) is used to predict *activities* of Ca^{2+} and SO$_4^{2-}$ ions in solution if gypsum is present. Since activities are related to concentrations of ions in solution, the solubility constant of gypsum allows for prediction of dissolved Ca^{2+} and SO$_4^{2-}$ concentrations. Or, using experimentally measured activities of dissolved Ca^{2+} and SO$_4^{2-}$ ions, the likelihood of precipitation or dissolution of gypsum can be calculated from **Eq. 4.7**. If the calculated value indicates disequilibrium, the activities of Ca^{2+} and SO$_4^{2-}$ must change by gypsum dissolution or precipitation until the aqueous species equilibrium distribution is achieved (i.e., the value of $K_s = 2.51 \times 10^{-5}$). Of course, depending on the environmental conditions, such as temperature and presence of other species in the system, this can take time, thus natural systems are often not at equilibrium.

Figure 4.2 compares Ca^{2+} and SO_4^{2-} activities in soil solution collected from several arid region soils in Algeria to the *theoretical* activities (solid line) predicted from the thermodynamic equilibrium constant (**Eq. 4.7**). The soils where the water samples were collected have high concentrations of dissolved ions, and thus the activity of water must be calculated using relationships not covered in this textbook. The predicted axes variables for the solid line are derived by taking the logarithm of **Eq. 4.7** (assuming gypsum has unit activity) and rearranging:

$$\log\left(Ca^{2+}\right)+\log\left(H_2O\right)$$
$$=-1\left(\log\left(SO_4^{2-}\right)+\log\left(H_2O\right)\right)+\log\left(K_s\right) \quad (4.8)$$

where the left-hand side of the equation is plotted on the y-axis, $\log(SO_4^{2-}) + \log(H_2O)$ is the x-axis variable, the slope is negative one, and $\log(K_s)$ is the intercept. Most of the data points are close to the solubility line, indicating equilibrium saturation with gypsum, suggesting that gypsic horizons are present in the soils.

Points below the line indicate undersaturation, and points above indicate supersaturation with respect to gypsum. The authors propose that dehydrated calcium sulfate mineral (anahydrite) controls the calcium and sulfate concentrations in soils that are supersaturated with respect to gypsum.

As shown in the gypsum solubility example, a main point of calculating thermodynamic activity relationships is to allow prediction of activity of species in solution and evaluation of stability of solid phases (i.e., are solids dissolving or precipitating). In aqueous solutions, this approach works well; particularly in well-controlled systems such as reaction vessels in the laboratory. In nature, aqueous systems are more complex because of the many fluxes and presence of species that are not pure, such as poorly crystalline minerals. Thus, thermodynamic equations are often limited for predictions in soil systems. However, despite the limitations, the thermodynamic approach allows for prediction of the theoretical *energy* of the system with respect to reactants and products, which is useful for understanding the directions reactions will go and the possible species in the system.

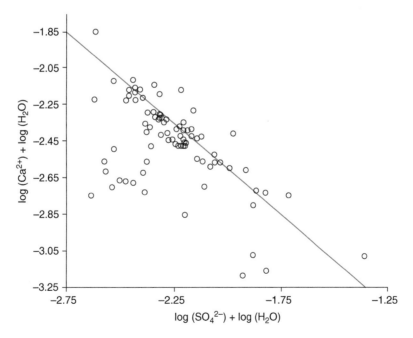

Figure 4.2 Comparison of soil solution activities in Algerian soils to theoretical gypsum solubility (solid line) derived from Equation 4.7 (plotted in linear form $\log(Ca^{2+}) + \log(H_2O)= -1$ ($\log(SO_4^{2-}) + \log(H_2O) - \log(K_s)$)). Plotted ion and water activities were calculated from empirical relationships with measured molar concentrations of ions in the soil solution as input. Adapted from Hamdi-Aissa et al. (2004).

Special Topic Box 4.1 A brief accounting of the history of thermodynamics of aqueous solutions

J. Willard Gibbs (1839–1903), a professor of chemistry at Yale University, is considered one of the founding fathers of physical chemistry. Albert Einstein referred to him as "the greatest mind in American history." One of his most note-worthy contributions to physical chemistry and science was the development of the ideas that underlie quantification of chemical energy in a system, which is evident in the equation named in his honor: *Gibbs free energy*. Gibbs's ideas were formally presented in 1876 in a series of papers total-ing 300 pages, "On the Equilibrium of Heterogeneous Substances," which forms the basis for predicting chemical reaction dynamics in systems at equilibrium.

Gilbert N. Lewis (1875–1946) furthered Gibbs's ideas of thermodynamics by measuring free energies of formation for many substances. Lewis was a chemistry professor at many institutions, including University of California–Berkeley and Massachusetts Institute of Technology. Lewis is also noted for developing some of the fundamental ideas behind the covalent bond. Lewis's work greatly contributed to the application and acceptance of Gibbs's ideas on thermody-namics of chemical systems. Lewis also became interested in the electrolytic properties of solutions, for which he pro-posed the concept of ionic strength. Peter Debye and Erich Hückel used Lewis's idea of ionic strength to develop the aqueous ion activity concept, first proposed in a paper they published in a German research journal in 1923.

The discipline of aqueous geochemistry to describe chemical processes in natural waters is credited to a handful of scientists who published works on ocean and natural water chemistry in the mid-twentieth century, including Robert Garrels (Northwestern University and Harvard University) and Lars Sillen (Swedish Royal Institute of Technology). Garrels and Charles Christ published the textbook *Solutions, Minerals and Equilibria*, in 1965, which is a foundational book in geochemistry. Werner Stumm, professor at Harvard from 1955–1969, and subsequently professor at the Swiss Federal Water Resources Centre

(EAWAG), is noted for conducting foundational research on modeling aqueous geochemical reactions and educating many scientists who contributed to the discipline. The text-book, *Aquatic Chemistry*, by Stumm and James Morgan (last edition published in 1997) is an invaluable text for advanced students and professionals.

Soil *solution* chemistry is a cross-disciplinary melding of soil science and aqueous geochemistry. Much of the scien-tific approach used in soil solution chemistry is the same fundamental approach used in aqueous geochemistry. Differences in soil solution chemistry compared to classical aqueous chemistry occur from the role that soil biota play in chemical reactions, solid–solution interface reactions that are more omnipresent because of the high solid to solution ratio in soils, high ion concentrations of soil-pore waters by evapotranspiration, and colloidal reactions. In 1979, Willard Lindsay, a soil chemist at Colorado State University, pub-lished a textbook titled *Chemical Equilibria in Soils*, which presented details of research on using classical aqueous thermodynamics to predict the extent of reactions in soil solutions. Like most textbooks, Lindsay's text draws on the work of many scientists; however, it is noteworthy because it brought the ideas under a single volume. The novelty of the text is its focus on reactions relevant to soil systems, such as formation of aluminosilicates, calcium phosphates, and poorly crystalline mineral phases. The principles described in Lindsay's text provide a solid foundation for use of aque-ous chemistry to understand soil solution chemistry.

Great advances over the years have allowed for mode-ling aqueous geochemical reactions that occur in natural systems. However, the assumption made in aqueous and soil equilibrium modeling is that the activity of solid phases is unity, but natural minerals are often mixtures that have nonunit activity. Methods to account for this non-ideality have been developed, and incorporations of such models into computer programs will improve our ability to predict the composition of natural waters.

4.2.2 Types of equilibrium constants

Aqueous chemistry includes several types of reactions, such as acidity, hydrolysis, complexation, and precipi-tation. Equilibrium constant symbols (K) for most

reaction types use a subscript to indicate the type of reaction. **Table 4.3** lists several different types of equi-librium constants covered in this chapter.

Equilibrium constants are typically derived from thermodynamic data. However, in some cases

Table 4.3 Different types of equilibrium constants for reactions in aqueous solutions. Alternate forms are given in parentheses.

Equilibrium constant	Reaction type	Example reaction and notes
K (K_{eq})	Nonspecific, can be used for any type of equilibrium constant	$A + B = C$
K_f	Equilibrium constant for formation of a species from elements at standard state temperature pressure and concentration	$H_2(g) + \frac{1}{2}O_2(g) = H_2O(g)$
K_a (K_{HA}, K_{OH}, K_{a1}, K_{a2})	Acidity constant	$H_2PO_4^- = H^+ + HPO_4^{2-}$ Negative log relationship is often used to express K_a of acids: $pK_a = -\log K_a$
K_1, K_2 ...K_n	Hydrolysis reactions	$Fe^{3+} + H_2O = FeOH^{2+} + H^+$
K_β	Summary hydrolysis constant	
$K_{complex}$	Stability constants for complexation	$Al^{3+}(aq) + SO_4^{2-}(aq) = AlSO_4^+(aq)$ Constant: K_{AlSO4+}
K_H	Henry's law for gas dissolution.	$CO_2(g) + H_2O = H_2CO_3(aq)$
K_s (K_{dis})	Solubility reaction, often dissolution	$CaSO_4 \cdot 2H_2O(s) = Ca^{2+} + SO_4^{2-} + 2H_2O$ $K_s = \dfrac{a_{Ca}a_{SO_4}a_{H_2O}^2}{a_{CaSO_4 \cdot 2H_2O}}$ Alternative notation: K_{CaSO4}
K_{sp}	Solubility product for dissolution reactions; assumes water and pure solids have unit activity.	$K_{sp} = a_{Ca}a_{SO_4} = \left(Ca^{2+}\right)\left(SO_4^{2-}\right)$
Q	Reaction quotient are derived using concentrations measured in system.	$Q = \dfrac{\left(Ca^{2+}\right)\left(SO_4^{2-}\right)\left(H_2O\right)^2}{\left(CaSO_4 \cdot 2H_2O\right)}$ where parenthesis are calculated activities based on concentrations in the system.
IAP	Ion activity product for dissolution reactions that uses concentrations measured in system; similar to Q except specifically for dissolution reactions.	$IAP_{CaSO_4} = \left(Ca^{2+}\right)\left(SO_4^{2-}\right)$ where parentheses are activities calculated from concentrations in the system.
K_{cond}	Concentrations are used instead of activities.	$K_{cond} = \left[Ca^{2+}\right]\left[SO_4^{2-}\right]$ where brackets are *concentrations* in the system
K_w	Hydrolysis of water	$H_2O = H^+ + OH^-$ $K_w = 10^{-14}$

conditional constants (K_{cond}) are used, which are typically only valid for systems very similar to the ones from which they were derived, and thus are less useful for prediction of aqueous chemistry. For example, in the gypsum dissolution reaction in **Eq. 4.1**, the conditional equilibrium (K_{cond}) constant is

$$K_{cond} = \left[Ca^{2+}\right]\left[SO_4^{2-}\right] \qquad (4.9)$$

where the brackets indicate the *concentration* of species as opposed to the *activities* that were used in **Eq. 4.7**. In a system with calcium and sulfate present, the concentrations of the ions can be input into **Eq. 4.9** to compare to a conditional equilibrium constant for

gypsum solubility. However, conditional equilibrium constants depend on solution conditions; thus, if other ions are present in solution, the value of K_{cond} may not be very accurate. To overcome this, equilibrium constants that use activity instead of concentration are used and are accurate across a range of solution conditions.

4.3 Calculation of ion activity

Activities of ions in solutions are not typically measured directly. One exception is pH, which is assumed to be the activity of H^+, or more specifically hydronium

(H_3O^+). The relation of activity to measured concentrations (c) is

$$a_i = \gamma_i c_i \qquad (4.10)$$

where γ_i is an *activity coefficient* for an ion (*i*). Activity coefficients are *correction factors* to account for the nonideality of ions in solution caused by ion–ion and ion–water interactions, which are related to valence, ionic size, and solution properties. Activity of an ion is the *effective concentration* and is unitless (the units of γ are thus L mol^{-1}).

As concentration of ions in solution approaches zero, ion interaction decreases because the chances of ions approaching each other becomes small, and thus, the ion activity coefficient approaches one:

$$\lim_{c \to 0} \gamma = 1 \qquad (4.11)$$

This relationship implies that as a solution is diluted, ions behave more ideally. Thus, in an ideal solution, $\gamma = 1$ and $a_i = c_i$.

As discussed in Chapter 2, ions in water are not free and unattached. The interactions of ions with water molecules and other ions affect the concentration-dependent (colligative) properties of solutions such as osmotic pressure, boiling-point elevation, freezing-point depression, and **chemical potential**. Chemical potential is analogous to ion activity – it describes the concentration-dependent chemical energy of species in solution, and thus quantifies the driving force of reactions. In predicting equilibrium reactions, activity is used instead of chemical potential to represent the effective *energy* of a species. **Equation 4.10** shows a direct relationship between activity and concentration in solution, which are related using an activity coefficient. Ion activity coefficients in aqueous solutions are less than one and decrease as the total concentration of ions in solution increases. Using **Eq. 4.10**, the relationship between the thermodynamic equilibrium constant and concentration can be solved using *Equation IV* in **Table 4.1**. For example, for a *generic* reaction

$$bB + cC = dD + eE \qquad (4.12)$$

$$K = \frac{(D)^d (E)^e}{(B)^b (C)^c} = \frac{\gamma_D^d \gamma_E^e}{\gamma_B^b \gamma_C^c} \times \frac{[D]^d [E]^e}{[B]^b [C]^c} \qquad (4.13)$$

where lowercase letters are stoichiometric coefficients, parentheses are *activities* of species B, C, D, E, and brackets represent concentrations. Thus, **Eq. 4.13** allows comparisons of analytically measured concentrations and theoretically predicted values calculated using the equilibrium constant (K), as shown in **Figure 4.2**, by using activity coefficients.

4.3.1   Use of ionic strength to calculate activity coefficients

Equation 4.13 requires activity coefficients to be known, which can be calculated using empirical relationships derived to account for the nonideal properties of ions in solution. In 1924, researchers Peter Debye and Erich Hückel proposed that the activity coefficient of an ion (i) is

$$\log \gamma_i = -A Z_i^2 \sqrt{I} \qquad (4.14)$$

where γ_i is the ion's activity coefficient, A is a constant (= 0.511), Z_i is the ion's charge, and I is the **ionic strength**. Ionic strength combines the effects of concentration and ion charge:

$$I = \frac{1}{2} \sum c_i Z_i^2 \qquad (4.15)$$

where c_i is the molarity (mol L^{-1}) and Z_i is the charge of each ion, *i*. The ionic strength estimates the *effective concentration of charge in a solution*. By convention, ionic strength is unitless, although units (e.g., mol L^{-1}) are sometimes indicated to be clear on the concentration units used in the calculation. A solution has only one ionic strength, but each ion may have a different activity coefficient (**Figure 4.3**).

Ions in solution conduct electricity through the solution; thus, not surprisingly, ionic strength is correlated to electrical conductivity (EC). Electrical conductivity is the inverse of resistivity. Electrical conductivity is measured using an electrical conductivity probe and is discussed in more detail in Chapter 13. Empirical equations relating EC to ionic strength have been derived, and in many cases are accurate over a broad range of solution conditions. This enables calculation of activity coefficients from estimates of ionic strength made using an EC probe, which is much easier than measuring all the cation and anion concentrations in solution. While it may be accurate in many cases, the most reliable approach is to calculate ionic strength from solution composition (i.e., **Eq. 4.15**).

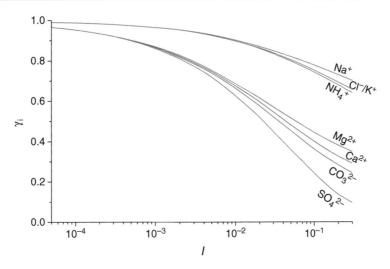

Figure 4.3 Ion activity coefficients versus ionic strength. Activities were calculated from the extended Debye-Hückel equation.

Whenever complete chemical analyses are provided for solutions (e.g., soil extracts or irrigation waters), the sum of cation charges in units of mmol(+) L^{-1} should equal the sum of anion charges (mmol(–) L^{-1}) to within ~10% difference. Exact agreement, however, often requires that one ion is determined by difference. In soil solutions and natural waters, this is often sulfate or organic acids due to difficulty in accurately measuring their concentrations and species. Concentrations of bicarbonate plus carbonate may be reported as alkalinity. Use of electrical conductivity is a workaround that allows for estimation of ionic strength when total cation and anion balance is not available (see Chapter 13).

Equation 4.14 was a great breakthrough in the understanding of ion behavior in very dilute solutions because it relates solution concentration to activity, which, in turn, is related to chemical potential and thermodynamic equilibrium constants. The **Debye-Hückel equation** assumes that ions interact electrostatically like charged particles of zero size. But ions have varying ionic radii, which affects their interactions. These and other assumptions make **Eq. 4.14** unreliable at ionic strengths greater than about 0.01 and limit its use in soil solutions, which can have *ionic strengths* on the order of 0.1 or greater. To calculate activities in solutions with I > 0.01, **Eq. 4.14** has been modified to include individual ion traits (referred to as the *extended Debye-Hückel equation*):

$$\log \gamma_i = -AZ_i^2 \left(\frac{\sqrt{I}}{1 + B\alpha_i \sqrt{I}} \right) \tag{4.16}$$

where $B = 0.33$ for aqueous solutions at 25 °C and α_i is an individual ion parameter determined experimentally.

Table 4.4 gives α_i values for some common ions. Values of α_i are correlated to the diameters of ions plus their associated water molecules (i.e., hydrated radius), and are called *the distance of closest approach*.

The **Davies equation (Eq. 4.17)** slightly modified the extended Debye-Hückel equation with empirical constants, alleviating the need for ion-specific α_i values, and yields satisfactory values for individual ion activity coefficients in solutions up to I ~ 0.1:

$$\log \gamma_i = -AZ_i^2 \left(\frac{\sqrt{I}}{1 + \sqrt{I}} - 0.3I \right) \tag{4.17}$$

Figure 4.3 shows the relationship of activity coefficient to ionic strength for several ions common in soil solutions. Note that as ionic strength decreases, activity coefficients increase towards one, in accordance with **Eq. 4.11**.

Table 4.4 Values of α_i in the extended Debye-Hückel equation. From Kielland (1937).

Ions	α_i
NH_4^+	2.5
Cl^-, NO_3^-, K^+	3
F^-, HS^-, OH^-	3.5
HCO_3^-, $H_2PO_4^-$, Na^+	4–4.5
HPO_4^{2-}, PO_4^{3-}, SO_4^{2-}	4
CO_3^{2-},	4.5
Cd^{2+}, Hg^{2+}, S^{2-}	5
Li^+, Ca^{2+}, Cu^{2+}, Fe^{2+}, Mn^{2+}, Zn^{2+}	6
Be^{2+}, Mg^{2+}	8
H^+, Al^{3+}, Fe^{3+}	9
COO^-	3.5

4.3.2 *Example* calculation of activity coefficient

Calculate the activity of Ca^{2+} in a soil solution that has a major cation composition of 0.009 M Ca^{2+}, 0.004 M Mg^{2+}, 0.0007 M K^+ and 0.008 M Na^+, assuming the anions are all monovalent (e.g., Cl^-, NO_3^-, HCO_3^-).

Step 1: To calculate activity, ionic strength is needed, which requires knowing the anion concentration. In solution, the anion charge ($[A^-]$) must equal the cation charge. Assuming all the anions are monovalent:

$$\left[A^-\right] = 2 \times \left[Ca^{2+}\right] + 2 \times \left[Mg^{2+}\right] + K^+ + Na^+ = 0.0347 M \tag{4.18}$$

Substitute this into **Eq. 4.15**.

$$I = \frac{1}{2}\left(2^2\left[Ca^{2+}\right] + 2^2\left[Mg^{2+}\right] + \left[K^+\right] + \left[Na^+\right] + \left[A^-\right]\right)$$
$$= 0.0477 \tag{4.19}$$

Step 2: At ionic strength 0.0477, the activity coefficient can be calculated using the Davies equation (**Eq. 4.17**).

$$\log \gamma_{Ca} = -0.511 \times 2^2 \left(\frac{\sqrt{0.0477}}{1 + \sqrt{0.0477}} - 0.3 \times 0.0477\right) = -0.337 \tag{4.20}$$

$$\gamma_{Ca} = 0.460 \tag{4.21}$$

Step 3: Calculate Ca^{2+} activity using **Eq. 4.10**.

$$a_{Ca} = \gamma_{Ca}\left[Ca^{2+}\right] = 0.460 \times 0.009 = 0.00414 \tag{4.22}$$

The Ca^{2+} activity in this example is much different than concentration (0.009 M) because aqueous solutions containing dissolved ions are nonideal. The activity coefficient corrects for the nonideality, thus yielding an *effective concentration* that can be used in activity-based calculations, such as solubility calculations.

4.4 Acids and bases

Substances that liberate H^+ in solution are ***Bronsted acids***. Strong acids like HCl, HNO_3, $HClO_4$, and H_2SO_4 completely dissociate into protons and anions (conjugate base). ***Weak acids*** such as $H_2PO_4^-$, HCO_3^-, HF, and acetic acid (CH_3COOH) deprotonate at higher pH than strong acids. For a generic weak acid called *HA*, the deprotonation reaction is

$$HA = H^+ + A^- \tag{4.23}$$

Its acidity constant is

$$K_a = \frac{\left(H^+\right)\left(A^-\right)}{\left(HA\right)} \tag{4.24}$$

For acid reactions, the equilibrium constant is often expressed in negative logarithmic form (**pK$_a$**) because this converts hydrogen activity in the equilibrium expression to pH:

$$-\log K_a = pK_a \tag{4.25}$$

$$pK_a = -\log\left(H^+\right) - \log\frac{\left(A^-\right)}{\left(HA\right)} \tag{4.26}$$

$$pK_a = pH - \log\frac{\left(A^-\right)}{\left(HA\right)} \tag{4.27}$$

This can be rearranged to solve for pH:

$$pH = pK_a + \log\frac{\left(A^-\right)}{\left(HA\right)} \tag{4.28}$$

Equation 4.28 is the ***Henderson-Hasselbalch equation***. From this equation the ratio of protonated and deprotonated acids as a function of pH can be calculated:

$$\log\frac{\left(A^-\right)}{\left(HA\right)} = pH - pK_a \tag{4.29}$$

Figure 4.4 shows that one pH unit above or below the pK$_a$ the ratio of protonated or deprotonated acids will change by an order of magnitude (i.e., 10 times). When the pH of a weak acid solution equals the pK$_a$ of the weak acid, the concentration of the protonated and deprotonated species will be equal. For example, acetic acid (CH_3COOH) is a weak acid with pK$_a$ = 4.76; at pH 4.76 the activities of CH_3COOH and CH_3COO^- in solution are equal and the log activity ratio in **Eq. 4.28** is zero making the equality correct. **Figure 4.4** shows the relationship for acetic acid species according to **Eq. 4.29**.

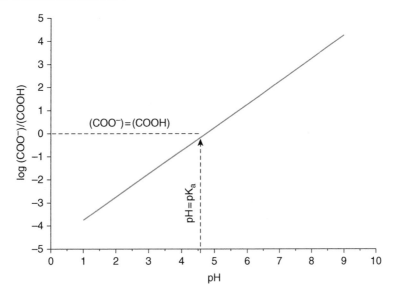

Figure 4.4 Ratio of acetate anion to acetic acid activity as a function of pH plotted according to **Eq. 4.29**. When pH equals the pK_a, activities of the acetate species are equal. At pH above the pK_a, the acetate anion predominates. At pH below the pK_a, acetic acid predominates.

The functional group on acetic acid is a carboxyl group, which is a common functional group in soil organic matter. Thus, the acidic behavior of soil organic matter (SOM) is similar to the much simpler acetic acid. The carboxylic functional groups on SOM cause the surface charge of SOM to vary with pH. At low pH, the carboxyl functional groups in SOM will be protonated and uncharged, while at higher pH the functional groups are negatively charged (e.g., R-COO⁻). The pK_a of SOM functional groups varies up to two pH units from the pK_a of acetic acid because the carboxylic acid moieties on SOM are bonded to different molecular entities than acetic acids (discussed further in Chapters 8 and 9).

4.4.1 Bases

Just as acids differ in strength, bases similarly differ in their ability to give up hydroxyl ions. The strong bases NaOH and KOH dissociate completely in solution. Bases such as $Ca(OH)_2$ and $Mg(OH)_2$ are less soluble than either NaOH or KOH. This lower solubility accounts for the lower pH of $Ca(OH)_2$ and $Mg(OH)_2$ solutions more accurately than does weak dissociation.

Ammonium hydroxide (NH_4OH) is a weak base with a dissociation reaction:

$$NH_4OH(aq) = NH_4^+(aq) + OH^-(aq) \quad (4.30)$$

$$K_b = \frac{(NH_4^+)(OH^-)}{(NH_4OH)} = 1.8 \times 10^{-5} \quad (4.31)$$

where K_b is the basicity constant. The NH_4OH molecule is the hydrate of NH_3, $NH_3\text{-}H_2O$, which is very soluble in water. Most of the solute exists in solution as NH_3 surrounded by a sheath of water molecules, but a small portion forms NH_4OH, which dissociates into NH_4^+ and OH^-, as shown in the reaction in **Eq. 4.30**. The NH_4OH activity in the denominator of **Eq. 4.31** is really both hydrated NH_3 molecule plus NH_4OH molecule.

4.4.2 Weak acids

The acidity and basicity of acids and bases in soil affects soil pH, one of the most characteristic chemical parameters of soils. **Table 4.5** provides acidity constants for acids found in soils. Weak acids with pK_a values between 3 and 9 are amongst the most interesting acids from the perspective of soil chemistry because their charge varies with soil pH (they are amphoteric), and because they contribute to the pH buffering in soils.

Knowing the pH when an acid gains or loses a proton is useful for understanding adsorption in soils. Protonation of weak acids can be predicted using

Acid	Formula	pK$_{a1}$	pK$_{a2}$	pK$_{a3}$
Hydrochloric	HCl	–3		
Sulfuric	H$_2$SO$_4$	–3	1.99	
Nitric	HNO$_3$	0		
Phosphoric	H$_3$PO$_4$	2.15	7.20	12.35
Hydrofluoric	HF	3.18		
Acetic	COOH	4.76		
Carbonic	H$_2$CO$_3$	6.35	10.33	
Silicic	H$_4$SiO$_4$	9.82	13.10	

equilibrium constants. For example, phosphoric acid (H_3PO_4) can exist as $H_2PO_4^-$, HPO_4^{2-}, and PO_4^{3-} in solutions. Sequential phosphoric acid deprotonation can be written using the following series of reactions and mass action equilibrium expressions:

$$H_3PO_4(aq) = H_2PO_4^-(aq) + H^+(aq) \tag{4.32}$$

$$K_{a1} = \frac{(H_2PO_4^-)(H^+)}{(H_3PO_4)} \tag{4.33}$$

$$H_2PO_4^-(aq) = HPO_4^{2-}(aq) + H^+(aq) \tag{4.34}$$

$$K_{a2} = \frac{(HPO_4^{2-})(H^+)}{(H_2PO_4^-)} \tag{4.35}$$

$$HPO_4^{2-}(aq) = PO_4^{3-}(aq) + H^+(aq) \tag{4.36}$$

$$K_{a3} = \frac{(HPO_4^{2-})(H^+)}{(PO_4^{3-})} \tag{4.37}$$

The fraction of phosphoric acid species ($f(H_nPO_4^{-3+n})$), where n is the number of protons, can be calculated using the ratio of the species to the total phosphate in solution; e.g.,

$$f_{H_2PO_4^-} = \frac{[H_2PO_4^-]}{[H_3PO_4]+[H_2PO_4^-]+[HPO_4^{2-}]+[PO_4^{3-}]} \tag{4.38}$$

where brackets indicate concentrations. For deriving theoretical trends in acid species, it is convenient to

assume that activity equals concentration; doing so causes small shifts in the predicted protonation deprotonation trends, but the relative amounts of acid species are accurate. For example, by substituting **Eqs. 4.33, 4.35,** and **4.37** into **Eq. 4.38**, the fraction of $H_2PO_4^-$ can be solved as a function of pH:

$$f_{H_2PO_4^-} = \frac{1}{\dfrac{(H^+)}{K_{a1}}+1+\dfrac{K_{a2}}{(H^+)}+\dfrac{K_{a2}}{K_{a3}}} \tag{4.39}$$

To better understand the derivation of **Eq. 4.39**, students should do the algebra of this substitution. Using **Eq. 4.39**, the fraction of $H_2PO_4^-$ in solution can be calculated from solution pH and the three acidity constants (**Table 4.5**). A similar calculation of $f_{H_3PO_4}$, $f_{HPO_4^{2-}}$, and $f_{PO_4^{3-}}$ can be done. **Figure 4.5** shows the trends of the fraction of phosphoric acid species as a function of pH (the fraction was converted to percent by multiplying it by 100). Within the pH range of most soils (pH 4 to 9), the predominant phosphoric acid species is either $H_2PO_4^-$ or HPO_4^{2-}. The species distribution information is useful for predicting the reactivity of phosphate in soil because the anion's charge causes it to be more, or less, attracted to charged surfaces. Thus, pK$_a$ values are needed to predict the adsorption behavior of phosphate in soils, which will be covered in detail in Chapter 10. The weak acid concept can also be used to predict the protonation and deprotonation of functional groups on mineral surfaces and on organic matter (discussed in Chapter 9).

Another multi-protic acid common in soils is carbonic acid (H_2CO_3), and its deprotonated ions bicarbonate (HCO_3^-) and carbonate (CO_3^{2-}). Carbonic acid is an important weak acid in soils because it can be present in high concentrations and buffers the pH of the soil solution. The concentration of carbonic acid and its two conjugate base anions in soil solution is a function of the **partial pressure** of CO_2 gas. CO_2 concentrations in the *soil atmosphere* are variable and may range from ambient tropospheric pressure in surface horizons, to up to 100 times greater in rhizospheres and at depth due to microbe and root respiration that produces CO_2 gas that slowly diffuses out of the soil.

In a closed system (no fluxes) containing a CO_2 gas-solution, acid speciation modeling uses a similar approach

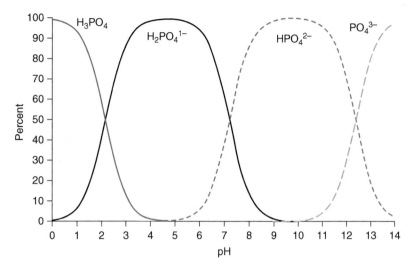

Figure 4.5 Percent phosphate species distribution as a function of pH. The lines cross where pH equals the pK_a values. In the range of most waters, phosphate exists primarily as $H_2PO_4^-$ or HPO_4^{2-}.

as used for phosphoric acid. **Figure 4.6** shows the relationship between percent H_2CO_3, HCO_3^-, CO_3^{2-} as a function of pH. Above pH 6.4 (pK_{a1}), the predominant species expected in solution is bicarbonate. CO_3^{2-} predominates at pH > 10; uncommon in soils. Below pH 6.4, carbonic acid becomes the predominant species in solution.

Protons added to neutral-to-alkaline soil solutions, such as by a plant root to maintain charge balance within the root tissue, are absorbed/neutralized by bicarbonate in solution, which is buffering the soil solution pH. In natural waters, bicarbonate is the major anion that is measured in **alkalinity** measurements, which is a quantitative measure of the ability of a solution to neutralize acid by consuming protons on the conjugate bases (deprotonated acids) in the solution.

4.5 Gas dissolution

Gas exchange from the soil solution to the atmosphere is an important process controlling aqueous chemistry of carbonate, ammonia, nitrogen, oxygen, and sulfur. An example is oxygen gas dissolution in water:

$$O_2(g) + H_2O(l) = O_2(aq) \tag{4.40}$$

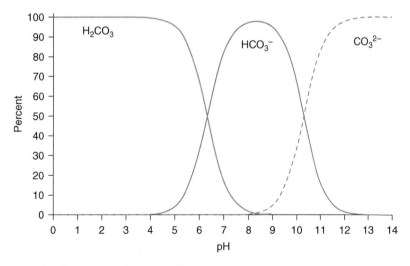

Figure 4.6 Carbonate species distribution as a function of pH.

Note the phase of the oxygen in this reaction changes from gas to aqueous. Dissolved gases are hydrated, i.e., surrounded by water, and in some cases have molecular bonds with water molecules, thus forming distinct species from the hydrated gas molecule. For example, CO_2 gas in solution hydrates to form aqueous H_2CO_3 ($H_2O + CO_2$).

The equilibrium distribution for gas dissolution reactions is predicted using **Henry's law**:

$$K_H = \frac{a_i}{P_i} \tag{4.41}$$

where K_H is Henry's law constant, a_i is the aqueous activity of dissolved species i, and P_i is the partial pressure of the gas. Partial pressure is the pressure of an individual gas in the volume occupied by all gases in the mixture; total pressure is the sum of partial pressure of each gas in the mixture. The SI unit for partial pressure is Pascal (Pa), but in aqueous chemistry the unit atmospheres (atm) is commonly used (1 atm = 101.325 kPa). Partial pressures can be converted to concentrations using the ideal gas law: $n/V = P_i/RT$, where n/V is the number of moles of gas per L and R and T are universal gas constant and temperature of the gas. Thus, gas concentration depends on temperature and partial pressure of the gas. As the partial pressure of a gas increases, its aqueous activity also increases. K_H for some important gases in soils are reported in **Table 4.6**.

Many biological processes in soils produce or consume gases in soil solution, such as sulfide, methane, oxygen, and nitrous oxide. The gases volatilize from the dissolved to the gas state. Since gas exchange between soil pores and the overlying atmosphere is controlled by soil porosity, pore structure, and water content, soil gas

Table 4.6 Henry's law constants for gases at 25 °C. From Stumm (1996).

Gas species	K_H (M atm^{-1})
CO_2	3.39×10^{-2}
N_2	6.61×10^{-4}
NH_3	57
O_2	1.26×10^{-3}
H_2S	1.05×10^{-1}
CH_4	1.29×10^{-3}
NO	1.9×10^{-3}
N_2O	2.57×10^{-2}

concentrations can be much greater or less than atmospheric concentrations. One of the greatest modern-day challenges in soil science is to understand how soil processes affect sequestration and release of greenhouse gases. Greenhouse gases include CO_2, N_2O, NO, NO_2, and CH_4. Henry's law is a basic tool needed to study and predict greenhouse gas fluxes into and out of soil solution. Two examples of gas dissolution reactions important for soil properties are shown next.

4.5.1 Predicting dissolution of ammonia in water

Ammonia (NH_3) has the highest solubility of the common soil atmospheric gases (**Table 4.6**) and is an important species in soil nutrient management. The ammonia gas dissolution reaction is

$$NH_3(g) + nH_2O = NH_3(nH_2O)(aq) \tag{4.42}$$

where the $NH_3(nH_2O)$ is the dissolved ammonia molecule surrounded by n water molecules. The Henry's gas law equation for ammonia dissolution is

$$K_H = \frac{(NH_3)}{P_{NH_3}} = 57 \, M\,atm^{-1} \tag{4.43}$$

The large value of K_H for ammonia dissolution indicates that the partitioning of ammonia is greatly in favor of the aqueous phase over gaseous phase. Thus, gaseous ammonia is short-lived in the environment; despite this, ammonia emissions and distribution as *dissolved* ammonia or ammonium are important environmental processes.

Below pH~9, aqueous ammonia is mostly converted to ammonium through a hydrolysis reaction:

$$NH_3(aq) + H_2O = NH_4^+ + OH^- \tag{4.44}$$

The ammonium and hydroxide are often written as a weak base pair (NH_4OH), which is another way to write NH_3H_2O. Writing it with the hydroxide shows that ammonia hydration is a base producing reaction (c.f., **Eq. 4.30**).

Ammonium (NH_4^+) is an important species of dissolved nitrogen for nutrient uptake by plants and microbes, and an important species in soil nitrogen redox processes (see Figure 5.20). Ammonium in solution is a weak acid:

$$NH_4^+\left(aq\right) = NH_3\left(aq\right) + H^+ \qquad (4.45)$$

The pK_a for ammonium is 9.25, indicating that under most conditions in soils (i.e., pH<9), it is protonated.

Anhydrous ammonia gas ($NH_3(g)$) is a popular nitrogen fertilizer because it can be applied pre-plant and, if soil moisture conditions are optimum, has minimal loss from the soil plant-root zone. The anhydrous ammonia is injected into the soil and quickly reacts with water associated with soil clays and soil organic matter, converting it to aqueous ammonium (**Eqs. 4.42** and **4.44**). There is a slight pH increase in the zone of anhydrous ammonia application from the hydrolysis of water (**Eq. 4.44**), but the pH typically quickly reverts to the natural soil pH because of soil buffering reactions and nitrification of the ammonium. This example shows that soil chemistry of ammonia is important for managing soils, and Henry's gas law is one of the basic equations used to predict its chemistry.

4.5.2 Predicting pH of water due to CO_2 dissolution

Another important use of Henry's gas law (**Eq. 4.41**) is the prediction of carbonic acid activity from carbon dioxide dissolution:

$$CO_2\left(g\right) + H_2O = H_2CO_3\left(aq\right) \qquad (4.46)$$

$$a_{H_2CO_2^*} = P_{CO_2}K_H \qquad (4.47)$$

where H_2CO_3 (aq) may be both a hydrated carbon dioxide molecule or the molecular species carbonic acid. In practice, the hydrated carbon dioxide and carbonic acid species in solution are considered as one species, often indicated using an asterisk, $H_2CO_3^*$. The activity of carbonic acid in solution is a function of the partial pressure of CO_2 (P_{CO2}). Using atmospheric P_{CO2} ($10^{-3.41}$ atm), K_H (3.39×10^{-2}) (**Table 4.6**), and carbonic acid deprotonation equilibrium constants (**Table 4.5**), rainwater is predicted to have a pH of 5.6. To calculate rainwater pH, first substitute values for Henry's gas law constant (K_H) and P_{CO2} in **Eq. 4.47** to calculate the activity of carbonic acid:

$$a_{H_2CO_3^*} = \left(H_2CO_3^*\right) = \left(10^{-3.41}\,atm\right) \times 3.39 \times 10^{-2}\,M\,atm^{-1}$$
$$= 1.32 \times 10^{-5} \qquad (4.48)$$

The carbonic acid is what contributes the proton to the water to increase water acidity. The equilibrium expression for carbonic acid disassociation to bicarbonate and a proton is

$$H_2CO_3^* = H^+ + HCO_3^- \qquad (4.49)$$

$$K_{a1} = \frac{\left(H^+\right)\left(HCO_3^-\right)}{\left(H_2CO_3^*\right)} = 10^{-6.352} \qquad (4.50)$$

Using the activity of $H_2CO_3^*$ at atmospheric P_{CO2} calculated in **Eq. 4.48**, and assuming the bicarbonate is equal to (H^+) by the stoichiometry of the **Eq. 4.49** (i.e., (HCO_3^-) = (H^+)), (H^+) in **Eq. 4.50** can be solved for:

$$\left(H^+\right) = \sqrt{\left(H_2CO_3^*\right)K_{a1}} = \sqrt{1.32 \times 10^{-5} \times 10^{-6.352}} = 10^{-5.62} \qquad (4.51)$$

Since pH is $-\log(H^+)$, the pH of rainwater in equilibrium with CO_2 in the atmosphere is 5.62, which generally matches observed rainwater pH. This assumes no other gasses like sulfur dioxide or nitrogen oxides are present, and that dust that affects solution pH is minimal. Using the approach described above, the general expression for calculating the pH of pure water in presence of CO_2 is

$$pH = -\log \sqrt{P_{CO_2} K_H K_{a1}} \qquad (4.52)$$

Due to root respiration and microbial activity, the partial pressure of CO_2 in the soil can be up to 100 times greater than atmospheric P_{CO2}, in which case, using **Eq. 4.52**, the predicted pH of pure soil water would be 4.7. When $CaCO_3$ is present in a soil, the system has three phases (solid, liquid, and gas) that must be considered in predicting pH and species distributions. The carbonate mineral dissolves at low pH releasing carbonate anions. The carbonate anions will become protonated, decreasing hydrogen activity and increasing pH. This reaction buffers pH to ~8.2, depending on the partial pressure of CO_2 gas. Carbonate mineral dissolution is covered in more detail in Section 4.6.3.

4.6 Precipitation and dissolution reactions

Formation of solids from ions in solution is a precipitation reaction; the reverse process is a dissolution reaction. Precipitates in soils are either minerals or salts.

Weathering processes include precipitation and dissolution reactions, which are integral to soil development. For example, consider the gypsum dissolution reaction shown in **Eq. 4.1**. The dissolution constant (K_{dis}) is written in **Eq. 4.7**. Assuming the gypsum solid phase and water have unit activity, the revised solubility expression is

$$K_{sp} = \left(Ca^{2+}\right)\left(SO_4^{2-}\right) = 2.51 \times 10^{-5} \tag{4.53}$$

where K_{sp} is the **solubility product** for gypsum dissolution. If Ca^{2+} and SO_4^{2-} concentrations and ionic strength are known, the **saturation state** of a solution with respect to gypsum can be calculated using **Eq. 4.53**. **Figure 4.2** shows the solubility line for (Ca^{2+}) and (SO_4^{2-}) calculated using **Eq. 4.53** (ignoring the water activity). When gypsum is present in a system, the solid maintains activities of calcium and sulfate ions in solution such that their product is constant. Here are two example scenarios to show how dissolution and precipitation of gypsum controls solution activity:

1. If calcium or sulfate is consumed by another process in the system, such as plant root uptake, gypsum dissolves to reestablish equilibrium activities of the ions.
2. If a flux such as rainwater infiltration into a lower soil horizon or evapotranspiration increases calcium or sulfate concentrations in the soil, their activities will exceed the solubility product of gypsum. To reestablish equilibrium, the aqueous activities of these ions are reduced by precipitation of gypsum. In arid regions soils, the reaction often results in formation of small gypsum crystals in the soil pores.

Additional details and example of precipitation and dissolution reactions are given below.

4.6.1   Solubility of minerals

Precipitation and dissolution reactions can be predicted using solubility equilibrium constants (K_s or K_{dis}). For example, consider the lead sulfate mineral anglesite ($PbSO_4$), which can occur in soils impacted by mining activities. The dissolution reaction for anglesite is

$$PbSO_4(s) = Pb^{2+} + SO_4^{2-} \tag{4.54}$$

The solubility constant for this reaction is

$$K_s = \frac{\left(Pb^{2+}\right)\left(SO_4^{2-}\right)}{\left(PbSO_4\right)} = 10^{-7.6} \tag{4.55}$$

If the anglesite is pure and crystalline it has unit activity and the solubility product (K_{sp}) is

$$K_{sp} = \left(Pb^{2+}\right)\left(SO_4^{2-}\right) = 10^{-7.6} \tag{4.56}$$

The reaction in **Eq. 4.54** shows that anglesite dissolution and precipitation has a 1:1 relationship for Pb^{2+} and SO_4^{2-} ion release or removal from solution. Thus, if anglesite is present in a system, the activities of Pb^{2+} and SO_4^{2-} ions can be calculated by substituting the mass balance expression ((Pb^{2+}) = (SO_4^{2-})) into **Eq. 4.56**:

$$K_{sp} = \left(Pb^{2+}\right)^2 = 10^{-7.6} \tag{4.57}$$

Solving for (Pb^{2+}) activity:

$$\left(Pb^{2+}\right) = \sqrt{10^{-7.6}} = 1.6 \times 10^{-4} \tag{4.58}$$

The relationship between lead and sulfate activities is shown graphically in an **activity diagram**, which plots the log of the activity of dissolved ions against each other. Activity diagrams typically are plotted in logarithmic form. Thus, taking the logarithm of the solubility product in **Eq. 4.56** yields

$$\log K_s = \log\left(Pb^{2+}\right) + \log\left(SO_4^{2-}\right) = -7.6 \tag{4.59}$$

Rearranging to solve for log(Pb^{2+}):

$$\log\left(Pb^{2+}\right) = \log K_s - \log\left(SO_4^{2-}\right) \tag{4.60}$$

which can be plotted as a straight line, shown in **Figure 4.7**. In a solution with only anglesite present, the activities of Pb^{2+} and SO_4^{2-} are equal (circle on the line **Figure 4.7**). Solution concentrations of lead and sulfate can be measured, and their activities calculated to compare to the theoretical activities that would occur at equilibrium if anglesite is controlling its solubility, which allows assessing if the solution is saturated, supersaturated, or unsaturated with respect to anglesite. When

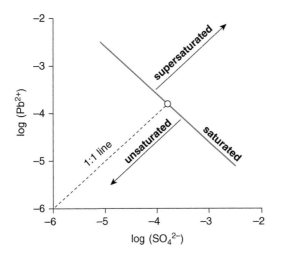

Figure 4.7 Activity diagram for lead and sulfate activities controlled by the precipitation and dissolution of the mineral anglesite. Where the 1:1 line intersects with solubility line the aqueous ion activities are those expected if there are no other phases present or competing common ions affecting the solution activities.

the product of activities of Pb^{2+} and SO_4^{2-} are above the line, the solution is supersaturated with respect to anglesite. Activities below the line are undersaturated. At equilibrium, the activities will occur on the line (saturated).

The discussion of anglesite solubility may appear to suggest that in a system with lead and sulfate present, anglesite precipitation and dissolution reactions control aqueous activities, and the activities would always equal 1.6×10^{-4}, as predicted by **Eq. 4.58**. However, this may not be the case because (1) there are several other lead and sulfate-containing solid phases that may control Pb^{2+} and SO_4^{2-} ion concentrations, including adsorption reactions; and (2) kinetics of reactions may be slow, causing disequilibrium.

When other reactions input lead or sulfate ions in solution, a **common ion effect** occurs. Consider, for example, if a calcium sulfate mineral such as gypsum ($CaSO_4 \cdot 2H_2O$) dissolves from an overlying horizon and the ions are transported into a horizon containing anglesite. The activity of SO_4^{2-} in the pore water will be much greater than Pb^{2+}. The *circle* on the line **Figure 4.7** would no longer apply, but the solid line would. The equilibrium condition would be represented by a point to the right on the diagram (down the line to the right). Anglesite still controls Pb^{2+} activity, but the increased SO_4^{2-} activity due to dissolved gypsum ion flux into

the horizon would decrease the Pb^{2+} activity so that the IAP is equal to the K_{sp}. In this scenario, the activity equilibrium point on the graph would lie *on the line* below 1.6×10^{-4} (open circle), and the Pb^{2+} activity would be less than 1.6×10^{-4}. Thus, due to the common ion effect, activities of Pb^{2+} and SO_4^{2-} in a solution can be unequal.

An important aspect of solubility modeling is that solubility reactions only predict the activities of *species* in the reactions. Other ion species dissolved from the mineral may occur in the solution, such as hydrolysis or protonated species, thus total ion concentration may be greater than predicted by mineral solubility. For example, in the anglesite example, lead can hydrolyze in solution to form $PbOH^+$ and other aqueous species making *total* lead in solution greater than just Pb^{2+} ($Pb_T = Pb^{2+} + PbOH^+ + \ldots$). This topic is addressed in Section 4.7.

4.6.2 Iron and aluminum dissolution from oxides and hydroxides

For oxide and hydroxide minerals, precipitation and dissolution reactions are pH dependent. **Table 4.7** lists the solubility constants for iron and aluminum oxide minerals common in soils. The dissolution of the iron oxide mineral goethite, for example, is

$$FeOOH(s) + 3H^+ = Fe^{3+} + 2H_2O \qquad (4.61)$$

The solubility constant for this reaction is

$$K_{sp} = \frac{(Fe^{3+})}{(H^+)^3} = 10^{-1} \qquad (4.62)$$

Taking the logarithm and rearranging:

$$\log(Fe^{3+}) = \log K_{sp} - 3pH \qquad (4.63)$$

Thus, the activity of Fe^{3+} in solution of a system in equilibrium with goethite will be controlled by goethite solubility (K_{sp}) and pH. If soluble iron is added to the system, and (Fe^{3+}) increases, more goethite will precipitate. If iron is removed from the system, and (Fe^{3+}) decreases, goethite will dissolve.

Table 4.7 Solubility products of iron and aluminum oxides at 25 °C. K_{sp} data are from Drever (1997).

Mineral	Dissolution reaction	K_{sp}	$\log K_{sp}$
Gibbsite (crystalline)	$Al(OH)_3(s) + 3H^+(aq) = Al^{3+}(aq) + 3H_2O(aq)$	$K_{sp} = \dfrac{(Al^{3+})}{(H^+)^3}$	8.11
Goethite	$FeOOH(s) + 3H^+(aq) = Fe^{3+}(aq) + 2H_2O(aq)$	$K_{sp} = \dfrac{(Fe^{3+})}{(H^+)^3}$	–1
Fe(OH)$_3$ (amorph)[1]	$Fe(OH)_3(s) + 3H^+(aq) = Fe^{3+}(aq) + 3H_2O(aq)$	$K_{sp} = \dfrac{(Fe^{3+})}{(H^+)^3}$	4.89
Hematite	$Fe_2O_3(s) + 6H^+(aq) = 2Fe^{3+}(aq) + 3H_2O(aq)$	$K_{sp} = \dfrac{(Fe^{3+})^2}{(H^+)^6}$	–4.01

[1] A ferrihydrite-type mineral that has variable crystallinity, mineral formula, and solubility.

Another important example of mineral dissolution in soils is aluminum hydroxide (gibbsite) dissolution:

$$Al(OH)_3(s) + 3H^+ = Al^{3+} + 3H_2O \qquad (4.64)$$

The solubility constant for this reaction is

$$K_{sp} = \frac{(Al^{3+})}{(H^+)^3} = 10^{8.11} \qquad (4.65)$$

Taking the logarithm and rearranging:

$$\log(Al^{3+}) = \log K_{sp} - 3pH \qquad (4.66)$$

Iron and aluminum ion activities as a function of pH are shown in **Figure 4.8**. An important aspect of solubility is that regardless of what additional reactions the dissolved ions undergo (e.g., complexation), the activity of the free ions is fixed when the system is in equilibrium with a solid phase. For example, if a dissolved Fe^{3+} ion released from goethite is complexed by an organic ligand to become *Fe³⁺-organic complex*, more iron will dissolve from goethite to replace the complexed Fe^{3+}, thereby maintaining a constant aqueous *free-Fe³⁺* activity (free Fe^{3+} is iron coordinated by water molecules only, no aqueous complexes).

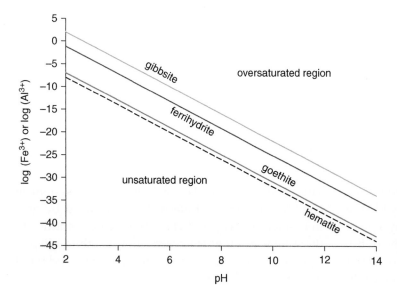

Figure 4.8 Activity of Fe^{3+} and Al^{3+} at equilibrium as a function of pH. If Fe^{3+} and Al^{3+} activity is above the solubility-equilibrium line, the solution is oversaturated with respect to the mineral phase and the metal ions should precipitate. If Fe^{3+} and Al^{3+} activity in a system is below the solubility-equilibrium line, the solution is unsaturated with respect to the mineral phase and ions will dissolve from the solid phase.

Table 4.8 Reactions, equations, and constants used in solid, liquid, and gas phase modeling of carbonate systems. Equilibrium constants from Stumm (1996).

Equation	Reaction	Equilibrium Equation	Equilibrium Constant at 25 °C
I	$CaCO_3(s) = Ca^{2+} + CO_3^{2-}$	$K_{sp} = (Ca^{2+})(CO_3^{2-})$	$10^{-8.480}$
II	$CO_2(g) + H_2O = H_2CO_3^*$	$K_H = \dfrac{(H_2CO_3^*)}{P_{CO_2}}$	$10^{-1.468}$
III	$H_2CO_3^* = H^+ + HCO_3^-$	$K_{a1} = \dfrac{(H^+)(HCO_3^-)}{(H_2CO_3)}$	$10^{-6.352}$
IV	$HCO_3^- = H^+ + CO_3^{2-}$	$K_{a2} = \dfrac{(H^+)(CO_3^{2-})}{(HCO_3^-)}$	$10^{-10.329}$

4.6.3 Calcite and carbon dioxide in soils

Carbon dioxide gas exists in the soil atmosphere and dissolves into soil solution to form carbonic acid, bicarbonate, and carbonate (see **Figure 4.6**). In soils, calcium carbonate readily interacts with aqueous carbonate species to affect soil pH and buffering, and the partial pressure of CO_2 in the soil atmosphere. Figure 1.11 illustrates the impacts of inorganic carbon reactions on pedogenesis.

The chemistry of carbonate systems includes reactions between solid, liquid, and gas phases. In natural systems, carbonate reactions are typically at equilibrium. Because equilibrium constants are well known for many carbonate reactions of interest, the carbonate chemistry can be modeled using equilibrium reactions. The principle equations and equilibrium constants for a carbonate system are listed in **Table 4.8**. In the carbonate system described in this table it is assumed that crystalline calcite is the only solid present, and no aqueous complexes occur.

Seven chemical species are listed in the carbonate system described in **Table 4.8** (including H^+). Modeling carbonate systems typically attempts to predict total concentration of inorganic carbon in solution, saturation state of the system with respect to a mineral, activity of hydrogen (pH), or concentration of calcium in the soil solution. Since there are seven species and only four equilibrium expressions in **Table 4.8**, solving the equilibrium system for a particular species requires introducing additional equations, or making assumptions about species. Two equations used to solve carbonate equilibrium systems are the electroneutrality condition (**charge balance**), and total dissolved inorganic carbon balance (T_{CO2}). The electroneutrality equation for the carbonate system is

$$2\left[Ca^{2+}\right]+\left[H^+\right]=\left[HCO^-\right]+2\left[CO_3^{2-}\right]+\left[OH^-\right] \quad (4.67)$$

The two times multiplier on calcium and carbonate concentrations in **Eq. 4.67** accounts for the fact that these ions are divalent. The mass balance equation for total inorganic carbon in solution is

$$T_{CO_2} =\left[H_2CO_3\right]+\left[HCO^-\right]+\left[CO_3^{2-}\right] \quad (4.68)$$

These equations are termed *conservative* because they are true so long as another aqueous component containing carbonate species is not introduced (i.e., the system is as defined in **Table 4.8**). Note that **Eqs. 4.67** and **4.68** use concentrations and not activities.

To solve the carbonate system, the concentration of a species can be fixed, or the concentration of the species can be assumed insignificant relative to other species. The partial pressure of CO_2 is a species activity that is commonly fixed. In the soil atmosphere, P_{CO2} is controlled by respiration of soil organisms and plant roots, soil porosity, and water content.

In soils with calcium carbonate present, the partial pressure of carbon dioxide and dissolution and precipitation of carbonate minerals buffer soil pH. If the soil carbonate system is described by the reactions in **Table 4.8**, pH as a function of P_{CO2} can be algebraically solved by making three assumptions:

1 $[CO_3^{2-}]$ is much less than $[HCO_3^-]$ because the system pH is below pH 9 in most soils (see **Figure 4.6**).

2 Proton and hydroxide activities are much less than calcium and bicarbonate activities. This is true when the soil pH is between 4 and 9; thus, their contribution to the electroneutrality balance equation (**Eq. 4.67**) is negligible.

3 Activities and concentrations are equal (i.e., solutions are ideal).

The assumption of an ideal solution for this scenario allows for determination of relative behavior of the ions in the carbonate system. Solving the system using activities is possible but requires additional assumptions or use of computers. Using the stated assumptions, **Eq. 4.67** becomes

$$2\left[Ca^{2+}\right] = \left[HCO_3^-\right] \tag{4.69}$$

The calcium concentration in **Eq. 4.69** is controlled by dissolution of calcite, thus using *Equation I* in **Table 4.8** to substitute for $(Ca^{2+}) = K_{sp}/(CO_3^{2-})$ into **Eq. 4.69** gives

$$\frac{2K_{sp}}{\left(CO_3^{2-}\right)} = \left(HCO_3^-\right) \tag{4.70}$$

Either parenthesis or brackets can be used here because activity and concentration are assumed equal. Using *Equations II–IV* in **Table 4.8**, the (CO_3^{2-}) and (HCO_3^-) needed for **Eq. 4.70** can be solved for as a function of $(H+)$ and P_{CO2}:

$$\left(HCO_3^-\right) = \frac{K_{a1}K_H P_{CO_2}}{\left(H^+\right)} \tag{4.71}$$

$$\left(CO_3^{2-}\right) = \frac{K_{a1}K_{a2}K_H P_{CO_2}}{\left(H^+\right)^2} \tag{4.72}$$

Substituting **Eqs. 4.71** and **4.72** in **Eq. 4.70** yields

$$\frac{2K_{sp}\left(H^+\right)^2}{K_{a2}K_{a1}K_H P_{CO_2}} = \frac{K_{a1}K_H P_{CO_2}}{\left(H^+\right)} \tag{4.73}$$

Rearranging **Eq. 4.73** to solve for (H^+) yields

$$\left(H^+\right)^3 = \frac{K_{a2}K_{a1}^2 K_H^2 P_{CO_2}^2}{2K_{sp}} \tag{4.74}$$

Taking the negative logarithm and solving for pH provides the following relationship:

$$pH = -\frac{1}{3}\log\frac{K_{a2}K_{a1}^2 K_H^2 P_{CO_2}^2}{2K_{sp}} \tag{4.75}$$

Substituting in constants from **Table 4.8** and isolating P_{CO2}, **Eq. 4.75** becomes

$$pH = 5.93 - \frac{2}{3}\log P_{CO_2} \tag{4.76}$$

Eq. 4.76 solves pH as a function of the P_{CO2} in a carbonate mineral system. This relationship is graphed in **Figure 4.9**, which shows that when P_{CO2} is $10^{-3.41}$ atm (2013 troposphere concentration) soil pH will be 8.2. At 10 times greater P_{CO2}, pH decreases to 7.5. Thus, soil becomes more acidic as P_{CO2} in soil pores increases. Experimental pH data from calcareous soils incubated under varying P_{CO2} are shown in **Figure 4.9**. There is very good agreement between the data and equilibrium model, suggesting that calcite dissolution and P_{CO2} is a major phase controlling pH of the soils. Given the expected range of P_{CO2} in soils, soils with calcite in them will likely have pH values ranging from 7.5 to 8.5. pH will be less than 7.5 if P_{CO2} exceeds 10 times tropospheric concentrations.

In soils, partial pressures of CO_2 and carbonate reactions are highly variable, influenced by the tropospheric concentrations, soil properties, climate, and especially soil biota. **Eq. 4.76** indicates that increased atmospheric P_{CO2} from fossil fuel burning will make soils slightly more acidic. Acidity accelerates weathering and mineral dissolution. For example, olivine dissolution by carbonic acid is

$$\begin{aligned} Mg_2SiO_4\left(s\right) + 4H_2CO_3\left(aq\right) \\ = 2Mg^{2+}\left(aq\right) + H_4SiO_4\left(aq\right) + 4HCO_3^- \end{aligned} \tag{4.77}$$

The aqueous bicarbonate, magnesium, and silicic acid products can either leach or produce secondary minerals. While such a reaction suggests that accelerated weathering will occur with increasing atmospheric CO_2, the full influence of increased atmospheric CO_2 to soil processes is not yet understood.

CO_2 equilibrium distribution in soils is a good example of how thermodynamic equations can be used to predict concentrations of different species. In this

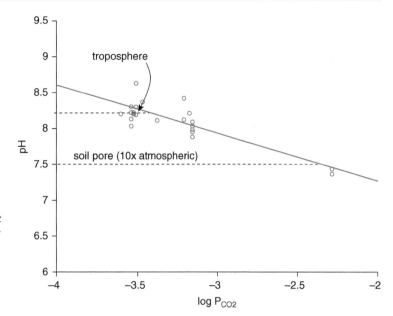

Figure 4.9 pH of a solution in equilibrium with calcite as a function of partial pressure of CO_2 plotted according to **Eq. 4.76** at 25 °C. Tropospheric (2013) P_{CO2} ($10^{-3.41}$ atm) and a value 10 times greater to represent soil P_{CO2} are shown. Points are experimental data from 10 Western US caclareous soils incubated at different P_{CO2} concentrations; pH measured in 1:1 soil solution. Observed data are from Whitney and Gardner (1943).

example, the system was solved for pH as a function of partial pressure of CO_2 (**Eq. 4.76**). Equations to solve for soluble calcium or total dissolved inorganic carbon as a function of pH or P_{CO2} can be solved using the reactions and equations in **Table 4.8** in a similar process as described in this example.

Several simplifying assumptions were required to solve the carbonate system equations that may not be possible or appropriate in other aqueous equilibrium problems. In addition, the assumption that activity and concentrations are equal (ideal solution) is fine for showing trends, but activity corrections can cause significant changes in the predicted pH or concentrations of the species. To solve equilibrium chemistry in complex systems where assumptions are not used requires use of numerical methods, which are best done using computers. Computer programs exist that can solve aqueous equilibrium problems and output accurate species concentrations (see Section 4.9).

4.6.4   Solubility of minerals in soils

In the systems discussed thus far, solubility has been considered an equilibrium process between solution and a pure solid phase. Soil systems are often not at equilibrium, and there are many different minerals that can participate in precipitation and dissolution reactions, each controlled by its solubility reaction rate and equilibrium constant. Furthermore, soil minerals

are not usually pure crystals with unit activity, and the solubility of solids changes, depending on the purity of the mineral, structural order, particle size, and hydration state of the mineral. For example, crystalline gibbsite has a solubility product of $10^{8.11}$, while poorly crystalline gibbsite is reported to have a solubility product constant of $10^{9.35}$.

Poorly crystalline phases are nonideal, and thus their activities cannot be assumed to be equal to unity, and the nonideality must be accounted for in the equilibrium expression. Soil-formed minerals tend to be particularly small and amorphous and contain many impurities. For example, common iron oxide minerals that exist in soils are ferrihydrite, goethite, and hematite that typically contain Al^{3+} and other metal cations within mineral structures (isomorphic substitution). Details of calculations of nonideal solid-phase activities are not covered in this book; however, a special case of nonideal solids that precipitate on mineral surfaces is discussed in Chapter 11.

To evaluate the status of aqueous and mineral systems with respect to solubility, measured ion concentrations can be converted to activities, and these values can be compared to solubility products. Using actual solution activities in solubility product equations gives the **ion activity product** (IAP). For example, in the goethite solubility reaction in **Eq. 4.61**, using actual solution measured concentrations to calculate Fe^{3+} and H^+ activities in **Eq. 4.62** yields

$$IAP = \frac{\left(Fe^{3+}\right)'}{\left(H^+\right)'^3} \qquad (4.78)$$

where the primes notate that the activities are calculated from measured concentrations. This equation assumes unit activities of the goethite; that is, the solid is pure, well crystalized goethite. At equilibrium, the IAP value equals the solubility product (K_{sp}). However, in systems not at equilibrium, IAP values will be greater than or less than the theoretical K_{sp}. To make comparisons of IAP to equilibrium of minerals, the ratio of IAP over K_{sp} is calculated. The logarithm of this ratio is called the **saturation index** (SI):

$$SI = \log \frac{IAP}{K_{sp}} \qquad (4.79)$$

When SI < 0, the system is **unsaturated**, when SI > 0 the system is **oversaturated**, and when SI = 0 the system is at equilibrium.

Ion activity products in soil solutions and equilibrium constants (solubility products) will agree only if the following system properties occur: (1) solubility of a single phase dominates the system; (2) competing reactions are insignificant; (3) the solid phase is pure; and (4) the system is close to equilibrium. Measurements of the ion activity product of iron and hydroxide (IAP = $(Fe^{3+})(OH)^3$) in many soil suspensions agree fairly well with the solubility product of the **poorly crystalline** iron oxide called ferrihydrite. Thus, soil iron appears to meet most of the criteria for using a solubility product – apparently Fe^{3+} reacts rapidly with OH^- so that single solid controls its solubility. Aluminum concentrations, on the other hand, appear to be controlled by slow processes, including weathering of aluminosilicates. Thus, aluminum mineral dissolution and precipitation reactions rarely meet the above criteria for using solubility products to predict aqueous ion activities and solution saturation states.

Phosphate mineral dissolution and precipitation reactions in soils are notoriously difficult to predict because they often do not follow thermodynamic solubility predictions. Instead, phosphate in soils is often precipitated as mineral phases that are not the least soluble species. Furthermore, the speed that phosphate minerals react may in fact control aqueous soil phosphate concentrations. For example, **Figure 4.10** shows the saturation index calculated for the Ca-P mineral octacalcium phosphate ($Ca_4H(PO_4)_3 \cdot 2H_2O$), abbreviated as OCP, in a manure-amended calcareous soil as a

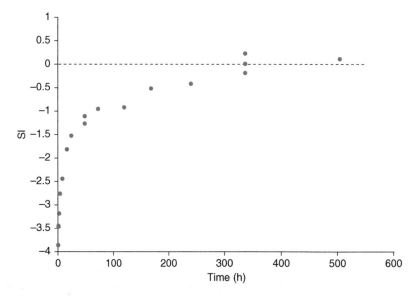

Figure 4.10 Octacalcium phosphate ($Ca_4H(PO_4)_3 \cdot 2.5H_2O$) saturation index (SI) of solutions extracted from a calcareous soil as a function of time. Initially the soil solution is undersaturated with respect to OCP. After 300 hours, the solution comes to equilibrium with respect to OCP. OCP appears to be the mineral phase controlling phosphate concentrations in this soil solution. Adapted from Hansen and Strawn (2003).

function of time. In calcareous soils, the least soluble mineral hydroxyapatite should theoretically control phosphorus solubility. However, the solutions are oversaturated with respect to hydroxyapatite, and appear to come to equilibrium with the more soluble OCP mineral. This suggests that solution phosphorus is controlled by precipitation and dissolution of a **metastable** phase (OCP). Theoretically, with time the OCP should *ripen* to the more stable and less soluble hydroxyapatite, but fluxes in the soil may disrupt progress to equilibrium, causing OCP to persist.

If a steady state existed between soil solids and dissolved phosphate ions, meaning the phosphate concentration in solution is constant with time, it might be described by a reaction such as

$$\text{soil-OH(s)} + \text{H}_2\text{PO}_4^-\,(\text{aq}) = \text{soil-H}_2\text{PO}_4 + \text{OH}^-\,(\text{aq})$$
$$(4.80)$$

where all the many phosphate interactions with soil particles are controlled by a single generalized species called *soil-H$_2$PO$_4$*; the equilibrium "constant" for this reaction is

$$\text{K}_{\text{soil-P}} = \frac{(\text{soil-H}_2\text{PO}_4)(\text{OH}^-)}{(\text{soil-OH})(\text{H}_2\text{PO}_4^-)} \qquad (4.81)$$

where $K_{\text{soil-P}}$ is a **reaction coefficient** rather than an equilibrium constant. While **Eqs. 4.80** and **4.81** are accurate, they are misleading – they have the form of equilibrium equations, but an equilibrium *constant* applicable to all soils does not exist. Also, the activities of soil-adsorbed ions such as soil-OH and soil-H$_2$PO$_4$ cannot be precisely defined, although models exist to predict them. Thus, reaction coefficients like **Eq. 4.81** describe several reaction processes rather than a single equilibrium. As such, constants for soil species are not available unless they are experimentally derived for a specific soil.

The reality of modeling soil mineral dissolution processes is that due to the complexity of soils, quantitative application of thermodynamic constants is often difficult. This is because natural systems are in a constant state of flux and thus generally are not at equilibrium with respect to mineral precipitation and dissolution; and because of the many competing reactions in the heterogenous soil environment. In laboratory incubations, some soil minerals require several years to come to equilibrium, and in natural systems, soil mineral dissolution and precipitation reactions may never achieve equilibrium. However, to predict whether minerals are dissolving or precipitating, equilibrium processes can be used to compare the solution and solid phase conditions in soils and to develop models of the processes controlling the steady-state movement of ions between the solid and solution phases.

Special Topic Box 4.2   Phosphate fertilizers

Next to nitrogen, phosphate is the most commonly applied fertilizer to increase crop growth. Phosphorus fertilizer is mined from phosphate mines. Some of the largest are in Kazakhstan, China, Morocco, Tunisia, and the United States (Florida and Idaho). The phosphorus containing rocks that are removed from the mines have mixed mineralogy. One of the most common phosphorus-bearing minerals in the rocks is fluorapatite, which has a low solubility that limits phosphorus available to plants after it is applied. The dissolution reaction of fluorapatite is

$$\text{Ca}_5(\text{PO}_4)_3\text{F}_1(\text{s}) + 6\text{H}^+ = 5\text{Ca}^{2+} + 3\text{H}_2\text{PO}_4^- + \text{F}^- \qquad (4.82)$$

This reaction shows that solubility increases as pH decreases, which is an important consideration for managing fertilizer application rates. Fluorapatite can be a good source of slow, long-term phosphorus, which may be beneficial in perennial cropping systems.

To produce more readily available phosphorus, phosphorus rock is processed to more soluble phosphate compounds. **Table 4.9** lists common phosphate fertilizers. Among these, triple super phosphate (TSP) is the most common phosphate fertilizer used. The dissolution reaction of TSP is

$$\text{Ca}(\text{H}_2\text{PO}_4)_2\cdot\text{H}_2\text{O}(\text{s}) = \text{Ca}^{2+} + 2\text{H}_2\text{PO}_4^- + \text{H}_2\text{O} \qquad (4.83)$$

When TSP dissolves in the soil, three important soil chemical processes control its availability for uptake by plants:

Table 4.9 Phosphorus fertilizers most commonly used in agriculture.

Fertilizer	Abbreviation	Mineral formula	Percent P_2O_5 by weight	Notes
Rock P (fluorapatite)	RP	$Ca_5(PO_4)_3F$	25–36	Very low solubility
Triple super phosphate (monocalcium phosphate)	TSP	$Ca(H_2PO_4)_2 \cdot H_2O$	44–52	Good solubility, can acidify in soil zone surrounding granule
Single super phosphate	SSP	$Ca(H_2PO_4)_2$ and $CaSO_4 \cdot 2H_2O$	16–22	Manufacturing produces a mixture of gypsum and TSP supplying a source of sulfate fertilizer
Monoammonium phosphate	MAP	$NH_4H_2PO_4$	48–62	Readily soluble source of N and P, slightly decreases soil pH
Diammonium phosphate	DAP	$(NH_4)_2HPO_4$	46–53	Twice as much N, slightly increases soil pH

1 The solution near the dissolved TSP granule becomes oversaturated with respect to calcium phosphate minerals such as dicalcium phosphate dihydrate (DCPD) (a.k.a. brushite):

$$Ca^{2+} + H_2PO_4^- + 2H_2O = CaHPO_4 \cdot 2H_2O(s) + H^+ \qquad (4.84)$$

which creates low pH near the zone of fertilizer application (note proton produced in precipitation reaction).

2 Calcium already in the soil impacts the mineral dissolution via the common ion effect.

3 Phosphate reacts with other soil ions and minerals, which changes its availability in the soil solution.

Precipitation of calcium phosphate minerals is often kinetically limited, thus meta-stable phases like DCPD precipitate, but over time (weeks to months), the least-soluble phase (hydroxyapatite) may predominate. Because of these factors, modeling availability of phosphorus from TSP and other fertilizers requires using numerous reactions such as **Eqs. 4.83**, **4.84** and others, including calcium, iron and aluminum mineral precipitation reactions, and calcium and phosphate adsorption reactions.

Application of phosphate is commonly done using mixed nitrogen and phosphate fertilizers such as MAP or DAP. A major advantage is the ease of handling and storage, and decreased shipping costs compared to applying the fertilizers separately. MAP causes slight acidification of the soil that is quickly counteracted by soil buffering, while DAP increase soil pH.

Unlike nitrogen, which is captured from the atmosphere in the Haber Bosch process, rock phosphorus mines are a finite resource. Some predictions of the lifetime of current phosphorus mines are they will become depleted in the coming decades. While such predictions are being debated, the real challenge in future phosphorus mining may be in the costs of extraction and distribution of the phosphorus fertilizers to regions throughout the world. For example, Europe does not have a major source of rock phosphorus, making it dependent on phosphorus from other regions. To alleviate Europe's reliance on imported phosphorus, there is considerable interest in recycling phosphorus recovered from municipal wastewater treatment plants, thereby closing the loop in the phosphorus life cycle (see the European Sustainable Phosphorus Platform website). While this would reduce phosphorus importation requirements, the recovery has its own set of economic and product quality challenges that must be overcome.

4.6.5 Solubility of contaminant metals from minerals

In soils impacted by mining activities, concentrations of metals such as lead, zinc, mercury, copper, cobalt, and cadmium are often elevated. Common species of these minerals in mine-impacted soils are solid phase minerals, including metal sulfates, phosphate, sulfides, carbonates, and oxides. A controlling reaction for metal contaminant fate and transport in the soil environment is precipitation and dissolution reactions of mineral phases. Some minerals, such as sulfides have very low solubility, while other phases, such as carbonates are more soluble. Prediction of the metal contaminant mineral solubility is done using solubility constants, as described above.

An example that is relevant to remediation of lead-contaminated soils is the use of phosphate amendment to immobilize lead. The immobilization mechanism is the transformation of lead ions from *available* phases to lead phosphate minerals that have low solubility, such pyromorphite $(Pb_5(PO_4)_3Cl)$. Upon dissolution of pyromorphite, the phosphate becomes protonated, thus the reaction has a pH dependency:

$$Pb_5(PO_4)_3Cl(s) + 6H^+ = 5Pb^{2+} + 3H_2PO_4^- + Cl^- \quad (4.85)$$

The solubility constant is

$$K_s = \frac{(Pb^{2+})^5 (H_2PO_4^-)^3 (Cl^-)}{(H^+)^6 (Pb_5(PO_4)_3Cl)} = 10^{-25.05} \quad (4.86)$$

Assuming unit activity for the solid phase and rearranging to solve for lead activity:

$$(Pb^{2+}) = \left(\frac{(H^+)^6 K_s}{(H_2PO_4^-)^3 (Cl^-)} \right)^{\frac{1}{5}} \quad (4.87)$$

which can be arranged to solve for (Pb^{2+}) as a function of pH by taking the logarithm:

$$\log(Pb^{2+}) = \frac{1}{5}\log(K_s) - \frac{3}{5}\log(H_2PO_4^-) \\ - \frac{1}{5}\log(Cl^-) + \frac{6}{5}\log(H^+) \quad (4.88)$$

Simplifying and substituting $-\log(H^+) = pH$, yields

$$\log(Pb^{2+}) = \frac{1}{5}\log(K_s) - \frac{3}{5}\log(H_2PO_4^-) \\ - \frac{1}{5}\log(Cl^-) - \frac{6}{5}pH \quad (4.89)$$

This equation can be used to plot an activity diagram of lead as a function of pH if phosphate and chloride concentrations are fixed (**Figure 4.11**). **Figure 4.11** also shows the lead activity predicted from solubility of PbO and cerussite ($PbCO_3$). Lead activity controlled by pyromorphite is much less than lead activity controlled by lead oxide or cerussite. **Figure 4.11** shows that increasing phosphate concentration in soil solutions will decrease Pb^{2+} activity because of the common ion effect.

In soils, phosphate concentrations are controlled by other solid phases, including dissolution and

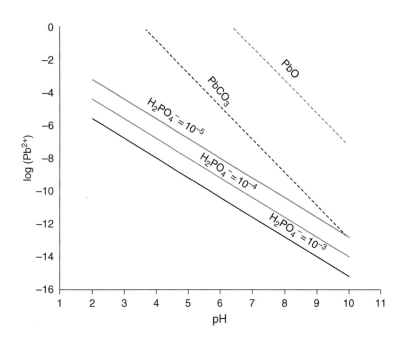

Figure 4.11 Activity diagram for lead activity predicted as a function of pH when pyromorphite, cerussite ($PbCO_3$) or PbO is controlling the solubility of lead. Phosphate activity ($H_2PO_4^-$) was fixed at three concentrations. Chloride activity was fixed at 0.001 M. Partial pressure of CO_2 was fixed at 0.001 atm. K_s (T= 25 °C) for pyromorphite is $10^{-25.05}$, for cerussite is $10^{4.65}$, and for PbO is $10^{12.72}$.

precipitation reactions of other phosphate-bearing mineral phases and adsorption reactions that need to be included in the equilibrium model to accurately predict lead solubility. However, **Equation 4.89** shows that if phosphate ion activity in soil increases, Pb^{2+} activity decreases. Also, as pH increases, Pb^{2+} activity decreases. Three important factors need to be considered in interpreting the pyromorphite activity diagram:

1 The diagram predicts lead species activity, but total Pb(II) in solution can be greater than Pb^{2+} activity due to formation of aqueous complexes with anions, ligands, and chelates.
2 Pyromorphite solubility across the pH range is calculated for a fixed $(H_2PO_4^-)$, but in natural systems there are many competing reactions that need to be considered.
3 The activity diagram is for a system at equilibrium under standard state conditions (1 atm and 298 K) and must be corrected for different temperatures.

A complete model, including all the reactions that affect pyromorphite solubility, is not given in this text. As systems become more complex because of numerous species present, computer programs are used to calculate equilibrium concentrations. However, the pyromorphite solubility example shows how phosphate affects solubility of lead and explains why phosphate works as a remediation strategy for lead-contaminated soils.

Solubility of metal carbonate, sulfate, oxide, phosphate, and sulfide minerals depends on the metal and the mineral species, and environmental factors such as presence of other ions. In addition, coprecipitation of metals creates mixed phases that have greater solubility. Another issue in soils is that mineral solubility does not account for sorption of metals on soil mineral and organic matter surfaces, which can be a dominant interface for solid-solution reactions. Thus, while modeling metal-mineral solubility is a useful tool to manage and predict contaminant metal availability for biological uptake or transport, the complexity of nature must be considered. Despite the complexities, there are metals and soils where solubility reactions are reasonably accurate. Additionally, mineral solubility models are useful for evaluating possible mineral species that control contaminant solubility in soil systems, thus providing some knowledge of system processes controlling availability.

4.7 Cation hydrolysis

For some cations, the attraction to the oxygen in water is so strong that the cation's charge completely repels the hydrogen ions resulting in a cation-hydroxide complex. This reaction is called **hydrolysis** because it splits a water molecule. Hydrolysis is most commonly observed in highly charged cations, or cations of small ionic radius. The charge on some cations such as P^{5+}, N^{5+} and S^{6+} is so great that they *permanently* bond with oxygen to form the oxyanions PO_4^{3-}, NO_3^-, and SO_4^{2-}. Cation hydrolysis is an acidic reaction because it produces a proton, thereby increasing acidity. For example, Al^{3+} hydrolysis is one of the most important reactions in creating soil acidity:

$$Al^{3+}(aq) + H_2O = Al(OH)^{2+}(aq) + H^+(aq) \qquad (4.90)$$

In soils, common cations that participate in hydrolysis reactions are Fe^{3+}, Al^{3+}, Pb^{2+}, Cu^{2+}, Zn^{2+}, and most other transition metals. The cations are surrounded by closely associated water molecules, such as $Fe(H_2O)_6^{3+}$, and the water molecules can progressively lose H^+ ions. Hydrolysis reactions change speciation of the cation, and thus affect total concentrations of the cations in solution. For example, in a soil with the iron oxide mineral goethite present, the Fe^{3+} species activity in solution is controlled by goethite dissolution, which is pH dependent, and can be predicted using the solubility constant (**Table 4.7**). While the Fe^{3+} activity in solution is fixed by dissolution of goethite, the total iron in solution is a function of the Fe^{3+} ion concentration plus the hydrolysis species:

$$Fe_T = \left[Fe^{3+}\right] + \left[Fe(OH)^{2+}\right] + \left[Fe(OH)_2^+\right]$$
$$+ \left[Fe(OH)_3\right] + \left[Fe(OH)_4^-\right] \qquad (4.91)$$

Equation 4.91 does not include iron hydrolysis species that are polymeric, that is, complexes containing more than one iron atom (e.g., $Fe_3(OH)_4^{5+}$). Polymeric metal hydrolysis complexes can be important in solutions, particularly for aqueous aluminum. Concentrations of iron hydrolysis species can be predicted from thermodynamic hydrolysis constants. For example, the first hydrolysis reaction for Fe^{3+} is written as

$$Fe^{3+}(aq) + H_2O = Fe(OH)^{2+}(aq) + H^+(aq) \qquad (4.92)$$

In this reaction, the shorthand notation of Fe^{3+} and $Fe(OH)^{2+}$ is used; however, in reality, the iron is coordinated by six water and/or hydroxide ligands (e.g., $Fe(H_2O)_5OH^{2+}$). The hydrolysis constant for the reaction in **Eq. 4.92**, omitting the water because its activity is one, is

$$K_1 = \frac{\left(Fe(OH)^{2+}\right)\left(H^+\right)}{\left(Fe^{3+}\right)} = 10^{-2.19} \tag{4.93}$$

Fe(III) hydrolysis continues progressively at still higher pH:

$$Fe(OH)^{2+}(aq) + H_2O = Fe(OH)_2^{+}(aq) + H^+(aq) \tag{4.94}$$

$$K_2 = \frac{\left(Fe(OH)_2^{+}\right)\left(H^+\right)}{\left(Fe(OH)^{2+}\right)} = 10^{-3.48} \tag{4.95}$$

Subsequent hydrolysis produces $Fe(OH)_3$ and $Fe(OH)_4^{-}$. Uncharged aqueous complexes can be noted with a superscript zero to signify that it is an aqueous complex and not a solid, for example, $Fe(OH)_3^{0}$. The negatively charged $Fe(OH)_4^{-}$ ion has a much different behavior than the positively charged or nonionic forms of aqueous iron because it is repelled from negatively charged surfaces and attracted to positively charged surfaces.

Activity of iron hydrolysis species can be predicted as a function of pH and Fe^{3+} activity using hydrolysis constants. If the solution is considered ideal so that concentrations equal activities, then **Eq. 4.91** can be solved algebraically for total iron by substitution of the hydrolysis equilibrium expression (**Eqs. 4.93, 4.95**, etc.) for the aqueous iron species:

$$Fe_T = \left[Fe^{3+}\right] + \frac{K_1\left[Fe^{3+}\right]}{\left(H^+\right)} + \frac{K_2K_1\left[Fe^{3+}\right]}{\left(H^+\right)^2} + \frac{K_3K_2K_1\left[Fe^{3+}\right]}{\left(H^+\right)^3}$$
$$+ \frac{K_4K_3K_2K_1\left[Fe^{3+}\right]}{\left(H^+\right)^4}$$

$$\tag{4.96}$$

where brackets indicate concentrations. **Equation 4.96** can be solved for $[Fe^{3+}]/Fe_T$ to allow calculation of the fraction of total iron present as Fe^{3+} ions as a function of pH:

$$\frac{\left[Fe^{3+}\right]}{Fe_T} = \left(1 + \frac{K_1}{\left(H^+\right)} + \frac{K_2K_1}{\left(H^+\right)^2} + \frac{K_3K_2K_1}{\left(H^+\right)^3} + \frac{K_4K_3K_2K_1}{\left(H^+\right)^4}\right)^{-1} \tag{4.97}$$

The fractions of the remaining hydrolysis species of Fe(III) can be calculated similarly. **Figure 4.12** shows the distribution of Fe(III) hydrolysis species as a function of pH. The predominant aqueous species in

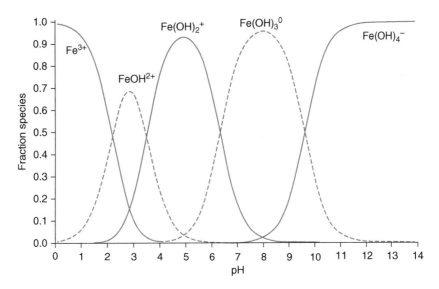

Figure 4.12 Distribution of iron hydrolysis species in solution as a function of pH. At the pH range of most soils (4 to 9), $Fe(OH)_2^{+}$ and $Fe(OH)_3^{0}$ are the predominant species in solution.

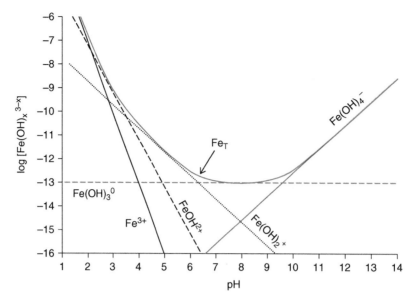

Figure 4.13 Molar concentrations of Fe(III) hydrolysis species in solution as a function of pH at 25 °C when goethite controls the Fe^{3+} concentration in solution. Ideal solution assumed (i.e., no activity corrections).

solution changes as a function of pH. For example, between pH 6.5 and 9 $Fe(OH)_3^0$ is predicted to be the predominant iron species in solution.

When a solid is present, Fe^{3+} concentration or activity is fixed, and the activity of the hydrolysis species is thus fixed by the solid's dissolution reaction. For example, if goethite is present, Fe^{3+} activity can be calculated using **Eq. 4.62**:

$$\left(Fe^{3+}\right) = \frac{\log K_{sp}}{\left(H^+\right)^3} \qquad (4.98)$$

Subsequently, $FeOH_2^+$ activity can be calculated by substitution of (Fe^{3+}) from **Eq. 4.98** into **Eq. 4.93**:

$$K_1 = \frac{\left(FeOH^{2+}\right)\left(H^+\right)}{\dfrac{\log K_{sp}}{\left(H+\right)^3}} \qquad (4.99)$$

$$\left(FeOH^{2+}\right) = \frac{\log K_{sp}}{K_1\left(H^+\right)^4} \qquad (4.100)$$

Continuing this algebraic substitution, the activity of the remaining hydrolysis species can be calculated. These expressions are only valid for the goethite dissolution reaction defined in **Eq. 4.61**. **Figure 4.13** shows the *concentration* of iron hydrolysis products when goethite controls the Fe^{3+} activity in an ideal solution (that is, all species have a unit activity).

Table 4.10 Hydrolysis constants of metal ions at 25 °C. Data are from Baes and Mesmer (1986).

Ion	log K_1	log K_2	log K_3	log K_4
Ca^{2+}	−12.85			
Mn^{2+}	−10.59	−11.61	−12.6	
Fe^{2+}	−9.5	−11.1	−10.4	−15
Ni^{2+}	−9.86	−9.14	−11	
Cu^{2+}	−8	−9.30	−10.5	
Zn^{2+}	−8.96	−7.94	−11.50	
Cd^{2+}	−10.08	−10.27		
Hg^{2+}	−3.40	−2.77		
Pb^{2+}	−7.71	−9.49	−10.86	
Al^{3+}	−4.95	−4.35	−5.70	−8.00
Fe^{3+}	−2.19	−3.48	−6.33	−9.60

Constants are for successive hydrolysis: $K_x = \dfrac{\left(M(OH)_x^{z-x}\right)\left(H^+\right)}{\left(M(OH)_{x-1}^{z-x+1}\right)}$

where M is the cation, z is the charge on the nonhydrolyzed cation, and x is the number of hydroxides on the metal hydrolysis product.

Total iron concentration in solution equals the sum of the concentration of each species (i.e., **Eq. 4.91**). The lowest total iron concentration occurs between pH 7 and 8.5, when the aqueous $Fe(OH)_3^0$ species predominates.

Table 4.10 gives the hydrolysis constants of several common soil cations. The smaller the hydrolysis constant (the more negative its exponent), the weaker the acid; meaning it is less likely to hydrolyze. For example, Fe(III) is a stronger acid and will hydrolyze at a

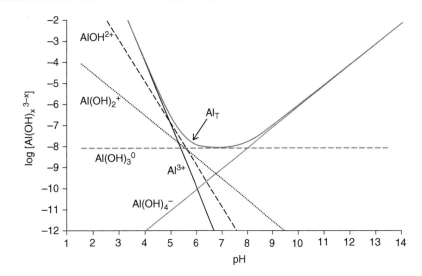

Figure 4.14 Molar concentration of Al(III) hydrolysis species in solution as a function of pH at 25 °C when gibbsite controls the Al^{3+} concentration in solution. Ideal solution assumed (i.e., no activity corrections).

lower pH than Al(III). Aluminum hydrolysis is an important reaction in soils. The hydrolysis reactions of aluminum are analogous to those of iron. The loss of the first H^+ from $Al(H_2O)_6^{3+}$ starts at approximately one pH unit below the pK and increases as pH increases. In solutions, the hydrolysis of the second and third protons is complicated by polymerization of the hydrolysis products into multi-nuclear Al(III) polymers, which were omitted from this Al-equilibrium system. The distribution of aluminum amongst the different hydrolyzed species can be calculated in the same manner as shown above for iron. The concentrations of Al-hydrolysis complexes as a function of pH, assuming gibbsite controls $[Al^{3+}]$ and disregarding Al(III) polymers, are shown in **Figure 4.14**. Minimum aluminum solubility occurs from pH 5 to 8, when $Al(OH)_3^0$ is the predominant hydrolysis species.

Changing signs of ions from positive to negative as a function of pH, as shown in **Figure 4.13** and **Figure 4.14**, is called amphoterism. **Amphoteric ions** can form positive, negative, or neutral hydroxyl complexes, depending on solution pH. The hydrolysis illustrated for Fe^{3+} and Al^{3+} occurs with other cations as well, but the extent of hydrolysis varies widely (**Table 4.10**).

4.8 Complexation

Closely interacting ions and molecules tend to lose their separate identities and become complex ions or ion pairs. **Complex ions** are the combination of a central cation with one or more ligands (**Figure 4.15**). A **ligand** is any ion or molecule in the coordination

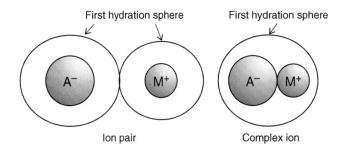

Figure 4.15 Illustrations of ion pairing where the hydration spheres surrounding each ion are maintained (left image), and a complex ion pair that changes the hydration sphere surrounding each ion (right image).

sphere around the central ion. Ligands are typically anions that have available filled electron orbitals that cations are attracted to, often forming ionic or covalent bonds. Water is technically a ligand (e.g., $Fe(H_2O)_6^{3+}$), but is not a complexing ion.

Complexing ligands can replace one or more of the water molecules in the primary hydration sphere or may exist in the second solvation sphere. When a ligand is in the second solvation shell, the cation and anion form an **ion pair**. Electrostatic attraction between ion pairs is weaker than the bonds between complexed ions. Complex ions and ion pairs are called **inner-sphere** and **outer-sphere complexes**, respectively. Many alkaline earth and transition metal cations occur in soil solutions as complex ions or ion pairs. To associate with a central ion, ligands must compete with the water molecules in the central ion's solvation sphere and must lose some of the water molecules in their own solvation sphere.

4.8.1   Predicting equilibrium for complexation reactions

Stability constants, or formation constants, refer to equilibrium constants for complex ions and ion pair reactions. The formation of ion complexes is the result of cation–anion attractive forces winning out in the competition between cations and water molecules or hydroxides in the cation's hydration sphere. An example is the formation of the mono-fluoro-aluminum ion complex:

$$Al(H_2O)_6^{3+}(aq) + F^-(aq) = Al(H_2O)_5 F^{2+}(aq) + H_2O$$

$$(4.101)$$

The water molecules are left off the F^- anion for simplicity. For clarity, this reaction can be written without the hydration waters on the cation:

$$Al^{3+}(aq) + F^-(aq) = AlF^{2+}(aq) \qquad (4.102)$$

Formation of complexes increases the overall solubility of ions, for example, total Al(III) in a fluoride solution is

$$Al_T = \left[Al^{3+}\right] + \left[Al(OH)^{2+}\right]...\left[AlF^{2+}\right] + \left[AlF_2^+\right].... \qquad (4.103)$$

where Al_T is the sum of the different species in solution, including hydrolysis species. In soils, aqueous Al^{3+} activity, and therefore concentration, is controlled by Al^{3+} dissolution from minerals such as gibbsite or aluminosilicates, and is thus fixed. The formation of hydrolysis and ligand complexes causes total aqueous aluminum species to increase above the concentration that would occur if only the uncomplexed Al^{3+} was present in solution.

Fluoride complexation of Al^{3+} is exploited to measure reactive Al^{3+} from soils. Adding fluoride creates AlF^{2+}. This temporarily lowers the activity of Al^{3+} in the water, which is replaced by slow dissolution or desorption of solid-phase aluminum. Measuring total dissolved aluminum in a fluoride extract is a relative measure of the readily soluble (presumably bioavailable) Al^{3+} in the soil. The stability constant for **Eq. 4.102** is

$$K_{AlF^{2+}} = \frac{\left(AlF^{2+}\right)}{\left(Al^{3+}\right)\left(F^-\right)} = 10^7 \qquad (4.104)$$

Equation 4.104 shows that the AlF^{2+} activity increases with increasing Al^{3+} or F^- activity. Increasing

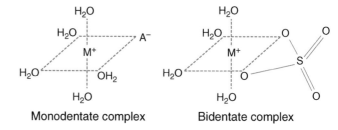

Figure 4.16 Structures of ligands bonding in monodentate and bidentate coordination. A^- is an anion ligand coordinated to a cation (M^+). The bidentate ligand is shown as a sulfate anion bonded to a metal cation (e.g., Ca^{2+}).

F^- concentration encourages more F^- ligands to replace water ligands around Al^{3+}, to a limit of AlF_6^{3-}. The hexa-fluoro-aluminum ion is the complex ion removed during fluoride extraction of aluminum in soils. In this system, protons also compete for the fluoride ion, not represented in **Eq. 4.104**; therefore, to accurately predict the activity of AlF^{2+} complexes, a HF protonation reaction and equilibrium constant are required.

Ligands such as H_2O, OH^-, F^-, and CN^- form only single bonds with cations, and are called **monodentate ligands** (or unidentate ligands) (**Figure 4.16**). **Bidentate ligands,** such as CO_3^{2-}, SO_4^{2-}, and PO_4^{3-} can form two bonds with cations. **Polynuclear complexes** (nuclear referring to the central cation) are created when ligands bridge two or more cations into a single molecule.

4.8.2   Chelate reactions with metals

Polydentate ligands occupy two or more positions around a cation (e.g., **Figure 4.17**). Such ligands are usually large organic molecules called **chelates**; from the Greek word for claw. Some enzymes, for example, are polydentate ligands occupying several positions around a central cation. Soil organic matter strongly adsorbs metal cations such as Cu^{2+} and Zn^{2+} through polydentate bonding.

Ethylenediaminetetraacetic acid (**EDTA**) is a chelating ligand (**Figure 4.17**) and is considered a good model of organic chelates that occur in soils and natural waters (although the ratios of nitrogen to carboxyl functional groups in natural organic matter are typically much less than in the EDTA molecule). In an EDTA-metal complex, the six ligand positions around the central cation are occupied by two amines and four carboxyl groups. Chlorophyll and hemoglobin are also chelates. Chelated metal cations are soluble, and tend

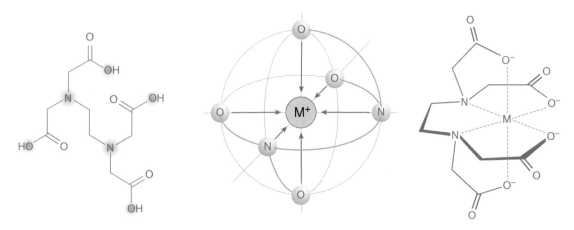

Figure 4.17 Schematic illustrating EDTA structure and metal coordination. EDTA has six ligands capable of bonding (highlighted in EDTA structure). The middle drawing shows a metal cation fully coordinated (octahedrally) by the ligands of an EDTA molecule (four oxygen and two nitrogen atoms). The molecular structure of the complex is shown in the drawing on the right. Left and middle images from Generalic (2014). Right image is public domain (http://image.absoluteastronomy.com/images/encyclopediaimages/m/me/metal-edta.png).

to keep metals, such as Fe^{3+}, Pb^{2+}, Zn^{2+}, Al^{3+}, and Cu^{2+} in solution. Chelates are used to extract micronutrients from soils in plant availability extraction tests and have been used to remediate metal-contaminated soils. The extent of cation-chelate complexation can be modeled using a stability constant; for example, the reaction and stability constant of the Fe(III)-EDTA complex are

$$Fe^{3+}(aq) + EDTA^{4-}(aq) = FeEDTA^-(aq) \qquad (4.105)$$

$$K_{FeEDTA^-} = \frac{(FeEDTA^-)}{(Fe^{3+})(EDTA^{4-})} = 10^{27.7} \qquad (4.106)$$

Protons effectively compete with iron for the EDTA sites and each acidic functional group on the chelate may have different complexation constants (pK_a values). For simplicity, the deprotonation reactions can be written as a summary reaction that includes all four protons:

$$H_4EDTA(aq) = 4H^+(aq) + EDTA^{4-}(aq) \qquad (4.107)$$

which is described by a deprotonation (or acidity) constant

$$K_{H_4EDTA^-} = \frac{(H^+)^4(EDTA^{4-})}{(H_4EDTA)} = 10^{23.76} \qquad (4.108)$$

If an iron oxide such as goethite controls Fe^{3+} activity, the solubility product (as shown in **Eq. 4.62**,) can be used

to predict Fe^{3+} ion activity. Substituting **Eqs. 4.62**, and **4.108** into **Eq. 4.106**, and rearranging gives the solubility of the Fe(III)-EDTA complex ion in equilibrium with goethite as a function of pH and H_4EDTA activity:

$$(FeEDTA^-) = (K_{H_4EDTA} K_{sp} K_{FeEDTA^-}) \frac{(H_4EDTA)}{(H^+)} \qquad (4.109)$$

Thus, in a system with goethite and EDTA, **Eq. 4.109** predicts that FeEDTA⁻ activity increases with increasing H_4EDTA activity and pH (higher pH corresponds to smaller (H⁺) values). However, **Eq. 4.109** cannot be easily solved without another equation because there are two unknowns (H_4EDTA and H⁺). One approach to solving the problem uses a mass balance for the total ligand concentration (e.g., $EDTA_{total}$ = [H_4EDTA] + [$EDTA^{4-}$]) and fix the $EDTA_{total}$ concentration in solution. Such problems are best solved using computer programs. Adding the chelated Fe(III) to the total Fe(III) solution concentration yields the total dissolved concentration of a metal in a EDTA-goethite system:

$$Fe_T = [Fe^{3+}] + [Fe(OH)^{2+}]...[FeEDTA^-]... \qquad (4.110)$$

Where the ellipses indicate the other hydrolysis and Fe(III)-EDTA complexes that occur in solution. **Figure 4.18** shows the concentrations of metal EDTA ligands and hydrolysis complexes expected in solution in equilibrium with goethite (calculated using a computer program).

Total iron concentration predicted with and without EDTA ligands as a function of pH is also shown. EDTA increases the total iron concentration in solution well above total iron concentration without EDTA. Plants take advantage of the *chelate phenomenon* by releasing Fe-specific chelates called siderophores to increase iron micronutrient availability. The complexed iron mobilizes to make Fe^{3+} more available to the plant roots. Specialized plant-root proteins can directly absorb Fe-chelate complexes.

In soil solutions, chelates and other **soluble organic compounds** are typically dissolved organic carbon (DOC). DOC concentrations in soils can be much greater than DOC in surface waters. DOC compounds in soils are soluble partially decomposed biomolecules with poorly defined structures and compositions. Some low-molecular-weight acids, organic bases, and neutral aliphatic or hydrophobic alcohols are also present in DOC; especially in rhizospheres.

Special Topic Box 4.3   Aluminum species and solubility in soils

Aluminum is the third most common element in Earth's lithosphere, after oxygen and silicon; and is extremely important in soils. Aluminum strongly influences soil acidity, is an important element in pedogenic processes, and can be an ecological toxicant for plants and animals. The soil chemistry of aluminum is complex because of the many species that exist in solid and solution phases (**Table 4.11**). Aqueous concentrations of Al^{3+} are controlled by precipitation and dissolution of Al-containing minerals such as aluminosilicates and oxides or hydroxides. In soils, aluminum minerals exist with varying degrees of crystallinity and surface areas causing varying equilibrium concentrations and reaction rates.

Characterization of what is termed *available* aluminum in soils is difficult, and often involves a method designed to measure relative availability of aluminum, such as exchangeable aluminum, bioavailable aluminum, and so on, which are not species specific. Measuring speciation of aqueous aluminum typically involves running solutions through a series of chromatographic and titrimetric methods that group it into different categories (**Table 4.11**); such as monomeric aluminum species, organic monomeric aluminum species, and total dissolved aluminum. Other phases of soluble aluminum are calculated by taking the differences between the different aqueous phases and total dissolved aluminum.

Spectroscopic methods, such as NMR, allow for the observation of aluminum species in solution, but are most useful for analyzing aqueous solutions that are not too complex (i.e., few interfering ions present), which limits ability for speciation of *soil solution* aluminum. Thermodynamic calculations can be used for speciation of aluminum in soil waters; however, caution must be used in interpreting the results because of the lack of accurate equilibrium constants for many of the aqueous aluminum complexes and polymers that are known to occur in solutions.

Table 4.11 Species of aluminum that exist in soils.

Al Species	Examples
Solution	
Free Al^{3+} (aquo-aluminum)	$Al(H_2O)_6^{3+}$
Hydrolysis complexes	$Al(OH)^{2+}$, $Al(OH)_2^+$ ... $Al(OH)_4^-$
Monomeric aqueous complexes	AlF^{2+}, $AlSO_4^+$
Monomeric organic complexes	Al complexed by organic acids (e.g., acetate, citrate)
Polymeric Al	$Al_{13}O_4(OH)_{24}(H_2O)_{12}$
Al nanoparticles	Polymer/mineral colloids
Strong organic complexes	Al-citrate, Al-SOM*
Solids	
Exchangeable Al	Al adsorbed on mineral or SOM phases
Multinuclear precipitates	Polynuclear Al sorbed/precipitated on mineral surfaces
Interlayer hydroxides	Gibbsite-like sheets or islands in interlayer of clay minerals
Primary minerals	Muscovite $K((Si_3Al)Al_2O_{10}(OH))_2$, orthoclase $KAlSi_3O_8$
Secondary minerals crystalline	Gibbsite $Al(OH)_3$, clay minerals (montmorillonite $M^+_{0.33}$ Si_4 $Al_{1.67}(Fe^{2+},$ $Mg)_{0.33}O_{10}(OH)_2$)
Secondary minerals poorly crystalline	Allophane, imogolite, poorly crystalline gibbsite

*SOM = soil organic matter

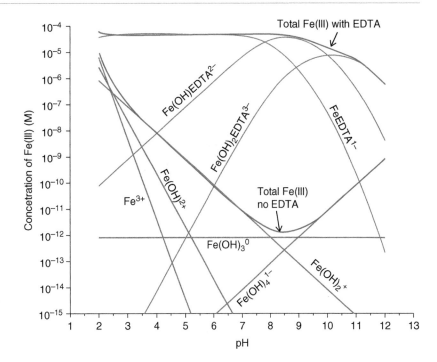

Figure 4.18 Concentration of Fe(III) hydrolysis and EDTA species in solution as a function of pH, based on goethite solubility with and without 5×10^{-5} M EDTA. Calculated using the program Mineql at 25 °C, using activity corrections.

Figure 4.19 A Spodosol from Michigan, USA, showing the leached E horizon and the spodic Bs horizons. The soil is classified as an Entic Haplorthod. Complexation of iron and aluminum by organic ligands in the surface horizon increases Fe^{3+} and Al^{3+} solubility and promotes movement down profile to the Bhs horizon where they accumulate. Photo credit: http://web2.geo.msu.edu/soilprofiles/. Source: Reproduced with permission of Dr. R. Schaetzl. See **Figure 4.19** in color plate section for color version.

The increased solubility of minerals in the presence of ligands such as chelates is called **ligand-promoted dissolution**. In addition to being important for trace element nutrient availability, ligands are also important for mobility and availability of metals that are contaminants. For example, organic ligands increase solubility and mobility of aluminum in soils and leaching potential of aluminum to surface waters. Color plate **Figure 4.19** shows a soil profile with a spodic horizon created from enhanced iron and aluminum mobilization caused by soluble organic ligands.

4.8.3   Trends in cation ligand affinity

At equal cation concentrations, transition metal cations compete more effectively for ligands than alkali and alkaline earth cations. Transition metal ions have the advantage of being able to shift some electrons to better accommodate ligand configurations (called **polarizability**). The ability of ligands to shift electron orbitals, and thus form stronger complex ions, increases in the order: $I^- < Br^- < Cl^- < F^- < C_2H_5OH < H_2O < NH_3 < EDTA < CN^-$. The strength of nitrogen-containing ligands is noteworthy in this list.

The relative ability of divalent cations to form ion complexes is $Mn^{2+} < Fe^{2+} < Co^{2+} < Ni^{2+} < Zn^{2+} < Cu^{2+}$. For trivalent cations, the order is $Al^{3+} < Fe^{3+} < Mn^{3+} =$

Table 4.12 Hard and soft Lewis acids and bases.

Lewis acids	Lewis bases
Hard acids	**Hard bases**
H^+, Li^+, Na^+, K^+ Mg^{2+}, Ca^{2+}, Sr^{2+} Al^{3+}, Be^{2+}, Si^{4+}, Ti^{3+}, Ti^{4+}, Mn^{2+}, Fe^{3+}, Co^{3+}, Cr^{3+}, lanthanides	H_2O, OH^-, O^{2-}, CO_3^{2-}, PO_4^{3-}, SO_4^{2-}, SiO_4^{4-}, F^-, NH_3
Intermediate acids	**Intermediate bases**
Fe^{2+}, Co^{2+}, Ni^{2+}, Cu^{2+}, Zn^{2+}, Pb^{2+}	NO_2^-, SO_3^-, Cl^-, Br^-, organic N-functional groups
Soft acids	**Soft bases**
Cd^{2+}, Hg^{1+}, Hg^{2+}, Cu^+, Ag^+	S^{2-}, CN^-, I^-, organic-SH

Cr^{3+}. The strongest complexing divalent cation is Cu^{2+}. Fe^{3+} is the weakest complexing trivalent transition metal ion but is stronger than other trivalent cations such as Al^{3+} and the lanthanides.

4.8.4 Predicting complexation using the hard and soft acid-base (HASB) concept

In aqueous solutions, cations react strongly with oxyanion ligands such as CO_3^{2-}, SO_4^{2-}, PO_4^{3-}, and NO_3^-. The metal cations Cu^{2+}, Cd^{2+}, and Hg^{2+} preferentially interact with non-oxygen-containing anions such as S^{2-}, CN^-. Cations that prefer to interact with oxygen-dominated anions (*hard* Lewis bases) are called *hard* Lewis acids (**Table 4.12**), whereas the cations that interact with non-oxygen-containing ligands (e.g., S^{2-}, CN^-) are called *soft* Lewis acids. Pearson suggested the following general rule to predict cation-anion complexation affinity:

Hard Lewis acids tend to associate with hard Lewis bases; soft Lewis acids tend to associate with soft Lewis bases.

Intermediate acids and bases can associate with both hard or soft bases and acids.

The classification of cations and anions as hard or soft acids and bases can be explained by considering the relative polarizability of the ions, which is related to ionic size and valence of the ion. Hard Lewis acids and bases are small ions and/or have high valence. Their electron orbitals are inflexible and thus, they form ionic bonds. Soft Lewis acids and bases are large ions that have low valence, and the electron orbitals are more polarizable and more likely to form covalent bonds. Soft Lewis acid cations are called covalent-bonding

ions. Chalcophiles are a group of soft Lewis acids that preferentially form bonds with sulfide-containing ligands as opposed to oxygen. Organic ligands and soil organic matter range from hard to soft Lewis bases.

The Lewis hard and soft acid–base concepts are useful to categorize cation and anion complexation. For example, reduced oxidation states tend to be softer Lewis acids and bases, and thus Fe^{3+} reacts differently than Fe^{2+} in soils. In this section, the HSAB concept is applied to aqueous complexation; however, HSAB concepts have also been shown to be useful for predicting metal toxicities, geochemical trends for mineral deposits, and anion and cation sorption affinities on solid surfaces.

Many permutations of the HSAB theory have been generated over the years, with the intention of providing a framework for predicting complexation. An important outcome of the concept is that ion properties, that is, ionic radius and valence, are important for predicting bonding between cations and anions. However, factors such as steric attraction and repulsion forces, hydration energies of ions, and energy involved in rearrangement of the ion constituents influence cation and anion attraction and complexation. The HSAB concept is a qualitative concept for categorizing cations and anions according to their potential to interact. Modern computing, however, makes HSAB less necessary because aqueous cation and anion speciation can be modeled using stability constants, as long as they are available.

4.9 Using software to predict soil solution equilibrium

Solving aqueous chemical equilibrium for simple systems is possible using algebra. Solving the equilibrium models in more complex, natural systems, however, is tedious, and requires many simplifying assumptions that may limit the accuracy of the results. For example, to solve equilibrium problems using activity corrections, numerous assumptions to gain initial estimates of species activities are typically required. Then using the initial estimates, the species activities are repeatedly recalculated until the species activity does not change (i.e., it converges). Such a calculation makes manually solving chemical species equilibria time consuming.

Computer programs make solving aqueous equilibrium problems easier. Programs such as PHREEQC, Geochemist Workbench, Mintec, or Mineql, to name a few, are powerful and user-friendly programs that can solve simultaneous equilibrium problems. In addition, many programs can couple the geochemical models with hydrologic flow models, collectively referred to as reactive transport (see Chapter 11), making them especially powerful for predicting fate and transport of chemicals in soils. However, as with most programs, getting the correct output requires correct input. In particular, input that defines the system species, reactions, and conditions is required.

Most aqueous systems can be solved with respect to concentration of a species of interest if the total concentration of ions, possible solid species, pH, and temperature are known. Modeling requires equilibrium constants, and, if temperature corrections are necessary, reaction enthalpies. Programs can output aqueous species distribution and concentration, pH relationships (e.g., titrations), and saturation state for solids. **Table 4.13** summarizes input and output of geochemical modeling.

Accuracy of computer output for prediction of species in a system depends on accuracy of the equilibrium constants used. Most programs contain a database of thermodynamic data and equilibrium constants for common

species. In addition, many programs allow manual input of equilibrium constants and species not included in the database. The user, however, should always verify and report equilibrium constants used by the program.

A challenge to using computer programs to solve *soil* equilibrium problems is that solid-solution interface reactions, such as adsorption and desorption, are typically not included in programs; or when added, the species and constants for the solid–solution reactions are not well defined. This is not a problem of the computer program *per se*, but a general problem of modeling soil reactions. Another issue is (as discussed above) that many natural systems are not at equilibrium, so equilibrium-based calculations do not accurately represent the species activities in a nonequilibrium system. As knowledge of soil species increases, and computer programs become more sophisticated, predictions of soil chemical processes will become more accurate.

4.10 Kinetics of chemical reaction in soil solution

Soil systems are often not at equilibrium, which means that reactions in the system are changing or are in a state of flux, thus they are continuously reacting to achieve the lowest energy state (Figure 1.13). Time-dependent chemical processes are called **kinetically controlled reactions**. Some reactions in nature take tens of thousands of years to come to equilibrium, whereas others take only seconds. **Figure 4.20** shows the rates of reactions typical of chemical processes that occur in soils and natural systems.

Many soil chemical reactions occur on time scales of minutes to days, which is a similar time scale to water flowing through soil macrospores during rain or irrigation input (**Figure 4.20**). Two important scenarios for chemical transport and reaction dynamics in soils are:

1 If the time a *unit* of water exists in a pore is less than the time it takes for a *solid phase* ion to dissolve, the ion concentration in the pore water will be less than equilibrium.
2 If a solution in a pore is saturated with respect to a solid, but precipitation is slower than the pore water is transported out of the system, little or no solid phase will precipitate. In this case, the saturated ions move out of the soil pores (e.g., leaching into

Table 4.13 Inputs and outputs of an aqueous geochemical system model.

Possible input	Possible output
Possible species	Dissolved concentration of species
Reactions	pH of system
Equilibrium constants for reactions	Saturation index of solids
Enthalpy of reaction (for temp. corrections)	Charge balance
Solid phases present	Ion activity products
Solid phase control (whether precipitation is allowed)	Equilibrium status of system
Activity coefficient correction model	Titration curve
Concentration of known species	
Total concentration of species	
pH	
Partial pressure of gases	
Temperature	
System fluxes (open or closed system)	
Modeling parameter variables (titration)	
Activity of solid phase	

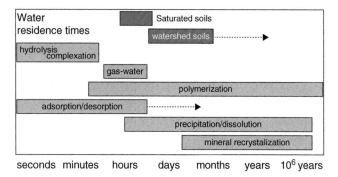

seconds minutes hours days months years 10^6 years

Figure 4.20 Reaction times of chemical processes in the environment, and average water residence times. Water residence times are for a saturated soil column and for water flow through a watershed (as reported in McGuire, A.D., et al. 2002). Dashed arrows on adsorption/desorption reactions and watershed soils indicate that processes may continue for longer times. Adapted from Langmuir (1997).

the lower horizon). Thus, even though the soil pore water is saturated with respect to a mineral phase, a precipitate does not form.

In the previous two scenarios, equilibrium modeling allows for predicting the state of the system with respect to reaction direction (i.e., precipitation or dissolution), but it cannot be used to quantify the activity or concentration of chemical in a solution. Predicting concentration change in a nonequilibrium system uses reaction kinetic models. Kinetic models are **rate equations** that quantitatively describe changes in concentration as a function of time:

$$r_{rxn} = \frac{\Delta C}{\Delta t} = \frac{\text{Change in concentration}}{\text{Change in time}} \qquad (4.111)$$

where r_{rxn} is the *overall* rate that the chemical reacts over a given time (Δt), and ΔC is the change in concentration of the chemical over the time. For some reactions, the rate of chemical reaction changes as the reaction proceeds, and r_{rxn} represents the average rate of reaction from the beginning and end of the reaction time increment.

A substance put in conditions in which it is unstable will sometimes not react at all until a stimulus is added. A mechanical example of such a **metastable equilibrium** is a rectangular block standing on end. It will not reach the more stable state of lying on its side until it is pushed so that its center of gravity is beyond its edge. In chemical reactions, pushing the block corresponds to **activation energy** (illustrated in Figure 1.13). Another example is a mixture of H_2 and O_2 gases, which will not

react until a spark or high temperature provides sufficient activation energy to perturb the metastable equilibrium and allow the gases to react. Photosynthesis is another example, where sunlight provides the activation energy that creates an activated state (glucose), which is metastable. The activated state returns to the stable initial states, CO_2 and H_2O, through a path of metabolism and decay that is as intricate as photosynthesis.

Nitrogen fixation, denitrification, soil weathering, phosphate fixation and release, clay mineral dissolution, and potassium fixation are soil chemical processes that are typically kinetically controlled in natural systems. Soil chemical kinetics has been the topic of intense study for the past four decades. It can even be argued that kinetics is intrinsic to thermodynamics because in reversible reactions, equilibrium is the condition where the *rates* of forward and reverse reactions are the same (i.e., forward and reverse reactions are occurring at an equivalent rate).

Small amounts of some substances increase reaction rates enormously. These substances, when left unchanged by the reaction, are called **catalysts**. One of the simplest types of catalytic action occurs when chemicals are adsorbed on surfaces causing them to have close proximity for relatively long periods. The probability of forming a new compound from the reactants is much greater than if they merely collide and rebound in a gaseous or liquid phase. Soil surfaces act as catalysts in this way. Catalysts lower the activation energy barriers that hinder reactions. The activation energy requirement can arise from many chemical and physical factors, and the mechanisms that catalysts use to lower the activation energy are just as numerous. Iron, manganese, and other transition metal organic catalysts such as enzymes speed up electron transfers for many reactions in living organisms.

Reaction inhibitors slow reaction rates. Nitrogen mineralization rates in soils, for example, can be slowed temporarily by chemicals that specifically slow or stop the microorganisms involved. Toxic metals can operate as enzyme inhibitors by replacing the metal coenzyme portion of an enzyme and thereby inactivating it.

Current research employs reaction kinetics to study the soil chemistry of carbon, nitrogen, potassium, phosphate, and trace metals. Soil reactions of these elements are often slow enough to be experimentally measurable. Because fluxes of carbon and nitrogen between soil, water, plants, animals, and the atmosphere are faster than rates at which they reach their most stable thermodynamic states, kinetics is the best approach to investigate and model changes of these elements.

For some ions, the reactions themselves can be very fast, but the ions may have to diffuse through soil pores before reaching a reaction site. The ions may also have to diffuse through the weathered surface of the mineral. With multiple diffusion and reaction processes occurring simultaneously, kinetic modeling can become very complex.

The general approach to model kinetic reactions starts with a reaction; for illustrative purposes, a generic reaction is used:

$$A + B \rightarrow C \tag{4.112}$$

which is a one-way reaction of *A* and *B* producing *C*. The mathematical expression used to predict the rate of the forward reaction is called a forward rate expression. The rate of production of product *C* at any time ($r_{rxn}(C)$) is proportional to the concentration of reactants in solution at that time:

$$r_{rxn}(C) = k[A]^{n_1}[B]^{n_2} \tag{4.113}$$

where k is the rate constant for the reaction; and n_1 and n_2 are reaction order coefficients for each of the species. Reaction order can be zero, one, two, or a fractional number, and is often set *a priori* based on experimental design or during modeling where different reaction orders are empirically fit to determine the best kinetic model for the time-dependent reaction. Rate constants are typically determined experimentally. Advanced study of reaction kinetics uses calculus to solve for the instantaneous rate of change in concentration of a species as a function of the instantaneous change in time.

If the reaction order is zero, the production rate of product *C* is linear with respect to time

$$r_{rxn}(C) = k \tag{4.114}$$

The production of product *C* in zero order kinetic reactions does not depend on any of the solution concentrations (**Figure 4.21**). The zero-order reaction can also be written in terms of the disappearance of a reactant, such as reactant A in the reaction in **Eq. 4.112**:

$$r_{rxn}(A) = -k \tag{4.115}$$

where the negative sign indicates that decreases as the reaction proceeds.

If the reaction is first-order with respect to each of the reactants, then the reaction rate depends on the *concentration* of reactants A and B. Using **Eq. 4.113**, the reaction rate with $n_1 = 1$ and $n_2 = 1$ is

$$r_{rxn}(C) = k[A][B] \tag{4.116}$$

In first-order reactions both the *rate of change in* concentrations of the reactants ([A] and [B]), and the products ([C]) decrease with time (**Figure 4.21**), thus the overall reaction rate changes as the reaction proceeds. Solving such kinetic equations is done using differential equations to relate the instantaneous reaction rate to the reactant and product concentrations at a specific time; this aspect of kinetic modeling is not covered in this textbook.

A useful aspect of measuring reaction kinetics instead of equilibrium concentrations is that characterization of time-dependent changes allows additional insight into reaction processes occurring in the system. It is

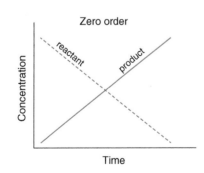
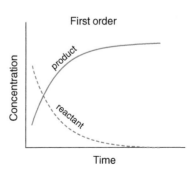

Figure 4.21 Theoretical shapes of zero- and first-order reactions showing relative concentrations of products and reactants. In zero-order reactions, the concentrations of the products and reactants change linearly with time. In first-order reactions, initially concentrations change fast, but as product concentration increases, concentration changes become slower.

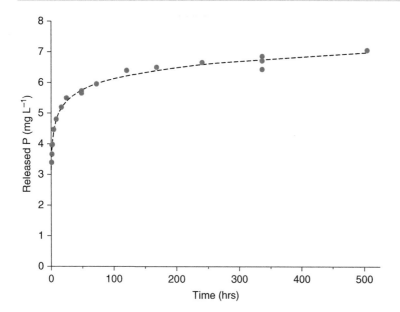

Figure 4.22 Concentration of phosphorus released from an alkaline surface soil amended with dairy manure as a function of time. Phosphorus release takes several days (>100 h) before approaching a steady state. Thus, if water is moving through the soil at a rate faster than phosphorus release rate, pore water phosphorus concentration is controlled by kinetic reaction processes. The dashed line is the best fit of the data using an empirical kinetic model. Data from Hansen and Strawn (2003).

analogous to *watching* a sports game as opposed to just seeing the final score. For this reason, kinetics of soil chemical processes are often studied by researchers to understand reaction processes occurring in a system.

Detailed treatment of modeling kinetics of soil chemical reactions is beyond the scope of this textbook. However, it is an active field of research, and kinetic equations are useful for predicting soil chemical processes. **Figure 4.22** shows the rate of dissolution of phosphate from an alkaline soil as a function of time. The data show that the reaction takes three to four days before equilibrium or *steady state* is achieved. Steady

state is a condition in a system when the rate of reaction is constant and often results when systems have fluxes in them removing reaction products. The rate-controlling reaction for the phosphate dissolution in **Figure 4.22** is likely the dissolution reaction of a Ca-phosphate mineral. The data were modeled using an empirical rate equation. The parameters from this equation predict the rate of desorption or dissolution from the soil and can be used to estimate leaching and off-site drainage in a scenario where water is moving through the soil faster than the chemical reactions comes to equilibrium with respect to phosphate release.

Special Topic Box 4.4   Kinetics of nitrogen loss from urea degradation

Many nitrogen reactions in soils are kinetically controlled, and prediction of their fate in soils requires use of kinetic equations. Urea ($CO(NH_2)_2$) is a common fertilizer added to provide nitrogen to crops. In soils, it is degraded by urease enzymes produced by plants and microbes. The summary degradation reaction is

$$CO(NH_2)_2 + 3H_2O = CO_2 + 2NH_4^+ + 2OH^- \qquad (4.117)$$

The release of the nitrogen from urea in soils is a relatively slow process, taking several days to weeks. Release

depends on soil temperature and water content, as well as soil physical, biological, and chemical properties. A problem with use of urea as a fertilizer is loss of the nitrogen through ammonia volatilization to the atmosphere. One estimate is that as much as 55% of the added urea-N may be lost through ammonia volatilization. **Figure 4.23** shows the many reaction processes that control the availability of nitrogen from urea addition to soil. All the reactions in this diagram are time-dependent reactions in soils, and thus kinetically controlled. For example, the loss of ammonia from soils occurs via volatilization and transport through soil pores to the atmosphere, which is a function of the soil porosity, temperature, water content (water decreases gas flux through the soil), and wind speed above the soil. While the ammonium-to-ammonia volatilization reaction is a relatively rapid process, the diffusion of the gas to the overlying atmosphere can be slow, making the overall volatilization process a rate-limited reaction. The rate of ammonia volatilization from a soil can be modeled using an apparent first-order rate coefficient and the concentration of ammonium in solution:

$$r_{rxn}\left(NH_3\right) = k_v\left[NH_4^+\right] \qquad (4.118)$$

where k_v integrates many time-dependent processes, including diffusion of gas through the soil to the overlying atmosphere, and the ammonia volatilization reaction (**Eq. 4.42**), which can be predicted by Henry's law constants. The full solution for ammonia volatilization reaction equation, and the associated reactions in **Figure 4.23**, is beyond the scope of this text. **Figure 4.24** shows the result of a computer-simulated model for the concentration of nitrogen in different soil species as a function of the reactions shown in **Figure 4.23**. Nitrogen transformation reaction constants were fit to nitrogen release data from laboratory microcosms. The model shows that urease and ammonium transformation reactions under these soil conditions are time dependent, with most of the reactions occurring within about 10 days. Different soil temperature or moisture conditions changes these reaction rates and are critical factors to consider when planning urease fertilizer application.

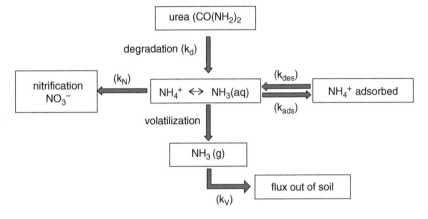

Figure 4.23 Model of urea degradation process shown as rate-controlled processes (indicated with k). Ammonium and nitrate fluxes into plants and microbes, or loss through leaching are not shown. Adapted from Bolado et al. (2005).

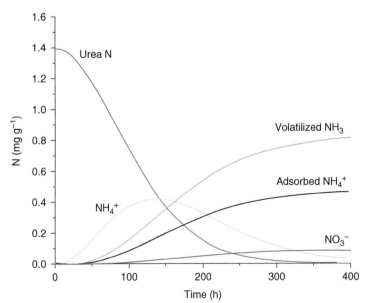

Figure 4.24 Results of simulation of urea degradation as a function of time. The degradation was modeled using kinetic constants derived from a soil incubation study. Adapted from Reddy et al. (1979) with permission.

To reduce loss of nitrogen from urea applied to soils, urease inhibitors have been shown effective. These block bacterial urease enzyme activity, thus preventing mineralization. Although urease inhibitors can be effective in reducing ammonia gas loss, their use in field application is limited because of costs and variability in their effectiveness. With increased research on best-use practices, urease inhibitors will become more cost effective and see broader use to prevent nitrogen loss through ammonia volatilization.

4.11   Summary of important concepts for soil solution chemistry

Soil solution chemistry uses classical thermodynamics to predict equilibrium of reactions within solutions and between solid and gas phases. Using thermodynamics, equilibrium constants can be calculated for reactions, and theoretical concentrations of reactants and species can be calculated. The calculations of concentrations include converting aqueous species activities to concentrations using activity coefficients. In many cases, reaction species are calculated as a function of pH, which is useful for soils since pH is a master variable and is easy to measure.

Although thermodynamic theory and its application to aqueous solution chemistry are well accepted, their application to soils is often tenuous because

1 There are multiple species present in soils.
2 Surface reactions on soil particles are not included, although they can be (covered in Chapters 10 and 11).
3 Biological processes prevalent in soils are often not accounted for in traditional aqueous thermodynamics.
4 The minerals precipitating in soils are not pure.
5 Soil reactions are often slow and do not come to equilibrium.

Despite these limitations, equilibrium modeling can successfully predict some soil reaction species concentrations. Even in systems not accurately represented by a set of thermodynamic models, use of the models to predict the equilibrium state improves our understanding of the processes controlling species concentrations. For example, ion activity products for a soil solution can be computed and compared to solid equilibrium constants to calculate if a system is saturated or unsaturated with respect to the solid phase (e.g., **Figure 4.1**). In cases where reactions are time-dependent, reaction kinetics can be used to predict species concentrations.

The utilization of reaction equilibrium constants and kinetics are core tools used to predict concentrations of species in soils. This chapter presented the basics of soil–solution equilibrium calculations. To completely predict soil chemical reactions, more advanced models are used, which, when coupled with molecular tools to investigate species present, is a powerful tool for predicting soil chemical processes. The development of more advanced computer programs is expected to make predicting soil chemical processes even more accurate. Such programs need to integrate chemical, biological, and physical processes, and should be flexible enough to work on a broad range of environmental systems.

Questions

1 Explain why $CaCO_3$ tends to accumulate at some depth in semi-arid and arid region soils, while salts that are more soluble tend to be in deeper horizons or absent. Why is the $CaCO_3$ layer thought to represent the average depth of wetting? Describe a soil-geochemical model that can be used to predict the precipitation of calcite; include the fluxes and relevant reactions. What are the dependent and independent variables in the model that are relevant to the soil system? List the information and assumptions needed to solve such a model.
2 Derive a solubility graph for gibbsite at 298 K. Label where SI > 0 and SI < 0; describe these two conditions with respect to mineral solubility. Show the equation for SI. If an organic acid complexed some of the aluminum, how would this change the Al^{3+} and total Al(III) concentrations?
3 Considering **Figure 4.2**, what are reasons that the activities of soil solution Ca^{2+} and SO_4^{2-} do not fall on the gypsum solubility line? Describe an example of a geochemical model that would represent the points off the line.

4 If the free energy of reaction for PbO dissolution is -7.26×10^4 kJ mol^{-1}, where Pb^{2+} and H_2O are the products, what is the solubility constant?

5 The tendency of minerals is to go to the lowest possible energy state. Why, then, do metastable soil minerals form?

6 What are two characteristics of ions that govern interactions with water?

7 What effect does one ion in solution have on another? How are these effects accounted for in aqueous chemistry?

8 A plant exudes a chelate to increase Fe nutrient availability. Write a chemical reaction for dissolution assuming $Fe(OH)_3$ is the solid phase in the soil with and without an organic acid. Derive the reaction equilibrium equation (i.e., mass action expression) for the organic dissolution reaction, and write another equation for total Fe^{3+} that includes free Fe^{3+}, the Fe-chelate, and two hydrolysis complexes. Describe how spending energy to make a chelate is a good strategy to increase Fe availability by the plant.

9 A soil solution extracted from a lysimeter was found to have 0.0065 M Na^+ and 0.0078 M Ca^{2+}, and the major anion is chloride. Calculate the ionic strength of the solution. What would the activity of the Na^+ and Ca^{2+} ions be?

10 Calculate the activity of Pb^{2+} in equilibrium with PbO at pH 4 and pH 7 (use the thermodynamic data from Question 4). What would happen if phosphate fertilizer was added to the soil.

11 Calculate the pH in (a) a 0.005 M solution of HCl, and (b) a 0.005 M solution of acetic acid. Explain why the pH values are different.

12 Considering the hard and soft acid–base theories, what ligands will complex with Ca^{2+}, and which will complex with Hg^{2+}?

13 If a soil solution is moving through a saturated soil profile at a rate of 0.15 m hr^{-1} with an average residence time of 40 minutes, and phosphate minerals are dissolving at a rate that takes 0.75 days to come to equilibrium, would it be more appropriate to use a kinetic equation or equilibrium equation to model the solution phosphate composition? Considering that phosphate adsorbs and desorbs from soil mineral surfaces, how might this affect the predictions?

14 List the following acids in the order of the pH that they will deprotonate: HF, H_3PO_4, H_2SO_4, COOH, HNO_3, and H_2CO_3 (include all protonated species).

15 Ion composition of a soil water from a wetland soil is listed in the following table.

Answer these questions:

a What is the ionic strength of the water?

b What is the activity coefficient for Fe^{3+}?

c What is the activity of Fe^{3+}?

d Is the soil water saturated, at equilibrium or undersaturated with respect to ferrihydrite (use the thermodynamic data in **Table 4.7**)?

Ion	Concentration (M)
Al^{3+}	2.5×10^{-7}
Ca^{2+}	1.6×10^{-3}
Fe (III)	1.8×10^{-4}
Fe(II)	2.0×10^{-4}
K^+	1.0×10^{-4}
Mg^{2+}	1.3×10^{-3}
Mn^{2+}	8.0×10^{-5}
Na^+	2.7×10^{-4}
Pb^{2+}	1.7×10^{-7}
Sr^{2+}	2.6×10^{-6}
Zn^{2+}	2.6×10^{-4}
pH	5.96
Cl^-	3.1×10^{-4}
SO_4^-	2.7×10^{-3}
HCO_3^-	1.4×10^{-3}

Data from Balistrieri et al. (2003).

Bibliography

Baes, C.F., and R.E. Mesmer. 1986. The Hydrolysis of Cations. 2 ed. Krieger Publishing Co., Malabar, FL.

Balistrieri, L.S., S.E. Box, and J.W. Tonkin. 2003. Modeling precipitation and sorption of elements during mixing of river water and porewater in the Coeur d'Alene River basin. Environmental Science & Technology 37:4694–4701.

Bolado, S., A. Alonso-Gaite, and J. Álvarez-Benedí. 2005. Characterization of nitrogen transformations, sorption and volatilization processes in urea fertilized soils. Vadose Zone Journal, 4(2):329–336.

Drever, J.I. 1997. The Geochemistry of Natural Waters: Surface and Groundwater Environments. 3rd ed. Prentice-Hall, Inc., Upper Saddle River, NJ.

Generalic, E. 2014. EDTA. Croatian-English Chemistry Dictionary & Glossary.

Hamdi-Aissa, B., V. Valles, A. Aventurier, and O. Ribolzi. 2004. Soils and brine geochemistry and mineralogy of hyperarid desert playa, ouargla basin, Algerian Sahara. Arid Land Research and Management 18:103–126.

Hansen, J.C., and D.G. Strawn. 2003. Kinetics of phosphorus release from manure-amended alkaline soil. Soil Science 168:869–879.

Kielland, J. 1937. Individual activity coefficients of ions in aqueous solutions. Journal of the American Chemical Society 59:1675–1678.

Langmuir, D. 1997. Aqueous Environmental Geochemistry. 1st ed. Prentice-Hall, Inc., Upper Saddle River, NJ.

McGuire, A.D., Wirth C, Apps M, Beringer J, Clein J, Epstein H, Kicklighter DW, Bhatti J, Chapin III FS, De Groot B, Efremov D. 2002. Environmental variation, vegetation distribution, carbon dynamics and water/energy exchange at high latitudes, *Journal of Vegetation Science* 13:301–314.

Reddy, K.R., R. Khaleel, M.R. Overcash, and P.W. Westerman. 1979. Nonpoint source model for land areas receiving animal wastes, 2. Ammonia volatilization. Transactions of the Asae 22:1398–1405.

Stumm, W.M., James J. 1996. Aquatic Chemistry: Chemical Equilibria and Rates in Natural Waters. 3rd ed. John Wiley & Sons, Inc., New York, NY.

Whitney, R.S., and R. Gardner. 1943. The effect of carbon dioxide on soil reaction. Soil Science 55:127–141.

5 REDOX REACTIONS IN SOILS

5.1 Introduction

In reduction and oxidation reactions, atoms exchange electrons causing changes in oxidation states and chemical species, and thus dramatically change the behavior of chemicals in soils. Redox reactions can be abiotic or biotically driven. In most natural systems, biological reactions have the greatest influence on redox processes. Redox reactions are fundamental to biogeochemical processes in soils because they influence pedogenesis and change mobility and bioavailability of nutrients and contaminants. **Table 5.1** shows some of the most common elements involved in redox processes in soils, and their oxidation states.

Oxidation is the loss or donation of electrons by an element. **Reduction** is the gain or acceptance of electrons. For example, the reduction reaction of Fe^{3+} to Fe^{2+}, which is an Fe^{3+}–Fe^{2+} couple, is

$$Fe^{3+} + e^- = Fe^{2+} \tag{5.1}$$

where e^- is a free electron. The reaction in **Eq. 5.1** is a **half-reaction** because it does not show where the electron is coming from, and thus describes only half of a complete reaction. Although half-reactions imply that free electrons exist, the free electron is a theoretical concept that does not occur in nature; the electron is always associated with an atom and directly transferred between ions or molecules. Thus, reduction of one substance requires oxidation of another.

An example of an oxidation half-reaction that provides electrons is oxidation of sulfide to elemental sulfur (S^{2-}–S^0 couple):

$$S^{2-} = S^0 + 2e^- \tag{5.2}$$

This half-reaction produces two electrons that must be donated to another element in a reducing half-reaction, which is not shown.

Soil Chemistry, Fifth Edition. Daniel G. Strawn, Hinrich L. Bohn, and George A. O'Connor.
© 2020 John Wiley & Sons Ltd. Published 2020 by John Wiley & Sons Ltd.

Table 5.1 Oxidation state and speciation of some common elements that participate in redox reactions in soils. Figure 2.2 lists oxidation states for other elements.

Element	Oxidation states		Species and notes
	Most reduced	Most oxidized	
Oxygen	2–	0	O_2 is electron acceptor in respiration that yields the most energy; reduced O^{2-} occurs in H_2O, and oxidized O occurs as O_2.
Nitrogen	3–	5+	N takes on every oxidation state between N(–III) and N(V). NH_3 or NH_4^+ are common N(–III) species; also amines ($R-NH_2$). N_2 is the elemental N^0 gas. N_2O, NO, NO_2^- are intermediate oxidation states involved in nitrification and denitrification. NO_3^- is an oxidized nitrogen species.
Carbon	4–	4+	Organic C compounds have many intermediate oxidation states, e.g., alkyne has C^0, and carboxyl has C^{3+}; reducing environments produce methane (CH_4) as C^{4-}; CO_2 gas is the most oxidized carbon compound (C^{4+}).
Sulfur	2–	6+	Many intermediate S oxidation states exist: H_2S and FeS and thiols (RSH) are sulfide (S(–II)) species; sulfite (SO_2) is S^{4+}; sulfate (SO_4^{2-}) is the most common S^{6+} species in oxidized environments.
Iron	2+	3+	Fe(II) (ferrous iron) is more soluble than Fe(III) (ferric). Fe(II) exists in unweathered primary minerals and reducing environments. Fe(III) exists as the Fe^{3+} ion and its hydrolysis complexes, Fe^{3+} ion pair and chelate complexes, and pedogenic oxides in soils.
Manganese	2+	4+	Aqueous Mn^{2+} is more soluble than Mn(III) and Mn(IV) species. In reduced soils, Mn^{2+} species occur in solution or on exchange sites. In oxidized soils, Mn(III) and Mn(IV) exist as oxide minerals or isomorphically substituted in iron oxides.
Arsenic	3+	5+	As^{3+} exists as the oxyanion acid arsenite (H_3AsO_3) in reduced environments. As^{5+} exists as the oxyanion acid arsenate (H_3AsO_4), which is similar in behavior to phosphate; arsenate is less soluble than arsenite.
Selenium	2–	6+	Similar oxidation states as sulfate: (Se(II), Se(0), Se(IV), and Se(VI). Common Se species include H_2Se, H_2SeO_3 (selenite), and H_2SeO_4 (selenate). Selenate is more soluble than selenite.
Mercury	0	2+	Elemental mercury is a liquid metal. Hg(II) exists as Hg^{2+} and its complexes in solution, as solid phase minerals, or sorbed on surfaces of minerals and organic matter in soils. In reducing environments, microbes produce methyl mercury ($CH_3 Hg^+$), which is highly toxic.
Chromium	2+	6+	Cr^{2+} is not common in nature. Cr^{3+} is common and is not very soluble. Cr^{6+} exists as the chromate (H_2CrO_4) oxyanion and is soluble.

Reactions in **Eq. 5.1** and **Eq. 5.2** can be combined into a full redox reaction:

$$2Fe^{3+} + S^{2-} = S^0 + 2Fe^{2+} \tag{5.3}$$

In this full redox reaction, the transfer of electrons is not shown, but the elements are accepting and donating electrons as shown in their half-reactions in **Eq. 5.1** and **Eq. 5.2**. Although the reaction in **Eq. 5.3** is technically correct, it does not represent a reaction that occurs in nature. Writing reactions for redox processes that occur in nature is one of the more challenging aspects of predicting redox chemistry in soils; especially considering that many of the reactions are biochemically mediated and the exact processes are not well understood.

Figure 5.1 shows the effects of redox on the distribution of iron, arsenic, and pH in flooded rice paddy soils. As incubation time under flooded conditions increased, oxygen concentrations decreased, redox potential decreased, concentrations of soluble iron and arsenic increased, and pH increased. **Redox potential** measures the *tendency* of a system to oxidize or reduce chemicals. High redox potentials indicate an oxidizing environment, and low redox potentials indicate a reducing environment. The word *tendency* is emphasized because redox potential measured in a soil is a quantitative measure of some, but not all, the redox processes occurring within a soil. The solubility of iron and arsenic in the paddy soils in **Figure 5.1** is affected by redox potential of the system, which alters the availability of these elements for plant uptake or transport out of the soil. Details of redox reactions, effects on chemicals, and measurement of redox in soils are discussed this chapter.

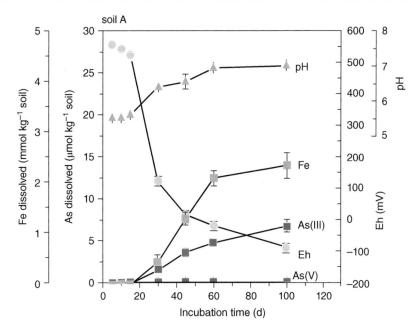

Figure 5.1 Eh, pH, and dissolved As(V), As(III), and Fe as a function of incubation time under flooded conditions for a rice paddy soil. Source: Yamaguchi et al. (2011).

5.2 Redox reactions in nature

Energy production and utilization processes in nature involve reduction and oxidation reactions. In soils, reduction reactions often occur in oxygen-limited environments, which include periodically or permanently flooded wetlands, rhizospheres, and soil regions that have high rates of microbial respiration and low gas permeability. Many of the most reactive redox environments are those that exist at the interface between oxidized and reduced zones (the **redox interface**). At the redox interface, chemical and mineral speciation is continuously changing. The redox interface is influenced by fluxes of temperature, water, nutrients, and gases through the soil. Oxygen fluxes into and out of the soil system are especially important because O_2 readily accepts electrons in abiotic and biotic processes and is the most common electron acceptor used for metabolism. When O_2 concentrations become low, biological organisms use alternative electron acceptors.

5.2.1 Photosynthesis redox reactions

Photosynthesis is a key redox process for fixing carbon and regulates CO_2 and O_2 gas cycling. Reduction of CO_2 by plants is done in chloroplasts that capture energy from the sun to fix CO_2 into biological carbon compounds:

$$CO_2 + 4e^- + 4H^+ = CH_2O + H_2O \tag{5.4}$$

where CH_2O represents a carbohydrate (such as glucose, $C_6H_{12}O_6$). The carbon in CO_2 accepts electrons, and its oxidation state changes from the C^{4+} in CO_2 to C^0 in carbohydrate. Simultaneously with a reducing reaction, another reaction must occur that provides the electrons. In photosynthesis, the O^{2-} in water gives up electrons in an oxidation reaction producing elemental oxygen (O_2 molecule):

$$2H_2O = O_2 + 4e^- + 4H^+ \tag{5.5}$$

The oxidation of the water molecule is done using energy from the sun coupled with enzymatic reactions that occur in chlorophyll. In the reactions in **Eqs. 5.4** and **5.5**, O^{2-} in H_2O is the **electron donor**, and C^{4+} in CO_2 is the **electron acceptor**. The overall reaction of photosynthesis is the sum of the half-reactions in **Eqs. 5.4** and **5.5**:

$$CO_2 + H_2O = CH_2O + O_2 \tag{5.6}$$

The process describes carbon fixation and oxygen production by plants, and is the reaction for energy transfer from the sun to Earth's biosphere. The sunlight energy input is not shown in the reaction.

Many redox reactions occur by enzymatic catalysis. Catalysis is necessary because most elements exchange electrons reluctantly. Enzymes lower the activation energy of electron transfer and increase reaction rates enormously. For example, the activation energy required for oxidation of many C, N, and S compounds to reach equilibrium creates the metastability that prevents you, the reader and the paper of this page from immediately oxidizing to CO_2.

5.2.2 Electron donors in nature

Oxidation of reduced carbon compounds by plants, animals, and microbes completes the carbon cycle by releasing the CO_2 and electrons used to make ATP (adenosine triphosphate):

$$CH_2O + H_2O = CO_2 + 4e^- + 4H^+ \qquad (5.7)$$

To obtain the energy and complete the reaction, organisms must find an electron acceptor to take the electrons released from the reduced carbon compounds. If oxygen is available, the half-reaction of aerobic electron acceptance is the reverse of the reaction in **Eq. 5.5**:

$$O_2 + 4e^- + 4H^+ = 2H_2O \qquad (5.8)$$

The reaction in **Eq. 5.7** summarizes many steps of the Krebs or citric acid cycle that organisms use to obtain energy, and the reaction in **Eq. 5.8** oversimplifies the intricate mechanism of electron acceptance by oxygen in living organisms.

In soils, carbon compounds in roots, microbes, dead plant matter, and soil organic matter (SOM) are the major electron donors. **Table 5.2** shows the approximate C, H, and O contents of the two largest plant components, cellulose and lignin, and of typical SOM. For simplicity, **Table 5.2** ignores the amounts of N, S, P, and other elements in these materials. Oxidation of plant matter does not go to completion immediately. Some carbon remains as SOM, microbial biomass, and partially metabolized byproducts.

Table 5.2 Approximate percent mass C, H, and O composition of lignin, cellulose, and soil organic matter (nitrogen, sulfur, and other elements are excluded).

	C(%)	H(%)	O(%)	Empirical formula
Lignin	61–64	5–6	30	$C_{2.8}H_{2.9}O$
Cellulose	44.5	6.2	49.3	$C_{1.2}H_2O$
Soil organic matter	58	5	36	$C_{2.2}H_{2.2}O$

Because cellulose oxidizes faster than lignin, SOM tends to have more aromatic groups that represent the accumulation of aromatic carbon from unreacted lignin. All these materials eventually oxidize in soils, but each succeeding oxidative step is slower. The half-reaction for the oxidation of *theoretical* soil organic matter is

$$C_{2.2}H_{2.2}O = 2.2C^{4+} + H_2O + 0.2H^+ + 9e^- \qquad (5.9)$$

The reaction in **Eq. 5.9** proposes that nine electrons come from oxidizing one mole of SOM to CO_2 (C^{4+}). This reaction is facilitated by microorganisms and other organisms (e.g., earthworms), who use the oxidation reaction to obtain electrons for respiration. Soil organic matter also contains amino (-NH_2) and sulfhydryl (-SH) groups, which are also electron donors.

Inorganic electron donors in soils typically occur in much smaller amounts than organic compounds, and include sulfide (S^{2-}), sulfur (S^0), Fe^{2+}, Mn^{2+}, and Mn^{3+}, and ammonia (N^{3-}). Oxidation of inorganic chemicals can occur biotically or abiotically. Organisms (called chemotrophs) can use electrons by oxidizing these elements. Nitrification is an example of an important soil process in which the chemotrophic microbes utilize the electrons in ammonia as an energy source. The reduced oxidation states of the trace elements Cr, Cu, Mo, Hg, As, and Se are also electron donors in soils, and most will oxidize in absence of biological oxidizers (i.e., abiotically). Oxidation of reduced inorganic elements can have important influences on speciation and solubility of chemicals in soil (**Table 5.1**).

5.2.3 Electron acceptors in nature

A key to obtaining energy from organic compounds, and thus to sustaining life, is availability of an electron acceptor. All heterotrophic aerobes oxidize carbon and reduce O_2 as the **terminal electron acceptor** (TEA), which produces CO_2 and H_2O. Oxygen is the strongest electron acceptor in nature yielding the most energy from reduction of O_2 to water (**Eq. 5.8**).

In soils, O_2 diffuses through pores to plant roots and soil microbes, where it can be utilized as a TEA. Some O_2 is also dissolved in soil water. Until a soil becomes quite wet, or the oxygen demand is high, the oxygen diffusion rate through soil is usually fast enough to maintain adequate O_2 availability. Even if only the larger soil pores are open to the atmosphere, the O_2 supply can be sufficient because O_2 gas diffusion through air is 10000 times faster than through water. Soils and waters that have available O_2 are called **oxic**. Soils that have no available O_2 are called **anoxic**. **Aerobic** and **anaerobic** are synonymous with oxic and anoxic, although they are more specific references to

respiratory processes of organisms in environments with and without oxygen.

If the diffusion path length from the soil surface through the soil is long, combined with a high O_2 demand from actively metabolizing roots and microbes, oxygen in the soil may be lacking. Also, many soils have restricted water drainage and are flooded for part or all of the year, further restricting O_2 diffusion. **Figure 5.2** shows the effects of water table fluctuations in a wetland on total O_2 in the soil.

Demand for O_2 by plant roots is constant during active growth, while microbial demand fluctuates widely in response to organic inputs and temperature. Oxygen consumption by plants and microbes can create zones of O_2 deficiency. Even in oxidized soils, the interiors of aggregates may be depleted in O_2 concentrations compared to the surrounding soil atmosphere

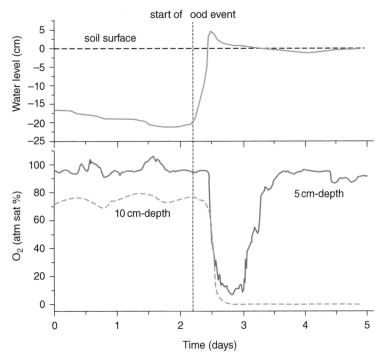

Figure 5.2 Wetland soil aqueous and gas O_2 concentrations at two soil depths during a five-day flood. The O_2 concentrations are percent saturation, where 100% represents equilibrium with tropospheric O_2 (g). Preceding the flood event, O_2 concentration at the 5-cm depth was near saturation, suggesting equilibrium with overlying atmospheric O_2 concentrations. At the 10-cm depth during the preflood event, O_2 concentrations were depleted relative to the atmospheric O_2 concentrations. The onset of flooding rapidly decreased O_2 concentrations at both the 5-cm and 10-cm depths. Approximately 1 day after the flood, O_2 concentrations at the 5-cm depth increased to levels indicating equilibrium with the overlying atmosphere; however, the 10-cm depth remained anoxic. The authors proposed that under flooding conditions, O_2 transfer to the 5-cm depth was facilitated by transport through aeraenchymous roots of the dominating vegetation cover (Phalaris arundinacea). At the 10-cm depth, respiration was faster than O_2 transport, preventing oxic recovery. Source: Jorgensen and Elberling (2012). Reproduced with permission of Elsevier.

because the O_2 respiration rates by microbes on the interior of the aggregates is greater than the gas diffusion rates, creating micro-sites of reduced redox potential. **Figure 5.3** shows the oxygen and redox profile of an aggregate incubated just below water saturation state in a laboratory microcosm. The slow O_2 diffusion created an anoxic zone in the interior of the aggregate causing the redox potential to also decrease.

Some soils contain dense subsoil horizons known as fragipans. Fragipan are characterized by large structural peds (typically > 5–10 cm in diameter) with low porosity and extremely slow permeability. As a result, water and roots occur primarily in the cracks between the large peds. Saturated conditions within these cracks limit oxygen availability and create sufficiently low redox potential conditions to reduce Fe(III) to more soluble Fe(II) (**Figure 5.4**). The mobilized iron then moves from the cracks toward the drier ped interiors where it precipitates due to the higher redox potential in these regions. As a result, exteriors of fragipan peds often become depleted of iron, and on the interior, zones of iron accumulation occur.

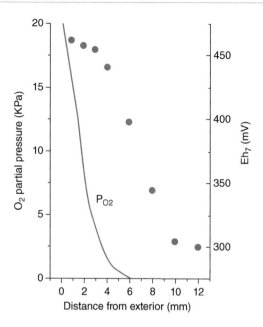

Figure 5.3 Oxygen gas partial pressure and redox potential in a 24-mm diameter soil aggregate incubated in the laboratory under near-saturated conditions (–1 kPa) for 15–25 days. The aggregate was made from soil collected from the B2 horizon of an Entisol in Germany (sandy clay loam texture). Data source: Zausig et al. (1993).

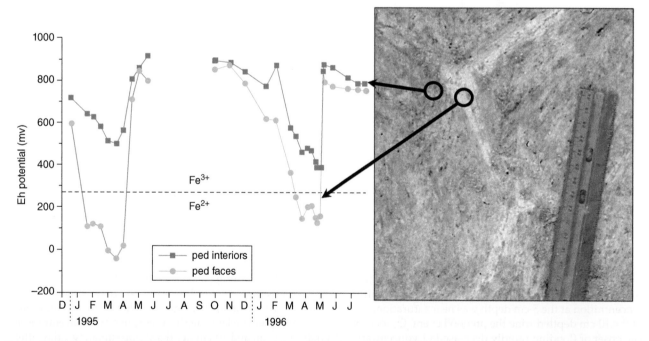

Figure 5.4 Redox potential (Eh) over a two-year period for a fragipan ped interior and exterior. The fragipan is formed in the subsoil of a Fragixeralf from northern Idaho. Because of its very low permeability, the cracks between fragipan peds are saturated for much of the winter and spring, while the ped interiors remain mostly unsaturated. Eh measurements represent an average of five Pt-electrodes placed in either the interior or the exterior region of the ped. Eh was corrected to the standard H_2 electrode. The horizontal line indicates the reduction potential of Fe^{3+} to Fe^{2+} at pH 5.5. Data are from P. McDaniel, University of Idaho. (See **Figure 5.4** in color plate section for color version.)

Agricultural practices can change the soil's ability to supply oxygen. Agronomic irrigation, cultivation, tillage, introduced crops, plant densities, and allowing fields to go fallow change the soil's water content and, therefore, change the pore space available for gas transfer between the root zone and the overlying atmosphere. Cultivation also destroys organic matter content, which maintains an open soil structure and improves permeability. In Midwestern North America, and other areas with high water tables, a common practice is to install tile drains under agricultural fields to remove the water that accumulates and limits oxygen availability to roots.

When O_2 availability is low, soil microorganisms utilize the oxidized states of nitrogen, sulfur, iron, manganese, and other elements as TEAs. Organisms that can use alternative TEAs are either facultative anaerobes or strict anaerobes, depending on their ability to selectively use a TEA other than O_2, or if they are *required* to use an alternative TEA for respiration. For plants, oxygen deprivation (anaerobic conditions) slows the rates of root metabolism and ion uptake and weakens the plant's resistance to soil pathogens. Lack of oxygen also promotes production of methane, nitrous oxide, and sulfide, as well as other reduced chemicals, which can have phytotoxic as well as environmental effects.

For anaerobic heterotrophic respiration, use of an alternative TEA does not release as much of the energy in carbon compounds as aerobic O_2 reduction. Thus, respiration by O_2 is more energetically favorable than other TEAs. The general order of preference for TEAs based on energy produced is:

$$O_2 > NO_3^- > Mn^{4+} > Mn^{3+} > Fe^{3+} > SO_4^{2-} > CO_2 > H^+$$

The exact order varies, depending on the species of the electron acceptors and environmental conditions. For example, Fe^{3+} present as goethite is a weaker TEA than sulfate, thus reduction of sulfate will yield more energy for an anaerobe. But Fe^{3+} present as ferrihydrite is a stronger TEA than sulfate, thus reduction of ferrihydrite will be energetically more favorable than sulfate. In addition, pH is a major environmental factor that influences the relative order of the preference for TEAs; explained in Section 5.6.

The most common **secondary electron acceptors** in soils include iron and manganese oxides, sulfate, and oxidized forms of nitrogen. Reducing half-reactions for these species are

$$FeOOH(s) + e^- + 3H^+ = Fe^{2+}(aq) + 2H_2O \tag{5.10}$$

$$2MnO_{1.75}(s) + 3e^- + 7H^+ = 2Mn^{2+}(aq) + 3.5H_2O \tag{5.11}$$

$$SO_4^{2-}(aq) + 8e^- + 8H^+ = S^{2-}(aq) + 4H_2O \tag{5.12}$$

$$NO_3^-(aq) + 2e^- + 2H^+ = NO_2^-(aq) + H_2O \tag{5.13}$$

$$2NO_3^-(aq) + 8e^- + 10H^+ = N_2O(g) + 5H_2O \tag{5.14}$$

$$2NO_3^-(aq) + 10e^- + 12H^+ = N_2(g) + 6H_2O \tag{5.15}$$

where $MnO_{1.75}$ is a *generic* mineral formula of a Mn(III-IV) soil oxide.

The products of the different nitrogen reduction reactions depend on soil properties and microbial biochemistry, and often are rate-limited reactions. The formation of N_2O (nitrous oxide) during denitrification is of interest because it is a long-lived greenhouse gas in the atmosphere. Nitrous oxide is often released initially after nitrate fertilizers are added to soils. By applying nitrate fertilizer when soils are not too wet, oxygen diffusion is adequate to prevent anaerobic conditions, and loss of nitrate by denitrification can be minimized.

Manganese and iron oxides contain manganese (Mn^{3+} and Mn^{4+}) and oxidized iron (Fe^{3+}) ions that some microbes can use as TEAs. The reduction of these metal oxides creates Mn^{2+} and Fe^{2+} ions. The reduced species are more soluble than their oxidized forms, leading to dissolution of the minerals in reducing environments. The reaction rates of mineral reduction are often slow because access to the solid-solution interface is limited by surface area, and because chemicals slowly diffuse through the porous mineral matrix. As a result, soils that undergo cyclic flooding are often at disequilibrium with respect to iron and manganese redox chemistry.

If both oxygen and secondary electron acceptors such as nitrogen, manganese, iron, and sulfur are absent, microorganisms in soils can still extract some energy from organic carbon compounds by fermentation. In fermentation reactions, microbes reduce carbon and release about 10% of the total energy in the initial compound. The products of fermentation include ethanol (C_2H_5OH), methane (CH_4), and some of the organic compounds that comprise peat. On a geologic time-scale, peat and buried organic matter are converted to coal and petroleum. Fermentation products retain about 90% of the original energy and are thus useful fuels.

Reduction of secondary electron acceptors produces compounds that are unstable in the presence of oxygen and convert back to the oxidized species through either abiotic or biotic processes. However, many abiotic oxidation reactions are slow. For example, soil organic matter stability is a result of incomplete oxidation due to lack of microbial activity, which may be caused by limited O_2 availability or low temperatures. The accumulation rate of SOM reflects the difference of the rates of organic matter addition vs. oxidation rates. The rate of addition is equivalent to the rate of net photosynthesis. The oxidation rate is governed by temperature and by the rate of oxygen supply. Subsidence (oxidation and carbon degradation) of drained peatlands occurs because of oxygen introduction to the massive store of incompletely oxidized carbon in peat soils.

Spatial variation of O_2 concentrations in soils provides zones of varying redox conditions, or redox gradients, where microbial communities establish their own niches based on the availability of TEAs. The concept of microbial niches was famously summarized by L.G.M. Baas Becking, a twentieth-century microbiologist noted for studying extremophiles: "Everything is everywhere, but the environment selects." This concept is observed in soils where there are distinct zones that microbes will thrive according to their ability to metabolize and outcompete other organisms because they are able to use the available food source (e.g., heterotroph vs chemotroph) and the available TEAs (e.g., aerobic vs anaerobic).

Because of the control that O_2 concentrations in soils have on soil redox processes, and the fact that O_2 concentrations can vary on scales of millimeters (e.g., ped interiors (**Figure 5.3**) to meters (e.g., a water table in lower horizon), soil redox gradients can occur on several scales. Zones in soils that have varying redox potential create redox gradients that alter soil development and create distinct pedogenic features called redoximorphic features (see Section 5.9).

5.3 Basic approaches for characterizing soil redox processes

It may seem that flooded soils such as wetland soils are anoxic and reduced, while drier soils are oxic and do not have reducing conditions. However, while many wetlands have soils with reducing conditions, a saturated soil is not necessarily an indicator of the *current* soil redox conditions. For example, temperate-region soils may flood in spring, but are not necessarily reduced because the temperature is too cold for biological activity to consume dissolved oxygen to create anoxic conditions needed for the soils to become reduced. In this scenario, if the landscape remains flooded and temperature increases, soil respiration will increase and eventually exceed available oxygen, and the soil will become reduced. An example of temperature dependence of redox in a wetland soil is shown in **Figure 5.5**. Thus, water status of soils is not a reliable indicator of redox processes.

The redox condition of soils is reported in numerous ways, which reflect the method of redox characterization. **Figure 5.6** summarizes descriptors of redox status in soils as a function of redox potential measured in volts (called Eh). While Eh measurement is based on well-established theory, its measurement in soils is subject to inaccuracies because many of the

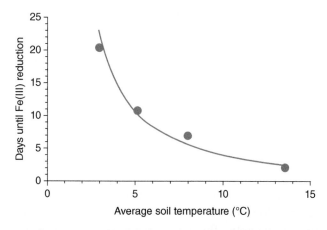

Figure 5.5 Effect of soil temperature on ferrihydrite reduction. As temperature increases, the number of flooding days until ferrihydrite reduction begins decreases. The exponential relationship indicates the strong influence that temperature has on microbial activity, and thus redox potential of soils. At less than 3 °C, ferrihydrite reduction in the wetland takes more than 20 days (~10 times longer than 13 °C). Source: Vaughan et al. (2009). Reproduced with permission of ACSESS.

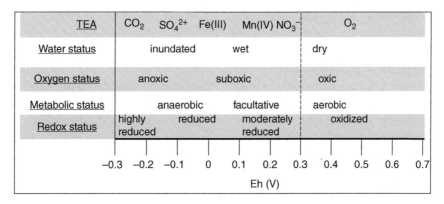

Figure 5.6 Correlation of soil redox potential with change in terminal electron acceptor (TEA), water, O_2, metabolism, and redox status category. Adapted from Reddy and DeLaune (2008).

requirements for Eh theory to be accurate are not valid in soil systems. This creates a problem when comparing field-measured redox potential with predictions of redox processes based on thermodynamics.

Despite the faults of Eh measurements in soils, it is a beneficial measurement to assess soil redox status because it is relatively easy to measure and is responsive to some dominant redox processes that occur in soils. Comparisons of Eh to redox processes such as presented in **Figure 5.6** are based on a combination of observed processes in soils and theoretical redox potentials. The ranges for most redox processes in the figure are estimates but are representative of redox processes observed in nature.

The redox condition where a chemical changes oxidation state is distinct for each species, and is related to its electronegativity and electron orbital configuration. For example, oxidation of Mn^{2+} to Mn^{4+} occurs at a lower redox potential than the oxidation of Fe^{2+} to Fe^{3+} because Mn^{4+} has a more stable electron configuration than Fe^{3+}.

The oxidized and reduced species of an element are called a redox couple. For example, one possible redox couple describing iron reduction is Fe^{3+}–Fe^{2+}; another is $Fe(OH)^{2+}$–Fe^{2+}. Every couple has a unique redox potential or thermodynamic constant that defines the relative order that chemicals will oxidize and reduce. To apply this fundamental aspect to soils requires prediction of soil redox processes and status. Three common methods are used to characterize soil redox condition:

1 Monitor changes in chemical species in the system that result from electron transfer processes.

2 Characterize redox reactions using a chemical reaction approach and the concepts of chemical activity, thermodynamics, and electrochemistry.
3 Measure the redox potential in a system using an electrode.

All three approaches have strengths and weaknesses and are used in combination to measure and predict redox status and processes in soils.

5.3.1 Using chemical species in soils to monitor redox status of soils

Redox properties in soils can be characterized by measuring changes in *concentrations* of chemicals, which change because redox affects chemical speciation and thus changes the solid, liquid, or gas phase distribution of the chemical (**Table 5.1**). For example, changes in concentration of species of oxygen, nitrogen, sulfur, methane, iron, or manganese in soils can be monitored and related to the redox status of a soil. Oxygen is one of the most important of these measures because it is a TEA for many organisms, and when absent, promotes use of secondary TEA. In-situ methods for measuring O_2 concentrations in saturated soils are fairly well developed, and allow for direct measure of the oxygen status of a soil in the field; see, for example, **Figure 5.2** and **Figure 5.6**. O_2 concentration measurement becomes more difficult as the soils dry out, which may make the sub-oxic delineation less accurate.

Ferrous iron (Fe(II)) is another species that can be monitored to assess redox status in soils. The presence of ferrous iron is highly indicative of reduced soils and can be assessed in-situ using α-dipyridyl solution, which forms a red ferrous complex when sprayed on the soil with dissolved Fe^{2+} in it.

When a soil undergoes redox cycling, it causes iron and manganese to form unique pedogenic solids called redoximorphic features. Redoximorphic features include iron and manganese **masses, concretions**, or **nodules**. Soil redox **depletions** are light gray because the iron and manganese have been reduced, solubilized, and leached away. Evaluating the distribution of redoximorphic features is an accurate method for characterizing soils that have undergone reduction and oxidation cycles (see Section 5.9).

5.3.2 Predicting redox processes in soil using chemical reactions

The chemical reaction approach to understanding redox processes in soils involves writing reactions and measuring or predicting chemical species within the reactions using thermodynamics and electrochemistry. For example, the Fe^{3+}–Fe^{2+} redox couple reaction is

$$Fe^{3+} + e^- = Fe^{2+} \tag{5.16}$$

The activity of the free electron (e^-) in half reactions is a useful concept for evaluating redox reactions. Free electrons do not exist in solutions, and thus their real activity is zero. However, the concept is beneficial for

modeling the redox potential at which electron acceptors and donors transfer electrons, and thus allows relative comparisons of redox couples.

For a half-reaction, a theoretical reaction constant (K_R) can be derived using the ratio of activities and products in a similar manner as an equilibrium constant for other types of reactions. For example, for the Fe^{3+}–Fe^{2+} half-reaction in **Eq. 5.16**, the equilibrium constant is

$$K_R = \frac{\left(Fe^{2+}\right)}{\left(Fe^{3+}\right)\left(e^-\right)} \tag{5.17}$$

where K_R is the redox equilibrium constant that is available in tables of thermodynamic data (e.g., **Table 5.3**) and (e^-) is the **electron activity** of the hypothetical free electron. In Chapter 4, methods to calculate the aqueous activity of species from their concentrations were presented; however, the activity of the e^- in **Eq. 5.17** cannot be calculated in the same manner, primarily because there is no corresponding concentration that can be measured. Electron activity is related to the amount of work done by the transfer of electrons in a redox reaction. It can be measured in an electrochemical cell (discussed below) as electron potential, measured in units of volts (Eh). By writing redox reactions as half reactions, the electrons can be tracked and related to thermodynamic reaction constants (K_R) for predicting the reaction progress relative to system conditions.

Table 5.3 Equilibrium constants for selected reduction half-reactions at 25 °C. The reactions are written as reduction reactions consuming a single electron. K_R source: Pankow (1991).

Reaction	log K_R	
$1/2Cl_2 + e^- = Cl^-$	23.6	Strong oxidizer (accepts electrons)
$1/5NO_3^- + 6/5H^+ + e^- = 1/10N_2 + 3/5H_2O$	21.05	
————oxygen stability————		
$1/4O_2 + H^+ + e^- = 1/4H_2O$	20.78	
$1/2NO_3^- + H^+ + e^- = 1/2NO_2^- + 1/2H_2O$	14.15	
$Fe^{3+} + e^- = Fe^{2+}$	13.0	
$1/8SO_4^{2-} + 5/4H^+ + e^- = 1/8H_2S + 1/2H_2O$	5.13	
$1/8CO_2 + H^+ + e^- = 1/8CH_4 + 1/4H_2O$	2.87	
$Cu^{2+} + e^- = Cu^+$	2.7	
$1/2S^0 + H^+ + e^- = 1/2H_2S$	2.4	
$1/6N_2 + H^+ + e^- = 1/3NH_3$	1.58	
$H^+ + e^- = 1/2H_2$	0	
————hydrogen stability————		
$1/2Fe^{2+} + e^- = 1/2Fe$	−7.45	Strong reducer (gives electrons)

Theoretical electron activity is often transformed to pe by taking the negative of the logarithm of its activity (analogous to pH):

$$pe = -\log(e^-) \qquad (5.18)$$

For example, the electron activity in the Fe^{3+}–Fe^{2+} half-reaction in **Eq. 5.16** can be transformed to pe by taking the logarithm of both sides of **Eq. 5.17**:

$$\log K_R = \log \frac{(Fe^{2+})}{(Fe^{3+})} - \log(e^-) = \log \frac{(Fe^{2+})}{(Fe^{3+})} + pe \qquad (5.19)$$

which can be rearranged to solve for pe:

$$pe = \log K_R - \log\left(\frac{(Fe^{2+})}{(Fe^{3+})}\right) \qquad (5.20)$$

Since K_R is a constant, the equation shows that pe is a function of the ratio of Fe^{2+} to Fe^{3+} activities.

Redox equilibrium constants can be used to predict the relative direction of reduction or oxidation of species in a system as a function of redox potential (represented by pe). For example, in the reduction reaction of Fe^{3+} shown in **Eq. 5.16**, as electron activity increases or decreases, the iron oxidation state will change (i.e., reaction will go to the left or right). The K_R for the iron reduction equilibrium constant expression in **Eq. 5.17** is 10^{13}; thus, the distribution of the Fe^{3+} and Fe^{2+} as a function of pe can be predicted by substituting the constant into **Eq. 5.20** and rearranging:

$$\log\left(\frac{(Fe^{2+})}{(Fe^{3+})}\right) = 13 - pe \qquad (5.21)$$

This equation shows that Fe^{3+} activity is greatest when redox potential is high (high pe), and vice versa. **Figure 5.7** graphically shows the $(Fe^{2+})/(Fe^{3+})$ distribution as a function of pe. The stable redox state of iron with respect to the system's redox potential can be predicted from this figure. For example, when pe is above or below log K_R (13 for Fe^{3+}–Fe^{2+} couple), the activity of one species exceeds the other, and when pe equals log K_R, the activity of the Fe^{3+} and Fe^{2+} are equal. When the activities of species in redox couple are equal ((Fe^{3+}) = (Fe^{2+})), the pe is called the *standard state* pe, abbreviated $pe°$.

As another example, consider the half-reaction for the reduction of Cu^{2+}:

$$Cu^{2+} + e^- = Cu^+ \qquad (5.22)$$

for which the equilibrium constant is $10^{2.6}$, or log $K_R = 2.6$. The corresponding equilibrium expression is

$$pe = \log K_R - \log \frac{(Cu^+)}{(Cu^{2+})} = 2.6 - \log \frac{(Cu^+)}{(Cu^{2+})} \qquad (5.23)$$

This equation is plotted together with iron reduction in **Figure 5.7**. Notice that at a given redox potential, log $(Cu^{1+})/(Cu^{2+})$ is much less than log $(Fe^{2+})/(Fe^{3+})$, thus Fe^{2+} is predicted to be stable at greater redox potentials than Cu^{1+}. This illustrates a general rule for determining the relative order that redox reactions will occur:

In comparing reducing half-reactions for redox couples, the couple with the lowest redox equilibrium constant (K_R) will oxidize or reduce at a lower redox potential than a redox couple with a higher K_R.

The rule is valid for only single-electron transfer reducing reactions. Reactions that include multiple electron transfers must be rewritten as single electron transfer reactions. To illustrate how to use K_R for redox couples to evaluate relative order of redox reactions, consider a copper and iron system that suddenly becomes oxidized; the Cu^{1+} will oxidize to Cu^{2+} at a lower redox

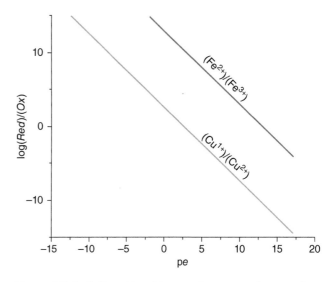

Figure 5.7 Activity ratios of iron and copper redox couples as a function of pe.

potential than Fe^{2+} oxidizes to Fe^{3+}. On the other hand, in a system that goes from oxidized to reduced, Fe^{3+} will be reduced to Fe^{2+} before Cu^{2+} is reduced to Cu^{1+}.

As another example of comparing redox couples, consider the H^+–H_2 couple. The hydrogen reduction reaction is

$$H^+ + e^- = \frac{1}{2}H_2 \qquad (5.24)$$

K_R for this reaction is 1, and log K_R = 0. Because the K_R value for the H^+–H_2 couple is less than the equilibrium constants for Fe^{3+} and Cu^{2+} reduction reactions, reduction of hydrogen will occur at a lower redox potential than Fe^{3+} and Cu^{2+}, and hydrogen oxidation will occur before Fe^{2+} and Cu^+. **Figure 5.8** shows the relative order of redox reactions discussed in examples in this section with respect to redox potentials.

Consumption of protons in many reducing reactions makes the redox reaction depend not only on pe but also pH. Because the pH-pe relationships are not always the same – that is, the stoichiometric coefficients on the (H^+) and (e^-) vary in the redox reactions – predicting the order of reduction or oxidation by comparing K_R alone can be misleading. The actual order for redox reactions that include proton reactants is a function of pH. Adjustment for pH will be discussed later in this chapter.

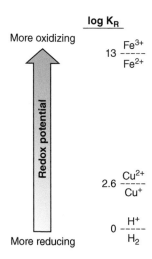

Figure 5.8 The relative order of Fe^{3+}, Cu^{2+}, and H^+ redox potentials and log K_R. If a highly reduced system is undergoing oxidization, the H^+–H_2 couple has the lowest K_R and will oxidize before the Cu^{2+}–Cu^+ couple, followed by the Fe^{3+}–Fe^{2+} couple. The order is reversed if an oxidized system is becoming reduced.

Table 5.3 provides the Log K_R values for many redox values that occur in soils. Comparing the values indicates the relative order of oxidation and reduction that *should* occur in soils. For most cases, redox processes in soils agree with thermodynamic predictions, but the complexities of natural systems make accurate characterizations difficult. For example, redox reactions in soils are driven by poorly understood biological processes, and thus actual redox reactions may not be included in a model developed to predict redox processes. Further, redox processes in soils are slow and rarely at thermodynamic equilibrium. The continuous fluxes occurring within soils change the activities of reactants and products faster than the reactions achieve equilibrium. For example, the Log K_R value for N_2 oxidation to nitrate is greater than oxygen reduction. Thus, in oxic conditions, N_2 should not be stable, which is not the case (troposphere is 78% N_2). The very high bond energy of the triple bond makes N_2 very unreactive, so oxidation is very slow. Another caveat of comparing log K_R values in **Table 5.3** is the assumption of standard-state conditions (1 atm, T = 25 °C, species activities equal one). Corrections to redox reactions for non-unit activity are discussed later in this chapter.

There are clear limitations to using theoretical redox chemistry to understand and predict soil redox processes, but the basic approach is nonetheless useful. Species that may occur in soils can be predicted, as well as the trends and direction reactions should go. For example, using redox reactions allows the relative order that chemicals will donate or receive electrons to be predicted. Knowing the order, and factors influencing it, is critical to understanding how redox reactions in natural systems occur as system fluxes change the availability of electron acceptors and donors.

5.3.3 Quantifying redox potential with a redox electrode

Redox potential measured for a soil or natural water is a quantitative indicator of the ability of the system to transfer electrons and is measured in volts. A specific definition of redox potential is:

Redox potential measures the ability of elements to accept and donate electrons.

In most natural systems, the measured redox potential represents a combination the redox potentials from several individual elemental redox potentials. In a system with a high redox potential there are few elements that will give up electrons, as most are already in their oxidized state. In a system with low redox potential, the electron orbitals of elements in the system have maximum *electron occupancy* and thus exist in reduced oxidation states.

In soils, the limits of redox potential vary with pH but are typically between –0.3 to 0.7 V (–300 to 700 mV). Soil redox potential is controlled by many system variables, including oxygen availability, temperature, microbial activity, availability of food, and availability of terminal electron acceptors (TEAs).

The voltage measured when a redox electrode is placed in a system is called **Eh**. In simple, well-controlled systems (such as a beaker with only aqueous ions), Eh is an accurate measure of redox potential. The redox probe used to measure Eh is an electrochemical cell (**Figure 5.9**). The cell measures the potential difference in electromotive force (EMF) between two separated platinum electrodes, with different redox half-reactions occurring at each electrode. One of the Pt electrodes is immersed in acid and serves as a reference electrode,

whereas the other is immersed in the solution containing the redox couple to be measured. The Pt electrodes accept or donate electrons to the redox couples in the half-cells, resulting in an electrical potential difference that is measured in volts. The voltage represents the difference in oxidation and reduction potential associated with the two redox half-reactions in the different cells. When the reference electrode is the **standard hydrogen electrode** (SHE), and the activities of the ions in the other cell are one, the measured redox potentials are the **standard electrode potentials** (Eh°). The SHE is a Pt electrode immersed in a solution of 1 M acidic solution in equilibrium with 1 atmosphere of H_2 gas; it is defined as having an Eh° of zero. Thus, in an electrochemical cell with a SHE as one of the electrodes, the measured redox potential, in volts or millivolts, is assigned the redox couple occurring in the solution surrounding the other Pt electrode. For example, in **Figure 5.9** the hydrogen gas oxidizes at the cathode and transfers the electron to the Pt electrode in the SHE. In the anode half-cell, Fe^{3+} accepts an electron from the Pt electrode and reduces to Fe^{2+}. The meter measures the electrochemical potential difference between the two half cells; for the Fe^{3+}–Fe^{2+} couple the meter would read Eh° = 0.77 V.

Because the SHE is not very practical, field and laboratory measurements use a reference electrode such as a solid-based calomel (mercury chloride) electrode, which is much easier to handle and maintain than a hydrogen electrode. The measured Eh from an electrode that includes an alternate reference electrode (i.e., non-SHE) must be converted to the SHE scale.

While the electrochemical cell and the standard state redox potential measurements in ideal systems are accurate, in less than ideal systems such as soils and natural waters, problems arise that make Eh measurement inaccurate. Nevertheless, soil Eh measured using a redox probe is a useful measurement because it is a relatively easy way to assess the *general* redox potential of a soil system, and, in some cases, such as an Fe(II)–Fe(III) redox system, measured Eh can be related to theoretical reactions. This is especially relevant for soils because iron is prevalent and may be controlling many of the electron transfer reactions occurring. Unfortunately, the exact iron species involved in the electron transfer reactions in soils may not be well known. Another use of measured Eh values is to provide a quantitative value to evaluate the *relative* redox

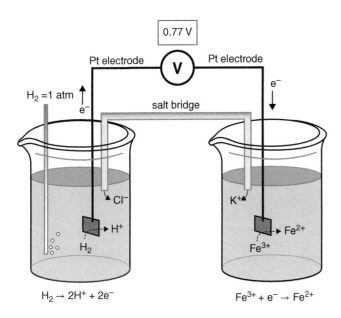

Figure 5.9 Electrochemical cell showing the standard hydrogen electrode and the Fe^{3+}–Fe^{2+} redox couple. The salt bridge maintains electroneutrality in solution by releasing ions. At standard state partial pressure of $H_2 = 1$ atm and (Fe^{3+}) and $(Fe^{2+}) = 1$. The cell is used to measure Eh°.

status of soils over time (e.g., seasons) or space (e.g., a transect across a wetland). Redox potential is pH- and temperature-sensitive, and these parameters should also be measured when measuring Eh.

5.3.4 Relating Eh to pe

Eh and pe are measures of redox potential and electron activity; where activity is synonymous with chemical potential (i.e., potential energy). Thus, one measure can be converted to the other. The relationship between redox potential measured using an electrochemical cell (Eh) and electron activity derived from a redox half reaction is made using the **Nernst equation**. The equation is named after Walther Nernst, a German chemist and physicist who made several notable contributions to the field of electrochemistry. The use of the Nernst equation to derive the mathematical relationship between pe and redox potential is not shown here; the result of the derivation is

$$Eh = \frac{RT}{F} \ln 10 \, pe \tag{5.25}$$

where R is the universal gas constant (8.3145 J mol^{-1} K^{-1}), T is absolute temperature (Kelvin) and F is Faraday's constant ($96\,485$ Coulombs mol^{-1}). The units of Eh are volts. At $25\,°C$ (298.15 K), **Eq. 5.25** simplifies to

$$Eh = 0.059 \, pe \tag{5.26}$$

Calculating theoretical Eh for redox couples and applying them to measured redox from soil has some limitations because of the inaccuracies with measured Eh in soils (discussed above).

Thus far, several concepts important to measuring and predicting redox reactions in soils have been discussed. In the remainder of this chapter, use of redox reactions to understand trends in redox processes in soils is discussed. To aid in organizing the concepts presented thus far, **Table 5.4** summarizes important redox terms.

Table 5.4 Terms and definitions of redox chemistry used in this chapter.

Redox term	Definition	Notes
Reducing half-reaction	A reaction where a chemical accepts an electron	An element gains one or more electrons.
Oxidizing half-reaction	A reaction where a chemical donates an electron	An element loses one or more electrons.
Reductant	Chemical that gives an electron	Chemical is in a reduced state.
Oxidant	Chemical that accepts an electron	Chemical is in an oxidized state.
Redox potential	The status of a system with respect to availability of electrons for transfer between chemical species	Most commonly measured as Eh. Can also be calculated as Eh or pe. High redox potential means chemicals will be oxidized, low redox potential means chemicals will be reduced.
(e$^-$)	Electron activity	A theoretical concept used in half-reaction equilibrium constants.
pe	$-\log$ (e$^-$), where (e$^-$) is the theoretical activity of an electron in solution	Similar mathematical concept as pH; related to redox potential measured by Eh via Nernst equation.
pe$^°$	pe when system is at standard state	At equilibrium, pe $^° = 1/n \ \log \ K_r$, where n = number of e$^-$ transferred.
Eh	Potential difference measured in volts between a hydrogen redox reaction and another redox reaction in a separate cell	Key concept in electrochemistry; measured in volts or millivolts; pe can be calculated from Eh.
K$_r$	Thermodynamic equilibrium constant for a redox reaction	Related to free energy (ΔG) of reaction for redox reaction Calculated from pe$^°$ and Eh$^°$.
Eh$^°$	Standard state redox potential	Eh of a half-reaction in which activities of all species are equal to one and T=25 °C. Related to redox potential measured by Eh via Nernst equation.
Eh$_{pH}$	pH-corrected redox potential	Redox potentials are commonly calculated for neutral (pH = 7) systems, indicated as Eh$_7$. Activities of oxidant and reductant are often included in calculations.
ORP	Oxidation/reduction potential	A general term referring to the redox potential of a system.

5.4   The role of protons in redox reactions

An important aspect of many redox reactions is the transfer of protons. For example, the half-reaction for reduction of Fe^{3+} in the iron oxide mineral goethite is:

$$FeOOH(s) + 3H^+(aq) + e^- = Fe^{2+}(aq) + 2H_2O \qquad (5.27)$$

This reaction is a **reductive-dissolution** reaction showing that for every mole of goethite reduced, three moles of protons are consumed. In nature, most redox reactions include transfer of protons, so redox reactions typically cause changes in soil pH. Oxidation of organic carbon in soils, for example, produces protons (see, e.g., SOM oxidation half reaction in **Eq. 5.9**), as does oxidation of ammoniacal fertilizers:

$$NH_4^+ + 3H_2O = NO_3^- + 10H^+ + 8e^- \qquad (5.28)$$

The *net* nitrification half reaction releases up to 10 H^+ per ammonium molecule oxidized. The actual number of protons released depends on the other *half* of the reaction (electron acceptor). Nitrification is a major soil acidification process in agricultural systems (see Chapter 12).

As a rule, oxidation reactions produce protons and decrease pH. Similarly, reducing reactions consume protons, increasing pH. A good way to remember this is that electrons and protons follow each other in redox reactions (in most cases). Thus, when characterizing or predicting redox processes in soils, simultaneous changes in redox potential and pH must be considered.

5.5   Redox potential limits in natural systems

The theoretical range of redox potentials possible in soils is limited by the stability of water. High redox potentials oxidize water to O_2:

$$H_2O = \tfrac{1}{2}O_2 + 2e^- + 2H^+ \qquad (5.29)$$

Water molecules are always present in natural systems, thus the oxidation of water to O_2 buffers the maximum redox potential. Oxidizing agents stronger than O_2 are unstable in water and are reduced via oxidation of water. Consider, for example, chlorine gas (Cl_2), for which the K_R for the Cl_2–Cl^- couple is greater than the water oxidation couple (**Table 5.3**). If Cl_2 gas is mixed with water, the chlorine takes electrons from oxygen atoms in water producing Cl^- and O_2.

At the reducing end of the redox scale, redox potential is buffered by reduction of H^+ to H_2:

$$2H^+ + 2e^- = H_2 \qquad (5.30)$$

Since H^+ is always present in water (even at high pH), the H^+–H_2 half-reaction is the lower limit of electrode potentials in aqueous systems and soils. As such, any species with a redox potential below the H^+–H_2 couple will theoretically be oxidized and H^+ will be reduced to hydrogen gas. For example, the redox potential for the Fe^0–Fe^{2+} couple is below the H^+–H_2 couple, and thus if Fe^0 is present in an aqueous system it will oxidize to Fe^{2+} by transferring electrons to H^+ and produce hydrogen gas. This is the initial reaction of the iron *rust* process.

The pe–pH relationship for the H_2O–O_2 couple in **Eq. 5.29** is derived from the equilibrium relationship (by convention the reaction is written as a reducing reaction, e.g., reverse of the reaction in **Eq. 5.29**):

$$K_R = \frac{(H_2O)}{(e^-)^2 (O_2)^{1/2} (H^+)^2} = 10^{41.54} \qquad (5.31)$$

This equilibrium expression can be transformed to include the terms pe and pH by taking logarithm of both sides:

$$\log K_R = (H_2O) - 2\log(e^-) - \tfrac{1}{2}\log(O_2) - 2\log(H^+) = 41.54 \qquad (5.32)$$

Substituting for pe and pH and rearranging **Eq. 5.32** becomes

$$pe = 20.77 - pH + \tfrac{1}{4}\log(O_2) \qquad (5.33)$$

where the (O_2) is the activity of oxygen gas that is equal to the partial pressure of oxygen gas. At 0.21 atm (partial pressure of O_2 at sea level), **Eq. 5.33** becomes

$$pe = 20.60 - pH \tag{5.34}$$

Similarly, for the $H^+ - H_2$ couple from reaction in **Eq. 5.30**, the pe–pH relation for hydrogen reduction is

$$pe = \log K_R - pH - 1/2 \log (H_2) \tag{5.35}$$

where $\log K_R$ is equal to zero (**Table 5.3**). The hydrogen gas (H_2) partial pressure can be set at one to represent the most reducing condition possible; thus, **Eq. 5.35** becomes

$$pe = -pH \tag{5.36}$$

The water pe–pH relationships described by **Eqs. 5.33** and **5.36** are plotted in **Figure 5.10**. The Eh axis is also included for reference (calculated using **Eq. 5.26**).

The graph shows the stability field for oxidation and reduction of water. In a system at equilibrium, redox potential will never become more oxidizing than water or more reducing than hydrogen gas. Since water is present in all aqueous systems, redox potentials of natural systems must be between the water stability lines in **Figure 5.10**.

Research done by Baas Becking and colleagues summarized Eh and pH data from 6200 measurements reported in the literature and plotted them on a graph to infer the pe–pH domain that occurs in nature (**Figure 5.11**). The data include microbial-rich environments, waters with extremophiles, natural waters, ocean sediments, bogs, and soils. The upper and lower limits are within the theoretical water limits. The mismatch at the upper limit for the $O_2 - H_2O$ couple is an artifact from the inefficiency of the platinum Eh electrode for measuring electron transfers with oxygen gas. The soils only Eh–pH domain

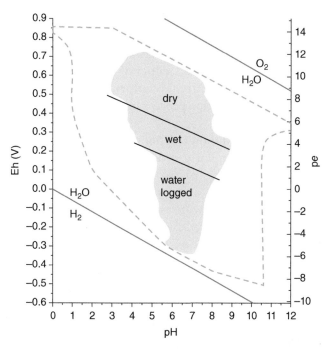

Figure 5.11 Eh–pH diagram showing the domain of 6200 measured Eh and pH from several different environments (dashed line). The subset of soil Eh–pH measurements is also shown (shaded region). The dry, wet, and water-logged conditions represent the environmental conditions at the various sites. Adapted from Baas Becking et al. (1960). Reproduced with permission of The University of Chicago Press.

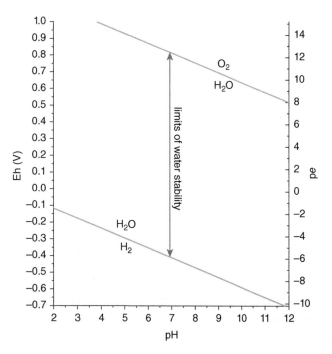

Figure 5.10 Eh–pH diagram showing the stability lines of water.

is also shown in **Figure 5.11**, which is different from the overall domain because of pH buffering that occurs in soils.

5.6   pe–pH diagrams

Because protons are included in most redox reactions, pe–pH phase diagrams (also called stability diagrams) are useful ways to represent stability fields for redox reactions. For example, in the goethite reductive dissolution reaction in **Eq. 5.27**, assuming H_2O and FeOOH have unit activity, the equilibrium constant for the half reaction is

$$K_R = \frac{\left(Fe^{2+}\right)}{\left(e^-\right)\left(H^+\right)^3} = 10^{13} \qquad (5.37)$$

This equilibrium expression can be transformed to include the terms pe and pH by taking logarithm of both sides:

$$\log K_R = \log\left(Fe^{2+}\right) - \log\left(e^-\right) - 3\log\left(H^+\right) = 13 \qquad (5.38)$$

Substituting in pe and pH, and rearranging **Eq. 5.38** becomes

$$pe = 13 - 3pH - \log\left(Fe^{2+}\right) \qquad (5.39)$$

This equation is plotted in **Figure 5.12** for Fe^{2+} activity fixed at 10^{-4}, which is the approximate activity of Fe^{2+} that occurs in many reduced soils. The line represents the pe–pH equilibrium conditions for the Fe^{3+}(goethite)–Fe^{2+} redox couple. If pe and pH of a system are above the line, then Fe^{3+} in goethite is predicted to be the stable species. If pe and pH are below the line, then Fe^{2+} is predicted to be the stable species.

An important use of pe–pH diagrams is comparison of redox stability lines from different redox couples. For example, at higher pH, the H^+–H_2 couple stability line becomes greater than the goethite-Fe^{2+} couple stability line. Using **Eq. 5.36**, the pe–pH relation for

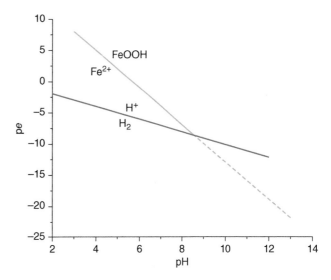

Figure 5.12 pe–pH diagram for goethite (FeOOH) reductive dissolution, with (Fe^{2+}) fixed at 10^{-4} ($T = 25$ °C). At pe and pH values above the line, goethite is stable. Below the line, goethite will reductively dissolve to Fe^{2+} (aq). The dashed region of the line represents the lower limit for pe in aqueous systems as controlled by the H^+–H_2 couple.

hydrogen redox couple can be compared to the goethite stability line. The hydrogen redox couple is the lower limit of the redox potential in an aqueous system so when its equilibrium line is greater than other redox couples, the hydrogen redox couple prevents the other redox couple from being active and the *system* redox potential is buffered. **Figure 5.12** shows that above pH 8.5, the H^+–H_2 redox couple buffers the pe, preventing it from decreasing enough to reduce goethite under the higher pH conditions. The upper H_2O–O_2 couple for the oxidation limit is off the pe scale of this diagram, so not relevant if goethite reduction is the redox process under consideration.

Because most redox reactions are pH dependent, pe–pH diagrams are useful for predicting the speciation of chemicals in a system. However, in natural systems, multiple different redox species exist, and redox equilibrium lines for each species must be considered. Predictions using multiple redox species can best be done using computer programs with appropriate redox reactions and equilibrium constants (K_R) as input.

Special Topic Box 5.1   Influence of iron sulfide mineral oxidation on soil properties

Iron sulfides occur in soils in several mineral types, but most commonly as pyrite (FeS_2). Iron sulfides occur in reduced soils such as wetlands and marshes or are input into soils as pyrite-bearing rock. Pyrite contains reduced ferrous iron ions (Fe^{2+}) and sulfide ions (S^2). Mine tailings often contain pyrite, as well as copper, zinc, cadmium, selenium, and arsenic. Weathering of pyrite in mine-impacted environments causes acid-mine drainage and releases the contaminant metals and metalloids associated with the pyrite minerals.

In oxygenated environments, iron sulfide mineral weathering oxidizes and transforms the iron and sulfur species. The oxidation reactions vary depending on the environment, include numerous electron transfer processes, and include microbiologically mediated reactions. A summary pyrite oxidative reaction is

$$FeS_2 + 2.5H_2O + 3.75O_2 = 2H_2SO_4 + FeOOH \qquad (5.40)$$

This reaction produces goethite mineral and two moles of sulfuric acid for every mole of pyrite oxidized. The sulfuric acid creates the low pH observed in acid mine tailings and in soils where pyrite is oxidized during weathering. Details of the reaction in **Eq. 5.40** are described below.

Step 1. Oxidation of sulfide in pyrite by oxygen (written as half reactions) to produce water and dissolved iron and sulfur:

$$FeS_2 + 8H_2O = 14e^- + 16H^+ + 2SO_4^{2-} + Fe^{2+} \qquad (5.41)$$

$$7/2O_2 + 14H^+ + 14e^- = 7H_2O \qquad (5.42)$$

Step 2. Oxidation of ferrous iron by oxygen (written as half reactions):

$$Fe^{2+} = Fe^{3+} + e^- \qquad (5.43)$$

$$1/4O_2 + H^+ + e^- = 1/2H_2O \qquad (5.44)$$

Step 3.1. Oxidation of pyrite by ferric iron (replace O_2 as TEA in reaction in **Eq. 5.42** with 14 Fe^{3+} electron acceptors and add to reaction in **Eq. 5.41**):

$$FeS_2 + 14Fe^{3+} + 8H_2O = 16H^+ + 2SO_4^{2-} + 15Fe^{3+} \qquad (5.45)$$

Step 3.2. Precipitation of ferric iron minerals (e.g., goethite):

$$Fe^{3+} + 2H_2O = FeOOH(s) + 3H^+ \qquad (5.46)$$

Step 4. Formation of secondary sulfate minerals:

$$3Fe^{3+} + 2SO_4^{2-} + 6H_2O + K^+ = KFe_3(SO_4)_2(OH)_6(s) + 6H^+$$
$$\text{(jarosite)}$$

$$(5.47)$$

$$Ca^{2+} + SO_4^{2-} + 2H_2O = CaSO_4 \cdot 2H_2O \quad \text{(gypsum)} \qquad (5.48)$$

The transfer of 14 electrons in Step 1 can occur abiotically but being a source of energy, microbes often catalyze the oxidation reaction, thus, speeding up the oxidation process. Likewise, in Step 2, microbial oxidation of the ferrous iron is an important process, typically the rate determining step in pyrite oxidation, and catalyzed by chemolithoautotrophic bacteria. Step 3.1 shows ferric iron as an oxidant (electron acceptor) for sulfide in pyrite oxidation (Step 1). The ferrous iron oxidant feedback creates self-oxidizing loop for pyrite in oxidizing environments. Step 3.2 shows an alternate fate of ferric iron where it hydrolyzes and precipitates as an iron oxide mineral (goethite or ferrihydrite); schwertmanite ($Fe_8O_8(OH)_6SO_4$) is another possible mineral phase. Step 4 shows two example sulfate mineral precipitates that may occur in soils where oxidative pyrite weathering has occurred. Jarosite is a distinctive mineral that occurs in low pH soils as bright yellow mineral coatings or masses (**Figure 5.13**). In the later stages of pyrite weathering, or when soil pH is neutralized by adding calcium carbonates, sulfate reacts with calcium to form gypsum.

The redox stability diagram for pyrite oxidation and precipitation is shown in **Figure 5.14**. The diagram shows that the stability field for pyrite occurs mostly in reduced soil conditions. But, as soil pH decreases, the upper limit for pyrite stability increases. An important consideration in interpreting the stability fields in **Figure 5.14** is that they are based on thermodynamic calculations, and do not necessarily include biochemical reaction processes. Abiotic reaction rates for some of the processes can be very slow. Microbes increase pyrite oxidation reaction rates by as much a million times (Taylor, Wheeler, and Nordstrom 1984).

Pyrite oxidation in soils causes low soil pH (<3.5) and formation of several secondary minerals such as iron oxides, jarosite, and gypsum. Soils impacted by iron-sulfide

Figure 5.13 Sulfuric horizon containing yellow jarosite in top horizon and a hypersulfidic horizon (black) below. The soil is located in a drained floodplain soil located along the Finniss River in S. Australia. See **Figure 5.13** in color plate section for color version. Photo courtesy of Rob Fitzpatrick, University of Adelaide; see Fitzpatrick et al. (2018) for site description.

minerals occur in drained tidal lands, wetlands, and soils developed from shale parent materials that have pyrite in them. Soils with pyrite-mineral oxidation occurring may

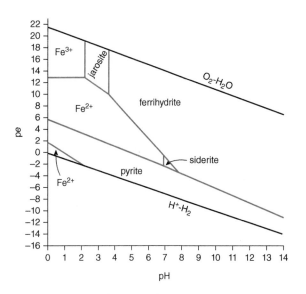

Figure 5.14 pe–pH diagram showing stability fields for ferrous Fe, pyrite (FeS_2), iron oxide mineral ferrihydrite ($Fe(OH)_3$), jarosite ($KFe_3(OH)_6(SO_4)_2$), and siderite ($FeCO_3$). $T = 25\ ^{\circ}C$, (Fe) $= 10^{-2}$, $P_{CO_2} = 10^{-2}$ atm, (S) $= 10^{-2}$, (K^+) $= 10^{-4}$. Source: Langmuir (1997). Reproduced with permission of Prentice-Hall, Inc.

develop sulfuric horizons, identified by the low pH and presence of jarosite minerals (**Figure 5.13**). In mine-impacted environments, pyrite-bearing rocks are brought to the surface and exposed to water and oxygen, creating acid-mine drainage problems that require remediation. Lime addition can be used to neutralize the acidity of pyrite oxidation in soils.

5.7 Prediction of oxidation and reduction reactions in soils

To compare the relative order that redox couples will oxidize and reduce, redox potentials are often placed on a linear scale, called a **redox ladder** for a given set of conditions. **Figure 5.15** shows the redox ladder for the half-reactions listed in **Table 5.5** together with their relative pe and Eh values. Eh is modified with the subscript (Eh_7) to indicate the value is for a system at pH 7. The Eh_7 values in **Table 5.5** were calculated by converting predicted pe values using **Eq. 5.26** that were calculated from the equilibrium constants for the listed redox half reactions together with activities of reactants typical for soils at pH 7. Thus, the values

include numerous assumptions on soil chemical conditions, but nevertheless are useful reference points to understand how redox potential influences transformation of species by electron transfer processes.

One use of a redox ladder is to compare relative positions of redox couples to predict if one redox couple can oxidize or reduce another couple. For example, the O_2–H_2O couple is above the other couples in **Figure 5.15**. Thus, any O_2 present in a system will be the preferred electron acceptor, and reduced species of couples lower down on the ladder should oxidize giving their electrons to oxygen to create water. As an oxidant is depleted, the oxidized species in the redox couple below it on the redox ladder becomes the dominant electron acceptor until it is depleted, and so on. So long

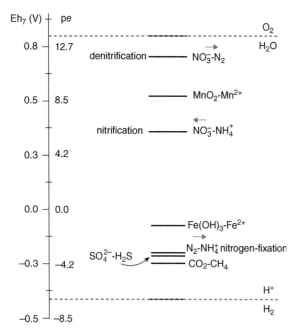

Figure 5.15 Redox ladder for several soil processes at pH 7. Redox potentials for the couples are listed in **Table 5.5**. Right arrows on nitrogen reactions indicate irreversible reducing reactions; left arrows indicate irreversible oxidizing reactions.

Table 5.5 Theoretical reduction half-reaction redox potentials at pH 7 for several common soil redox couples. Reactions are net reactions; for many of the reactions, actual processes include multiple intermediate redox couples.

Reaction	Eh_7 (V)*	Assumptions
O_2 disappearance $O_2 + 4e^- + 4H^+ = 2H_2O$	0.8	$P_{O2} = 0.21$ atm pH = 7
Aqueous Fe^{3+} reduction $Fe^{3+} + e^- = Fe^{2+}$	0.77	(Fe^{3+}) and $(Fe^{2+}) = 10^{-4}$
NO_3^- disappearance via denitrification $NO_3^- + 6H^+ + 5e^- = \frac{1}{2}N_2 + 3H_2O$	0.70	$(NO_3^-) = 10^{-4}$ $P_{N2} = 0.78$ atm pH = 7
MnO_2 reduction to Mn^{2+} formation $MnO_2 + 2e^- + 4H^+ = Mn^{2+} + 2H_2O$	0.52	$(MnO_2) = 1$ $(Mn^{2+}) = 10^{-4}$ pH = 7
NH_4^+ loss via nitrification $NO_3^- + 8e^- + 10H^+ = NH_4^+ + 3H_2O$	0.36	$(NO_3^-) = 10^{-4}$ $(NH_4^+) = 10^{-4}$ pH = 7
$Fe(OH)_3$ reduction to Fe^{2+} $Fe(OH)_3 + e^- + 3H^+ = Fe^{2+} + 2H_2O$	−0.071	$(Fe(OH)_3) = 1$ $(Fe^{2+}) = 10^{-4}$ pH = 7
NH_4^+ production via N_2 fixation $N_2 + 8H^+ + 6e^- = 2NH_4^+$	−0.19	$(NH_4^+) = 10^{-4}$ $P_{N2} = 0.78$ atm pH = 7
H_2S formation $SO_4^{2-} + 10H^+ + 8e^- = H_2S + 4H_2O$	−0.21	$(SO_4^{2-}) = 10^{-4}$ $P_{H2S} = 10^{-4}$ atm pH = 7
Goethite reduction to Fe^{2+} $FeOOH + e^- + 3H^+ = Fe^{2+} + 2H_2O$	−0.23	$(FeOOH) = 1$ $(Fe^{2+}) = 10^{-4}$ pH = 7
CH_4 production $CO_2 + 8H^+ + 8e^- = CH_4 + 2H_2O$	−0.24	$P_{CO2} = 3.2 \times 10^{-4}$ $P_{CH4} = 10^{-4}$ atm pH = 7
H_2 formation $2H^+ + 2e^- = H_2$	−0.41	$P_{H2} = 1$ atm pH = 7

*Values from James and Bartlett (1999).

as there is an oxidant present, the system's redox potential is maintained (**poised**) at the redox potential for that couple. Once the oxidant poising the system is depleted, the redox potential decreases to the next oxidant couple below it on the ladder.

In laboratory experiments, where the redox conditions are carefully controlled, and incubation time is long enough for the system to come to equilibrium, theoretical redox trends predicted in **Figure 5.15** generally agree with observed redox trends. The stability and progress of redox reactions in nature, however, depend on the kinetics of the reaction and microbial activities, which cannot be inferred from a redox ladder because redox ladders are based on thermodynamic equilibrium. In addition, many redox reactions (especially nitrogen redox reactions) are not reversible, so the redox sequence shown on the ladder may not occur in the order shown. Despite these limitations, viewing redox couples on a redox ladder is instructive as an indication of the relative sequence of redox reactions. For example, the position of the NO_3^-/N_2 couple indicates that denitrification occurs under anoxic conditions, and before Mn(III, IV)-oxide minerals are reduced. This behavior is observed in wetlands and rice paddies.

The redox couples between the stability limits of water in **Figure 5.15** can exist as both oxidized and reduced species in the environment, depending on the redox potential of the system. Under anoxic conditions, Fe^{2+}, S^{2-}, CH_4, and NH_4^+ can occur as stable species. If O_2 is available, Fe(III), SO_4^{2-}, CO_2, and NO_3^- are stable. For some elements, metastable intermediates may occur. For example, NO_2^-, N_2O, S^0, S^{4+} are intermediate forms that are not shown in **Figure 5.15**. Some of these metastable species can persist for very long times.

5.7.1 Reduction reactions on the redox ladder

Redox ladders are often used to conceptualize the order that microorganisms will use electron acceptors in respiration (TEAs). The higher the redox couple on the redox ladder, the more energy that can be derived by using the oxidized form of the couple as a TEA. Thus, the most competitive microorganisms are those that can utilize the strongest electron acceptors available, which in oxic soils is O_2. If the O_2 supply is insufficient,

the next strongest electron acceptor available is the NO_3^-–N_2 couple (denitrification). The NO_3^-–N_2 couple is a theoretical redox couple because it summarizes several steps in the denitrification process that have different redox potentials (see Section 5.10).

After all of the available oxygen and nitrate have been reduced, Fe(III) and Mn(IV,III) oxides are reduced, causing increases in the Fe^{2+} and Mn^{2+} concentrations in the soil solution. Mn(IV, III) and Fe(III) oxides are solids that cannot diffuse to the microbes, and thus biological reduction may be rate-limited. The redox potential of goethite in the ladder is lower than the poorly crystalline iron oxide ferrihydrite because goethite has less surface area and the overall activity of the mineral-bound iron is less. Thus, structural ferric iron ions in goethite do not accept and donate electrons as readily as the ferric iron ions in ferrihydrite, and goethite may persist in some soils even when the redox potential is low enough to reduce the structural goethite iron ions. As highly energetic electron acceptors like oxygen, nitrate, Mn(III, IV), and Fe(III) become more scarce, strong reducing conditions occur. Under these conditions, microbes can reduce sulfate to sulfur or sulfide, ferment organic matter to methane, and in extreme cases, reduce water to H_2.

As soils become reducing, changes in element oxidation states cause the concentrations of species in the soil solution or gases in the soil atmosphere to change. An idealized redox-solubility sequence for many common electron acceptors in soils is shown in **Figure 5.16**, where the *x-axis* is the incubation time corresponding to onset of an event that decreases availability of electron acceptors, such as occurs during flooding. If a reduced soil dries out, and becomes oxidizing, theoretically the concentrations would reverse. Because of the irreversibility and kinetics of redox reactions, however, and involvement of specific microbes, wetting and drying redox couple sequences are often not reversible.

The stepwise order of reduction in **Table 5.5** and **Figure 5.15** is idealized. In an oxygen-limited environment, anaerobe respiration is usually much faster than alternative electron acceptor availability. This is mostly due to the spatial discrepancy between electron donors and acceptors; thus, soil redox processes are diffusion limited. As a result, utilization of alternative TEA in soils rarely follow an idealized

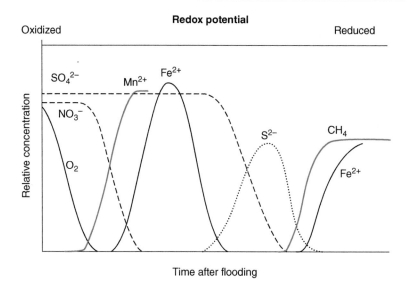

Figure 5.16 Theoretical concentrations of O_2, NO_3^-, SO_4^{2-}, Mn^{2+}, H_2S, and CH_4 in a flooded soil as a function of time after flooding; assuming there is a readily available food source. As O_2 is depleted, NO_3^- is consumed in a denitrifying reaction, and nitrate concentrations decrease. Subsequently, manganese and iron oxides are used as TEA producing aqueous Mn^{2+} and Fe^{2+}. As redox potential decreases further, SO_4^{2-} is used as a TEA, and H_2S is produced, which reacts with Fe^{2+} to produce insoluble FeS minerals, decreasing concentration of dissolved Fe^{2+}. Continued decrease in redox potential results in CO_2 being reduced to CH_4. When S^{2-} is completely used or lost from the system, dissolved Fe^{2+} concentrations increase because FeS is no longer a sink for Fe^{2+}. Adapted from Kehew (2001).

sequence, and instead appears to overlap, thus *smearing* the theoretical redox boundaries shown in **Figure 5.15**. This artifact is not because the redox theory is inaccurate but because of the limitations of measuring redox in soils. Accordingly, arrays of redox probes are used in soils to obtain an average redox potential, but the procedure ignores the precise processes occurring in the soils that vary on a millimeter scale. The best approach is to use multiple methods to evaluate redox processes in soils, including measurement of soil physical characteristics.

5.7.2   Oxidation reactions on the redox ladder

When a reduced soil system becomes oxidized, many oxidation reactions occur abiotically because the reduced species are unstable in the presence of oxygen. Oxidation of reduced compounds are important processes determining the fate of (NH_4^+), Mn(II), Fe(II), Se(IV), Se(II), As(III), and organic carbon compounds. The reduced elements are sometimes suddenly introduced into oxidizing environments from

changes in the water table in soils, or as fluxes from a reduced region of soil. The reduced elements can also enter soils via various land-applied wastes, sediment deposition, or fertilizer application.

An important example of reduced element oxidation processes in soils occurs in wetland soils where reductive dissolution of iron oxides solubilizes Fe(II) and the soluble Fe(II) moves with water toward the surface, driven by evapotranspiration. When the mobilized Fe(II) reaches a zone in the surface of the soil that is more oxidized than the lower profile, the Fe(II) oxidizes, and precipitates as ferric iron oxide. The result is an iron-oxide enriched horizon in the upper soil profile, and an iron-depleted horizon in the lower profile. **Figure 5.17** shows a soil that has an iron-enriched horizon caused by this process. At the centimeter-to millimeter-scale in soils, reduction and oxidation of iron and manganese create redox concentrations and depletions called root linings, nodules, and concretions.

Redox couples lower on the redox ladder can be used as electron sources by chemotrophic organisms. The lower on the redox ladder, the more energy yielded by oxidizing the reduced chemical. Some soil

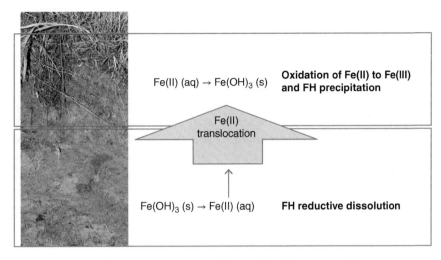

Figure 5.17 Soil redox processes that create iron enrichment of the surface horizon are shown next to a wetland soil profile. FH indicates ferrihydrite, which has a hypothetical mineral formula Fe(OH)$_3$ in the reactions shown. (See **Figure 5.17** in color plate section for color version.)

microorganisms can oxidize methane or hydrogen gas as sources of energy. Russian microbiologist Sergei Winogradsky discovered that in oxic environments, microbes could oxidize ammonia to form a series of nitrogen oxide species and pass off the electrons to O$_2$ to gain energy. The final product of ammonia oxidation is nitrate, and the process is called nitrification:

$$NH_4^+ + 2O_2 = NO_3^- + H_2O + 2H^+ \qquad (5.49)$$

Some heterotrophic organisms in soils also oxidize ammonia (anammox); probably to dissipate low redox potentials, although the exact reason is unclear.

In flooded soils, organic matter oxidation is slow, thus creating a buildup of SOM. The apparent resistance to degradation is not because the organic matter is thermodynamically stable in the reduced conditions; the redox equilibrium for most organic molecules is below hydrogen, and thus SOM is thermodynamically not stable in even the most reduced flooded soils. Rather, the persistence of SOM in flooded soils is because of the extremely slow abiotic electron transfer reactions between reduced carbon and electron acceptors other than oxygen. In aerobic environments, degradation by heterotrophic aerobes speeds up the degradation rate of organic matter. The slow organic matter degradation in flooded soils is the reason for the buildup of large quantities of SOM that occurs in wetlands soils, such as Histosols.

5.8 Redox measurement in soils

Measured redox potentials for redox couples often differ from theoretically predicted ones like those listed in **Table 5.5**. Reasons for the discrepancies have been discussed above (kinetics, irreversibility, redox probe reactivity, microbial population, etc.). In some cases, such as in highly reduced soils, a steady state may be reached approximating equilibrium, thus making measured Eh an acceptable measure of the overall redox status of the system.

Under oxic conditions, measured redox potentials often deviate from the potentials of soil redox couples. In anoxic soils, redox potentials may be more quantitatively related to ion activities because Fe^{2+} and Mn^{2+} concentrations are high, and both these ions are reactive on Pt electrodes. The range of redox potentials that have been measured in soils is shown in **Figure 5.11**. The envelope drawn around soils data (shaded region) was considered by the investigators to be the limits of redox potentials and pH values in soils.

5.8.1 Other methods to assess redox status of soils

Because of the problems associated with accuracy of redox probes in soil measurements, researchers have developed other methods to quantify redox potentials in the environment, including:

1 Measure concentrations of a redox couple and use the Nernst equation and standard electrode potential to calculate Eh.[1]
2 Measure the concentrations of predominant *alternative* electron acceptors or donors, multiply them by the number of electrons involved in their redox couple half-reaction, and sum the results (**oxidative capacity (OXC)**).[2]
3 Use an iron oxide–based paint on a plastic pipe and monitor removal of the paint after inserted into the soil.[3]
4 Assessment of anaerobic microbe activity in soils (e.g., Nitrosomonas, Desulfuromonas), or presence of enzymes involved in secondary TEA processes.

While all four methods are useful, none is ideal. The alternative methods offer insight into the redox potential of a system but should be used in conjunction with a redox probe to increase the confidence in soil redox potential.

5.9 Soil redoximorphic features and iron reduction in wetlands

Saturation or flooding of soils leads to anaerobic conditions that create distinct soil properties important for determining best management practices. Soil properties indicative of water saturation include redoximorphic features created by the oxidation and reduction of iron and manganese in the soil. Because oxygen status within soils is variable over time and space, reduced soluble iron and manganese move from reduced zones in the soil to more oxidizing zones, creating

redoximorphic features such as iron masses, concentrations, depletions, or coatings; and Mn/Fe-cemented soil particles called concretions, nodules, and pore linings. Redoximorphic features occur as mottles, or entire horizons may become enriched or depleted in iron (see for example **Figure 1.14** and **Figure 5.17**). Soil that is depleted in iron is typically gray. However, a soil that has high concentrations of Fe(II) may also appear gray because reduced Fe(II) compounds have a light bluish tint as opposed to the red color of Fe(III) compounds. A soil with abundant Fe(II) will turn red or orange when exposed to air because the Fe(II) oxidizes to Fe(III) and out of solution. Alpha-alpha-Dipyridyl is a colorimetric chelate that turns reddish-pink when applied to soils containing Fe(II).

Redoximorphic features in a soil can serve as *indicators* that a high-water table affects the soil. This is particularly important for determining agricultural uses, establishment of wildlife sanctuaries and wetlands (hydric soil delineation) and assessing soil drainage characteristics for septic or building purposes. Iron reduction is the most important factor in development of redoximorphic features, and redox potentials for Fe(III)–Fe(II) redox couples can complement Eh values in determining the redox potential at which iron reduction will occur in a soil. A common iron solid species in wetland soils is ferrihydrite (modeled as $Fe(OH)_3$); thus, the ferrihydrite-Fe^{2+} redox couple is a good system to evaluate soil redox potential with respect to iron reduction. The ferrihydrite-Fe^{2+} reduction reaction is

$$Fe(OH)_3 + e^- + 3H^+ = Fe^{2+} + 2H_2O \qquad (5.50)$$

The equilibrium expression for this reaction is

$$K_R = \frac{(Fe^{2+})}{(e^-)(H^+)^3} = 10^{16.2} \qquad (5.51)$$

This equilibrium expression can be transformed to include the terms pe and pH by taking logarithm of both sides:

$$\log K_R = \log(Fe^{2+}) - \log(e^-) - 3\log(H^+) = 16.2 \qquad (5.52)$$

Substituting for pe and pH, and rearranging, **Eq. 5.52** becomes

$$pe = 16.2 - 3pH - \log(Fe^{2+}) \qquad (5.53)$$

---

[1] *Note:* Theoretically accurate, but it is often difficult to accurately measure activities (concentrations) of both redox couples. It is also difficult to a priori predict the relevant redox couple to measure to use in the Nernst equation. The equilibrium assumption of the Nernst equation makes this measurement tenuous.

[2] *Note:* Requires measurement of concentrations of relevant TEA, which can be difficult for some elements.

[3] *Note:* Called IRIS, which stands for iron reduction in soils. IRIS tubes are an indicator of the ferrihydrite-Fe^{2+} couple, which is an important redox reaction in soils, and thus IRIS tubes are good relative indicators of redox status. Manganese oxide-based paints are also useful redox indicators.

The ferrihydrite reduction pe–pH relationship is plotted as a stability line in **Figure 5.18** (pe was converted to Eh using Eq. **5.26**). To assess if reduction of iron in a soil is occurring, soil pH and Eh are measured and compared to the ferrihydrite-Fe^{2+} redox couple stability line. If the measured soil Eh and pH values are below the line, iron should exist as reduced soluble Fe(II) species. Soil Eh and pH values plotted above the line suggest that iron in the soil should exists as Fe(III) in ferrihydrite.

The wet soils region plotted in **Figure 5.18** shows that in many *wet* soils, redox potentials are above that for the ferrihydrite-Fe^{2+} redox couple; shaded dark blue in **Figure 5.18**. Thus, under these anoxic conditions (where iron reduction should occur) the ferrihydrite-Fe^{2+} couple may not be the appropriate redox couple to predict iron reduction. Ferrihydrite reduction is only one model reaction for iron reduction in soils that can be used for assessing redox status. Ferric iron redox

couples with higher redox potentials than the ferrihydrite-Fe^{2+} redox couple include iron hydrolysis complexes (e.g., $Fe(OH)^{2+} + e^- + H^+ = Fe^{2+} + H_2O$) and organic Fe(III)–$Fe^{2+}$ redox couples, such as soil organic-matter complexed Fe(III). Determining an accurate Fe(III)–Fe(II) redox couple in natural systems is difficult because the correct iron species to model are not well known, and the system is usually not at equilibrium. In addition, microbial-facilitated reactions are involved in most iron transformation reactions, and the abiotic iron redox reactions used to model stability fields may not account for the iron biochemistry occurring in the system.

Given the limitations of the theoretical redox couples, empirical limits on redox potential have been established. These relationships are based on observed Eh–pH conditions and are not associated with a specific redox couple. **Figure 5.19** shows such a relationship. Soils with Eh–pH values below the line are deemed to be reducing and thus likely influenced by

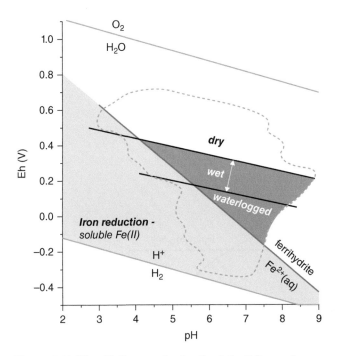

Figure 5.18 Eh–pH diagram for ferrihydrite-Fe^{2+} couple assuming Fe^{2+} activity is 10^{-4}. The domain of observed soil Eh–pH values from Baas Becking (1960) (**Figure 5.11**) is also shown. Since ferrihydrite is a common iron oxide species in saturated soils such as occur in wetlands, the Fe^{2+}-ferrihydrite redox line indicates the limits of Eh and pH conditions where soil iron is reduced and redoximorphic features should occur. The dark-blue shaded region highlights the region of Eh–pH domain observed in *wet* soils that are above the ferrihydrite-Fe^{2+} redox stability line.

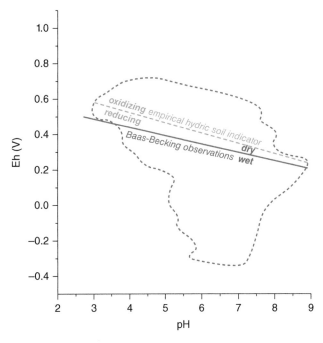

Figure 5.19 Two empirical cutoff lines for distinguishing soil redox status based on Eh–pH measurements. The empirical hydric soil indicator is used by the US Natural Resources Conservation Service to relate soil redox potential measurements to hydric soil classification. The line separating observed redox potentials of wet and dry soils is based on observed soil conditions, as reported by Baas Becking (see **Figure 5.11**).

high-water tables and become anoxic (hydric). Soils with values above the empirical line are not deemed hydric based on redox potential measurements. Even if the redox potential is not below the empirical redox line, other factors may qualify the soil as having hydric conditions. The empirical line is similar to the *dry soil–wet soil* Eh–pH boundary conditions observed by Baas Becking, which is provided in **Figure 5.19** for comparison. This boundary line is drawn from actual observations of soil conditions and Eh–pH measurements and supports the empirical hydric soil indicator line as being close to observed field conditions.

5.10   Nitrogen redox reactions in soils

Within soil, nitrogen simultaneously exists as gas, liquid, and solid states, with multiple species occupying each state. For example, solid nitrogen species include ammonium adsorbed on mineral surfaces, organic nitrogen compounds (e.g., proteins), amine functional groups in soil organic matter, and nitrate salts that may occur in arid soils. Redox reactions are one of the most important processes for controlling nitrogen species.

Table 5.6 shows the oxidation states of the different nitrogen species. **Figure 5.20** shows the different nitrogen species and the general redox transformation reactions that occur in nature. There are many other nitrogen redox reactions that occur in nature that are less well studied than those listed in **Figure 5.20**. Transformation of nitrogen in soils via redox reactions

Table 5.6 Nitrogen molecules, names, oxidation states, and common phase of molecule in most environments (gas, aqueous, solid).

Molecule	Name, common state (gas, aqueous, solid)	Oxidation state
HNO_3, NO_3^-	Nitric acid, nitrate (aq)	+5
NO_2	Nitrogen dioxide (g)	+4
HNO_2, NO_2^-	Nitrous acid, nitrite (aq)	+3
NO	Nitric oxide (g)	+2
N_2O	Nitrous oxide (g)	+1
N_2	Dinitrogen (elemental nitrogen gas)	0
NH_2OH	Hydroxylamine	−1
N_2H_4	Hydrazine	−2
NH_3, NH_4^+	Ammonia, ammonium (g, aq, s)	−3
$C-NH_2$	Organic nitrogen (e.g., amine) (s)	−3 (varies)

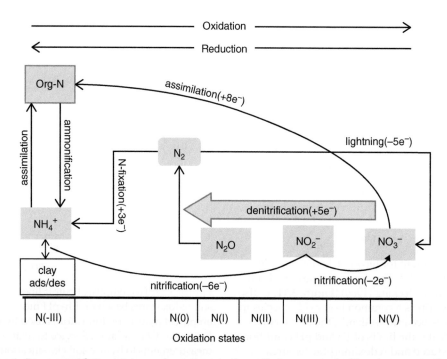

Figure 5.20 Common nitrogen oxidation states and reactions in the soil nitrogen cycle. The electrons in the parentheses indicate the number of electrons required to be added or subtracted from the reactant nitrogen species to produce the corresponding product.

is highly varied, depending on the soil properties, management factors, and environmental conditions (e.g., temperature, microbial population, and water content). Nitrogen is an important agricultural fertilizer, and it is estimated that as much as 50% of the applied nitrogen is not utilized by plants; transformation of the fertilizer to different species by redox reactions are processes partially responsible for nitrogen fertilizer use inefficiency.

Redox reactions of nitrogen in soils are dominated by biological processes. They include reactions to make the nitrogen available for biological uptake, and reactions that are part of an organism's metabolism; i.e., electron sources or terminal electron acceptors. The major nitrogen redox reactions that occur in soils are nitrification, denitrification, nitrogen fixation, ammonification, and assimilation (**Figure 5.20**). The Eh_7 values for these reactions are listed in **Figure 5.15** and **Table 5.5**.

5.10.1 Nitrogen assimilation

Assimilation redox reactions describe the uptake of nitrate or ammonium into a plant or microbe and incorporation into proteins (amino acids), nucleic acids, and other biomolecules within the organism. Absorbed nitrate is quickly reduced to N(-III) oxidation states within organisms, so nearly all nitrogen biomolecules occur as reduced N(-III) in amine function group (R_{3-x}-NH_x) or amide (R-CO-NH_2) groups.

5.10.2 Ammonification

Ammonification is a mineralization reaction where the organic nitrogen is converted to NH_4^+. It is a degradation reaction done by heterotrophic bacteria and fungi, who use the organic carbon associated with the nitrogen compounds. There are several intermediates in the breakdown of the organic nitrogen compounds; for example, protein is first broken down to amino acids, and then to amine and urea compounds, which are subsequently broken down to ammonia that hydrolyzes to form ammonium. A summary reaction for ammonification is

$$Protein(NH) + H_2O = NH_4^+ + OH^- + CO_2 \quad (5.54)$$

Soil factors that affect ammonification include temperature, water availability, and the C/N ratio in the organic compound being degraded. The fate of the ammonium is either assimilation, loss as ammonia volatilization, conversion to NO_3 (nitrification) and leached out of the soil, or adsorption on a clay particle. If soils contain illite or vermiculite clays with high tetrahedral layer charge, ammonium will become fixed in the clay by the collapsed interlayer.

5.10.3 Nitrification

Nitrification is a two-step process that involves chemoautotrophic bacteria that use the electrons in ammonium as an energy source and CO_2 as a carbon source. In the first step, Nitrosomonas oxidizes NH_4^+ to nitrite (NO_2^-):

$$NH_4^+ + 2H_2O = NO_2^- + 8H^+ + 6e^- \quad (5.55)$$

The second oxidation reaction, done by Nitrobacter, quickly oxidizes the nitrite to nitrate:

$$NO_2^- + H_2O = NO_3^- + 2H^+ + 2e^- \quad (5.56)$$

The reaction in **Eq. 5.56** is fast and prevents poisonous nitrite concentrations from increasing too much in the soil; except in the spring when the soil temperature is low and oxidation of nitrite is slow. The net change in oxidation state in nitrification is loss of eight electrons by ammonium. The reaction occurs in oxidizing environments, and thus oxygen is the ultimate reducing half-reaction. The summary reaction for nitrification is

$$NH_4^+ + 2O_2 = NO_3^- + H_2O + 2H^+ \quad (5.57)$$

The overall nitrification reaction is rapid in soils when conditions are optimal. Critical factors for nitrification include availability of inorganic carbon (bicarbonate and CO_2), temperature, and types of clay minerals present.

The protons produced in nitrification reactions are a major source of acidity to soils; especially when ammoniacal fertilizers are added (see Chapter 12). In addition to Nitrobacter and Nitrosomonas, other microbes have been identified as capable of oxidizing ammonium. The nitrate produced is either leached from the soil, taken up by an organism, or undergoes denitrification and is lost as gaseous nitrogen.

5.10.4 Denitrification

Denitrification converts nitrate to nitrogen gas when oxygen is limiting, and anaerobic microorganisms use nitrate and the various intermediates as terminal electron acceptors. The intermediates are nitrite (NO_2^-) and nitrous oxide (N_2O). Nitric oxide (NO) may be produced abiotically. There are many different types of nitrate reduction reactions carried out by several different anaerobic microorganisms. If complete denitrification occurs, the final reduced nitrogen species is nitrogen gas (N_2). A summary reducing half-reaction for denitrification is

$$NO_3^- + 6H^+ + 5e^- = \frac{1}{2}N_2(g) + 3H_2O \qquad (5.58)$$

Losses of nitrous oxide during denitrification are a major concern because it is a potent greenhouse gas. The half-reaction for nitrous oxide production from nitrate is

$$2NO_3^- + 10H^+ + 8e^- = N_2O(g) + 5H_2O \qquad (5.59)$$

Nitrous oxide may also be produced directly from nitrite (NO_2^-), depending on the organisms present and the conditions of the soil. The complete reduction of NO_3 to N_2 yields the most energy, which occurs in multiple biochemical steps that go to completion so long as intermediate gases are not lost from the soil.

Organic compounds in soils are the source of electrons for denitrification. A key soil condition for denitrification is low oxygen availability, which occurs when soils are saturated, or in the interior of aggregates that have slow oxygen diffusion (replenishment). Nitrogen gas losses as N_2 and N_2O are common, while NO gas loss is less prevalent because of its low concentration buildup. **Figure 5.21** shows the relative concentrations of the different nitrogen species in a system that goes from oxic to anoxic. As the soil becomes more reducing, nitrogen progressively reduces to the various nitrogen species shown;

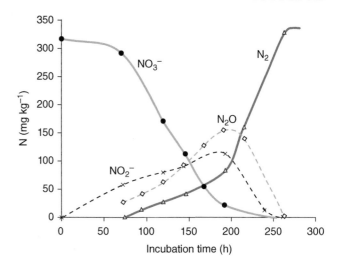

Figure 5.21 Amount of nitrogen in soil incubated anaerobically as a function of time. As the soil becomes more reduced, nitrite, nitrous oxide, and nitrogen gas are produced. Data from Leffelaar and Wessel (1988).

after 250 days, all of the nitrogen in the soil is converted to N_2 gas.

5.10.5 Biological nitrogen fixation

Fixation of N_2 gas is a source of nitrogen for some plants. Nitrogen fixation is carried out by N-fixing bacteria containing the nitrogenase enzyme. Most of the bacteria are symbiotic in the roots of plants, especially legumes. The product of nitrogen fixation is ammonium. A net half-reaction is

$$N_2(g) + 8H^+ + 6e^- = 2NH_4^+ \qquad (5.60)$$

The nitrogenase enzymes are actually two enzymes; one contains iron and the other molybdenum as their reactive centers. The breaking of the triple bond in the nitrogen gas molecule requires a great deal of energy. The N-fixing microbes are typically associated with plants that supply energy in exchange for the ammonium nutrient. The nitrogen reduction reaction is halted when oxygen is present; thus, microbes produce locally anaerobic conditions. Microbes further protect the nitrogenases from oxygen in the root zone by producing oxygen scavenger proteins called leghemoglobin, which is a plant analogue to the hemoglobin animals use for O_2 transport.

5.10.6 Anammox and dissimilatory nitrogen reduction to ammonium

Anaerobic ammonium oxidation (anammox) is a process that is a predominant N_2 production process in anoxic marine sediment environments. It is responsible for up to 50% of the dinitrogen produced from oceans. Unlike other prominent processes in the nitrogen cycle that have been studied since the late nineteenth century, the anammox process is a more recent discovery. The overall reaction is summarized as an NH_4^+–NO_2^- redox couple:

$$NH_4^+ + NO_2^- = N_2 + 2H_2O \tag{5.61}$$

where the nitrite is accepting electrons from the ammonium. The nitrite can be produced as an intermediate species in nitrification processes when an anaerobic–aerobic interfacial environment occurs, or a denitrification intermediate in an anaerobic environment. Use of the nitrite as an electron acceptor instead of oxygen is not energetically favorable and requires hydrazine (a component of rocket fuel) as an intermediate electron shuttle species. Thus, the anammox process is slow.

Anammox has been observed in soils, including wetlands, rice paddies and even some nonflooded forest soils. However, it is typically responsible for less than 15% of the total N_2 production in soils.

Dissimilatory nitrate reduction is similar to denitrification, except the nitrogen is reduced all the way to ammonia rather than dinitrogen (N_2). Ecologically it is advantageous because it conserves nitrogen rather than losing it as N_2 gas flux from the soil, such as may occur at the end of typical denitrification. It is only operative in anoxic environments like sediments and wetlands; especially anoxic environments that have very high microbial activities (warm temperatures and abundant organic carbon), such as coastal wetlands.

5.10.7 Limitations to theoretical nitrogen redox reaction predictions

Nitrogen redox reactions are biologically driven, and thus include enzymes and cellular processes. The reactions listed in **Eqs. 5.54–5.61** are misleading to the extent that they do not include many of the intermediate electron transfer processes. Although the reactions are summary reactions, thermodynamic processes still apply. Thus, redox potentials for the half-reactions can be calculated and redox diagrams or redox ladders can be developed to relate the environmental conditions that favor the various reactions (such as shown in **Figure 5.15**). While this thermodynamic approach is instructive for understanding the energetics involved in the different redox processes, actual modeling and predictions from such reactions is tenuous. Instead, researchers typically base predictions of nitrogen cycling on the type of nitrogen fertilizer added to soil (e.g., organic, ammoniacal, or nitrate fertilizers), temperature, organic carbon available (C/N ratio), soil pH, organisms present, and aerobic or anaerobic state of the soil; the latter can be evaluated with the redox probe.

5.11 Summary of important concepts in soil redox reactions

The concepts of electrochemistry are elegant applications of physical chemistry that support measurements and predictions of redox chemistry. **Table 5.4** summarizes many of the terms used in redox chemistry. A point of confusion by many students is the mixed use of Eh, pe, and K_R, which are key concepts used in electrochemistry. These and some other important concepts in redox chemistry are summarized here:

1 Redox potential is a quantitative measure of the degree to which a system will oxidize or reduce chemicals.
2 Redox potential can be quantified using an electrode to measure Eh in units of volts.
3 Theoretical redox potentials can be calculated for redox reactions as pe.
4 Equilibrium constants for redox reactions (K_R) can be found in tables.
5 Eh°, pe°, and K_R are linearly related by theoretical relations, and can be interconverted.
6 Most redox reactions in soils and waters involve transfer of protons (protons typically follow electrons).
7 Thermodynamically based redox potentials of half-reactions can be compared to determine the oxidation and reduction sequence for redox couples.
8 The limits of redox potential in natural environments are constrained by reduction and oxidation of water.

Redox is one of the most important processes controlling chemical speciation and mobility in the environment and is thus an important topic in managing natural systems. Despite the many limitations in application of redox theory to natural systems, prediction of redox processes is possible in at least some cases. As capabilities for modeling natural processes improve and a better understanding of redox reactions and species is gained, the application of redox chemistry to natural environments will undoubtedly improve.

Questions

1 Why do redox potentials measured in soils differ from theoretical redox potentials?

2 What are some soil properties impacted by redox processes?

3 What are some of the major *drivers* for redox processes in soils?

4 Write the balanced reactions for the oxidation of SOM (**Eq. 5.9**) by oxygen.

5 Derive a half-reaction for the reductive dissolution of MnO_2 to Mn^{2+}. Derive an equation to predict this reaction as a function of pH using an equilibrium approach (log K_R= 43.6). What is the Eh_7 for this reaction? *Hint:* First calculate pe at pH 7 (assume (Mn^{2+}) = 10^{-4}) and then convert to Eh. How does it compare to iron mineral reduction and nitrate reduction on the redox ladder (i.e., describe which redox couple would reduce first in a reducing environment, and which would oxidize first in an oxidizing environment)?

6 Assuming equilibrium, what is the Eh_7 value of the O_2–H_2O couple at P_{O2} = 0.21 (normal air), 0.20 (typical soil air), and 0.01 (possible P_{O2} in an anaerobic soil aggregate)? *Hint:* First calculate pe at pH 7 and then convert to Eh.

7 Why do the redox potentials differ from the electrode potential?

8 Describe the likely chemical changes in a soil as it becomes increasingly anaerobic.

9 Why are quantitative predictions of nitrogen redox reaction using thermodynamic redox equilibrium often inaccurate?

10 Describe the environmental conditions that promote nitrification and denitrification.

11 Why is temperature important for redox reactions?

12 Why is a standard hydrogen half-cell used in a redox electrode? Why is it more practical to use a calomel reference electrode?

13 What are some of the limitations of the redox probe for soil redox measurements? What are some alternatives for characterizing soil redox?

14 Why should pH and temperature always be measured when measuring redox potential of a soil?

15 Describe the soil factors that prevent redox reactions from coming to equilibrium.

16 Characterization of wetlands uses several soil and site parameters. Soil redox potential is one parameter that is used on a relative basis. Why are measured redox potentials only good as "relative" indicators of redox status of wetlands? What other parameters might be useful for characterizing a soil impacted by a high-water table?

17 How does redox affect arsenic and phosphate distribution in soils?

18 Compare the redox reactions and species of arsenic, selenium, and sulfur. Include a description of how redox might impact environmental mobility of the species.

19 Write the half reactions for anammox.

20 What are the major electron donors and acceptors in aerobic soils and how does their relevance change under anaerobic conditions?

Bibliography

Baas Becking, L.G.M., I.R. Kaplan, and D. Moore. 1960. Limits of the Natural Environment in Terms of pH and Oxidation-Reduction Potentials. Journal of Geology 68:243–284.

Fitzpatrick, R.W., P. Shand, and L.M. Mosley. 2018. Soils in the Coorong, Lower Lakes and Murray Mouth Region. *In* L. Mosely, et al. (eds.) Natural History of the Coorong, Lower Lakes and Murray Mouth Region. Royal Society of South Australia, Adelaide, South Australia.

James, B., and R. Bartlett. 1999. Redox Phenomena, p. B169–B194. *In* M. E. Sumner (ed.) Handbook of Soil Science. CRC Press, Boca Raton, FL.

Jorgensen, C.J., and B. Elberling. 2012. Effects of flooding-induced N_2O production, consumption and emission dynamics on the annual N_2O emission budget in wetland soil. Soil Biology & Biochemistry 53:9–17.

Kehew, A.E. 2001. Applied chemical hydrogeology. Prentice Hall, Upper Saddle River, NJ.

Langmuir, D. 1997. Aqueous Environmental Geochemistry. 1st ed. Prentice-Hall, Inc., Upper Saddle River, NJ.

Leffelaar, P.A., and W.W. Wessel. 1988. Denitrification in a homogeneous, closed system – experiment and simulation. Soil Science 146:335–349.

Pankow, J.F. 1991. Aquatic Chemistry Concepts. Lewis Publishers, Chelsea, MI.

Reddy, K.R., and R.D. DeLaune. 2008. Biogeochemistry of Wetlands: Science and Applications. CRC Press, Boca Raton, FL.

Taylor, B.E., M.C. Wheeler, and D.K. Nordstrom. 1984. Stable isotope geochemistry of acid-mine drainage – experimental oxidation of pyrite. Geochimica Et Cosmochimica Acta 48:2669–2678.

Vaughan, K.L., M.C. Rabenhorst, and B.A. Needelman. 2009. Saturation and temperature effects on the development of reducing conditions in soils. Soil Science Society of America Journal 73:663–667.

Yamaguchi, N., T. Nakamura, D. Dong, Y. Takahashi, S. Amachi, and T. Makino. 2011. Arsenic release from flooded paddy soils is influenced by speciation, Eh, pH, and iron dissolution. Chemosphere 83:925–932.

Zausig, J., W. Stepniewski, and R. Horn. 1993. Oxygen concentration and redox potential gradients in unsaturated model soil aggregates. Soil Science Society of America Journal 57:908–916.

6 MINERALOGY AND WEATHERING PROCESSES IN SOILS

6.1 Introduction

The solid–solution interface is the most reactive part of soils and controls the composition and availability of nutrients and contaminants in the soil solution. The reactivity of the solid surface is highly variable and depends on composition and properties of the solids. This chapter focuses on inorganic solids. Organic solids are also extremely important for solid–solution interfacial reactions in soils; they are discussed in Chapter 8.

Inorganic soil particles range in size from colloidal (<2 μm) to gravel and rocks (>2 mm), and consist of minerals, salts, and amorphous solids. Early workers assumed that minerals formed in soils were all amorphous. X-ray diffraction revealed that many soil clay minerals are in fact crystalline. A common definition of a mineral is:

A *mineral* is a naturally occurring inorganic solid that has an ordered atomic arrangement, and definite, although not fixed, chemical composition.

This definition covers most of the solids in the earth that are of interest for geologic studies. In soils, however, many solids are amorphous or lack long-range order, and thus would not technically be considered minerals. A more inclusive definition for soil scientists is:

Soil minerals are inorganic solids that occur in soils, including well-ordered minerals, amorphous phases, and salts.

Sand and silt particles result from physical breakdown of igneous and metamorphic rocks in soils and are often composed of **primary minerals** (minerals formed from melt) (**Figure 6.1**). Clay-sized particles (<2 μm) are responsible for most chemical reactions in soils. A **clay particle** is defined only by size (<2 μm). A **clay mineral** is a group of minerals for which there are several different species, mostly phyllosilicates. In soils, most clay minerals are clay-sized particles. Often, the term *soil clay* is used by soil scientists and

Soil Chemistry, Fifth Edition. Daniel G. Strawn, Hinrich L. Bohn, and George A. O'Connor.
© 2020 John Wiley & Sons Ltd. Published 2020 by John Wiley & Sons Ltd.

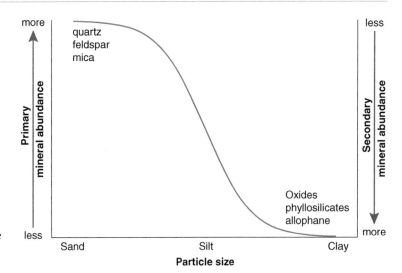

Figure 6.1 Typical abundance of primary and secondary minerals in soils as a function of particle size. Adapted from McBride (1994). Reproduced with permission of Oxford University Press.

(a)   (b)

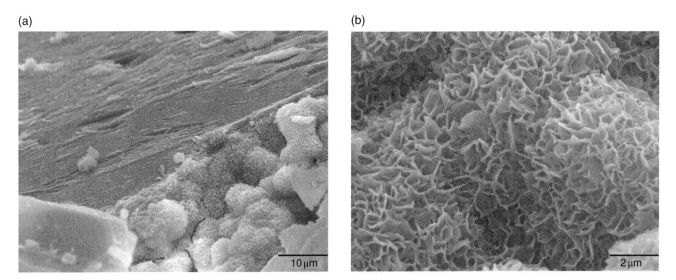

Figure 6.2 SEM micrographs of weathered augite (a pyroxene mineral group). Left image shows the augite crystals with the smaller smectite minerals formed on the weathered edge. The right image shows a magnified view of the smectite minerals and their tactoid morphology. From Velbel and Barker (2008). Reproduced with permission of Clays and Clay Minerals.

refers to secondary minerals, including phyllosilicates and secondary oxides that predominate in the clay-size fraction. The specific surface area of clay particles is much greater than sand and silt particles (see Figure 1.9), creating high surface charge that dominates the solid–solution interface.

The clay fraction of soils is mostly composed of **secondary minerals** formed during low-temperature weathering reactions. **Figure 6.2** shows a microscopic view of a primary mineral pyroxene grain that is weathered to a secondary mineral smectite on its exterior. In some soils, secondary minerals are inherited

from deposited sediments or sedimentary rocks. Soils that have not undergone extensive weathering may contain primary minerals in the clay fraction.

The secondary minerals in soils include phyllosilicates, aluminum and iron oxides, carbonates, phosphates, and sulfate minerals. Phyllosilicates are very common in soils. Oxides in soils, which are Al, Fe, Mn, and Ti oxides that accumulate in the soil as weathering removes silicon, are common in highly weathered soils. Oxides range from amorphous to crystalline, and often exist as coatings on larger soil particles.

6.2   Common soil minerals

Minerals are organized by their structural and chemical characteristics. One system categorizes minerals into classes based on their anion group. Common mineral classes in soils are **silicates**, **oxides**, and carbonates, with lesser amounts of phosphates, borates, chlorides, sulfates, and sulfides. Minerals can also be categorized based on similarities of their structure. For example, phyllosilicates (e.g., kaolinite and montmorillonite) have layered structures composed of sheets of tetrahedral cations bonded to sheets of octahedral cations (discussed in detail in the Chapter 7). **Table 6.1** lists common **mineral classes**, groups, and species observed in soils.

The occurrence and type of primary and secondary minerals in soils depends on the **parent material** and degree of weathering. Soil parent materials are rocks and minerals from which a soil develops. Rocks are conglomerates of minerals. Minerals are composed of different elements. The most common elements in Earth's crust are oxygen, silicon, iron, and aluminum (Figure 3.3), which combine to form iron- and aluminum-bearing silicates or oxides (**Figure 6.3**). Calcium,

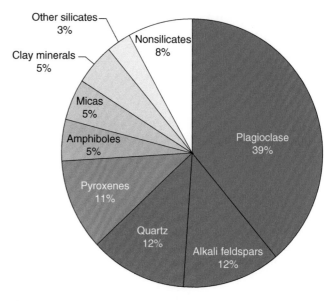

Figure 6.3 Estimated percentage by volume of minerals in Earth's crust. 92% are silicates (blue wedges). Most of the silicates are tectosilicates (62%, dark blue wedges), which are the most resistant to weathering. The second most common type of minerals are inosilicates. The values are averages for the whole Earth's crust. Adapted from Klein et al. (2002). Reproduced with permission of Wiley.

Table 6.1 Common mineral classes in soils.

Mineral Class	Types	Examples	Comments
Silicates	Nesosilicates, inosilicates, phyllosilicates, and tectosilicates	Pyroxene, amphibole, feldspar, mica, montmorillonite, chlorite, vermiculite, quartz, zeolites	Phyllosilicates are common secondary minerals in soils; quartz and feldspars are common primary minerals in sand fraction.
Oxides	Common: iron and aluminum oxides Less common: Mn, and Ti oxides	Fe oxides: ferrihydrite, goethite (FeOOH), hematite (Fe_2O_3) gibbsite ($Al(OH)_3$)	Common in all soils, predominant minerals in well weathered soils; high surface areas; can be anion or cation exchangers.
Carbonates	Ca and Mg carbonates	Calcite ($CaCO_3$), aragonite ($CaCO_3$), dolomite ($CaMg(CO_3)_2$)	Calcite and aragonite are polymorphs; low to moderate solubility; common in semi-arid and arid region soils; important carbon sink in soils; buffers pH.
Sulfates	Ca and Fe sulfates	Gypsum ($CaSO_4 \cdot 2H_2O$), jarosite ($KFe_3(SO_4)_2(OH)_6$)	Soluble; occur in arid region soils.
Evaporites (not a class)	Chlorides and soluble sulfates	NaCl (halite), Na carbonates, Na and Mg sulfates	Occur in arid region soils that have excessive evapotranspiration; promote halophytic vegetation.
Sulfides	Iron sulfides	Pyrite (FeS_2), marcasite, mackinawite	Occur in reduced soils or soils developed from parent materials containing sulfides; oxidizes and creates acidic soils.
Phosphates	Ca, Mg, Fe, and Al phosphates	Fluorapatite ($Ca_5(PO_4)_3F$), brushite ($CaHPO_4 \cdot 2H_2O$), strengite ($FePO_4 \cdot 2H_2O$), struvite ($MgNH_4PO_4 \cdot 6H_2O$)	Multiple Ca phosphate minerals occur in alkaline soils
Others	Poorly crystalline minerals	Allophane, imogolite, ferrihydrite, opal silicate glass	Very common in soils and nature; these minerals often coat larger mineral particles.

magnesium, sodium, and potassium are also common elements that occur in minerals.

Soils have been called *Earth's weathering engine,* and *clay factories.* Secondary soil minerals originate from weathering of rocks and sediments, and thus are composed of the same elements as crustal minerals. In well-weathered soils, the oxide mineral composition in soils increases because desilication removes silicon from the profile. CO_2 gas in the soil leads to formation of carbonates in some soils (e.g., calcite). Because of desilication and carbonate formation, the relative amounts of elements in soil minerals are different from crustal minerals.

Types of minerals and concentrations of the elements vary in soils because of different parent materials, climate, vegetation, landscape, and weathering time (soil formation factors). Thus, soils acquire unique mineralogical compositions that give them distinct physical and chemical properties. **Table 6.2** lists soil orders and predominant mineral characteristics.

6.3 Crystal chemistry of minerals

The properties of minerals are controlled by their physical and chemical characteristics (**physicochemical properties**). Physical characteristics of minerals include: particle size, surface area (specific surface), habit (mineral shape), porosity, surface charge, and aggregation. Chemical characteristics of minerals include: types of molecular bonds between atoms, elemental composition, stability of atoms within the mineral to chemical reactions and oxidation, and the arrangement of atoms in the mineral lattice.

Minerals have varying degrees of crystallinity. A **crystal** is an arrangement of ions or atoms that is repeated at regular intervals in three dimensions. The smallest repeating three-dimensional array of atoms in a crystal is called the **unit cell**. Unit cells contain specific atom arrangements representing the basic mineral composition and structure of the mineral. A perfect crystal structure occurs when several unit cells are combined. Minerals with foreign atoms in them, or

Table 6.2 Soil orders and their typical mineral and organic matter characteristics.

Soil order	Mineral, organic matter, and weathering characteristics
Andisol	Soils typically formed in volcanic ejecta, which weathers to small, high-surface-area minerals, including ferrihydrite, allophane, and imogolite; metal-humus complexes are present in the colloidal fraction.
Gelisol	Permafrost inhibits mineral weathering and typically leads to accumulation of high amounts of soil organic matter.
Histosol	Soils composed of organic matter in varying states of decomposition, from peat (slightly decomposed) to muck (highly decomposed); typically occur in wetland or bog environments; may contain sulfides and reduced iron minerals.
Aridisol	Soils developed in an arid environment, with little organic matter accumulation and varying degrees of weathering; have subsoil accumulations of $CaCO_3$, gypsum, silica, and/or evaporites; salts occur as small white crystals in cracks and voids or on soil particle surfaces.
Entisol	Soils with little weathering or organic matter accumulation; thus, mineral composition strongly reflects parent material.
Inceptisol	Soils with only minimal subsoil horizon development; mineral composition generally reflects parent material; some formation of phyllosilicates and iron oxides as coatings on primary mineral grains.
Vertisol	Soils characterized by high concentrations of high shrink-swell clays such as montmorillonite (a 2:1 smectite phyllosilicate) that cause the soil to crack when dry and expand when moist; deep incorporation of SOM.
Mollisol	Slightly to moderately weathered soils with a thick, dark A horizon enriched in organic matter; a variety of 2:1 phyllosilicates may be present.
Alfisol	Moderately weathered soils with a clay-enriched B horizon resulting from illuviation of clay minerals; a variety of phyllosilicates may be present.
Spodosol	Slightly to moderately weathered soils; the presence of plant-derived soluble organic compounds and large water flux lead to eluviation of upper profile to form a light-colored E horizon; illuviation forms a B horizon enriched in organic matter complexed with Al and Fe ions and iron oxides such as ferrihydrite.
Ultisol	Moderately to highly weathered soils with a clay-enriched B horizon; desilication leads to formation of kaolinite, chlorite, and iron oxides; iron oxides cause subsoil to have characteristic red, yellow, or orange colors.
Oxisol	Highly weathered soils composed of predominantly iron and aluminum oxides (hematite and gibbsite) and kaolinite; few or no primary minerals present; relatively uniform mineral distribution throughout profile.

atoms not in the correct unit cell position, are less than ideal and called poorly crystalline or amorphous.

A basic tenet of mineralogy is: *properties of minerals can be predicted from their chemical and physical characteristics.* By extension, mineral properties strongly influence the physicochemical properties of soils. In this section, details of mineral chemistry are presented to provide a framework for studying soil mineral properties.

6.3.1   Bonds in minerals

When atoms combine to form minerals, the bonds between the atoms change the distribution of electrons. The type of bond depends on the electronic structure of the combining atoms (i.e., electron orbitals and electron affinity or electronegativity, discussed in Chapter 2). Following is a brief review of bond types that are important in minerals:

- *Ionic or electrostatic bonding* occurs between oppositely charged ions such as Na^+ and Cl^-. Ionic bonding is strong, and not constrained to any specific bond angle (undirected). Ionic-bonded compounds tend to be hard solids and have high melting points but dissolve more readily in aqueous solutions than covalently bonded compounds because charged ions in ionic bonds are more easily solvated than covalently bonded ions.

- *Covalent bonding* (shared electron pairs) is common between atoms having similar electrical properties, such as in H_2O, F_2, CH_4, and C (diamond). In covalent bonding, the electrons are shared between atoms to form strong bonds. Bond angles in covalently bonded structures are directed, determined by the geometric positions of the electron orbitals involved. Covalent bonds are not as *physically* strong as ionic but are more resistant to aqueous reactions.

- *Hydrogen bonding* occurs between molecules containing H^+ and ions of high electronegativity, such as F^-, O^{2-}, and N^{3-}. The hydrogen bond is a weak electrostatic bond and is important in crystal structures of minerals that have OH^- and water molecules in their structure, such as the layer silicates. Summed over many atoms, the individually weak hydrogen bonds can impart an overall strong attachment of minerals to adjacent mineral components. Kaolinite and goethite are examples of minerals with hydrogen bonds.

- *van der Waals forces* are weak electrostatic forces between residual charges on molecules that occur from polarization dipoles in molecules, or vibrational (instantaneous) dipoles created when atoms approach each other.

Each type of chemical bond imparts characteristic properties to a substance. If more than one type of bond occurs in a crystal, the weakest bonds typically determine the physical properties such as hardness, mechanical strength, and melting point. In addition, the type of bond dictates the susceptibility of atoms in minerals to chemical weathering, which is an aqueous reaction. The physical and chemical properties of minerals are influenced by the types of elements in a mineral and the bonds within the structure.

Although bond types appear to be distinct, in most crystals bonding is a mixture of ionic and covalent bonds. For example, the Si-O bond in silicates is 51% ionic and 49% covalent. The degree of ionic nature of the Si-O bond is sufficient however to apply the rules for ionic bonding to most silicate mineral structures, thus, directional aspects of the covalent bonds can be disregarded. Ion size and charge are therefore the most important characteristics in determining crystal structure. The bond type of Al^{3+}, Fe^{3+}, and Mg^{2+} cations with oxygen in minerals is like that of Si^{4+}, so ionic size is equally important in determining the structure of minerals composed of these cations.

6.3.2   Rules for assembling minerals

Mineral structure and properties are controlled by atomic composition and geometric arrangement of ions in the crystal structure. At the beginning of the twentieth century, development of X-ray diffraction allowed scientists to discover how atoms are arranged in minerals (see Special Topic Box 6.1). In the 1940s, Linus Pauling, two-time Nobel Prize winner, summarized the observations that mineralogists were making about atomic structure in minerals into a set of rules that are useful guidelines for how minerals are assembled:

Rule 1: Cations in crystal structures are surrounded by anions. The distance between cations and anions determines the atomic arrangement of ions in the mineral lattice.

Special Topic Box 6.1   History of the use of X-ray diffraction for studying soil mineralogy

Soil mineralogy began around 1850 with the discovery of cation exchange in soils. However, little was known about the properties of the soil clays, and a debate on whether the clays were amorphous or crystalline substances ensued. Many studies on chemical and physical properties were pursued through the latter half of the nineteenth century.

A major breakthrough in understanding mineral properties was the invention of X-rays in 1895 by Wilhelm Conrad Röntgen, for which he was awarded the first Nobel Prize in Physics. In 1912, Max von Laue and colleagues in Munich showed that crystals diffract X-rays, proving that minerals were repeating arrangements of atoms. In 1912, father and son, W.H. Bragg and W.L. Bragg in England, discovered the first crystal structure determinations using X-ray diffraction (XRD) (**Figure 6.4**) to calculate the distance between planes of atoms (Bragg's law).

Phyllosilicates were noted for their interesting properties (e.g., cation exchange), and researchers soon began

studying clay minerals using XRD. Scientists such as Hadding in Sweden, Rinne in Germany, and Ross in the United States were among the first to study clay minerals by XRD in the early 1920s. They proposed that micron-sized clay particles were in fact crystalline. In 1930, Pauling published a foundational paper on the crystal structure of micas determined using XRD. Following Pauling's methodology, in 1932 Gruener published the structure of kaolinite. In 1933, Hofmann, Endell, and Wilm published the structure of montmorillonite, in which the mineral structural properties were linked to the expanding nature of clay minerals. Other clay mineral structures followed. In 1951, Brindley, and in 1953 Grim published texts detailing clay mineral structures.

Nearly 70 years after the discovery of cation exchange by soil clays, S.B. Hendricks and W.H. Fry (1930) and W.P. Kelley, W.H. Dore, and S.M. Brown (1931) used XRD and chemical analysis to discover that soil clays were crystalline. For the first time, properties of soils could be understood based on their atomic-level mineralogical properties, and a mineralogical basis for cation exchange and shrink-swell behavior of soils was discovered (discussed in detail in Chapter 7).

The structures of hematite, goethite, and gibbsite were worked out in the early days of using X-ray patterns. In their 1930 paper, Hendricks and Fry were amongst the earliest researchers to report the presence of bauxite (aluminum oxide) in soil colloids examined by X-ray diffraction. In a 1932 paper, Kelley and co-workers reported that iron oxides were present as coating on many clay minerals in soils.

Much of the discovery of soil mineralogy is linked to X-ray diffraction analysis of phyllosilicates and oxides. Many soil clays, however, are poorly crystalline, such as iron and aluminum oxides and allophane. Poorly crystalline clays lack long-range atomic ordering, and thus X-ray diffraction studies of their structure are less useful. Advanced tools (see Section 6.14) are allowing for more detailed study of soil mineralogy, as well as new knowledge of how soil minerals interact with SOM and chemicals in soil solutions.

Figure 6.4 X-ray diffractometer (1912) used by the W.H and W.L. Bragg to investigate the crystal structure of minerals. Adapted from https://en.wikipedia.org/wiki/William_Henry_Bragg#/media/File:X-ray_spectrometer,_1912._(9660569929).jpg.

Rule 2: The charge of a cation is dissipated equally to the first coordination sphere of anions.

Rule 3: Cations and anions arrange themselves so that like charges are as far apart as possible.

Rule 4: Cations having large valences and low coordination numbers tend not to share anions because the high cationic charges would be too close.

Rule 5: The most stable minerals are those with the fewest number of elements.

Pauling's rules are a useful set of guidelines for how *ideal* minerals are assembled, and for understanding natural minerals in soils (which are often *non ideal*). The rules are interdependent, meaning that during mineral formation, the crystal structures will achieve an overall balance of atomic composition and arrangement that best satisfies the five rules. Nonideality creates unique characteristics in minerals, such as variable crystallinity and weatherability, and creation of surface charge. The first two of Pauling's rules are especially helpful for understanding how minerals are assembled and are discussed in more detail in the following sections.

Pauling's Rule 1 states that cations and anions are bonded to each other such that the space surrounding the cation has as many anions as will fit. The number of anions that pack around a central cation is called the **coordination number** of the central ion. This rule is a geometry problem constrained by the ionic radii of the cations and anions. The most common anion in minerals is oxygen (O^{2-}), which has an ionic radius of 0.14 nm (**Table 6.3**). The ion radius of oxygen is much larger than that of most cations, so oxygen comprises $\geq 90\%$ of the volume of most common silicates and oxide minerals. Accordingly, mineral structures are determined by the structural arrangement of oxygen ions around cations.

In the mineral lattice, the space between the oxygen ions that accommodates cations depends on the arrangement of the oxygen ions. **Figure 6.5** shows examples of two different voids created by four and six oxygen atoms, which represent **tetrahedral** and **octahedral** coordination, respectively (so named because of the number of sides in the polyhedral representation). If the central cation is too large, the surrounding anions are forced apart, creating too much space for a stable molecular coordination. If the cation is too small, the void is incompletely filled, and the cation size is insufficient to keep the anions at their optimal separation (anions repel anions).

Several possible coordination structures surrounding voids can be created from the three-dimensional **closest packing** of *hard* spheres (oxygen ions): trigonal, tetrahedral, octahedral, cubic, and cuboctahedron. For silicates and oxides, tetrahedral and octahedral coordination (**Figure 6.6**) are the most common coordination structures. The ball-and-stick models in **Figure 6.6** greatly exaggerate bond lengths, and the size of the O^{2-} ions is reduced to show the relative positions of each

Table 6.3 Ionic radii for common mineral ions and their predicted and observed coordination numbers with O^{2-}. Calculated ionic radius varies as a function of ion charge, coordination number, and molecular orbital structures; values given here are the most commonly used in crystal structures. Data from Shannon (1976).

Ion	Coordination number	Pauling ionic radius (nm)	Observed coordination number
Al^{3+}	6	0.054	4, 6
Ca^{2+}	8	0.11	6, 8
Fe^{2+}	6	0.078	6
Fe^{3+}	6	0.065	6
K^+	8	0.15	8, 12
Mg^{2+}	6	0.072	6
Mn^{2+}	6	0.083	6
Mn^{4+}	6	0.053	6
$N^{3-}(NH_4^+)$	4	0.15	8, 12
Na^+	6	0.10	6, 8
O^{2-}	–	0.14	–
Si^{4+}	4	0.026(0.04)[a]	4
Ti^{4+}	6	0.061	6

[a] For predicting Si coordination in minerals the value 0.04 is more commonly used (Shannon and Prewitt, 1969).

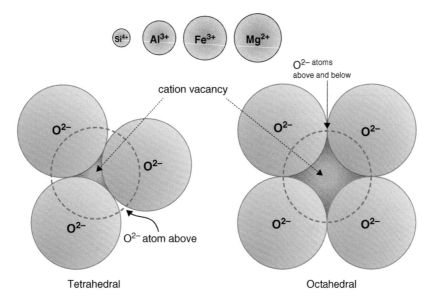

Figure 6.5 Illustration of how different arrangements of oxygen ions create different-sized voids. In the tetrahedral structure, the fourth oxygen atom (dashed circle) sits above the cation site. In the octahedral structure, the fifth and sixth oxygen atoms sit above and below the cation vacancy. Silicon fits in the void in the center of the tetrahedral coordination. Iron and magnesium fit well in oxygen octahedra. Aluminum fits in both.

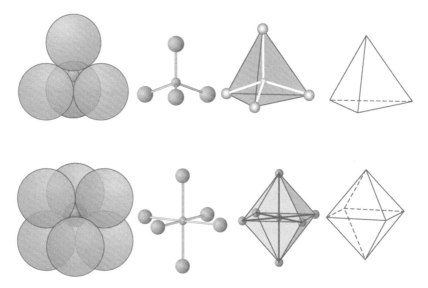

Figure 6.6 Different representations of tetrahedral (four-sided) and octahedral (eight-sided) coordination.

ion. The space-filling models in **Figure 6.6** more accurately represent the cation and anion relative sizes.

Assuming ions act as rigid spheres, the stable cation-anion polyhedral coordination can be calculated using **radius ratio** of the cation and O^{2-}. **Table 6.4** shows the limits of the radius ratio for the varying coordination structures. The Si^{4+} cation has a Si/O radius ratio of 0.29 and occurs in tetrahedral coordination. In tetrahedral coordination, the radius of the void between the four O^{2-} ions is only 0.22 times the O^{2-} radius, or 0.031 nm (0.14 nm × 0.22). Thus, cations must be at least that big to effectively shield the anion charges from each other. In minerals, cations ranging from 0.031 to 0.056 nm radius occur in four-fold (tetrahedral) oxide coordination. The radius of Si^{4+} in four-fold coordination is about 0.04 nm.

Table 6.4 Spatial arrangement of rigid spheres in relation to radius ratio and coordination number. The radius ratios are the limits for the size of the cations to fit in the void in the polyhedra. For example, cations in a tetrahedron must be no smaller than 22% and no bigger than 40% of the surrounding anions.

Radius ratio (r_{cation}/r_{anion})	Arrangement of anions around cations	Coordination number of central cation
0.15–0.21	Corners of an equilateral triangle	3
0.22–0.40	Corners of a tetrahedron	4
0.41–0.72	Corners of an octahedron	6
0.73–0.99	Corners of a cube	8
1	Closest packing	12

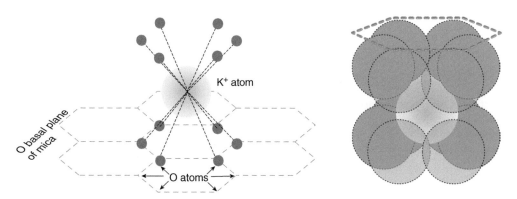

Figure 6.7 Illustration of a potassium atom coordinated by 12 oxygen atoms in the interlayer of a mica phyllosilicate mineral.

If the central cation size increases beyond a certain limit, additional anions can fit in the coordination sphere, thus increasing the oxygen coordination number. After tetrahedral coordination, the next largest cation void is octahedral coordination, where six oxygens surround the central cation.

Aluminum has an oxygen radius ratio of 0.39 and occurs in either six-fold octahedral coordination or tetrahedral coordination. When the radius ratio is near the boundary between two types of coordination, the cation may occur in either of the coordination structures, depending on conditions during crystallization. High crystallization temperatures favor low coordination numbers. Thus, in minerals that solidified from high temperatures, Al^{3+} tends to assume four-fold coordination, and substitutes for Si^{4+}. At lower temperatures, Al^{3+} tends to occur in six-fold coordination.

Potassium has a larger ionic radius than many other cations, and its O^{2-} radius ratio is 1.1. The large ionic radius of K^+ can fit 8–12 oxygen anions in its first-shell coordination environment. In mica, a phyllosilicate mineral, K^+ exists in the interlayer forming ionic bonds with up to six oxygens from a layer above and up to six oxygens from a layer below; thus, satisfying its structural coordination requirements (**Figure 6.7**).

Pauling's Rule 2 explains that cation charges are balanced by charges from first-shell coordinated anions such that the overall cation charge sums to zero. In some cases, however, charge balance does not occur, and the mineral gains a net structural charge. The charge deficit transfers to the mineral exterior as permanent surface charge and is balanced by ions adsorbed on the surface. The charge distribution around a cation or anion in a mineral is calculated from its **electrostatic bond strength** (s), which is the ratio of the valence of the ion (Z) divided by its coordination number (CN):

$$s = \frac{Z}{CN} \tag{6.1}$$

A Si-tetrahedral structure (**silicate**) has four O^{2-} ligands coordinated around one Si^{4+}, thus, the electrostatic bond strength for the Si is $+4/4 = +1$. The silicon ion gives one positive charge to each of its four oxygen ligands (**Figure 6.8**). From the oxygen's perspective, each of the four O^{2-} anions give one negative charge to balance the Si^{4+} cation charge. Thus, each of the four O^{2-} ions in the silicate molecule has a remaining net charge of -1, creating the negative four valence of the SiO_4^{4-}. The oxygen in the $Si-O^{1-}$ atomic pair bonds to adjoining cations to satisfy its negative one charge. Electrostatic bond strength calculation is a useful way to keep track of charges in a mineral's structure. This concept will be used later to show how **permanent surface charge** is calculated.

Pauling's rules provide a logical framework for understanding how the geometric and charge characteristics of cations and anions constrain the arrangement of atoms in minerals. For example, the linkages of cation-ligand units, such as Si^{4+} tetrahedra or Al^{3+} octahedra, are a consequence of the charge distribution and coordination number (**Figure 6.9**). Understanding how minerals are assembled is useful for understanding mineral and soil properties.

6.3.3 Isomorphic substitution

Isomorphic substitution, isomorphism, atomic substitution, and solid solution all refer to the substitution of one ion for another without significantly changing the structure of the crystal. Isomorphic substitution in mineral structures is common. In primary minerals, isomorphic substitution takes place during crystallization and occurs between ions of the same or different charges. The size of the ions, rather than the charge, determines the potential to isomorphically substitute (hence, the name – *iso* means same, *morphic* means shape). When an ion of different charge isomorphically substitutes, electrical neutrality is maintained by simultaneous substitution of ions elsewhere in the structure, or by adsorption of ions on the outside of the structure. Isomorphic substitution typically takes place between ions that are within 10–15% difference in ionic radii.

In silicate and oxide minerals, common isomorphic substitutions are Al^{3+} or Fe^{3+} for Si^{4+} in tetrahedrally coordinated cation positions, and Mg^{2+}, Fe^{2+}, and Fe^{3+} for Al^{3+} in octahedrally coordinated cation positions. Substitution between ions of unequal charge in phyllosilicates leaves negative or positive charges within the crystal that are neutralized by ions in the interlayer of the mineral. Isomorphic substitution is common in phyllosilicates, creating a net negative charge, which is primarily responsible for cation exchange capacity (CEC) in soils.

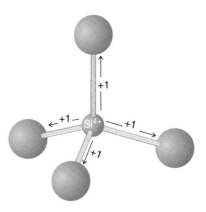

Figure 6.8 Charge distribution (*s*) for a Si^{4+} tetrahedron.

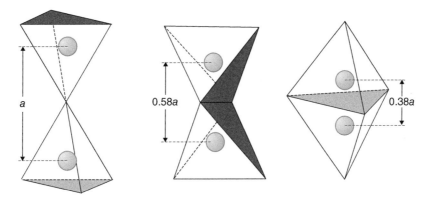

Figure 6.9 Three possible ligand sharing modes for two tetrahedrally coordinated cations. The cation separation distance (a) is greatest for corner sharing ligands (left). High charge cations such as Si^{4+} do not like to share more than one ligand with each other due to increased electrostatic repulsion created by shorter separation distances (Pauling Rule #4). Thus, Si^{4+} tetrahedra in minerals have corner sharing with other ligands as opposed to edge (middle) or face sharing (right).

Some areas of soil clay particles have a net positive charge, and an anion exchange capacity. For example, Al^{3+} can substitute for a Mg^{2+} in an octahedral position of a phyllosilicate, creating a localized negative-charge deficiency (i.e., a positive charge surplus). However, in phyllosilicates, the positive octahedral charge is typically less than the negative charge created by isomorphic substitution of Al^{3+} for tetrahedral Si^{4+}, and thus the clay particle has an overall negative charge.

6.3.4 Mineral formulas

The chemical composition of a mineral is represented by the number of atoms in its unit cell. There are three basic *guidelines* for writing mineral formulas:

1 The net charge of the formula should be zero.
2 All atoms in the unit cell must be represented.
3 The number of each of the atoms in the formula is stoichiometric with respect to their proportions in the unit cell.

For example, the mineral quartz has a formula of SiO_2:

Guideline 1. The mineral formula is charge balanced (4+ from Si^{4+} ion, plus 2 × 2- from the two O^{2-} ions equals zero).
Guideline 2. Pure, or ideal, quartz consists of only Si and O.
Guideline 3. Within the mineral, the ratio of silicon to oxygen is 1:2.

Quartz is a simple mineral formula, but the rules also work for more complex minerals. A good way to keep track of the charge balance of a mineral formula is to make a table to add up the charges. **Table 6.5** shows

Table 6.5 Charge balance for the mineral talc $(Si_4Mg_3O_{10}(OH)_2)$.

Element	Number in mineral formula	Charge	Total charge in mineral formula
Mg	3	2+	6+
Si	4	4+	16+
O	10	2-	20-
OH	2	1-	2-
		Sum	0

the charge balance for talc $(Mg_3Si_4O_{10}(OH)_2)$. The charge from the talc mineral formula sums to zero.

Mineral formulas are commonly written as ideal formulas, which represent the major elemental composition of the mineral species, and ignores irregular, minor, or nonideal element compositions. Minerals formed in nature, however, are typically nonideal. For example, the *ideal formula* for the clay mineral montmorillonite is

$$M^{+}_{0.66}Si_8Al_{3.34}Mg_{0.66}O_{20}(OH)_4$$

where M^+ represents cation charges in the interlayer. The measured composition and mineral formula for a montmorillonite specimen acquired from Wyoming, USA (SWy-2) is

$$\left(Ca_{0.12}Na_{0.32}K_{0.05}\right)\left(Al_{3.01}Fe(III)_{0.41}Mn_{0.01}Mg_{0.54}Ti_{0.02}\right)$$
$$\left(Si_{7.98}Al_{0.02}\right)O_{20}(OH)_4$$

where the parentheses isolate cation groups that occupy the same atomic positions in the mineral structure. SWy-2 is a common mineral used to test clay mineral properties. **Table 6.6** shows the structural atom occupation in *ideal* montmorillonite and *real* SWy-2 montmorillonite.

One aspect of mineral formulas of clay minerals that is often confusing is the listing of both full- and **half-unit cell formulas** that is used when writing phyllosilicate formulas. A half unit cell for a phyllosilicate has stoichiometric units of one half the **full unit cell**.

Table 6.6 Comparisons of the mineral formulas for ideal montmorillonite and a montmorillonite collected from Wyoming, USA (SWy-2). Even though the tetrahedral and octahedral sheet cation compositions of the two formulas are different, the total number of cations in each sheet is the same.

Mineral	Interlayer cations	Tetrahedral sheet cations	Octahedral sheet cations	Anions
Ideal montmorillonite	$M^+_{0.66}$	Si_8	$Al_{3.34}Mg_{0.66}$	$O_{20}(OH)_4$
Wyoming montmorillonite (SWy-2)	$Ca_{0.12}Na_{0.32}$ $K_{0.05}$	$Si_{7.98}Al_{0.02}$	$Al_{3.01}$ $Fe(III)_{0.41}$ $Mn_{0.01}$ $Mg_{0.54}$ $Ti_{0.02}$	$O_{20}(OH)_4$

For example, the half unit cell of ideal montmorillonite is $M^+_{0.33}Si_4Al_{1.67}Mg_{0.33}O_{10}(OH)_2$ (the stoichiometry is ½ the full unit cell formula given in **Table 6.6**). The listing of full- and half-unit cell formulas for clay minerals originates from differences in how the unit cell is defined for different phyllosilicates. In this text, most phyllosilicate formulas are given as half-unit cell formulas.

6.4 Common primary mineral silicates in soils

Primary minerals are igneous rock minerals formed from melts either in Earth's interior (**magma**) or at Earth's surface (**lava**). **Table 6.7** lists the common primary minerals that occur in soils.

Table 6.7 Common primary minerals present in soil environments in order of increasing stability. Source: Feldman and Zelazny (1998). Reproduced with permission of ACSESS.

Name	Mineral class	Environment	Ubiquity in soils[a]	Importance
Pyrite	Sulfides	Tidal marshes (reducing conditions and hard-rock mine tailings (coal and shale beds)	C	Primary mineral (oxidizing conditions) but secondary phase forms in reducing environments; large metal and acidity output during weathering
Dolomite	Carbonates	Shallow, young soils formed in limestone	R	Constituent of limestone parent material; fertilizer source
Talc	Phyllosilicates	Ultramafic rocks	R	No layer charge; little chemical reactivity; unstable in soils
Olivine	Nesosilicates	Basic and acidic igneous rocks	R	Source of Fe, Ca, Mg, and Mn; unstable in highly leached soil
Pyroxenes	Inosilicates	Igneous and contact metamorphic rocks	R	Source of Fe, Ca, Mg, and Mn; unstable in highly leached soil
Amphiboles	Inosilicates	Igneous and metamorphic rocks	R	Source of Fe, Ca, Mg, and Mn; vermiculite precursor
Chlorite (Mg-interlayer)	Phyllosilicates	Metamorphic or igneous rocks	R	Important precursor for 2:1 soil clay minerals
Biotite (Fe^{2+}-bearing micas)	Phyllosilicates	Granitic and high-grade metamorphic rocks	R	Stable in only the youngest or least weathered soils; precursor of other 2:1 soil clay minerals and Fe-oxides; source of K
Feldspars (Plagioclase and Alkali-feldspars)	Tectosilicates	Wide variety of igneous/metamorphic rocks; persistence in soils and geologic deposits is related to weathering intensity, and the duration of exposure to weathering	C	Source of Ca, Na, and K; weathering products can be kaolinite, mica, gibbsite, halloysite, smectite, or amorphous materials
Muscovite	Phyllosilicates	Granitic and high-grade metamorphic rocks	C	Vermiculite, smectite, and interstratified 2:1 precursor, K source
Epidote	Sorosilicates	Medium-grade metamorphic and mafic igneous rocks	R	Source of Fe, Ca, Mn; very resistant to weathering
Quartz	Tectosilicates	Nearly all soils and parent materials	U	Concentrated in sand- and Si Lt-fractions; soluble in clay fraction
Garnets	Nesosilicates	High-grade metamorphic/acid igneous rocks	R	Source of Fe, Mg, Ca, Al, and/or Mn precursor of Fe-oxides
Tourmaline	Cyclosilicates	Granites and pegmatites; detrital sediments	R	Stable in soils, but may alter to secondary 2:1 clay minerals
Rutile	Oxides	Igneous and metamorphic rocks; detrital sediments; quartz inclusion	C	Very stable in soils, but some mobility of Ti; rarely alters to other oxides
Zircon	Nesosilicates	Acid and basic plutonic, metamorphic rocks	C	Very resistant; used as index minerals in pedologic studies

[a] U, ubiquitous; C, common; R, rare.

Table 6.8 Structural classification of silicates. The Si:O ratio refers to the number of Si atoms divided by the number of O atoms associated with the Si tetrahedra in the mineral structure. In determining Si:O ratio, isomorphic substitution of Al^{3+} in the Si^{4+} positions (underlined Al in table) counts as a silicon atom.

Classification (group)	Structural arrangement	Si/O ratio	Examples
Nesosilicates (island silicates)	Single tetrahedra	1/4	Olivine ($(Mg,Fe)_2SiO_4$), garnet
Inosilicates (single-chain silicates)	Continuous single chains of tetrahedra sharing two corners	1/3	Pyroxene ($MgSiO_3$)
Inosilicates (double-chain silicates)	Continuous double chains of tetrahedra sharing alternately two and three oxygens	4/11	Amphiboles ($Mg_{3.5}Si_4O_{11}(OH)_1$)
Phyllosilicates (sheet or layer silicates)	Continuous sheets of tetrahedra, each sharing three oxygens	2/5	Micas, clay minerals (muscovite-$K_2Al_4Si_6\underline{Al}_2O_{20}(OH)_4$)
Tectosilicates (framework silicates)	Continuous framework of tetrahedra, each sharing all four oxygens	1/2	Quartz, feldspars, zeolite (quartz-SiO_2; orthoclase-$K(Si_3\underline{Al}O_8)$)

The most common primary minerals in soils and Earth's crust are the silicates listed in **Table 6.8**. Different silicate structures arise from the various ways in which the SiO_4 tetrahedra combine with one another. There are four O^{2-} ligands on a silicate unit. The Si-O bond leaves a negative one charge on the oxygen, so it must bond to other cations to balance the remaining charge. If the bonding cation is another Si^{4+} atom, a corner-sharing Si-tetrahedral configuration results (Si-O-Si). A silicate unit (SiO_4^{4+}) can form zero, one, two, three, or four corner-sharing bonds to other tetrahedral silicate units. Because Si-O bonds are highly covalent, and thus strong, the more Si-O-Si bonds there are in a mineral, the more resistant it is to weathering reactions. Thus, a mineral with four Si-O-Si bonds is more resistant to weathering than a mineral with three, two, or one corner-sharing Si-O-Si bonds. In quartz, for example, all four of the oxygen atoms in the Si tetrahedra are shared with other Si atoms (**Table 6.8**), so it is very resistant to weathering, and persists in soils. In the mineral olivine, on the other hand, the four oxygen atoms in the Si tetrahedra are not shared with any other Si atoms, so it is easily weathered.

The different silicate groups (**Table 6.8**) are distinguished by the number of Si-O-Si bonds for each tetrahedron, and are represented by the silicon-to-oxygen ratio (**Si/O ratio**). Aluminum sometimes substitutes for Si^{4+} atoms, and the Al^{3+} in this position counts as a *Si* in the Si/O ratio. Details of four of the most common silicate groups are discussed below.

6.4.1 Nesosilicates

Nesosilicates are island silicates typified by olivine. Nesosilicates are common in mafic, Fe^{2+}- and Mg^{2+}-rich rocks such as basalt. The silicate unit in nesosilicates does not bond to any other silicate units, but instead bonds to other cations, such as Fe^{2+} or Mg^{2+}, to satisfy the negative one oxygen charge. Nesosilicates are common parent materials for secondary minerals that occur in soils because of their ease of weathering.

6.4.2 Inosilicates

Inosilicates include two groups, the single- and double-chain silicates, called **pyroxenes** and **amphiboles**, respectively. As the name implies, inosilicates are distinguished by the continuous chains of Si tetrahedra (**Figure 6.10**). Hornblende is a common amphibole observed in some soils, especially ones that are weakly weathered.

6.4.3 Phyllosilicates

Phyllosilicates have sheets of silicon tetrahedra where three of the oxygen anions coordinating a Si^{4+} cation are bonded to three separate Si^{4+} cations within a plane called the tetrahedral sheet. The fourth silicate oxygen anion is bonded to Al^{3+}, Mg^{2+}, or Fe^{3+} cations located in a sheet below the tetrahedral sheet. The sheet structure results in a platy morphology. Isomorphic substitution

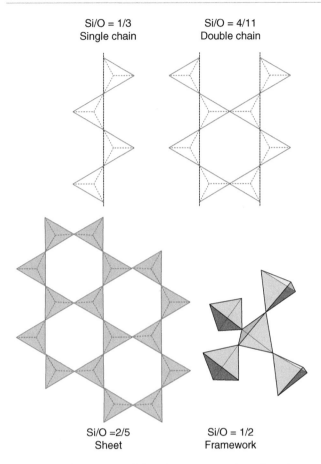

Si/O = 1/3
Single chain

Si/O = 4/11
Double chain

Si/O =2/5
Sheet

Si/O = 1/2
Framework

Figure 6.10 Structures of the silicate groups based on the silicon polymerization through sharing of the oxygen ligands. The higher the Si/O ratio, the more Si-O-Si bonds per tetrahedral silicate unit.

of Al^{3+} for Si^{4+}, or divalent cations such as Mg^{2+} for Al^{3+} octahedral cations, creates permanent negative surface charge. The phyllosilicate mineral group includes micas and secondary clay minerals that occur in soils. Muscovite and biotite are two common micas that occur in sand and silt fractions of soil. If mica minerals weather to clay-size particles, they are called illites, which are common in soils. Because of their importance in soils, the mineral structures of phyllosilicates will be presented in detail in the next chapter.

6.4.4 Tectosilicates

Tectosilicates are minerals in which all four of the oxygen atoms of the silicon tetrahedron are bonded to another silicon atom by corner-sharing linkages, creating a framework of silicate structural units. Quartz is a

good example of a tectosilicate. If Al^{3+} isomorphically substitutes for Si^{4+} in a tectosilicate, the framework structure is maintained, but Na^+, K^+, or Ca^{2+} cations are included in the mineral structure to balance the charge deficit on the aluminum-tetrahedral oxygen anions. Common aluminum-containing tectosilicates in soils are **feldspars**. Tectosilicates, especially quartz, are resistant to weathering and commonly occur in the sand and silt fractions of soils.

6.4.5 Cations in primary silicates

In addition to Si^{4+}, primary silicates typically contain Al^{3+}, Fe^{3+}, Mg^{2+}, Fe^{2+}, Ca^{2+}, Na^+, and K^+ cations (**Table 6.9**). A mineral with *regular* isomorphic substitution is called a **solid–solution** series between two **end members**. Examples of end members in a solid–solution series are the Na^+ and K^+ pure end members of feldspars (albite-orthoclase) (**Table 6.9**). Alkali feldspar solid–solution

Table 6.9 Examples of common silicate mineral species and their mineral formulas.

Chemical formula	Name
Nesosilicates (olivine) Si/O = 1/4	
Mg_2SiO_4	Fosterite
Fe_2SiO_4	Fayalite
$Mg_{1.8}Fe_{0.2}SiO_4$	Chrysolite
Inosilicates	
Pyroxenes (single chain) Si/O = 1/3	
$MgSiO_3$	Enstatite
$FeSiO_3$	Orthoferrosilite
Amphibole (double chain) Si/O = 4/11	
$Ca_2Mg_5Si_8O_{22}(OH)_2$	Tremolite
$CaMg_4FeSi_8O_{22}(OH)_2$	Actinolite
$NaCa_2Mg_5Fe_2Si_7AlO_{22}(OH)_2$	Hornblende
Phyllosilicates (mica) Si/O = 2/5	
$K_2Al_4Si_6Al_2O_{20}(OH)_4$	Muscovite
$K_2Mg_4Fe_2Si_6Al_2O_{20}(OH)_4$	Biotite
$K_2Mg_6Si_6Al_2O_{20}(OH)_4$	Phlogopite
Tectosilicates Si/O = 1/2	
SiO_2	Quartz
Feldspaars	
$KAlSi_3O_8$	Orthoclase
↕ *alkali feldspar solid-solution*	
$NaAlSi_3O_8$	Albite
↕ *plagioclase solid-solution*	
$CaAl_2Si_2O_8$	Anorthite

minerals are feldspar minerals that have a mixture of Na^+ and K^+ atoms in their structure (e.g., anorthoclase $((Na,K)AlSi_3O_8)$). Plagioclase minerals are feldspar solid–solution minerals that have Na^+ and Ca^{2+} compositions that are between albite and anorthite.

6.5   Minerals and elements in rocks

Rocks are conglomerates of single or mixed minerals. They are classified as **igneous**, **sedimentary**, or **metamorphic**. Metamorphic rocks are formed from sediments that have been transformed by heat and pressure. The minerals in rocks occur in various grain sizes, ranging from fine to coarse grained. The composition and grain size impart most of the weathering characteristics of rocks.

The major minerals of igneous rocks, in decreasing order of general abundance, are feldspars, quartz (SiO_2), biotite $(K(Mg,Fe)_3AlSi_3O_{10}(OH)_2)$ and muscovite $(KAl_2AlSi_3O_{10}(OH)_2)$. The feldspars include orthoclase and the plagioclase series ranging from albite $(NaAlSi_3O_8)$ to anorthite $(CaAl_2Si_2O_8)$. Other minerals in igneous rocks are present in lesser amounts. **Figure 6.11** shows the mineral composition of two examples of **granite** and **basalt** rocks.

6.5.1   Elemental composition of rocks

Table 6.10 lists the elemental contents of several rock types expressed as weight percentage of the oxides.

Expressing the elements as oxides emphasizes the importance of oxygen as the predominant anion in rocks and soils. Even sulfates, phosphates, and carbonates are oxyanions in which the negative charge comes from oxygen ligands. The only other significant ligand in minerals is sulfide; the sulfur in sulfides is unstable in aerobic soils.

Granitic or acid igneous rocks (>66% SiO_2) are richer in silicon and potassium, and poorer in magnesium and iron, than basaltic or basic igneous rocks (**Table 6.10**) (45 to 52% SiO_2). Metamorphic rocks form from heating and compression of mineral bodies. Metamorphic rock compositions vary with the source material, but are similar to many igneous rocks, except for marble and a few others. For example, quartzite is composed of metamorphosed sandstone, and augen gneiss is metamorphosed granite. Metamorphic rocks are occasionally rich in Mg^{2+}-containing minerals, such as pyroxenes $(Ca(Mg, Fe)Si_2O_6)$. Serpentine metamorphic rocks are rich in Mg^{2+} because of the presence of antigorite $(Mg_3Si_2O_5(OH)_4)$, a phyllosilicate.

Sedimentary rock materials have already undergone some weathering before the rock is formed. Their composition represents depletion of weatherable elements, as in sandstone, or selective accumulation of minerals and elements, as in shale and limestone (**Table 6.10**). Sandstones are mostly accumulation of quartz grains. Shales are fine-grained sedimentary rocks that tend to form in regions of particulate accumulation, such as lake or sea floors. Thus, shales are richer than sandstones in K^+, Mg^{2+}, Fe^{2+} and Fe^{3+}, and

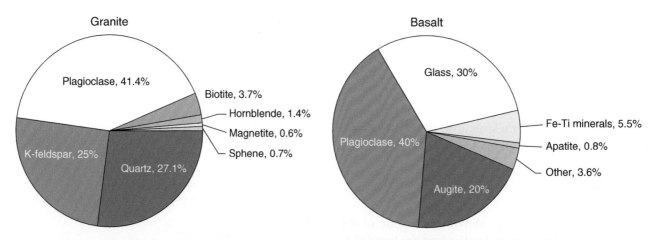

Figure 6.11 Mineral compositions of Half Dome granite and Columbia River basalt rocks. A large fraction of the Columbia River basalt is glass, which is an amorphous siliceous material rich in iron, magnesium, and aluminum. Augite is a pyroxene $((Ca, Na)(Mg, Fe, Al, Ti)(Si, Al)_2O_6)$. Data from Hausrath et al. (2009).

Table 6.10 Average elemental compositions of several rocks.

Element oxide	Granodiorite[a] (Granitic) (%)	Basalt[a] (%)	Shale[b] (%)	Sandstone[b] (%)	Limestone[b] (%)
SiO_2	65.1	49.3	58.1	78.3	5.2
K_2O	2.4	1.2	4.3	1.4	0.04
TiO_2	0.5	2.6	0.6	0.2	0.06
Al_2O_3	15.8	14.1	15.4	4.8	0.8
Fe_2O_3	1.6	3.4	4.0	1.1	0.5
FeO	2.7	9.9	2.4	0.3	–
MgO	2.2	6.4	2.4	1.2	7.9
CaO	4.7	9.7	3.1	5.5	42.6
Na_2O	3.8	2.9	1.3	0.4	0.05
H_2O	1.1	–	5.0	1.6	0.8
P_2O_5	0.1	0.5	0.17	0.08	0.04
SO_3	–	–	0.6	0.07	0.05
CO_2	–	–	2.6	5.0	41.5
Total	100	100	100	100	100

[a] From Tyrrell (1950).
[b] From Pettijohn (1957).

Ca^{2+}/Mg^{2+}-containing carbonates. The sodium content of sedimentary rocks is much lower than that of igneous rocks, although some marine sedimentary rocks contain entrapped sodium.

6.6 Stability of silicates to weathering

The major reaction process that weathers minerals is the dissolution of mineral-bound ions by water. In addition to the energy of hydration released when ions dissolve in water, an ion's entropy increases as it is freed from a rigid structure into aqueous solution because an ion's most stable state (lowest energy) is at infinite dilution in water. The amount of water in soils is limited to the thin film on soil particle surfaces and the water that exists in small pores, so the entropy increase due to dilution is limited to that extent.

The resistance of igneous minerals to weathering increases the same as the order of mineral crystallization from cooling magmas and is described by the **Goldich dissolution series (Figure 6.12)**. Minerals that are most stable at high temperatures (i.e., crystallize first from the molten magma or lava) are the least stable at low temperatures. Such minerals, including calcic feldspars, olivine, and hypersthene ((Mg, Fe)SiO_3), tend to be rich in the water-soluble alkali and alkaline earth ions, as well as Fe^{2+} ions. The weathering rate of

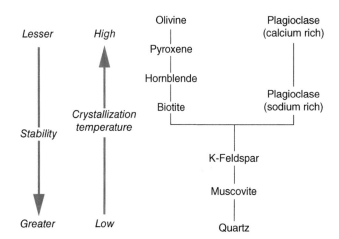

Figure 6.12 The Goldich dissolution series predicts that minerals that form at high temperature and pressure are the least stable at surficial temperatures. The left branch is a discontinuous series derived from ultra-mafic magma. The right is a solid–solution series that forms from mafic magma. The minerals lower in the diagram derive from more felsic magma. As magma temperature decreases, minerals with higher Si/O ratios begin to form, making them more stable to weathering.

minerals increases with increasing content of alkali and alkaline earth cations.

Another factor affecting mineral weatherability is the position of ions in the structure. For example, at room temperature, Al^{3+} is more stable in octahedral coordination than tetrahedral coordination. Isomorphic substitution also creates charge and strain in minerals, decreasing their stability and making them more soluble. For example, the tetrahedral position in calcium feldspars (e.g., anorthite) contains equal amounts of Al^{3+} and Si^{4+}. The charge deficit created by the Al^{3+} substitution is balanced by Ca^{2+} ions between the tetrahedra. The structural strain from the charge deficit in the tetrahedra and Ca^{2+} counter-charge weaken the calcium-feldspar structure compared to feldspar with less Al^{3+} isomorphic substitution, such as sodium and potassium feldspars that have only one-quarter of the tetrahedral positions occupied by Al^{3+}, which creates less strain and less charge deficit. Thus, calcium feldspars are more soluble than potassium and sodium feldspars. Sodium feldspars are more soluble than potassium feldspars because K^+ fits better between adjacent tetrahedra.

A third chemical factor affecting mineral weatherability is the degree to which the silicon tetrahedra are linked together. Increased linkage between tetrahedra means increased stability against weathering. Feldspars and quartz are three-dimensional networks of tetrahedra in which each of the four tetrahedral

oxygens is a corner of another tetrahedron. This maximum sharing of oxygen atoms produces considerable stability. Hence, quartz is very persistent in soils, especially as sand or silt-sized grains. Feldspars would be as resistant to weathering as quartz if not for the incorporation of Na^+, K^+, or Ca^{2+} in the feldspar mineral structure that creates instability that counteracts the stabilizing effect of tetrahedral linkage.

A fourth factor of mineral stability is the Fe^{2+} and Mn^{2+} contents. In aerobic soils, the presence of the reduced ions in mineral structures increases weathering rates because oxidation to Fe(III) and Mn(III, IV) creates charge imbalances within the minerals. In anaerobic soils, on the other hand, minerals containing oxidized Fe^{3+} and Mn^{3+} or Mn^{4+} are unstable.

Figure 6.13 summarizes the relative weathering rates of common primary silicates that occur in rocks. In soils, weathering rates depend on temperature and moisture, mineral particle size, and planes of physical weakness (cleavage) in the crystal. Moisture effects include both the rate of flow of soil solution past mineral surfaces, and the composition of the solution. Solids dissolve more slowly if the solution already

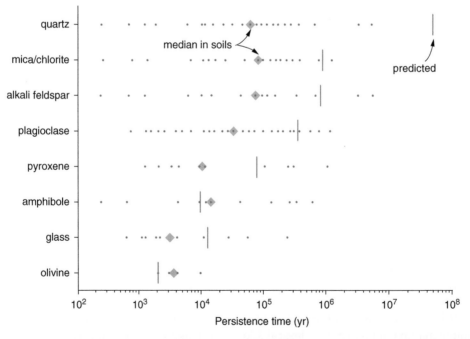

Figure 6.13 Median persistence times of silicate mineral in soils as judged from chronosequences. Diamonds are averages and gray circles are individual observations. Vertical bars show the persistence time of a 1-mm-sized pure mineral particle calculated from laboratory-measured mineral dissolution rates. Mineral persistence in soils varies with crystallinity and morphology, weathering factors such as climate, organisms (organic acids), and solute fluxes through the soil. Adapted from Hausrath et al. (2008).

contains ions held within the mineral. For example, if the solution contains high concentrations of Ca^{2+}, then calcite will not dissolve because of the common ion effect. High electrolyte concentrations, on the other hand, can maintain higher ion concentrations at equilibrium because of lower activity coefficients and because of ion-pair formation or complexation.

Physical weathering is a major weathering process at cleavage planes or stress fractures that allows particles to be broken apart. Feldspars and micas have well-defined cleavage planes that hasten the rate of mineral breakdown.

Particle size affects the susceptibility of the minerals to weathering reactions because of increased solid–solution interface. Thus, on an equal mass basis, smaller particles weather more rapidly than larger particles. The size effect is most significant when the particles are less than a few micrometers in size.

Figure 6.13 shows the *predicted* persistence of silicate minerals and shows the *observed* mineral persistence in several soils. Agreement is good, but, in many cases, mineral persistence is much less than theoretically predicted. This is because most minerals have less than perfect crystallinity, and microbial and organic acid reactions on minerals enhances their weathering rates.

6.7 Chemistry of soil weathering and mineral formation

6.7.1 Initial breakdown of primary minerals

Rocks formed beneath Earth's surface are unstable when raised to the surface where they contact water, O_2, CO_2, and organic acids. Soil is formed when the ions in rock minerals at Earth's surface change or weather to more stable chemical states. Soil development, in the chemical sense, is synonymous with weathering. This section discusses the general weathering reactions that are driven by reactions with water, O_2, and CO_2, which create soil minerals (secondary minerals).

Weathering of igneous and metamorphic rocks changes the dense solids into unconsolidated particles that are distinct from the chemical composition and structure of the parent minerals. The changes during weathering of sedimentary rocks are less striking than

when igneous and metamorphic rocks weather; thus, the boundary between sedimentary rocks and soil is often physically and chemically diffuse.

Table 6.10 shows the elemental composition of common rocks that are soil parent materials. The crystal structures and ion valences in rock minerals are stable at the conditions that the rocks formed. The physical processes of erosion, freezing and thawing, glaciation, heating and cooling, and root growth at Earth's surface breaks rocks apart and exposes more surfaces for chemical weathering. A bigger change in the rock minerals, however, results from the chemical conditions within the soils that dissolve the minerals, including exposure to water, oxygen, carbon dioxide, and organic acids and chelates.

Some ions that dissolve from primary minerals into the soil solution combine to create new solids; other ions are leached away. This means that the composition of the soil solution changes during soil development and may change enough to cause the first-formed soil minerals to re-dissolve and portions of them to re-precipitate as still other minerals.

Some minerals remain virtually unweathered because their dissolution rate in water is exceedingly slow. Quartz particles larger than several micrometers in size (fine silt) remain in soils for a long time because of their resistance to weathering. Feldspar disappears from the sand and silt fractions more rapidly than quartz (**Figure 6.13**). When finely divided into clay-sized particles, however, quartz persists only slightly longer in soils than clay-sized feldspar.

6.7.2 Formation of soil minerals

New solids in soils form by dissolution of the old mineral and subsequent precipitation of the solute ions. When only part of the dissolved mineral solute re-precipitates, the overall process is called **incongruent dissolution/precipitation** reaction. **Congruent dissolution/precipitation** is complete dissolution of the mineral and formation of a secondary mineral from the dissolved ions. Soil weathering reactions are commonly incongruent dissolution reactions because silicate minerals often do not completely dissolve and re-precipitate; instead, the silicate minerals partially dissolve and the ions released form secondary clay minerals and amorphous solids.

Dissolution of minerals and formation of new minerals (**neogenesis**) is an important pedogenic process that is summarized by the following concepts:

1 Weathering of igneous and metamorphic rocks releases considerable alkali and alkaline earth cations during the initial transition from rock to soil. Some of these cations are retained by plant absorption and cation exchange, but most are readily lost from the soil.
2 Weathering releases considerable silica to the soil solution, much of which leaches from the soil. The silica that is not leached reacts to form secondary minerals such as kaolinite, smectite, and chlorite, which are common in soils, but transitory on a geologic time scale.
3 Aluminum and iron hydroxides are insoluble and tend to accumulate in soils (titanium and manganese oxides are also insoluble but are not as prevalent in rocks and soils).
4 Weathering initially produces some alkalinity.
5 The second stage of weathering or soil development produces acidity.

The secondary minerals formed in soil tend to be small and have varying degrees of crystallinity – from crystalline to amorphous. Minerals formed in soils are primarily aluminosilicates, aluminum and iron oxides, and calcite in semi-arid and arid climates. The small secondary minerals have large surface areas and tend to be charged because of unsatisfied ion charges within the crystals and at crystal edges. Large surface areas and unsatisfied bonds make secondary minerals very reactive with ions in soil solution, including protons and hydroxides.

The weathering rates of smectites and chlorites in soils, and leaching of their ions, are similar to their growth rates, and thus smectite and chlorite crystals larger than 1 μm are rare. Crystal growth is more evident for kaolinite. **Figure 6.14** shows the effects of time on the number of clay-sized *particles* in a chronosequence of soils. The clay particles in the chronosequence soils are clay minerals such as smectites and chlorites, and iron and aluminum oxides.

Secondary minerals continue to weather because they are stable only within certain concentration limits of soluble silica, alkali and alkaline earth cations, and H^+ in the soil solution. As solutes are leached away, many of the secondary minerals (smectites, calcite, gypsum, etc.) dissolve. Thus, as weathering progresses, the secondary mineral suite progresses to more stable (less soluble) secondary minerals, which include iron and aluminum oxides and kaolinite.

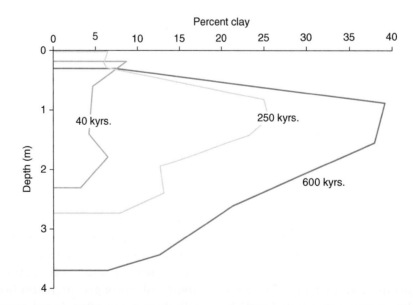

Figure 6.14 Clay content in a chronosequence of soils developed on granitic sands. Shallow argillic (clay accumulation) horizons develop as a result of mineral precipitation and illuviation processes. Data are from soils developed on a series of terraces of different age that span from the base of the Sierra Nevada mountains (California) west to the Merced River. Adapted from White et al. (2005). Reproduced with permission of Elsevier.

6.7.3 Weathering effects on element composition in soils

The effect of weathering reactions on the total composition of soils is illustrated by the data in **Table 6.11**. Three soils of increasing maturity, or degree of weathering, are compared to the average composition of igneous rocks. The comparisons are not exact because the parent materials of the soils differ from each other and from the average igneous rock.

The first stage of weathering releases large amounts of Ca^{2+}, Mg^{2+}, Na^+, and K^+ from the rock minerals, shown in **Table 6.11** by the change in composition from an igneous rock to the Barnes soil (Mollisol). Calcium and Na^+ cations form *pedogenic* minerals, such as carbonate minerals, in regions with limited precipitation. Most of the alkali and alkaline earth cations remaining after the first weathering stage are in large unweathered mineral grains. Smaller fractions of Ca^{2+}, Mg^{2+}, Na^+, and K^+ are retained by adsorption to negatively charged secondary mineral particles and change less than the total amounts of these ions. These fractions are significant because they supply these essential macro-elements to plants and soil microbes, they are subject to further leaching losses, and they influence soil pH.

The efflux of cations slows markedly after initial breakdown of rock minerals. The difference in total composition between the immature Barnes soil and the more mature Cecil soil is much less than that between the Barnes soil and igneous rock. The differences in soil maturity between the Barnes and the Cecil stages involve primarily the rearrangement of elements into secondary soil minerals. Rearrangements plus slight differences of total composition can create large differences in the availability of ions for plant growth. Phosphate is probably more available in the Barnes soil, for example, than in the Cecil soil, although the total amounts of phosphate are similar. Soluble plus exchangeable aluminum reaches phytotoxic concentrations in the Cecil soil. The slight differences in total elemental composition of the two soils can encompass a wide range of secondary mineral compositions.

The phosphate content of soils tends to remain constant during soil development. Phosphate is only slowly leached from soils, so the total phosphorus content of soils varies little with soil maturity. The form of phosphate, however, changes from predominantly apatite ($Ca_5(OH, F)(PO_4)_3$) in igneous rocks to aluminum and iron phosphates in moderate to strongly weathered soils.

Soil sulfur, nitrogen, and carbon are associated with the soil's organic fraction, and do not change with weathering as much as other elements. Their soil contents are related to biological activity and climate. Thus, while sulfates are potentially easily leached from soils, plant and microbial uptake of sulfate and its incorporation into organic compounds tends to maintain the sulfur content of soils. Ammonium is an exception because it can be fixed in clay mineral interlayers.

Table 6.11 Composition of average igneous rocks and of three surface soils of increasing maturity. Data from Byers et al. (1935).

Element oxide	Average of igneous rocks	Barnes loam (South Dakota) Mollisol	Cecil sandy clay loam (North Carolina) Ultisol	Columbiana clay (Costa Rica) Oxisol
SiO_2	60	77	80	26
Al_2O_3	16	13	13	49
Fe_2O_3	7	4	5	20
TiO_2	1	0.6	1	3
MnO	0.1	0.2	0.3	0.4
CaO	5	2	0.2	0.3
MgO	4	1	<0.1	0.7
K_2O	3	2	0.6	0.1
Na_2O	4	1	0.2	0.3
P_2O_5	0.3	0.2	0.2	0.4
SO_3	0.1	0.1	–	0.3
Total	100.5%	100.9%	100.6%	100.4%

Soil development involves a steady loss of silicon. However, this is not clear from the composition of the Barnes and Cecil soils in **Table 6.11**. The parent materials of these soils are apparently silica-rich. The loss of silicon is evident, however, from the low SiO_2 content of the highly weathered Columbiana soil. As the SiO_2 content of this soil decreased to less than half the average for igneous rocks, the Al_2O_3 and Fe_2O_3 contents increased threefold. The silica loss is from dissolution, not erosion.

The solubilities of iron, aluminum, titanium, and manganese oxides are much lower than the solubility of silica. These oxides are more stable than the secondary silicates that might have formed earlier in this soil. Aluminum accumulates in the clay mineral fraction because it precipitates as aluminosilicates and hydroxides. The aluminum remains behind in the soil as other ions leach away. Iron also accumulates in soils where it precipitates as oxides. Illite is altered parent material, and unlike secondary phyllosilicates such as kaolinite, montmorillonite, and allophane, is not precipitated from the soil solution.

Some weathered solutes reprecipitate in lower soil horizons. Examples include clay accumulation in the B horizon, silica pans (impermeable layers of soil particles indurated with silica), and the widespread caliche horizons of $CaCO_3$ accumulation in semi-arid and arid regions.

Although the net effect of weathering is the loss of soluble components from the soil, the effects of weathering on ion distribution in soils is not continuous or unidirectional. For example, during dry seasons, the soil solution can flow upward; and plant uptake and decay deposit ions on the soil surface. The rates of ion absorption by plants during nutrient cycling are far greater than the rates of weathering, and thus plant uptake reduces weathering losses of ions from the soil.

Large-scale losses of ions from soils usually occur only during periods of high rainfall or limited plant growth due to overgrazing, forest clearing, or fires. Chemicals are also moved physically in the soil profile by pedoturbation; by soil fauna such as worms, termites, and ants; and by uprooting of fallen trees. In montmorillonitic soils (Vertisols), pedoturbation occurs where extensive vertical cracks as deep as 1 m form in the dry season and soil particles break from the walls of the cracks and fall to the bottom. When a Vertisol soil is wetted, the fallen particles swell and force the overlying soil upward, causing a slow,

continual mixing of the surface meter of soil. Human influences such as amendments, pollution, and plowing also dramatically alter the distribution of chemicals within the soil profile.

Salt accumulation due to impeded soil drainage or seawater inundation enriches ions in the soil profile. Weathering under these conditions reverses in the sense that elements that are unstable under well-drained oxidative conditions are deposited in the surface horizon. Accumulation of salts decreases the productivity of soils for agricultural crops and is a severe problem in arid regions that have poorly drained soils and rely on irrigation (see Chapter 13). Better drainage and excess water are needed to leach salts out of the soil profile.

6.8 Formation of secondary minerals in soils

In the previous sections, factors that affect silicate mineral solubility, pedogenic processes, and effects of weathering on ion movement in soils were discussed. All these factors affect the formation of secondary minerals in soils. **Table 6.12** lists the common secondary minerals in soils. Formation of iron and aluminum oxides occurs from the precipitation of the poorly soluble ferric iron and aluminum ions. Prediction of these reactions based on thermodynamics was discussed in Chapter 4 (see for example Figure 4.13 and 4.14). Formation of secondary aluminosilicates, such as phyllosilicates, is much more complex to predict than oxide precipitation.

An early theory of soil clay mineral formation was that they form solely by differential migration of ions into and out of existing silicate structures (incongruent dissolution/precipitation reaction). While this type of mineral formation is not a predominant mechanism for soil mineral formation, some secondary 2:1 layer silicates are formed by solid-phase changes of mica fragments inherited from the parent material. Illite, for example, is a product of physical breakdown of mica particles to smaller particles with simultaneous loss of interlayer K^+ cations at the edges of the particle by cation exchange reactions. Illite can be further modified to vermiculite by incongruent weathering reactions. The illite-to-vermiculite weathering process is not completely understood, probably because it is highly variable in nature. But, it likely involves the outward diffusion of

Table 6.12 Common secondary minerals present in soils in order of increasing stability. Source: Feldman and Zelazny (1998). Reproduced with permission of ACSESS.

Name	Mineral class	Environment	Ubiquity in soils[a]	Importance
Halite	Halide	Arid, saline/sodic soils	C	Highly soluble; adverse osmotic effects on plants
Gypsum	Sulfate	Arid soils	C	Moderately high solubility; used to reclaim sodic soils
Jarosite	Sulfate	Add sulfate soils; mine overburden, coastal wetlands	C	Product of pyrite oxidation resulting in large production of acidity and toxic metals
Calcite	Carbonate	Arid soils; very limited leaching	C	May act as cementing agent; high P sorption
Pyrite	Sulfide	Tidal marshes (reducing conditions) and hard-rock mine tailings (coal and shale beds)	C	Primary mineral (oxidizing conditions) but poorly crystalline secondary phase forms in reducing environments
Ferrihydrite	Oxide	Wetland soils, soils derived from volcanic ash deposits	C	Product of rapid weathering and oxidation of Fe(II)
Allophane/Imogolite	Alumino-silicate	Soils derived from volcanic ash deposits	R	Short-range order, highly adsorptive
Sepiolite/Palygorskite	Phyllosilicate	Marine sediments, arid soils, high Si and Mg levels	R	Moderately high CEC, surface area, and adsorptive properties
Halloysite	Phyllosilicate	Volcanic ash; granitic (feldspathic) saprolite	R	Ephemeral in intensely weathered soils
Vermiculites (dioct.)	Phyllosilicate	Mica alteration in well-drained soils	R	Very high CEC; fixation of K^+; sink for solution Al
Smectites	Phyllosilicate	Mica and/or vermiculite alteration	C	High CEC; high surface area; high shrink-swell capacity
hydroxy-interlayered vermiculite	Phyllosilicate	Acid, highly weathered soil surface horizons	U	Variable CEC; (Al and Mg hydroxide interfilling); high anion adsorption
Kaolinite	Phyllosilicate	Desilication of 2:1 clays/feldspathic	U	Low charge but highly pH-dependent; high anion adsorption
Hematite & goethite	Oxide	Well-drained, near-surface soil	U	Low charge but highly pH-dependent; high anion adsorption
Gibbsite	Oxide	Old, stable soils or feldspar pseudomorphs	C	Low charge but highly pH-dependent; high anion adsorption
Anatase	Oxide	Dissolution of Ti-bearing parent minerals	R	Unreactive

[a] U, ubiquitous; C, common; R, rare.

K^+ from between the layer lattices, and a subsequent or simultaneous reduction of charge within the layer lattice. Further alteration of illite and vermiculite via congruent or incongruent reactions to montmorillonite or chlorite occurs with more advanced weathering.

Kaolinite, the most widespread of the crystalline clay minerals, has no structural counterpart in igneous minerals. The mechanism for kaolinite formation is the complete breakdown of other silicate minerals, such as feldspar or phyllosilicates, and the precipitation of kaolinite from Al^{3+} and Si^{4+} ions in solution. The smectites nontronite and montmorillonite are also likely formed in soils by precipitation from solution; as are poorly crystalline aluminosilicates such as allophane and imogolite.

Phyllosilicates have distinct chemical and physical characteristics that affect their mineral properties. Montmorillonite, for example, has a diffuse negative surface charge due to isomorphic substitution in its structure. The crystal habit of montmorillonite is platy, and the particles stack on top of each other like a stack of dish plates. The space between individual clay platelets (layers) creates **interlayers**, which contain water that is connected to, but distinct from, the bulk soil solution. Cations are attracted, and anions are repelled from the negatively charged interlayers. Because clay mineral interlayers are hydrated, cations can exchange via cation exchange reactions. In addition to cation exchange, the surface charge and hydrated interlayer in montmorillonite create high shrink-swell behavior.

6.8.1 Prediction of secondary mineral formation

Assuming the soil is at equilibrium, mineral formation can be predicted using thermodynamic properties of ideal minerals. Many mineral precipitation and dissolution reactions are slow, however, so the assumption of equilibrium may be incorrect. In addition, uncertainties in thermodynamic constants are often large, especially for mineral phases that are non-ideal with respect to ion composition and degree of crystallinity. These factors, and slow equilibrium of silicate minerals, make accurate thermodynamic prediction challenging. Despite such limitations, thermodynamic predictions can be used to generate mineral solubility diagrams that provide a general understanding of mineral weathering and soil development processes.

Activities of dissolved Si^{4+}, Al^{3+}, and H^+ are the predominant factors typically considered in aluminosilicate mineral dissolution and precipitation modeling. Activities of other ions such as Na^+, Ca^{2+}, K^+, and Mg^{2+} are also important, but these ion activities are often fixed at likely concentrations to reduce the number of variables in the model to less than three, allowing for plotting on a two-coordinate graph to show the conditions where the mineral is stable. The dissolution reaction of kaolinite is

$$Al_2Si_2O_5(OH)_4 + 6H^+ = 2Al^{3+} + 2H_4SiO_4 + H_2O \quad (6.2)$$

The corresponding kaolinite equilibrium reaction constant (solubility product) is

$$K_{sp} = \frac{\left(Al^{3+}\right)^2 \left(H_4SiO_4\right)^2}{\left(H^+\right)^6} = 10^{5.45} \quad (6.3)$$

Taking the logarithm of **Eq. 6.3**, and solving for (Al^{3+}) yields

$$\log\left(Al^{3+}\right) = 2.73 - \log\left(H_4SiO_4\right) - 3pH \quad (6.4)$$

Equation 6.4 shows that the Al^{3+} activity is a function of pH and the activity of the silicic acid in solution. This theoretical relationship can be plotted on a three-coordinate axis graph, called a stability diagram. Three-coordinate graphs, however, are difficult to interpret when other mineral phases are plotted on the same graph, which is one of the most useful aspects of a stability diagram. Instead of a three-axis graph, **Eq. 6.4**

can be transformed into a two-axis stability diagram by combining the variables (Al^{3+}) and pH:

$$\left[\log\left(Al^{3+}\right) + 3pH\right] = 2.73 - \log\left(H_4SiO_4\right) \quad (6.5)$$

The left-hand side of **Eq. 6.5**, $[\log(Al^{3+})+3pH]$, can be plotted as a dependent variable on the y-axis as a function of the independent variable $\log(H_4SiO_4)$ plotted on the x-axis (**Figure 6.15**). The line in **Figure 6.15** represents the equilibrium conditions for the kaolinite dissolution reaction in **Eq. 6.5**. If solution activities fall above the line, the solution condition is supersaturated with respect to kaolinite, indicating that kaolinite should precipitate from solution. If solution activities fall below the line, solution conditions are unsaturated, so kaolinite should dissolve. A simpler way to view the two-variable y-axis is to consider a single pH value; for example, $\log(Al^{3+})+3pH$ at pH = 6 is shown as a second y-axis in **Figure 6.15**. The numbers on the second y-axis are equal to $\log(Al^{3+})$ when the pH = 6. A second line can be plotted for another pH on the graph to show how the solubility of kaolinite changes as a function of pH.

Stability diagrams such as **Figure 6.15** can be used to evaluate mineral precipitation and dissolution trends. For example, if silicic acid activity decreases, the stability field diagram predicts that kaolinite will dissolve, and $\log(Al^{3+})$ will increase. Within soil, desilication causes increased Al^{3+} activity, which may exceed the solubility of gibbsite $(Al(OH)_3)$ causing it to precipitate from solution. This is the general weathering trend observed in soils.

A main use of a stability diagram is to compare solubilities of different minerals to determine the most stable phase in a system. For example, the dissolution reaction of ideal montmorillonite is

$$Mg_{0.33}\left(Al_{1.44}Mg_{0.66}\right)\left(Si_4\right)O_{10}\left(OH\right)_2 + 6H^+ + 4H_2O$$
$$= 0.99Mg^{2+} + 1.44Al^{3+} + 4H_4SiO_4 \quad (6.6)$$

From the dissolution reaction, a solubility product can be written and $\log(Al^{3+})+3pH$ can be solved for as a function of $\log(H_4SiO_4)$ in a similar manner as done for kaolinite above. However, (Mg^{2+}) is an added variable and pH will end up on both sides of the equation[1].

---

[1] Derivation of the montmorillonite dissolution equation is not covered here; details are presented in Kittrick (1971) and Kittrick (1969).

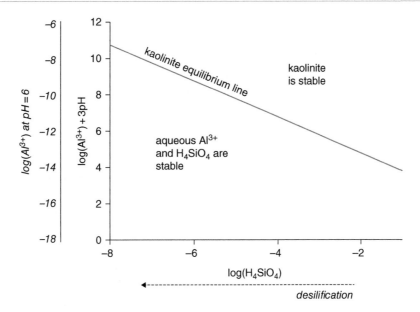

Figure 6.15 Stability diagram showing kaolinite solubility at 25 °C. One y-axis shows the $\log(Al^{3+})$ value in a system if pH is fixed at six; the other shows $(\log(Al^{3+}) + 3pH)$ values. Above the line (H_4SiO_4) and (Al^{3+}) are supersaturated with respect to kaolinite; below the line is undersaturation.

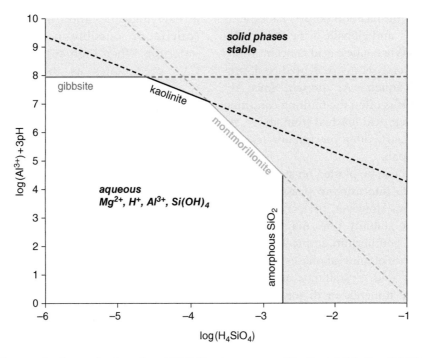

Figure 6.16 Stability diagram for four soil minerals at 25 °C. For the montmorillonite stability line (Mg^{2+}) is fixed at 10^{-3} M and pH is fixed at 7. The gray zone is the range where solid mineral phases are stable products. Adapted from Lindsay (1979).

By fixing Mg^{2+} activity and pH, a montmorillonite stability line can be plotted with the kaolinite diagram (**Figure 6.15**). This allows montmorillonite solubility to be compared to kaolinite (i.e., **Eq. 6.5**). **Figure 6.16** shows the montmorillonite dissolution stability line, along with the stability lines for three other minerals controlling Al^{3+} and Si^{4+} solubility in soils. Fixing pH for the montmorillonite solubility line and having it as a variable on the y-axis is counterintuitive; however, doing so facilitates comparison of solubilities for different aluminosilicate minerals such as kaolinite. With change in pH, the slope of the montmorillonite solubility line remains constant, but shifts up or down about 0.5 y-axis units for each one pH unit change.

The solid lines in **Figure 6.16** show the borders for aqueous transition from solution to the solid phases montmorillonite, kaolinite, gibbsite, or amorphous SiO_2. The shaded zone in **Figure 6.16** shows where the system is saturated with respect to at least one solid. Thus, if ion activities and pH conditions are within the shaded zone, precipitation from solution will occur. However, because rates of mineral dissolution and precipitation in soils are often slow, over- and undersaturated solutions often occur.

The order of mineral stability in **Figure 6.16** matches the weathering conditions observed in soils. As silicic acid is leached, its activity in the overlying soil decreases, causing montmorillonite to dissolve, and kaolinite is the stable mineral species. As weathering continues and further desilication occurs, kaolinite is unstable and dissolves, and gibbsite is the stable mineral phase predicted. When silicic acid concentrations are increasing, the reverse sequence is predicted, provided there is enough aqueous Al^{3+} present. Since Al^{3+} is not very soluble, this is often the limiting condition in zones receiving silicic acid leached from above. As a result, amorphous quartz often precipitates, leading to formation of a duripan.

Soil pore water data can be plotted on stability diagrams to gain insights into mineral phases that are controlling equilibrium. However, perfect agreement between data and the stability lines may not occur because of the lack of equilibrium, imprecise equilibrium constants, presence of mineral phases with varying degrees of purity and crystallinity, and lack of inclusion of important minerals or dissolved phases (e.g., aqueous complexes) that are reacting in the system. Despite these limitations, comparison of stability diagrams to soil pore-water composition is useful for predicting minerals that may occur in the soil, and whether minerals should be dissolving or precipitating. In-situ characterization of soil–solid phases increases the ability to predict mineral dissolution and precipitation processes in soils (see Section 6.14).

In this section, derivation and interpretation of weathering stability diagrams for secondary aluminosilicates was discussed. Similar diagrams can be developed for other mineral species of interest, including non-aluminosilicates. Depending on the mineral, the aqueous activities on the axes are changed to show their relationship and mineral solubility.

6.9 Soil carbonates

In regions of limited rainfall, carbonates (particularly $CaCO_3$) accumulate in soils. Where evapotranspiration exceeds precipitation, the downward flow of water through the soil profile is sufficient to remove only highly soluble weathering products, such as Na^+ salts. Even soils where the amount of percolating water is 1% or less of the total rainfall, intermittent rains flush out soluble salts. Salts and minerals that are less soluble, on the other hand, accumulate because of limited water flow and subsequent evaporation. Secondary silicates containing Ca^{2+} as a structural ion are rare, but Ca^{2+} remains instead as an exchangeable cation, or precipitates as the $CaCO_3$ (**calcite** or possibly polymorphs aragonite and vaterite). When high sulfate concentrations occur, Ca^{2+} precipitates as the more soluble **gypsum** ($CaSO_4 \cdot 2H_2O$). Figure 1.11 illustrates pedogenic carbonate formation processes, and the formation of a calcic horizon. Calcite minerals formed in soil do not substitute very much Mg^{2+} into their structure; thus, dolomite ($CaMg(CO_3)_2$) formation in soils is uncommon; however, it may occur if inherited from parent materials.

The carbonate anion in carbonate minerals consists of a carbon atom coordinated to three oxygen atoms forming a planar triangle. Due to the high charge of C^{4+} ion, the bonds between carbon and oxygen are strong, thus forming the molecular *carbonate unit*. Oxygen cannot share electrons with more than one carbon atom in carbonate minerals, so carbonate oxygen ligands are bonded to Ca^{2+} cations, which *glue* the carbonate triangles together. A small amount of isomorphic substitution of Mg^{2+}, Sr^{2+}, or metals such as Zn^{2+} into the calcite structure may occur.

The effect of chemical processes in oceans and freshwaters on $CaCO_3$ solubility has been considered at great length by geochemists. However, carbonate chemistry in surface waters is simpler than in soils, which have limited water content, varying partial pressures of CO_2 (P_{CO2}), and multiple sources of Ca^{2+}. In addition, P_{CO2} varies with the rate of upward diffusion of CO_2 to the atmosphere and with plant root and microbial respiration. In flooded soils and soils with high amounts of root and microbial activity, P_{CO2} can range from slightly less than the overlying atmospheric P_{CO2} to 100 times greater.

Although the environmental conditions that bring about carbonate accumulation in soils are varied, soil Ca^{2+} ion activity, which is controlled by dissolution of calcite and exchange of clay minerals, and soil pH and partial pressure of CO_2, are the main factors that control its formation:

$$CaCO_3(s) + 2H^+ = Ca^{2+} + H_2O + CO_2(g) \qquad (6.7)$$

Acidity favors $CaCO_3$ dissolution and increasing P_{CO2} favors $CaCO_3$ precipitation if Ca^{2+} ions are present. Dissolved Ca^{2+} and bicarbonate ions in soils may be moved out of one zone to another. In horizons that become enriched with these ions, calcite reprecipitates as pedogenic calcite (**calcification**):

$$Ca^{2+} + HCO_3^- = CaCO_3(s) + H^+ \qquad (6.8)$$

The high Ca^{2+} concentrations and limited water contents in soils favor calcite precipitation reactions, as described in **Eq. 6.8**. However, acidification via management practices (e.g., ammoniacal fertilizers) or acidic inputs (e.g., acid rain) favors dissolution reaction according to **Eq. 6.7**. The reversible reactions are describing the same process, but they are written differently for emphasis.

When the mass of $CaCO_3$ in soils exceeds several percent, it controls both soil pH and soil solution Ca^{2+} concentrations. The Ca^{2+} activity can be calculated from equilibrium expressions of the reactions in **Eq. 6.7** (or **Eq. 6.8**), and the solubility product of $CaCO_3$:

$$K_s = \frac{(Ca^{2+})(CO_2)}{(H^+)^2} = 10^{9.74} \qquad (6.9)$$

where (CO_2) is the partial pressure of carbon dioxide. Rearranging to solve for $\log(Ca^{2+})$:

$$\log(Ca^{2+}) = \log(K_s) - \log(CO_2) - 2pH \qquad (6.10)$$

The solubility of calcite as a function of pH at three different partial pressures of CO_2 is plotted in **Figure 6.17**. The solubility line shows that as pH decreases, (Ca^{2+})

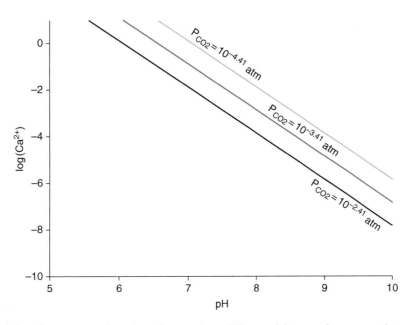

Figure 6.17 Calcite solubility diagram as a function of pH at three different CO_2 partial pressures (atmospheric, 10 times more, and 10 times less atmospheric).

increases from dissolution of calcite, and as P_{CO2} increases, (Ca^{2+}) decreases because of calcite precipitation. This model can be applied successfully to arid soils to predict calcite precipitation and dissolution reactions. In soils, an important consideration that may not be included in the models is the inhibitory effect of organic matter on carbonate precipitation because SOM-Ca^{2+} complexes decrease the Ca^{2+} species activity.

Calcite buffers soil pH (see example in Eq. 4.76 and Figure 4.9). If exchangeable sodium is low, the pH of calcareous soils when measured in the laboratory is typically around 8.3. Field-pH values of calcareous soils are usually less than 8.3 because the CO_2 concentration is higher in the soil's gas phase than in the atmosphere due to root and microbial respiration (see Figure 4.9). The average P_{CO2} in the pores of agricultural soils can be 10 to 100 times that of the atmosphere. Many workers use the value of 0.01 atm as the typical P_{CO2} in soil pores. In warmer and drier soils, lower P_{CO2} is likely. The actual CO_2 concentration depends on the rate of microbial and root respiration, and CO_2 diffusion to the atmosphere.

6.10   Evaporites

Soluble salts accumulate in soils where drainage water from surrounding soils accumulates, or where the amount of percolated water is small compared to the amount of water evaporated. The term *evaporite* refers to compounds more soluble than $CaCO_3$. Soil salinity is dealt with in more detail in Chapter 13; in this section, the types of mineral salts that occur in soils are discussed.

Salt minerals line pores of saline soils and can form extensive masses or crust on the soil surface in evaporative basins, salt flats, and playas (former and intermittent lake beds). The salts that form in soils include sodium salts such as **halite** (NaCl) and trona ($NaHCO_3$). Smaller amounts of sulfates, borates, potassium, and occasionally lithium salts precipitate in soils having high rates of evaporation. In alkaline salty soils, some secondary silicates also form, including the zeolite analcime ($NaAlSi_2O_6 \cdot 6H_2O$) and sepiolite ($Mg_4Si_6O_{15}(OH)_2 \cdot 6H_2O$). **Figure 6.18** shows a theoretical evaporative sequence for a salt solution. In an evaporative sequence,

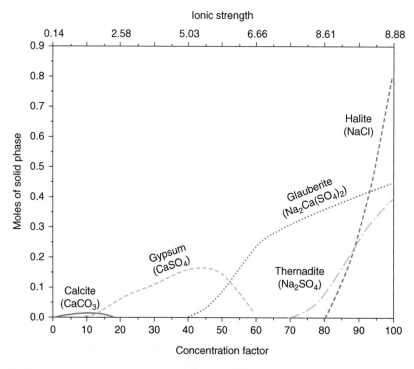

Figure 6.18 Sequence of salt precipitation from a hypothetical solution containing Na^+, Ca^{2+}, Mg^{2+}, SO_4^{2-}, Cl^-, HCO_3^- at pH 7.4 as a function of concentration increase caused by evaporation. Adapted from Doner and Grossl (2002). Reproduced with permission of ACSESS.

calcium carbonate minerals precipitate first, followed by sulfate minerals, and then, at the highest ionic strength, NaCl precipitates.

6.11 Soil phosphate minerals

Phosphate minerals are minor constituents of soils, but, given phosphorus's importance in plant nutrition and surface water quality, the mineral phases of phosphorus in soils are important to understand. The primary phosphorus molecule in the environment is orthophosphate, which strongly adsorbs to mineral surfaces. However, several phosphate minerals occur in soils that control its availability in the environment (**Table 6.13**). Below pH 6, iron and aluminum phosphate minerals with strengite- and variscite-like compositions are important. In alkaline soils, calcium phosphate minerals form (**Table 6.13**). Many calcium phosphate minerals formed in soils are metastable, such as OCP and β-TCP, which means they will transform to more stable phases over time. The solubility (and stability) of calcium phosphate minerals decreases in the order brushite > monetite > OCP > β-TCP > **hydroxyapatite** > fluorapatite. Magnesium phosphate minerals, such as struvite, may form in soils upon initial phosphate fertilization; however, they are more soluble than all the calcium phosphate minerals, and thus are typically transient. However, in soils with high amounts of magnesium, such as manure-amended soils, magnesium-calcium phosphates can persist because of continuous fluxes of the ions into the soils.

The reaction processes and rates of calcium phosphate mineral formation and dissolution in soils depend on pH, organic matter, ion composition, adsorbents, temperature, and ionic strength, which may be affected by soil management. Hydroxyapatite and fluorapatite minerals are the most thermodynamically stable, and least soluble calcium phosphate minerals in alkaline soils, but their formation is controlled by time-dependent processes (see Figure 4.10). Apatite may not precipitate directly from solution, but instead form from ripening of poorly crystalline or metastable phases. **Figure 6.19** shows the solubility of phosphorus from a soil as compared to theoretical phosphorus solubility of calcium phosphate minerals. The initially leached solutions from the soil appear to be in equilibrium with metastable calcium phosphate mineral phases, in agreement with the concept that such phases control phosphate availability in calcium-rich soils.

6.12 Sulfur minerals

Two common inorganic forms of sulfur occurring in soils are sulfide and sulfate. Sulfate is a large divalent oxyanion consisting of S^{6+} in tetrahedral oxygen coordination. Sulfide is a smaller S^{2-} ion that forms metal sulfide compounds, and hydrogen sulfide gas. In wetland and marsh areas, sulfide exists in the reduced soils as metal sulfides. In oxidized soils, the stable form of sulfur is sulfate (SO_4^{2-}). The most common soil sulfate mineral is gypsum ($CaSO_4 \cdot 2H_2O$), which forms white to clear crystals that fill pores and cracks in soils in semi-arid and arid regions. In extremely dry soils, anhydrite ($CaSO_4$) may form (see Figure 4.2).

In acid-sulfate soils (low pH), Fe(III)-sulfate minerals are common; including jarosite ($KFe_3(SO_4)_2(OH)_6$) and schwertmannite ($Fe_8O_8(OH)_6SO_4$). Schwertmannite is unstable and transforms to goethite or some other iron oxide at equilibrium (*ejecting* the sulfate). Jarosite is more common than schwertmannite, and occurs in

Table 6.13 Phosphate minerals that occur in soils. Relative solubility of variscite and strengite depend on the solid phases controlling Al^{3+} and Fe^{3+} activities, respectively.

Mineral	Formula	Relative solubility
Acidic soils		
Strengite	$FePO_4 \cdot 2H_2O$	Depends on Fe activity
Variscite	$AlPO_4 \cdot 2H_2O$	Depends on Al activity
Alkaline soils		
Struvite	$MgNH_4PO_4 \cdot 6H_2O$	Most soluble
Newberryite	$MgHPO_4 \cdot 3H_2O$	
Brushite	$CaHPO_4 \cdot 2H_2O$	
Monetite	$CaHPO_4$	
Octacalcium phosphate (OCP)	$Ca_4H(PO_4)_3 \cdot 2.5H_2O$	
β-tricalcium (β-TCP)	β-$Ca_3(PO_4)2$	
Hydroxyapatite (HA)	$Ca_5(PO_4)_3OH$	
Fluorapatite (FA)	$Ca_5(PO_4)_3F$	
		Least soluble

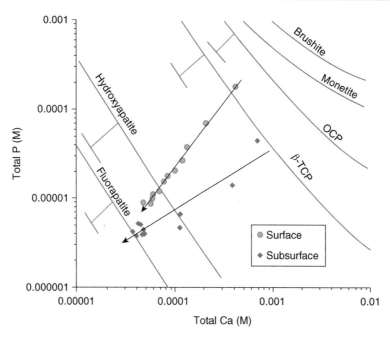

Figure 6.19 Total phosphorus in solutions as a function of total calcium concentration when the extraction solution (water) was sequentially replaced with a fresh solution. The curves are for several Ca-P minerals (pH = 7.5, [F$^-$] = 10^{-4} M). Arrows indicate sequence of solution replenishment: first replenishment at the beginning of arrows and last replenishment at the arrow tips. Offset bars represent calculated shift in solubility lines for pH = 8.0. OCP and β-TCP control phosphorus solubility in the initial leachates. As the soil is successively leached, the less-soluble minerals hydroxyapatite and fluorapatite control the phosphate solubility. Adapted from Hansen and Strawn (2003). Reproduced with permission of Wolters Kluwer Health.

soils as small, bright yellow masses or coatings (see Figure 5.13 in color plate inserts).

Iron sulfide minerals formed in soils are small, poorly crystalline, and unstable in oxidizing environments. This makes them difficult to study. The primary sulfides in soils are pyrite (FeS$_2$), mackinawite (FeS$_{0.9}$), marcasite (FeS$_2$), and greigite (Fe$_3$S$_4$). Pyrite forms in wet, sedimentary soil environments, where microbial reduction of iron and sulfur occurs (see Special Topic Box 5.1). Mackinawite and greigite are intermediate iron sulfides that should transform to pyrite. Marcasite is preferred at pH less than five, while pyrite is the most stable iron sulfide at pH greater than five.

Iron and other metal sulfides are often emplaced in soils as mining waste. The oxidation of pyrite produces sulfuric acid. For example, acid-mine drainage occurs when mining activity brings buried iron sulfides to the oxygen-rich surface environment. Pyrite introduced into a highly oxidizing and hydrated environment rapidly oxidizes. The overall acidification and oxidation reaction is

$$FeS_2(s) + 3.75O_2 + 2.5H_2O = FeOOH + 2SO_4^{2-} + 4H^+$$

(6.11)

Four moles of H$^+$ are produced for every mole of pyrite oxidized. The reaction is facilitated by bacteria that utilize the electrons from oxidation of Fe(II) and S(II-) as an energy source (chemotrophs). In soils undergoing an oxidation-reduction cycle, such as in marsh areas, pyrite oxidation creates acid-sulfate soils. Special Topic Box 5.1 covers pyrite oxidation in more detail.

6.13 Time sequence of mineral formation in soils

Weathering involves the movement of water through the soil profile and the gradual removal of silica and alkali and alkaline earth cations. The flow of water usually prevents accumulation of soluble salts, but is slow enough to permit silicon, aluminum, and some magnesium to reprecipitate as secondary minerals. The change in secondary soil clay mineral composition correlates to the amount of weathering and leaching in the soil (**Table 6.14**). The sand and silt fractions of soils are usually relics of the soil parent material.

Table 6.14 Sequence of clay mineral distribution with increasing soil development. Adapted from Jackson and Sherman (1953). Reproduced with permission of Elsevier.

Relative degree of soil development	Prominent minerals in soil clay fraction
Predominant in young soils	
1	Gypsum, sulfides, and soluble salts
2	Calcite, dolomite, and apatite
3	Olivine, amphiboles, and pyroxenes
4	Mica
5	Feldspars
Predominant in intermediately weathered soils	
6	Quartz
7	Muscovite
8	Vermiculite, illite, and chlorite
9	Smectite (montmorillonite)
Predominant in advanced weathered soils	
10	Kaolinite and halloysite
11	Gibbsite and allophane
12	Goethite, limonite, and hematite
13	Titanium oxides, zircon, and corundum

The mineral groups of **Table 6.14** indicate progressively increasing stages of soil maturity. A given suite of minerals may not necessarily dominate the clay fraction, but its presence in detectable amounts is a reliable indicator of the degree of soil weathering and development. The criterion for a mineral's presence normally is whether the mineral is detectable by X-ray diffraction; however, poorly crystalline minerals are not easily detected by XRD. The clay mineral groups that are characteristic of increasing soil maturity can be broken down into three stages of weathering: young, intermediate, and advanced. Minerals that distinguish *young* soils are inherited from the soil's parent material and are typically rapidly weathered (dissolved) when clay-sized. Typical mineral occurrences in young soils are:

- *Soluble salts.* Halite (NaCl) and gypsum ($CaSO_4 \cdot 2H_2O$), as well as sulfides (pyrite, FeS_2) in soils reclaimed from seas or swamps. These minerals readily dissolve in percolating water or, in the case of the sulfides, are readily attacked by oxygen. Saline and sodic (alkali) soils are examples of this category.
- *Calcite* ($CaCO_3$), *dolomite* ($CaMg(CO_3)_2$), *and apatite* ($Ca_5(F,OH)(PO_4)_3$). Carbonates of clay size are rapidly leached from humid soils. In arid regions, calcite accumulates. The phosphate as calcium phosphates in alkaline soils or aluminum and iron phosphates in acid soils.
- *Olivine* ($(Mg, Fe)_2SiO_4$) *and the feldspathoids* (amphiboles – mostly hornblende and pyroxenes). In these minerals, Fe(II) will oxidize and increase the weathering rate.
- *Primary-layer silicates.* These include biotite ($K(Mg, Fe, Mn)_3Si_3AlO_{10}(OH)_2$) from igneous and metamorphic parent materials, glauconite ($K(Fe, Mg, Al)_2Si_3AlO_{10}(OH)_2$) from marine sediments, and magnesium chlorite ($(Mg, Fe)_6(Si, Al)_4O_{10}(OH)_8$. Aluminum often substitutes for silicon in the tetrahedral sheets; up to one out of every four silicon atoms can be isomorphically substituted. At room temperature, tetrahedrally coordinated aluminum is less stable than octahedrally coordinated aluminum. Presence of Fe(II) in micas increases their instability because it oxidizes destabilizing the mineral's bond valence balance.
- *Feldspars.* Albite ($NaAlSi_3O_8$) and anorthite ($CaAl_2Si_2O_8$) are the end members of the plagioclase feldspar continuum, covering the whole range of Na^+ to Ca^{2+} mixtures. The greater the calcium content of plagioclase, the faster its rate of weathering. Orthoclase and microcline feldspars (polymorphs of $KAlSi_3O_8$) in the clay fraction weather at roughly the same rate as albite.

Minerals in young soils disappear over time with weathering. Minerals that distinguish *intermediately* weathered soils are more resistant to weathering than those that occur in young soils:

- *Quartz.* Clay-sized quartz (SiO_2) is more easily weathered than sand-sized quartz; thus, its predominance in sand.
- *Muscovite* ($KAl_2(Si_3Al)O_{10}(OH)_2$). Muscovite mica is more stable than biotite. Biotite mica contains oxidizable Fe(II) and some Mn(II), whereas muscovite does not. In addition, the dioctahedral aluminum layer of muscovite seems to fit much better between the two layers of silica/alumina tetrahedra, and hence is more stable than the trioctahedral layer of biotite that contains magnesium and Fe(II).
- *Interstratified or intermixed layer silicates.* These include vermiculite ($M^+(Mg, Fe)_3(Si_{4-n}, Al_n)O_{10}(OH)_2$, where M^+ is the interlayer cation) and the hydrous micas (illite). Whether the minerals of this category

are inherited from the parent material or are secondary products derived from inherited minerals is uncertain in many cases.

- *Montmorillonites or smectites* (M^+ (Al, Mg)$_2$Si$_4$O$_{10}$ (OH)$_2$). These are the secondary Mg- and Al-rich 2:1 layer silicates that can form in soils. The chemical composition varies greatly, but the basic structure remains essentially the same.

In soils at *advanced* stages of weathering, predominant minerals are all secondary; and few primary parent minerals remain in such soils. By the time these stages of weathering are reached, inherited minerals in the clay fraction have either disappeared or are present in only minor amounts:

- *Kaolinite and halloysite* (Al$_2$Si$_2$O$_5$(OH)$_4$). The 1:1 phyllosilicates (kaolins) are more stable with time than the 2:1 phyllosilicates. Kaolinite is a common component of soil clays and occurs in relatively high concentrations in intermediate to advanced weathered soils.
- *Aluminum oxides*. These include gibbsite (Al(OH)$_3$) and allophane. The loss of silicon from soils leaves an aluminum- and iron-rich residue in soil clays.
- *Iron hydroxyoxides*. These include goethite (FeOOH) and hematite (Fe$_2$O$_3$).
- *Titanium oxides*. These include anatase and rutile (TiO$_2$), leucoxene (hydrated, amorphous titanium oxide), and ilmenite (FeTiO$_3$), plus zircon (ZrO$_2$) and corundum (Al$_2$O$_3$). Titanium and zirconium are immobile in soils and are used as indicators of the amount of parent material that has weathered to produce soil.

The above sequence of minerals occurs as soils develop from young to advanced weathering stages; however, soil mineral weathering does not always progress from one weathering category to the next. The sequence in **Table 6.14** is a time sequence in the sense that soluble cations and silica are increasingly lost from soils with time. The secondary minerals in each successive weathering stage of **Table 6.14** are stable at lower Si(OH)$_4$ concentrations and lower soil pH than those in the previous stage. Relative weatherability of soil minerals that occur in soils is also shown in **Table 6.7** and **Table 6.12**.

6.14   Measurement of soil mineralogy

The greatest challenge in identifying soil minerals is that they occur as heterogeneous mixtures that are difficult to isolate. In addition, soil minerals are rarely pure, well-crystallized materials. Three basic strategies are used to study and identify soil minerals:

1 Isolate the fractions into size separates (sand, silt, clay). Sand can be sieved, and the silt and clay can be separated by suspended particle settling.
2 Chemically treat the soil to selectively remove organic matter, carbonates, poorly crystalline minerals, or oxides.
3 Analyze the soil or separated fractions using microscopic, spectroscopic, X-ray diffraction, or total dissolution methods.

Each of the methods can be applied independently or in sequence to learn about a soil's mineralogical composition. **Table 6.15** summarizes some common mineralogical identification methods. A rule of thumb in mineral analysis of soils is that it is good to use more than one method to determine soil mineralogy.

In addition to direct or indirect analytical methods for soil mineral analysis, matching solution chemistry (i.e., activity of ions and pH) with phase diagrams provides some knowledge of possible existing phases in the soils (e.g., see **Figure 6.16**).

6.14.1   Principles of X-Ray Diffraction (XRD) for clay mineralogy

XRD is one of the most useful tools for identification of minerals (see Special Topic Box 6.1). XRD uses a monochromatic beam of X-rays focused on a sample arranged at a known angle with the incident X-ray beam. Part of the beam is reflected (diffracted) by successive repeating planes of atoms; planes of atoms are atoms that occur in a two-dimensional plane within the mineral lattice. The diffracted X-rays from each of the planes combine. Conceptually, maximum diffracted beam intensity occurs wherever the diffracted beam travels a distance equal to an integral number (n = 1, 2, 3) of wavelengths (λ) (see **Figure 6.20**). This condition is described by the **Bragg equation**:

$$n\lambda = 2d \sin \theta \qquad (6.12)$$

Table 6.15 Methods for determining soil mineralogy.

Method	Use	Basic principle	Advantage	Limitation
Spectroscopy and Diffraction				
X-ray diffraction (XRD)	Determine long-range order of minerals.	X-rays are reflected from planes of minerals (d-spacing).	Accurate, easy access, long history of use to characterize minerals; advanced methods use high-energy X-rays and microprobes.	Mineral must have some long-range order (crystallinity); peaks are often not unique.
X-ray absorption spectroscopy (XAS, XAFS, XANES, EXAFS)	Determine the local atomic structure surrounding atoms in minerals.	X-rays excite central atom and electromagnetic waves are scattered off atoms surrounding central atom.	Element specific, gives local atomic structure surrounding atom, sensitive.	Difficult to get access; gives average atomic environment in mixed mineralogy samples; difficult to analyze data.
Infrared (IR, FTIR, Raman)	Identify minerals based on absorption spectra from known compounds.	Infrared radiation excites atoms and causes stretching and vibration characteristic of bonds and coordination.	Easy to use, sensitive, easy access, many reference spectra.	Mixed samples are difficult to deconvolute; some signal bands overlap.
X-ray fluorescence (XRF)	Identify the elemental composition of sample.	Each element gives off fluorescent energy at a known energy.	Provides concentration of atoms, nondestructive, easy access, simple to use.	Does not give any information on the species, only elemental composition.
Other spectroscopies (Mössbauer, EPR, NMR)	Oxidation states and molecular speciation.	Probing atomic structure use high energy and measuring emission from relaxation.	Mössbauer-Fe and Mn oxidation states and molecular structure; EPR and NMR-molecular structures via fingerprinting.	Mössbauer limited to Fe and Mn mineralogy; EPR is used for Cu, Cr, Mn, and organics; NMR used for P, Al, Si, and organics.
Microscopic Methods				
Electron microscopy (SEM, TEM, Microprobe)	Determine the morphology of submicron particles; elemental composition	Electron beam impinged on sample and detected.	Images of very small minerals (~0.5 nm); provides size and shape of reactive surface	Samples must be specially prepared; often requires a vacuum environment.
Light microscopy	Determine morphology and shape of minerals.	Light is reflected or passed through sample.	Inexpensive; easy access; mineral identification.	Limited to particles greater than 200–800 nm.
Atomic force microscopy	Determine surface topography.	Probe is scanned across the surface and computer logs height of probe.	In-situ analysis possible (no vacuum); good spatial resolution (~1 nm).	Sample preparation and analysis artifacts are common.
General Methods				
Color	Mineral color can be used for identification.	Can be semi-quantitative by employing a spectrometer.	Can be done in field, low cost.	Not all minerals have characteristic colors; accuracy subjective.
Dissolution and analysis of elemental composition	Concentration of elements in minerals are determined.	Mineral is dissolved and then concentration of elements is determined.	Readily available; can be used to identify minerals if assumptions are made.	Time consuming; extractions may not be mineral specific.
Thermal analysis	Determine the weight loss curve of minerals.	Samples are heated, and the weight loss is calculated at each temperature, which is characteristic of mineral.	Easy; inexpensive; can be accurate if sample is not of mixed composition.	Resolution of peaks may be difficult; distinction of minerals in mixed samples is difficult.
Extractions	Selectively extract mineral phases.	Minerals have variable solubility in different extracts.	Low overhead; easy; straightforward; reproducible.	Readsorption of dissolved phase; extracts are often not specific.

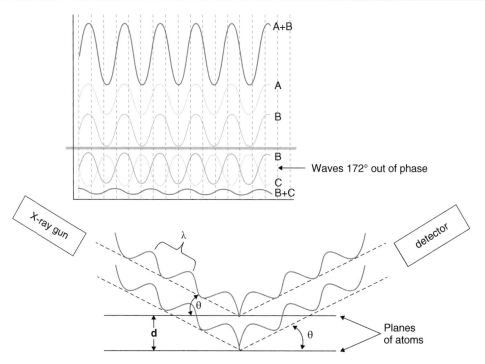

Figure 6.20 Basic concepts of X-ray diffraction. The X-rays of a known wavelength (λ) are directed onto the sample at a known angle (θ). The X-rays diffract from the planes of atoms at the same angle. The outgoing waves are shifted relative to the separation distance of the planes of atoms (d). When the waves diffracted from the different planes are in phase, they constructively sum yielding the maximum intensity in the detector. When the waves are completely out of phase (180°), they destructively sum, yielding minimum intensity (zero) in the detector. By changing θ, the phase of the diffracted wave is varied, allowing for determination of the angle and d-spacing at which maximum intensity (amplitude) occurs.

where n is an integer corresponding to a shift in the wavelength, λ is the wavelength of the X-radiation, d is the distance between successive layers of atoms within the crystal, and θ is the angle at which the radiation strikes the mineral. The right-hand side of **Eq. 6.12** (2d sin θ) equals the distance that the X-rays travel between planes of atoms. When this distance is equal to some integer (n) times the radiation wavelength, the outgoing wavelengths become aligned, and the diffracted beam intensifies. When the conditions of Bragg's equation are not met, the diffracted beams cancel each other (**Figure 6.20**). Maximum intensity in a **diffractogram** occurs when the incoming X-rays strike the surface of the mineral at a specific angle that causes the left-hand side of **Eq. 6.12** to equal the right-hand side when n = 1, 2, or 3.

In a crystal, the separation distance of planes of atoms (d) is fixed, thus the angle of the incident X-ray (θ) can be changed until diffraction is maximized, satisfying **Eq. 6.12**. The maximum diffraction for a given d occurs when the incoming X-rays impinge on the sample at angle θ = sin^{-1} (n λ/2d), where n is typically one or two. To find this, the sample is rotated through different angles (θ) and the counts are recorded. The angle of the maximum peak corresponds to a d-spacing that can be calculated using **Eq. 6.12** because θ, λ, and n are known. Differences in crystal d-spacing distances as small as 0.01 to 0.001 nm can be detected by X-ray diffraction. This allows the position of atoms in the structure to be determined, which is a distinct characteristic of the mineral species.

X-ray diffraction works well for identifying phyllosilicates in soil clay fractions that have been isolated and cleaned of organic matter and oxide coatings. The XRD distance of the planes of atoms in phyllosilicates is referred to as the **c-spacing**, which is indicative of the height of a clay platelet (a 2:1 or 1:1 layer), including its interlayer. They are also referred to as **d-spacing** because of the **d** parameter in the Bragg equation.

To aid in detection of the c-spacing, soil clays are often prepared as oriented films, which can be created because of the platy morphology of the clay minerals.

The oriented-clay film enables maximum diffraction signal from the **001 plane** (basal plane of O atoms), allowing the greatest sensitivity for c-spacing measurement. To differentiate clays, they are saturated with different interlayer cations. The interlayer spacing of phyllosilicates varies as a function of the degree of hydration of cations in the interlayer. For example, vermiculite interlayers collapse when K-saturated, but expand to 1.4 nm when Na-saturated. The interlayer of montmorillonite does not shrink as much as vermiculite when K-saturated; c-spacing is ~1.2 nm. Sodium-saturated montmorillonite has a c-spacing of 1.5–1.7 nm. Such variances are helpful for distinguishing between the different clay minerals.

6.14.2 Example calculation of d-spacing from a diffractogram

Figure 6.21 shows an X-ray diffractogram for a pure kaolinite sample. The units on the x-axis are degrees 2θ because the angle of the X-ray gun and detector is two times the X-ray's angle of incidence. The diffractogram shows multiple peaks that are a type of fingerprint for the kaolinite crystal structure. The two most intense peaks occur at 12.27 and 24.80 degrees 2θ. Using these peaks, the d-spacing of the mineral can be calculated by solving **Eq. 6.12**:

$$d = \frac{n\lambda}{2\sin\theta} \tag{6.13}$$

Thus, for the first peak where $\theta = 12.27/2 = 6.135°$ the calculated d-spacing is 0.721 nm (the wavelength for a Cu radiation source is 0.154 nm, and a value of one was used for n). Similarly, for the second peak, the d-spacing is 0.359 nm. These d-spacings correspond to the distances that planes of atoms in kaolinite are separated.

When there is only a single mineral phase, a few peaks may be enough to positively identify the mineral species. However, in samples with mixed mineralogy, such as soils, other methods are needed to help identify the minerals. As discussed above, treatment of the soil clays prior to XRD measurement can be used to enhance detection of clay minerals in mixed samples.

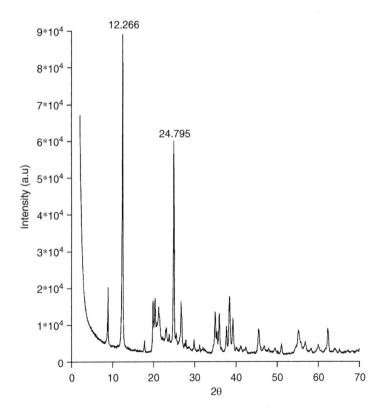

Figure 6.21 X-ray diffractogram from a kaolinite sample. The two dominant peaks are characteristic of kaolinite. Adapted from Zegeye et al. (2013), with permission.

6.14.3 Selective extraction of iron oxides and amorphous aluminosilicates from soils

Characterization of soils for iron oxides and poorly crystalline or amorphous aluminosilicates (e.g., allophane) is important because they are common minerals that occur in most soils, and thus, impart important characteristics to the soils. Use of spectroscopic and XRD characterization can provide some information of soil mineralogy. However, because iron oxides and poorly crystalline aluminosilicates have varying degrees of crystallinity, spectroscopic and XRD analysis of soils in many cases are limited. Furthermore, quantification of the amount of oxides in soils is difficult with spectroscopic and XRD analysis.

Selective extraction of soils using chemicals provides quantitative information on the amount of crystalline and poorly crystalline iron oxides and aluminosilicates in soils. Citrate, bicarbonate, dithionite (CBD) extractant solution is a reducing and complexing extractant that removes all secondary iron oxides from soils by reducing the Fe(III) to Fe(II), which has increased solubility and complexes to the citrate in solution to keep it dissolved. CBD is a common extractant for removing crystalline iron oxides goethite and hematite from soils for quantification; ferrihydrite is also extracted by CBD extraction. Ammonium oxalate extraction in the dark (AOD) solution can be used to remove poorly crystalline ferrihydrite and aluminosilicates from soils for quantification. By subtracting the iron concentration removed using AOD from the iron concentrations removed using CBD, the total concentration of goethite or hematite can be quantified.

Quantification of iron oxides and poorly crystalline aluminosilicates by extraction is a common mineralogical analysis method used in many soil science disciplines—from pedology to soil fertility. For example, by knowing the type and amount of iron oxide in a soil, phosphate adsorption behavior can be predicted. AOD extractable aluminum and silicon concentrations are used to characterize whether a soil has andic soil properties.

6.15 Important concepts in soil mineralogy

In this chapter, fundamental concepts of mineralogy were presented, including detailed discussion of how atomic properties can be used to understand how minerals are *assembled*. The arrangement of elements in minerals is governed by their ionic radius, ionic charge, and bond strength. Isomorphic substitution is an important factor in mineral characteristics. Properties of primary minerals and rocks were discussed, with detailed description of primary mineral silicate groups: nesosilicate, inosilicate, phyllosilicate, and tectosilicate. These minerals are the parent materials from which soils form.

Weathering in soils creates secondary minerals. The different types of secondary minerals and their weathering environment were discussed. Aluminosilicates (i.e., clay minerals) and iron and aluminum oxides are common secondary minerals that occur in soils. Formation of secondary minerals can be predicted using solubility modeling. More details of soil clay minerals and oxides are provided in the next chapter.

Questions

1 Distinguish between primary and secondary minerals and give examples of each. Which minerals are more important in determining soil properties?

2 Which minerals occur in the sand and silt fractions of soil? Which are commonly found in the clay-sized fraction? Why?

3 Determine the radius ratio and coordination environment (number) surrounding the following elements (assume O (r = 0.132 nm) is the coordinating ligand): Al, Fe^{3+}, Ca, Si, Na, K, Mg.

4 Give examples of ionic, covalent, hydrogen, and van der Waals bonding in minerals.

5 Why are high-charged ions less likely to form edge or face-sharing linkages in minerals?

6 What ion dominates silicate structures?

7 What is the dominant characteristic that determines whether ions may isomorphically substitute for one another?

8 What is the relevance of the number of Si to O atoms in a mineral formula?

9 Employ Pauling's electrostatic valence principle and show that the residual charge on O bound to Si in a tetrahedral environment can be satisfied by two Al atoms in octahedral coordination. Hint: Draw a Si tetrahedron and show the number of Al bonds needed when Al is in octahedral coordination to make the net charge on an oxygen ligand equal to

zero. How many octahedral coordinated Mg atoms would be required to satisfy the oxygen ligand charge on the Si tetrahedra?

10 Feldspars and quartz are both tectosilicates. Describe their differences and similarities and relate this weathering susceptibility.

11 Based on mineral composition, which rock would weather quicker in the same environment, granite or basalt? Explain your answer.

12 Why are observed mineral weathering rates in soils faster than predicted weathering rates (see **Figure 6.13**)?

13 Discuss the flux of ions in soils as weathering proceeds with respect to secondary mineral formation (use **Figure 6.16**).

14 Write a dissolution reaction for the aluminosilicate pyrophyllite ($Al_2Si_4O_{10}(OH)_2$) and plot a line on a stability diagram (similar to kaolinite in **Figure 6.15**). Log K_{sp} = 1.92. Hint: Reactants are pyrophyllite, six hydrogen ions, and four waters; products are four silicic acids and two aluminum ions.

15 What are the soil processes that lead to formation of calcite? What influences does calcite have on soils?

16 Discuss how the *d*-parameter in Bragg's law is used to identify a mineral.

17 How can vermiculite and smectite be differentiated from each other using XRD?

18 What are the shortcomings of a salt precipitation model (such as shown in **Figure 6.18**) for predicting the fate of ions (e.g., Na^+, Mg^{2+}, Cl^-) in soils?

19 Given a diffractogram of a clay mineral with a peak at 5.08 degrees 2θ that corresponds to its c-spacing, what is the d-spacing of this mineral? What clay mineral might this be?

Bibliography

Byers, H.G., L.T. Alexander, and R.S. Holmes. 1935. The Composition and Constitution of Colloids of Certain of the Great Groups of Soils. US Department of Agriculture, Washington D.C.

Doner, H.E., and P.R. Grossl. 2002. Carabonates and evaporites, p. 199–228. *In* J. Dixon and D. Schulze, (eds.) Soil Mineralogy with Environmental Applications. SSSA Book Series Ser. 7. Soil Science Society of America, Madison, WI.

Feldman, S.B., and L.W. Zelazny. 1998. The chemistry of soil minerals, p. 139–152. *In* P. M. Huang, (ed.) Future Prospects of Soil Chemistry. Soil Science Society of America, Madison.

Hansen, J.C., and D.G. Strawn. 2003. Kinetics of phosphorus release from manure-amended alkaline soil. Soil Science 168:869–879.

Hausrath, E.M., A. Neaman, and S.L. Brantley. 2009. Elemental release rates from dissolving basalt and granite with and without organic ligands. American Journal of Science 309:633–660.

Hausrath, E.M., A.K. Navarre-Sitchler, P.B. Sak, C.I. Steefel, and S.L. Brantley. 2008. Basalt weathering rates on Earth and the duration of liquid water on the plains of Gusev Crater, Mars. Geology 36:67–70.

Hendricks, S.B., and W.H. Fry. 1930. The results of X-ray and microscopical examinations of soil colloids. Soil Sci. 29:457–478.

Jackson, M.L., and G.D. Sherman. 1953. Chemical weathering of soils, p. 221–319. Advances in Agronomy. Ser. 5. Soil Science Society of America, Madison.

Kelley, W.P., W.H. Dore, and S.M. Brown. 1931. The nature of the base-exchange material of bentonite, soils, and zeolites, as revealed by chemical investigation and X-ray analysis. Soil Sci. 31:25–45.

Kittrick, J.A. 1969. Soil minerals in Al_2O_3-SiO_2-H_2O system and a theory of their formation. Clays and Clay Minerals 17:157–167.

Kittrick, J.A. 1971. Stability of Montmorillonites. 1. Belle Fourche and Clay Spur Montmorillonites. Soil Science Society of America Proceedings 35:140–145.

Klein, C., C.S. Hurlbut, J.D. Dana, and C. Klein. 2002. The 22nd Edition of the Manual of Mineral Science (after James D. Dana). 22nd ed. J. Wiley, New York.

Lindsay, W.L. 1979. Chemical Equilibria in Soils. John Wiley and Sons, New York.

McBride, M.B. 1994. Environmental Chemistry of Soils. Oxford University Press, New York.

Pettijohn, F.J. 1957. Sedimentary Rocks. 2d ed. ed. Harper, New York.

Shannon, R.D. 1976. Revised effective ionic–radii and systematic studies of interatomic distances in halides and chalcogenides. Acta Crystallographica Section A 32:751–767.

Shannon, R.D., and C.T. Prewitt. 1969. Effective ionic radii in oxides and fluorides. Acta Crystallographica Section B-Structural Crystallography and Crystal Chemistry B 25:925–945.

Tyrrell, G.W. 1950. The Principles of Petrology; An Introduction to the Science of Rocks. [11th ed.]. ed. Methuen, London.

Velbel, M.A., and W.W. Barker. 2008. Pyroxene weathering to smectite: conventional and cryo–field emission scanning electron microscopy. Koua Bocca ultramafic complex, Ivory Coast. Clays and Clay Minerals 56:112–127.

White, A.F., M.S. Schulz, D.V. Vivit, A.E. Blum, D.A. Stonestrom, and J.W. Harden. 2005. Chemical weathering rates of a soil chronosequence on granitic alluvium: III. Hydrochemical evolution and contemporary solute fluxes and rates. Geochimica Et Cosmochimica Acta 69:1975–1996.

Zegeye, A., S. Yahaya, C.I. Fialips, M.L. White, N.D. Gray, and D.A.C. Manning. 2013. Refinement of industrial kaolin by microbial removal of iron-bearing impurities. Applied Clay Science 86:47–53.

7 CHEMISTRY OF SOIL CLAYS

7.1 Introduction

Weathering produces secondary minerals in soils, including phyllosilicates, oxides, salts, carbonates, and amorphous phases. Most of the secondary minerals in soils occur in the clay-size fraction (see Figure 6.1). Soil scientists often use the term **soil clay minerals** to refer to the group of minerals that have important soil physical and chemical properties, but do not belong to any specific type or species of mineral. The most common soil clay minerals in moderately weathered soils are secondary phyllosilicates. In highly weathered soils and soils that have significant amounts of volcanic ash input, iron and aluminum oxides are common, as well as short-range ordered aluminosilicates.

To study mineralogy, minerals are separated into groups or classes based on their major anion type. While instructive for classifying them, in soils, minerals occur as mixtures that are physically and chemically associated with each other. For example, soil minerals often exist as mixtures of two or more mineral types within a single particle; such as clay minerals that are mixed-layer silicates rather than a single species, and oxides coating clay and sand grains. Additionally, many soil minerals are impure, and have widely varying degrees of crystallinity. The complexities are amplified by the presence of soil organic matter, which interacts with soil minerals by complexation or aggregation, including coatings on mineral surfaces (**Figure 7.1**).

The complex mixture of soil minerals makes them difficult to study in-situ. To overcome these complexities, minerals are typically physically or chemically separated from the soil, or synthetic minerals are produced and their distinct properties studied. For example, the mineral and surface properties of the iron oxide goethite (FeOOH) are usually gleaned from synthetic mineral samples produced in the laboratory. In fact, much of the understanding of soil mineral properties and behaviors comes from studying pure or

Soil Chemistry, Fifth Edition. Daniel G. Strawn, Hinrich L. Bohn, and George A. O'Connor.
© 2020 John Wiley & Sons Ltd. Published 2020 by John Wiley & Sons Ltd.

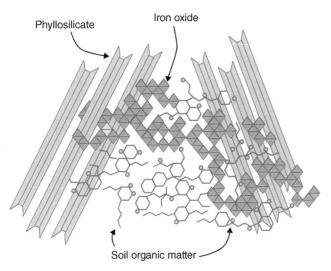

Figure 7.1 Illustration of pedogenic minerals and soil organic matter (SOM) mixtures. Rings, chains and spheres represent SOM constituents (C, O, OH, S, or N). Diamonds represent pedogenic Fe and Al oxides. 2:1 phyllosilicate minerals are represented by two trapezoids on either side of a rectangle.

Figure 7.2 Scanning electron micrograph of a kaolinite sample isolated from weathered sandstone showing crystal morphology and clay stacking. Image reproduced from Wilson et al. (2014), with permission.

ideal minerals. However, properties of purified soil minerals or isolated minerals are not necessarily the same as minerals mixed with each other and with soil organic matter, such as occurs within soils. Thus, care must be taken when extrapolating pure-phase mineral properties to soil clays.

7.2 Structural characteristics of phyllosilicates

Many soil clay minerals belong to the **phyllosilicate** group of the silicate class of minerals. **Figure 7.2** shows a micrograph of kaolinite, one of the most common clay minerals in soils. The terms *layer silicate, layered aluminosilicate, phyllosilicate,* **clay mineral**, and *sheet silicate* are used synonymously. Soil scientists define *clay minerals* as a specific type of phyllosilicate mineral found in nature, particularly soils. Secondary phyllosilicates formed in soils typically have maximum width of less than 2 μm; thus, they have large solid-solution interfaces that greatly influence soil properties (see Figure 1.9). The extensive reactivity of clay minerals makes them useful for many industrial processes (see Special Topic Box 7.1).

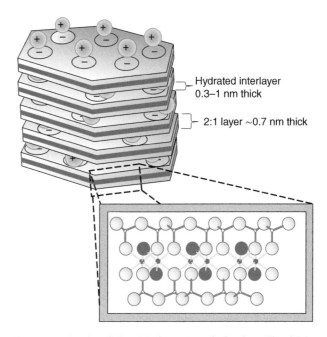

Figure 7.3 Stacks of clay 2:1 clay mineral platelets. The thicknesses of a 2:1 layer and the interlayer are shown. The close-up shows the atoms and their arrangement in a 2:1 crystal.

Phyllosilicates are a combination of sheets of Al-O, Mg-O, or Fe-O octahedra between sheets of Si-O tetrahedra (**Figure 7.3** and **Figure 7.4**). The tetrahedral and octahedral sheets bond together by sharing oxygen atoms. The characteristics that differentiate phyllosilicates are:

Special Topic Box 7.1   Uses of phyllosilicates

Phyllosilicates are integral components of soils and are used in a variety of manufacturing and industrial processes, and appear in many products used by humans. Acquisition of phyllosilicates is usually by mining, typically from sedimentary deposits. In 2017, clay mining companies in the United States sold 25 million tons of clay at a value of $1.54 billion. There are many uses of clay minerals (**Table 7.1**) – from their namesake use in pottery (primarily kaolins), to their use as catalysts in petroleum processing (catalytic cracking).

Table 7.1 Common uses of clay minerals.

Use	Notes
Paper production	Kaolin is the primary clay used in paper production as a surface treatment to add brightness and finish, and alter ink absorption properties, or it can be used as filler. Smectites are used in some papers to add distinct properties such as water drainage.
Ceramics	Kaolinite is used to produce ceramic wares and molds for casting.
Catalyst	Used in chemical production to aid in breaking organic bonds (cracking) or chemical synthesis. Clays are typically saturated with exchange cations or acid treated.
Antimicrobial	Clays exchanged with ions such as Cu^{2+} or $Ag+$ have antimicrobial properties.
Medicines	Clays are used to soothe stomach irritation; as fillers, binders and carriers of medicines; to sorb ingested toxic chemicals, including metals or foodborne toxins (e.g., aflatoxin).
Packing	Expanded clays such as vermiculite are used to pack hazardous materials because they have high liquid absorption potential.
Insulation	Clays are good thermal insulators, either combined with other materials or by themselves. They are used as both pressed and loose insulation materials in moldings or buildings, respectively.
Oil drilling and exploration	Clay slurries are used to seal and lubricate during ground drilling operations. Also used in water drilling operations.
Clay liners	Clays are used as liners in landfills because their low hydraulic conductivity prevents seepage of waste fluids offsite.
Plastics	Clays are added to plastic and rubber as fillers and to provide unique properties.
Cosmetics	Talc and kaolinite are used in many cosmetic products.
Pesticides	Pesticides can be sorbed on clay minerals to aid in dispersion, release timing, and prevent degradation. Clays (kaolinite) have been shown to directly inhibit attack of insects on plant tissue, and acts as a sunscreen for fruit.
Hazardous chemical treatment	Clays are used to encapsulate hazardous spills, and to remove hazardous chemicals from waters.
Absorbents	Absorbency is a valued attribute for cat litter.
Building materials	Fired into brick, or combined with straw or other fiber, clay building materials have been used in ancient and modern dwellings. Used in spackling for sheet rock.

---

1 The number and sequence of tetrahedral and octahedral sheets.
2 The layer charge per unit cell.
3 Interlayer cations and hydration state.
4 The cations in the octahedral sheet.

These four clay properties are used to categorize clay minerals into species types.

There are several ways to represent the phyllosilicate structure, including ball and stick and polyhedral (**Figure 7.4**). A common *shorthand* model used to represent 2:1 phyllosilicates is two trapezoids (**tetrahedral sheet**) sandwiching a rectangle (**octahedral sheet**) (**Figure 7.5**). A 1:1 clay is depicted as a single trapezoid on a rectangle.

The octahedral sheets in the phyllosilicates are either dioctahedral or trioctahedral, depending on whether the cation sites within the sheet are completely filled (trioctahedral) or only two-thirds filled (dioctahedral). In **trioctahedral sheets**, the divalent cations Mg^{2+} or Fe^{2+} ions occupy every octahedral sheet cation position (**Figure 7.6**). The magnesium sheet is sometimes called

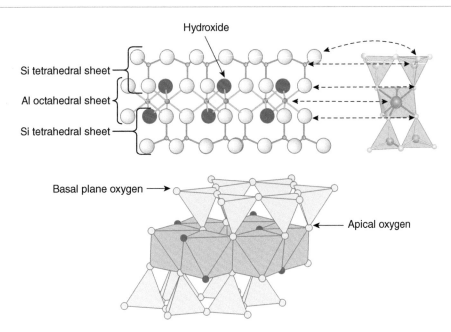

Figure 7.4 Basic phyllosilicate components showing the basal plane oxygens, apical oxygens, and tetrahedral and octahedral sheets. The octahedral sheet is di-octahedral when Al^{3+} or Fe^{3+} is the octahedral sheet cation. When Mg^{2+} or Fe^{2+} is present, the octahedral sheet is trioctahedral (not shown – the vacant position in the Al^{3+} octahedral sheet shown is filled with Mg^{2+} or Fe^{2+}).

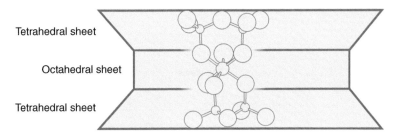

Figure 7.5 The generic illustration of a 2:1 phyllosilicate layer.

a *brucite-like sheet* after the magnesium hydroxide mineral brucite $(Mg(OH)_2)$. In **dioctahedral sheets,** the trivalent cations Al^{3+} or Fe^{3+} occupy two out of every three cation positions (**Figure 7.7**). The dioctahedral sheet is referred to as a *gibbsite-like sheet* because it is analogous in structure and composition to gibbsite $(Al(OH)_3)$.

In phyllosilicates, the tetrahedral sheet attaches to the octahedral sheet by sharing the upward- and downward-pointing apical oxygens in the tetrahedra with the octahedral cations $(Al^{3+}, Mg^{2+}, Fe^{2+},$ or $Fe^{3+})$. The **apical oxygen** effectively takes the place of two-thirds of the hydroxides in the brucite or gibbsite-like sheets. The apical oxygens and the octahedral hydroxides form a plane of atoms parallel to the basal plane of the clay structure (**Figure 7.4**). The hydroxide ions in the plane of oxygen atoms between the tetrahedral and octahedral sheet are positioned at the bottom of the

hexagonal (ditrigonal)-shaped voids created by the six corner-sharing silicate molecules in the **basal plane** of oxygen atoms (**Figure 7.8**). The resulting bonded tetrahedral and octahedral sheets form a **layer** composed of either one tetrahedral and one octahedral sheet (1:1, tet:oct), or an octahedral sheet sandwiched between two tetrahedral sheets (2:1, tet:oct:tet). The platy morphology facilitates stacking of the layers, and the spaces between are referred to as **interlayers**.

7.2.1   1:1 phyllosilicates

The **kaolin** group consists of minerals with a single tetrahedral sheet joined to an octahedral sheet, which are **1:1 phyllosilicates** (**Figure 7.9**). On the octahedral face of kaolin is a plane of closely packed hydroxide anions. On the silicate face of kaolin minerals is a **basal plane**

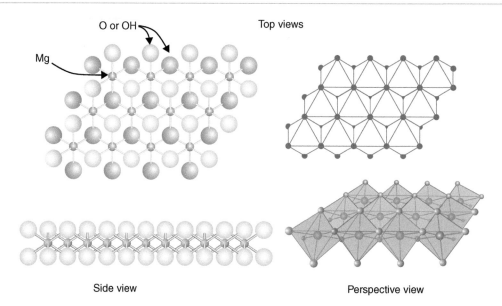

Figure 7.6 Four different illustrations of the trioctahedral Mg^{2+} sheet. The top two models show the planar view ((001) face). The bottom shows the side view ((110) face) obtained by rotating the planar ball and stick model 90°. The bottom right image shows a perspective view of the trioctahedral sheet.

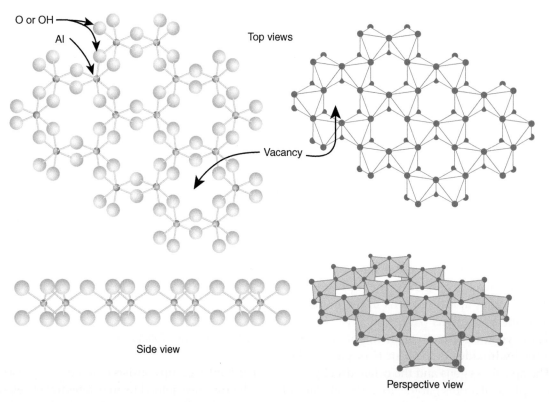

Figure 7.7 Four different illustrations of the dioctahedral Al^{3+} sheet. The top two models show the planar view ((001) face). The bottom shows the side view ((110) face) obtained by rotating the planar ball and stick model 90°. The bottom right image shows a perspective view of the trioctahedral sheet.

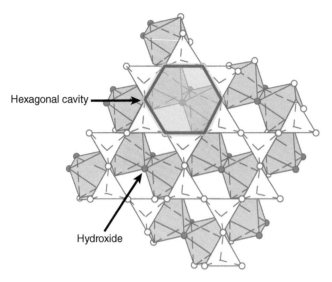

Figure 7.8 Top view of a dioctahedral phyllosilicate model showing the hexagonal cavity created by the rings of silicate anions that occur in the tetrahedral sheet. At the bottom of the hexagonal cavity is a hydroxide ion. The symmetry of the hexagonal cavity is often skewed creating ditrigonal symmetry and is thus called a *ditrigonal cavity*.

of oxygens associated with the silicon tetrahedron. The 1:1 phyllosilicates are apparently very inflexible in their structural requirements, and allow little or no isomorphic substitution, and the structure is overall electrically neutral. However, the edges of kaolin minerals have charge created by the terminus of the mineral lattice. The 1:1 layers of kaolinite are held together strongly by the many hydrogen bonds between the

OH groups on an octahedral face and the basal plane oxygen atoms of the next sheet.

7.2.2 2:1 phyllosilicates

The **2:1 phyllosilicates** are made up of an octahedral sheet with a tetrahedral sheet above and below. The 2:1 layers stack in the ***c* direction** (perpendicular to the clay sheets). Most 2:1 phyllosilicates have extensive isomorphic substitution in the octahedral and tetrahedral sheets (**Figure 7.10** and **Figure 7.11**), which leads to negatively charged surfaces, or **layer charge,** on the crystal. The charge is expressed on the surface of the mineral, and is called **permanent surface charge**. The layer charge is balanced by cations in the interlayer. The 2:1 minerals are classified according to layer charge and octahedral sheet composition. **Table 7.2** shows the classification and characteristics of some phyllosilicates.

Layer charge magnitude plays a dominant role in determining the strength and type of bonding between the 2:1 layers. **Figure 7.12** illustrates the differences in layer charge and interlayer composition for several different phyllosilicates. If the layer charge is zero, as in pyrophyllite and talc (not shown in **Figure 7.12**), the 2:1 layers are held together by weak van der Waals forces. If the layer charge is negative, the 2:1 sheets bond electrostatically to charge-compensating cations in the interlayers. The greater the layer charge, the more cations and the stronger the interlayer bond.

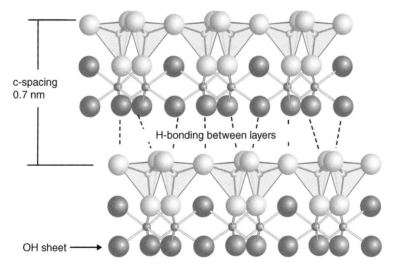

Figure 7.9 Two layers of kaolinite showing the interlayer H-bonds between the hydroxide plane of atoms and the basal plane of O atoms on the adjacent layer. The c-spacing is the distance between two equivalent planes of atoms within the layers. Compare this illustration to the kaolinite micrograph in **Figure 7.2**.

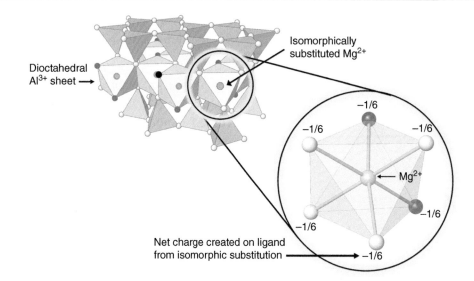

Figure 7.10 Layer charge caused by isomorphic substitution of Mg^{2+} in an Al^{3+} octahedral sheet. The overall charge created by one Mg^{2+} substitution for one Al^{3+} cation is –1. The charge deficiency is spread over six oxygen ligands, creating a net charge of –1/6 on each oxygen ligand.

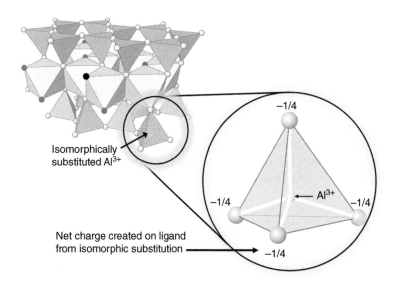

Figure 7.11 Layer charge caused by isomorphic substitution of Al^{3+} in a Si^{4+} tetrahedral sheet. The overall charge created by one Al^{3+} isomorphic substitution for one Si^{4+} cation is –1. The charge deficiency is spread over four oxygen ligands, creating a net charge of –¼ on each oxygen ligand.

Smectites have the lowest layer charge amongst the expandable phyllosilicates, and the bond between the layers is relatively weak. The low layer charge allows the interlayer to expand via entry of water. In soils, smectite minerals expand or swell when wet, and collapse when dry. In contrast, in high layer charge phyllosilicates, such as the **micas** muscovite and biotite, or **illite** (a slightly weathered mica of clay size), K^+ cations in the interlayer form strong electrostatic or ionic bonds with the basal plane oxygen atoms (see Figure 6.7). Potassium has a weak hydration sphere, and is just the right size to fit within the **ditrigonal cavity** created by six corner-sharing silica tetrahedra (**Figure 7.8**). The strong ionic bond prevents water molecules from entering the interlayer, thus, soils that have mica or **illite**-dominated clay fractions are nonexpanding and the soil does not shrink and swell with changing moisture content. **Vermiculites** are intermediate in layer charge

Table 7.2 Composition and nomenclature of phyllosilicates arranged according to layer type, layer charge (x), layer composition, and interlayer cations (M).

Type, group (layer charge (x))	Octahedral sheet arrangement	Name	Ideal half-cell chemical formula			
			Interlayer cations	Tetrahedral sheet	Octahedral sheet	Anions
1:1	dioctahedral	Kaolinite	none	Si_2	Al_2	$O_5(OH)_4$
Kaolin-serpentine		Halloysite	$2H_2O$	Si_2	Al_2	$O_5(OH)_4$
(x = 0)	trioctahedral	Antigorite	none	Si_2	Mg_3	$O_5(OH)_4$
2:1	dioctahedral	Pyrophyllite	none	Si_4	Al_3	$O_{10}(OH)_2$
Pyrophyllite-talc	trioctahedral	Talc	none	Si_4	Mg_3	$O_{10}(OH)_2$
(x = 0)						
Smectite-saponite	dioctahedral	Montmorillonite	$M_{0.33}, H_2O$	Si_4	$Al_{1.67}(Fe^{2+}, Mg)_{0.33}$	$O_{10}(OH)_2$
(x = 0.2–0.6)		Beidellite	$M_{0.33}, H_2O$	$Si_{3.67}Al_{0.33}$	Al_2	$O_{10}(OH)_2$
		Nontronite	$M_{0.33}, H_2O$	$Si_{3.67}Al_{0.33}$	Fe^{3+}_2	$O_{10}(OH)_2$
	trioctahedral	Saponite	$M_{0.33}, H_2O$	$Si_{3.67}Al_{0.33}$	Mg_3	$O_{10}(OH)_2$
		Hectorite	$M_{0.33}, H_2O$	Si_4	$Mg_{2.67}Li_{0.33}$	$O_{10}(OH)_2$
Vermiculite	dioctahedral	Dioctahedral vermiculite	$M_{0.74}, H_2O$	$Si_{3.56}Al_{0.44}$	$Al_{1.4}Fe^{3+}_{0.3}Mg_{0.3}$	$O_{10}(OH)_2$
(x = 0.6–0.9)	trioctahedral	Trioctahedral vermiculite	$M_{0.70}, H_2O$	$Si_{3.3}Al_{0.7}$	Mg_3	$O_{10}(OH)_2$
Illite	dioctahedral	Illite	$K_{0.6–0.85}$	$(Si,Al)_4 (Al<1)$	$(Al_{2-x-y}, Mg_x, Fe^{2+}_y)_2$	$O_{10}(OH)_2$
(x = 0.6–0.9)						
Mica	dioctahedral	Muscovite	K	Si_3Al	Al_2	$O_{10}(OH)_2$
(x = 1)	trioctahedral	Biotite	K	Si_3Al	$(Mg, Fe^{2+})_3$	$O_{10}(OH)_2$
		Phlogopite	K	Si_3Al	Mg_3	$O_{10}(OH)_2$
Chlorite	trioctahedral	Chlinochlore	$Mg_2Al(OH)_6^+$	$(Si,Al)_4$ ($Al = 0.8$-1.6)	$(Mg, Fe^{2+})_3$	$O_{10}(OH)_2$
(x = 0.8–1.6)						

between mica and smectite and are also intermediate between micas and the smectites in swelling properties.

Chlorites are clay minerals that have octahedral cation hydroxide sheets in the interlayers instead of only cations. The octahedral interlayer sheet is typically trioctahedral, and has Fe^{3+} and Al^{3+} cations isomorphically substituting for Mg^{2+} creating excess positive charge that balances the 2:1 layer charge. Because of blocked access to the interlayer surface charge, chlorites are non-expanding and have less cation exchange capacities than the expanding clays (smectites and vermiculites).

7.3 Relation of phyllosilicate structure to physical and chemical properties

7.3.1 Interlayer bond

The interlayer bond has a major effect on the physical and chemical properties of layer silicates. Bonding *within* the layers is much stronger than *between*

adjacent layers. When the mineral is subjected to physical or thermal stress, it fractures first between the unit layers, along the basal plane. This is the reason for the platy shape of most macroscopic layer silicate crystals (**Figure 7.13**). Also, the stronger the interlayer bond, the greater the crystal growth in the c dimension before fracture. Hence, the weatherability and size and shape of layer silicate crystals is a direct consequence of the strength of their interlayer bonds.

7.3.2 Surface area

The specific surface area of layer silicates is related to their expanding properties, and may be either external only, or, if the interlayer is hydrated, external plus internal. **External surface** refers to the faces and edges of the whole crystal. **Internal surface** is the area of the basal plane surfaces of each layer. Nonexpanding minerals exhibit only external surface; expanding minerals have both internal and external surface. The total

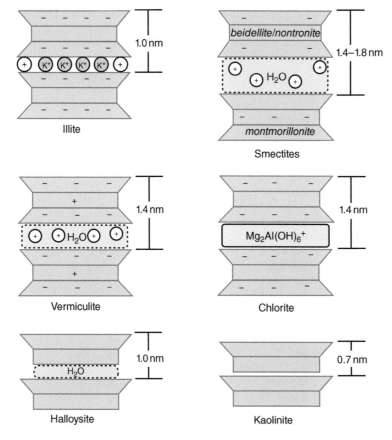

Figure 7.12 Polyhedral models of common clay minerals in soils showing charge origination (negative signs), interlayer composition, and c-spacing. Positive circles are mono or divalent cations. The smectite model shows the charge originating in the octahedral sheet in a montmorillonite layer, and tetrahedral sheets in a beidellite or nontronite layer. Illite shows hydrated cations on the edges that exchanged for interlayer K^+. Adapted from McBride (1994). Reproduced with permission of Oxford University Press.

specific surface area of montmorillonite can be as large as 800×10^3 m^2 kg^{-1}. The specific surface area of kaolinite, a nonexpanding and 1:1 layer silicate, is usually only 10 to 20×10^3 m^2 kg^{-1}. **Table 7.3** shows the specific surface area of some clay minerals.

7.3.3   c-spacing

The **c-spacing** of the layer silicates is determined by the number of tetrahedral or octahedral sheets per layer, and the presence of ions and/or polar molecules in the interlayer. This spacing is conveniently measured by X-ray diffraction (Chapter 6).

Mica's interlayer is collapsed, and the adjacent layers of this mineral approach one another closely, preventing water entry (**Figure 7.12**). The resulting constant c-spacing is 1 nm (**Table 7.3**). In minerals having weak interlayer bonds, such as montmorillonite, cations, and water or other polar molecules can enter between the basal planes, causing the c-spacing to increase (**Figure 7.14**). The expansion varies greatly with the amount and type of hydrated cation. In minerals with strong interlayer bonding, such as mica, chlorite, and kaolinite, water and hydrated cations cannot enter between the basal planes, and the c-spacing is fixed.

7.3.4   Cation adsorption and layer charge

The permanent negative charge on phyllosilicates resulting from isomorphic substitution is balanced by adsorbed cations, which creates **cation exchange capacity (CEC)**. The edges of crystals have additional charge created by unsatisfied bonds. The amount of **edge sites** is a function of crystal size. In 2:1 layer

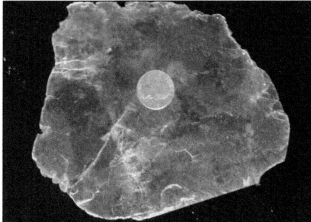

Figure 7.13 Pictures of mica showing the platy morphology of phyllosilicates. The top image shows ~1 cm tall stacks of muscovite mica. The bottom two images show the sheet-like, or platy, morphology. The coin (2.4 cm diameter) lies beneath the mica *sheet*. The piece of mica in the bottom image is ~1 mm thick and thus contains approximately 1 million 2:1 layers and interlayers. If the interlayers expanded by 0.5 nm due to weathering and hydrated cation exchange, the 1 mm thick sheet would increase its thickness to 1.5 mm—a 50% increase.

silicates, the largest surface charge occurs from isomorphic substitution.

Layer charge can be calculated using mineral formula. **Table 7.4** shows the calculation of layer and

interlayer charges for muscovite mica. Counter ions in the hydrated interlayers are exchangeable. If the layer charge is large and the interlayer cations are dehydrated, adsorbed cations cannot be removed and water cannot enter; thus, the *effective* CEC is reduced. **Table 7.3** shows the source of charge and effective or measured cation exchange capacities of several clay minerals. Special Topic Box 7.2 provides an example calculation of clay mineral formula, layer charge, and *theoretical* CEC.

7.3.5 Shrink and swell behavior and interlayer collapse

The number of layers of water molecules in clay interlayers can vary from none to three or more, which changes the c-spacing of the clay minerals (**Figure 7.14** and **Figure 7.15**) causing them to shrink and swell. Factors affecting the number of water layers in clay interlayers are the type of cations in the interlayer, the layer charge of the clay mineral, and hydration state of the clay mineral. Interlayer spacing is controlled by electrostatic forces, which are described by Coulomb's law (Eq. 2.2) – the force between charges is directly proportional to the magnitude of the charges, and inversely proportional to the square of the separation distance of the charges. There are three sources of electrostatic forces in clay interlayers that affect interlayer separation distance:

1 Cation valence is directly proportional to the force between the cation and the clay layer charge, thus higher charged cations shrink interlayer distances, and vice versa.
2 Hydrated cation size (Table 2.3) is inversely proportional to the force between the cation and the clay layer charge; thus, smaller hydrated cations shrink interlayer distances, and vice versa.
3 The force between an interlayer cation and clay layer charge is inversely proportional to the distance between isomorphic substitution charge in the 2:1 layer and the basal plane. Thus, a lower force between the octahedral sheet charge and the interlayer cations occurs when layer charge originates in the octahedral sheet compared to layer charge originating in the tetrahedral sheets (for illustration, see Figure 9.1).

Table 7.3 Selected properties of clay minerals.

Component	Layer charge (half cell)	CEC (mmol(+) kg⁻¹)	Specific surface area (m² g⁻¹)	c-spacing (nm)	Expansible	pH-dependent charge	Forces holding layers together
Kaolinite	~0	10–100	5–20	0.72	No	Extensive	Hydrogen bonds
Montmorillonite	0.2–0.6	800–1200	600–800	1.2–2.1	Yes	Minor	Electrostatic
Vermiculite	0.6–0.9	1200–1500	300–500	1.0–1.5	limited	Minor	Electrostatic
Illite	0.6–0.9	100–400	70–120	1.0	No	Medium	Ionic-type
Chlorite	~1	200–400	70–150	1.4	No	Extensive	Electrostatic
Allophane	–	100–1500	500–900	–	–	Extensive	

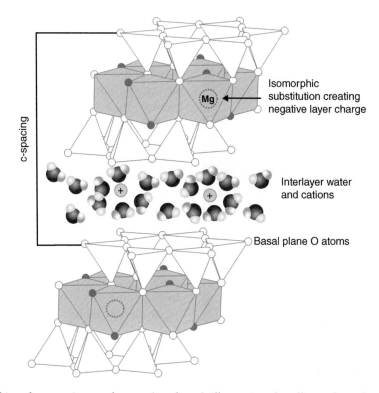

Figure 7.14 Clay mineral interlayer cations and water (one layer), illustrating the effects of interlayer spacing on c-spacing distance.

Decreased force of attraction between the interlayer cations and surfaces leads to increased swelling due to more water in the interlayer. **Figure 7.15** illustrates the change in interlayer spacing of sodium-saturated montmorillonite due to varying amounts of interlayer water. As the amount of water in the interlayer increases, the c-spacing increases, and vice versa. Thus, in soils that have a lot of montmorillonite in them, such as Vertisols, drying and wetting cause extensive changes in interlayer water causing extreme shrink and swell behavior. This shrink–swell behavior creates large cracks to form when the soils are dry, allowing for organic matter to fall deep into the soil, creating thick A horizons.

In mica, K^+ is dehydrated, and thus fits perfectly into the ditrigonal cavities in the basal plane. Thus, the cation to layer charge separation is smaller than if larger hydrated cations existed in the interlayer. The forces of attraction between K^+ and the clay layers causes the adjacent layers of mica to collapse, and the interlayer K^+ cations are not exchangeable. In montmorillonite, the layer charge originates in the octahedral sheet, and

Table 7.4 Calculation of layer charge for muscovite mica $(K(Si_3Al)Al_2O_{10}(OH)_2)$. The layer charge of -1 is balanced by the K^+ cation in the interlayer.

Element	Number in mineral formula	Ion valence	Total charge in mineral formula
Si	3	+4	+12
Al(tet)	1	+3	+3
Al(oct)	2	+3	+6
O	10	-2	-20
OH	2	-1	-2
Total layer charge (sum)			-1
Interlayer charge (K^+)			+1

the increased distance between the layer charge and the interlayer cations causes a weaker force than in mica, or even vermiculite. Thus, cations in the interlayer of montmorillonite remain hydrated, and are readily exchangeable.

Cations in the interlayer of vermiculite are predominantly Ca^{2+}, Mg^{2+}, and Na^+. These cations are exchangeable because the layer charge is not strong enough to collapse the layers and cause cation dehydration of the ions' solvation sphere (i.e., cation-layer charge force < cation hydration energy). However, if K^+ or NH_4^+ cations occupy the exchange sites of vermiculite, the interlayer will collapse because (i) K^+ and NH_4^+ cations

---

Special Topic Box 7.2 Determination of mineral formula

The measurement of the atomic composition of minerals is important for determining mineral identification and presence of impurities. Two common methods are: (1) digestion of the mineral in an acid followed by measurement of ion concentrations in the digest; and (2) X-ray fluorescence (XRF). Complete acid digestion requires special treatment to *liberate* the silicate by either fluxing in an alkali salt at high temperature prior to acid digestion, or using hydrofluoric acid as one of the digestion acids. Acid digestions are done with heat, pressure, and mixtures of acids.

Elemental composition is typically given in weight percent (Wt%) of the oxide component, such as Wt% SiO_2 or Fe_2O_3. For example, the mineral Wyoming montmorillonite sold by the Clay Mineral Repository as SWy-2 has a reported composition of 62.9% SiO_2, 19.6% Al_2O_3, 0.09% TiO_2, 3.35% Fe_2O_3, 0.32% FeO, 0.006% MnO, 3.05% MgO, 1.68% CaO, 1.53% Na_2O, 0.53% K_2O, 0.11% F, 0.049% P_2O_5, 0.05% S, which corresponds to a half cell mineral formula of

$$Ca_{0.06}Na_{0.16}K_{0.025}\left(Si_{3.99}Al_{0.01}\right)\left(Al_{1.55}Fe^{3+}_{0.20}Mg_{0.25}\right)O_{10}(OH)_2$$

With a layer charge of -0.25 (the Ti, S, F, and P elements are not included in the mineral structure). The fractional atomic composition of some atoms in the mineral formula is a result of charge balancing needed for the unit-cell mineral formula and is equivalent to a structural composition

averaging of the atoms in several unit cells to an individual unit cell – in the mineral structure there are, of course, no partial atoms.

To convert from Wt% oxide to mineral formula, or atoms per formula unit, dimensional analysis and charge balance in a mineral formula are used. The results of converting Wt% oxide to atoms per formula unit for a beidellite mined from Idaho, USA, are shown in **Table 7.5** to show how the measured atoms are distributed in phyllosilicate mineral formulas.

Table 7.5 Composition of a clay sample in weight percent oxide and atoms per formula unit (APFU). Sample from the DeLamar Mine (Bench 1) Post et al. (1997).

Element oxide	Wt.% oxide*	APFU (half cell)
SiO_2	61.34	3.63
Al_2O_3	31.97	2.24
Fe_2O_3	2.17	0.10
MgO	0.81	0.07
MnO	0.02	0.00
CaO	0.79	0.05
K_2O	2.53	0.19
Na_2O	0.32	0.04
BaO	0.04	0
Sum	100.00	

*0.8Wt% TiO_2 removed.

The mineral formula for the DeLamar clay sample shown in **Table 7.5** is

$$Na_{0.04}Ca_{0.05}K_{0.19}Si_{3.63}Al_{2.24}Fe^{3+}_{0.1}Mg_{0.07}O_{10}(OH)_2$$

The atomic formula can be rearranged to more clearly show the exchangeable, tetrahedral, and octahedral composition of the mineral by assuming a basic dioctahedral mineral formula and partitioning the atoms to the tetrahedral, octahedral, and interlayer positions in the mineral structure. For example, for the DeLamar beidellite, 3.63 Si^{4+} atoms are placed in the tetrahedral sheet and the remaining tetrahedral positions are occupied by 0.37 Al^{3+} atoms so that there are four total atoms in the tetrahedral position in the half cell formula. The remaining Al^{3+} atoms (1.87) are placed in the octahedral sheet along with 0.1 Fe^{3+} atoms, and 0.03 Mg^{2+} atoms so that the number of atomic positions in the octahedral sheet sums to two. The Na^+, K^+, Ca^{2+}, and remaining Mg^{2+} ions are placed as interlayer cations. The revised mineral formula is:

$$Na_{0.04}Ca_{0.05}K_{0.19}Mg_{0.04}(Si_{3.63}Al_{0.37})(Al_{1.87}Fe^{3+}_{0.1}Mg_{0.03})O_{10}(OH)_2$$

From the mineral formula, the layer charge can be calculated (**Table 7.6**). The mineral has a layer charge of 0.4 negative charges per formula unit of clay (half-cell). The mineral weight is 380 g mol^{-1}. Using the layer charge and mineral weight, the *theoretical* cation exchange capacity is

$$\frac{0.4\,mol(+)}{mol\,clay} \times \frac{mol\,clay}{380\,g\,clay} \times \frac{1000\,g}{kg} \times \frac{1000\,mmol}{mol}$$

$$= \frac{1053\,mmol(+)}{kg\,clay} \qquad (7.1)$$

Table 7.6 Calculation of layer charge for a beidellite. ($Na_{0.04}Ca_{0.05}K_{0.19}Mg_{0.04}(Si_{3.63}Al_{0.37})$ $(Al_{1.87}Fe^{3+}_{0.1}Mg_{0.03})O_{10}(OH)_2$). The difference in layer charge and interlayer cation charge (0.01) results from experimental errors and assumptions in mineral formula calculation methods.

Element	Number in mineral form	Ion valence	Total charge in mineral formula
Si^{4+}	3.63	+4	+14.52
Al^{3+}(tet)	0.37	+3	+1.11
Al^{3+} (oct)	1.87	+3	+5.61
Mg^{2+}	0.03	+2	+0.06
Fe^{3+}	0.1	+3	+0.30
		Cation charge	+21.6
O^{2-}	10	−2	−20
OH^-	2	−1	−2
		Total layer charge	−0.40
Interlayer cations			
Na^+	0.04	+1	+0.04
Ca^{2+}	0.05	+2	+0.10
Mg^{2+}	0.04	+2	+0.08
K^+	0.19	+1	+0.19
		Total interlayer charge	+0.41

The actual CEC is measured by summing the charge from the cations released in an exchange experiment, which will likely be different than the theoretical CEC.

have lower hydration energies and thus smaller hydrated radii than Ca^{2+}, Mg^{2+}, and Na^+; and (ii) vermiculite layer charge originates in the tetrahedral sheet; thus, the ion charge to layer charge distance is short, causing a stronger force between the cations and the layer. The force between the K^+ or NH^+ cations and the vermiculite layer charge is greater than the hydration energy of the cations causing dehydration and collapse of the vermiculite interlayer, rendering the interlayer cations effectively unavailable for exchange. This phenomenon is called **K^+-fixation**, or **NH_4^+-fixation**.

Figure 7.12 illustrates the effects of the amount of tetrahedral or octahedral charge on the size of the interlayer for various phyllosilicates. Since the magnitude and location of the charge is due to isomorphic substitution, the physical properties of clay minerals can be predicted from mineral structure and properties.

In some clay minerals, such as kaolinite and pyrophyllite, there is no layer charge, and bonding between layers occurs via weak hydrogen or van der Waals bonds between planes of oxygen or hydroxide atoms. Although individual hydrogen bonds or van der Waals forces are relatively weak, summed across the two

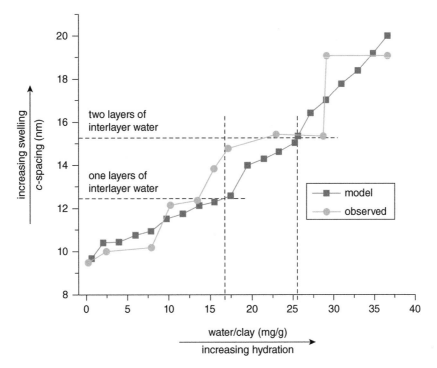

Figure 7.15 Effect of clay hydration state on c-spacing distance. Graph shows data from a Na-montmorillonite clay hydration experiment and results from a molecular modeled simulation that calculates theoretical c-spacing based on mineral and ion charge. The stable equilibrium points of one layer and two layers of interlayer water are shown. A completely dehydrated interlayer would have a c-spacing of ~ 1 nm. Data are from Meleshyn and Bunnenberg (2005), with permission.

adjacent layers of clay, the total hydrogen-bonding forces holding the layers together are significant.

7.4   Detailed properties of phyllosilicates

7.4.1   Kaolins

Kaolins are 1:1 layer silicate minerals (**Figure 7.9**) that occur in soils as hexagonal crystals with an effective size of 0.2 to 2 μm (**Figure 7.2**). Si^{4+} is the only cation in the tetrahedral sheet of kaolins, and Al^{3+} or Mg^{2+} occupy the octahedral positions. When Al^{3+} is in the octahedral sites, the mineral is **kaolinite** $(Al_2Si_2O_5(OH)_4)$ or one of its poorly crystallized polymorphs, dickite or nacrite. When Mg^{2+} is in octahedral sites (trioctahedral), the kaolin mineral is antigorite $(Mg_3Si_2O_5(OH)_4)$. **Halloysite** is a form of kaolinite where water is held between structural units in the basal plane. Kaolinite layers are held together by hydrogen bonding between oxygen ions of the tetrahedral sheet, and hydroxide anions of the adjacent octahedral sheet.

The hydrogen bonding between the 1:1 kaolin layers prevents expansion and entry of water and cations (halloysite is an exception because of formation conditions). The c-spacing of kaolinite and antigorite is 0.72 nm. Surface area is limited to external surfaces, and hence specific surface area ranges from only 10 to 20 m^2 g^{-1}. In fully hydrated halloysite, water in the interlayer causes a c-spacing of 1 nm. Kaolinite is coarse clay with low colloidal activity, including low plasticity and cohesion, and low swelling and shrinkage.

The ideal unit formula for kaolinite has a Si/Al ratio of one. Kaolinites have little or no isomorphic substitution. Any deviation of the Si/Al ratio from one could be due to surface coatings that were not removed during preparation of the sample for elemental analysis. Most of the cation exchange capacity of kaolinite (10 to 100 mmol(+) kg^{-1}) is attributed to dissociation of protons from OH groups at clay edges. Protonation and deprotonation of edge sites is pH dependent, making kaolinite a **variable charge mineral**.

7.4.2 Smectite

The layer charge of smectites is 0.2 to 0.60 charges per half-unit cell formula. Because of the relatively low-layer charge compared to mica and vermiculite, smectites expand freely. The c-spacing varies with the exchangeable cation and the degree of interlayer solvation. Complete dehydration yields spacing near 1 nm, and full hydration can swell the layers to greater than 2.0 nanometers.

The different types of smectite vary in octahedral cation composition and location of isomorphic substitution (**Table 7.2**). In soils, the predominant smectite is the dioctahedral mineral **montmorillonite**, where Al^{3+} is the major octahedral cation, and some Mg^{2+} isomorphic substitution occurs. A typical half unit-cell formula for montmorillonite is $Na_x(Al_{2-x}Mg_x)(Si_{4-y}Al_y)O_{10}(OH)_2$ where x ranges from 0.2 to 0.6, Na^+ is the charge-compensating exchangeable cation, Al^{3+} and Mg^{2+} are in the octahedral layer, and y is typically less than 0.33.

Soil smectites have extensive isomorphic substitution, with some Al^{3+} substituting for Si^{4+} in the tetrahedral sheet, and Fe^{2+} and Mg^{2+} substituting for Al^{3+} in the octahedral sheet. In montmorillonite, however, isomorphic substitution in the tetrahedral sheet is minor, and most of the layer charge arises from Mg^{2+} substitution for Al^{3+} in the octahedral sheet.

Beidellite is a smectite that has tetrahedral Al^{3+} (**Table 7.2**). Beidellite is less common in soils than montmorillonite. **Nontronite** is like beidellite but has an Fe^{3+} octahedral sheet instead of an Al^{3+} octahedral sheet. Nontronite is a common product in weathered basalt, but is not common in soils; probably because during weathering of rocks to soils, liberated iron forms iron oxides instead of phyllosilicates. Interestingly, nontronite is common in Mars *soil* samples (see Special Topic Box 7.3).

Cation exchange capacities for montmorillonite range from 800 to 1200 mmol(+) kg^{-1} (**Table 7.3**). The cation exchange capacity is only slightly pH dependent. The low layer charge allows the mineral to expand freely, exposing both internal and external surfaces to the soil solution. Low-charge smectites (0.2–0.4 charges per half unit cell formula) have greater hydration in the interlayer, and thus greater shrink-swell behavior. Due to the accessibility of the interlayer, total specific surface area is 600 to 800 m² g^{-1}, with as much as 80% of the total due to internal surfaces. Montmorillonite has high colloidal activity, including high plasticity and cohesion, and high swelling and shrinkage. Montmorillonite normally occurs as a fine or colloidal clay, with irregular crystals 0.01 to 1 µm in size. Smectites are common in Vertisols and in soils of alluvial plains.

7.4.3 Vermiculite

Vermiculites occur extensively in soils formed by weathering of micas. The layer structure of vermiculite resembles that of mica from which the mineral derives. Both trioctahedral and dioctahedral vermiculites exist. Weathering or alteration of the precursor micas replaces the interlayer K^+ mostly with Mg^{2+}, Ca^{2+}, and Na^+, and expands the c-spacing up to 1.5 nm.

Vermiculite includes several mineral compositions (**Table 7.2**), but generally Al^{3+} is substituted for Si^{4+} in the tetrahedral sheet to the extent of 0.6 to 0.9 per half unit-cell formula, and Mg^{2+} and Fe^{2+} occur as the octahedral cations. An idealized half unit-cell formula for trioctahedral vermiculite is $Mg^{2+}_{x/2}(Mg, Fe^{2+})_3(Si_{4-x}Al_x)O_{10}(OH)_2$, where the Mg^{2+} is the exchangeable cation in the interlayer and x ranges from 0.6 to 0.9.

An interesting property of some vermiculites is the *internal* balancing of some of the layer charge by substituting Fe^{3+} or Al^{3+} for Mg^{2+} and Fe^{2+} in the octahedral sheet, such as in the mineral formula $Na_{0.7}(Al_{0.2}Fe(III)_{0.4}Mg_{2.4})(Si_{2.7}Al_{1.3})O_{10}(OH)_2$. This creates a tetrahedral sheet charge of –1.3 and octahedral sheet charge of +0.6, and a resulting net charge of –0.7 per unit formula; which is balanced in this example by 0.7 Na^+ ions in the interlayer.

The layer charge in vermiculite gives rise to a cation exchange capacity ranging from 1200 to 1500 mmol(+) kg^{-1} (**Table 7.3**); considerably higher than the exchange capacity of montmorillonite. Some high charge vermiculites have very high CEC of up to 2100 mmol(+) kg^{-1}. As with montmorillonite, the cation exchange capacity is only slightly pH dependent.

Most of vermiculite's layer charge originates in the tetrahedral sheet, thus the distance between the interlayer cation and the surface charge is less than montmorillonite, which causes less swelling than vermiculite (**Figure 7.12**). When K^+ or NH_4^+ cation adsorb in the interlayers of vermiculite, they become *fixed* because they cannot be exchanged with ordinary salt solutions, and the

Special Topic Box 7.3   Mineral weathering on Mars

An objective of modern space exploration is discovery of possible life and water on our closest planetary neighbor, Mars. The Curiosity Rover launched by NASA in 2012 took approximately eight months to travel 560 million kilometers to Mars (**Figure 7.16**).

Figure 7.16 The surface of Mars in Gale Crater taken by the Curiosity Rover in 2013. The Mars Science Laboratory is equipped with several instruments and tools to survey the Martian environment. Many of the analyses are designed to detect the mineralogical characteristics of the Mars *soil*. Source: Courtesy NASA/ JPL-Caltech.

Compared to Earth, Mars's surface is dry and barren. However, recent scientific investigation suggests that Mars once had a hydrated surface. Water facilitates mineral weathering and formation of regolith akin to soil. Most of Earth's soils (at least those that we are usually interested in) contain organic matter from biological activity. The presence of water on Mars leads to the question: Could simple forms of life, such as microbes, have lived on Mars? Answering these questions and preparing for future exploration requires knowledge of the geologic environment on Mars.

Many Mars missions have focused on studying its mineral composition using satellites with spectrometers that can detect mineral identities on the surface, and rovers to make in-situ measurements of element and mineral composition. Such missions have discovered vast quantities of clays, including smectites, on the surface of Mars. Clay minerals are direct evidence of the presence of hydrous weathering processes, meaning water was present. Nontronite is a particularly abundant clay species observed on Mars's surface. In Earth's soils, nontronite occurs less commonly than its aluminum-rich smectite cousin montmorillonite. It is not clear what conditions favor nontronite formation. On Mars, nontronite likely formed from water alteration of Fe-rich basalt rocks.

Mars landscapes rich in clay minerals are prime environments to look for signatures of microbial life because the presence of clay minerals indicates water was once present. Such science is driving many fields of inquiry, and promises a bright future for planetary mineralogists ... and possibly exobiologists.

interlayer dehydrates and collapses. Collapsed vermiculite is nonswelling (with a fixed c-spacing of 1 nm). Other large alkali cations such as Cs^+ and Rb^+ can also become fixed in vermiculites when they are adsorbed. When Na^+, Mg^{2+}, or Ca^{2+} cations are in the interlayer, the c-spacing is 1.2–1.5 nm. The total specific surface area of vermiculite, when not K^+ or NH_4^+ saturated, ranges from 600 to 800 m^2 g^{-1}. The CEC of vermiculite when the interlayer is collapsed is considerably less because of lack of access to the interlayer cations for exchange.

7.4.4   Mica and Illite

Mica minerals are abundant in soils as primary minerals inherited from parent materials. Micas are precursors for other 2:1 clay minerals, notably vermiculites. In soils, mica is often interstratified with other expandable phyllosilicates. Altered mica containing less K^+ and more water than well-ordered mica is called illite or hydrous mica, and is the form of mica that usually occurs in the clay-size fraction.

A typical half unit cell formula for dioctahedral **muscovite** mica is $K(Al_2(Si_3Al)O_{10}(OH)_2$. Isomorphic substitution of Al^{3+} for Si^{4+} in the tetrahedral layer creates negative charge in the basal plane of oxygen atoms. The short distance between the layer charge and the interlayer cation results in strong coulombic attraction for charge-compensating cations. Interlayer K^+ is so strongly adsorbed that it is not exchanged in standard cation exchange capacity (CEC) measurements. Thus, despite the large layer charge (about –1 per half unit cell), the CEC of illite is only 100 to 400 mmol(+) kg^{-1} (**Table 7.3**). Total specific surface area of illite is about 70 to 120 m^2 g^{-1}, and is restricted to external surfaces. Micas are nonswelling, and are only moderately plastic in texture. The fixed K^+ is released slowly during weathering and is a source of K^+ for plants.

Biotite is dark-colored mica that has a trioctahedral sheet containing Fe^{2+} and Mg^{2+}. Biotite weathers much more rapidly than muscovite because Fe^{2+} oxidizes to Fe^{3+}, destabilizing the structure, and because the proton in the bottom of the ditrigonal cavity (**Figure 7.8**) is more strongly repelled from the Si^{4+} cations in a trioctahedral configuration than a dioctahedral configuration. This occurs because the vacancy in the dioctahedral sheets allows the proton to shift further away from the high-charge Si^{4+} cations, while in the trioctahedral sheets there is no vacancy and thus the proton remains surrounded by the high charge Si^{4+} ions making it less stable than the dioctahedral micas. As a result of biotite's weathering susceptibility, it is uncommon in the clay fraction of soils because it readily alters to vermiculite, chlorite, and other secondary minerals.

Special Topic Box 7.4   Potassium release from soil minerals

Potassium is a macronutrient required by plants, thus its availability for plant uptake is an important consideration for managing soil fertility. **Figure 7.17** shows a biogeochemical cycle for potassium in soils. Most plant available potassium is released from minerals in the soil to the soil solution where roots absorb it.

There are three main categories of mineral-bound potassium in soils: (1) exchangeable K^+ on clay minerals, (2) K^+ *fixed* on clay minerals such as muscovite and biotite (see **Table 7.2**), and (3) K^+ associated with primary minerals such as K-feldspar ($KAlSi_3O_8$). The availability of K^+ in soil for plant uptake is assessed using NH_4^+ solutions to release the exchangeable K^+ to the extraction solution. Potassium concentrations in the NH_4^+ extracts are correlated with plant uptake to make fertilizer recommendations. However, the NH_4^+ extraction method may underestimate the actual K^+ availability because it does not accurately assess K^+ release from mica and K-feldspar minerals; which may be especially important in the rhizosphere.

Potassium fixed on clays and K-feldspar are considered **nonexchangeable K^+** and are thus categorized as *unavailable*. Because nonexchangeable K^+ is released at a much slower time scale than exchangeable potassium, ammonium exchange to measure plant availability underestimates K^+ in the soils for plant nutrient availability. In soils, K^+ categorized as nonexchangeable can represent more than 90% of the total K^+. Studies have reported, however,

that the nonexchangeable K^+ may be more than 80% of the potassium that is eventually supplied to plants, and thus labeling it unavailable is not correct.

Fixed potassium in soils is associated with vermiculite, illite, and mica minerals. K-feldspars and micas are common in soils derived from granitic rock parent materials, occurring in the silt and fine sand fractions. As mica weathers, the interlayer potassium ions, which are the weakest bonds in the crystal, are exchanged for other cations like Na^+ and Ca^{2+}, starting first at the edges of the mica (**Figure 7.18**). This process creates frayed edges because the Na^+ and Ca^{2+} have larger hydrated radii than K^+; subsequently, the mica becomes illite. With continued weathering, the additional interlayer K^+ is exchanged for hydrated cations, and some dissolution of the 2:1 crystallite occurs, creating vermiculite. This weathering sequence is referred to as vermiculitization. Weathering of mica is enhanced in rhizospheres, which accelerates the release of K^+ by increased microbial activity and presence of organic acids that dissolve the minerals.

Adsorption of K^+ on clays by cation exchange is a common process and can be a sink for K^+ fertilizer. Potassium exchange on montmorillonite is reversible. However, due to the high layer charge associated with the tetrahedral sheets in illite and vermiculite (**Table 7.3**), adsorbed K^+ dehydrates when adsorbed to these clay minerals, causing the vermiculite and illite interlayers to collapse and K^+ to become fixed.

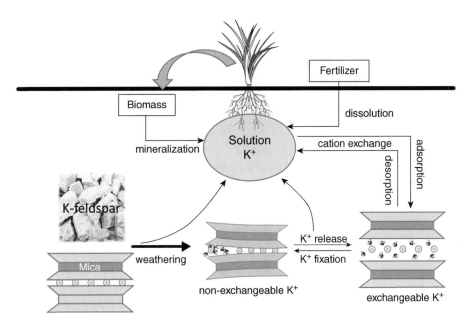

Figure 7.17 Biogeochemical cycle of potassium in soils.

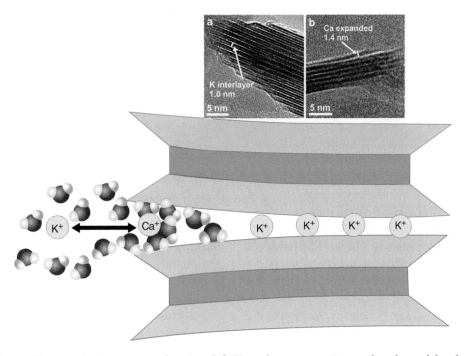

Figure 7.18 Mica to illite weathering process showing Ca^{2+}-K^+ exchange occurring on the edges of the clay mineral platelets, which causes the interlayer to become hydrated and edges to fray. Micrographs at the top show transmission electron microscopic image of a mica mineral before and after Ca^{2+} exchange. After Ca^{2+} exchange on the edges of the mica mineral, the edges become frayed and the c-spacing increases from 1 nm to 1.4 nm. TEM micrographs from Fuller et al. (2015), with permission.

7.4.5 Chlorite

Secondary phyllosilicates with sheets of octahedra in the interlayers instead of cations, like chlorite, are common in soils. In chlorite, a trioctahedral sheet is precipitated in the interlayer. The positively charged Mg-OH octahedral sheet between the negatively charged 2:1 sheets restricts swelling, decreases the effective surface area, and reduces the cation exchange capacity of the mineral. An idealized half unit-cell formula for trioctahedral chlorite (chlinochlore) is $(AlMg_2(OH)_6^+)_x$ $(Mg, Fe^{2+})_3$ $(Si_{4-x}Al_x)$ $O_{10}(OH)_2$, where x ranges from 0.8 to 1.6. The $AlMg_2(OH)_6^+$ component of the chlorite structure is a trioctahedral sheet that is cationic and balances the layer charge of the 2:1 layer. Isomorphic substitution in typical chlorites occurs in the tetrahedral sheet of the 2:1 layer. Compared to other phyllosilicates, the high tetrahedral Al^{3+} substitution creates a relatively large negative layer charge. The interlayer hydroxide sheet is typically trioctahedral. Dioctahedral chlorites do exist with the interlayer consisting of Al^{3+} or Fe^{3+} hydroxides, but dioctahedral sheets in the interlayer are typically discontinuous hydroxide *islands*, and are referred to as **hydroxy-interlayered smectites** (HIS) or **hydroxy-interlayered vermiculites** (HIV). Aluminum is the most common interlayer cation in hydroxy-interlayered minerals.

Chlorite and "chlorite-like" minerals are common in soils that form from sedimentary rocks. The elemental composition of chlorites varies widely. Some mafic- (Fe- and Mg-rich magma) derived chlorites contain chromium and nickel hydroxides in their interlayer. Soils derived from serpentine parent materials contain chlorite, and many are infertile because of their high Mg^{2+} and low Ca^{2+} contents. HIV and HIS occur in soils as intermediate weathering products that form under acidic conditions. With increased weathering, desilication dissolves the silicate from the phyllosilicate in the HIS and HIV.

The layer charge of the 2:1 portion of chlorite varies, but is similar to mica. Cation exchange capacity ranges from 200 to 400 mmol(+) kg^{-1} (**Table 7.3**). Total specific surface area ranges from 70 to 150 m^2 g^{-1}.

7.5 Allophane and imogolite

Most of the crystallized secondary minerals found in soils have passed through intermediate amorphous steps during chemical weathering. Amorphous is defined as being nondetectable by X-ray diffraction. During the reorganization of hydrous gels that precipitate from the soil solution, phases with varying degrees of crystallinity are produced and the distinction between amorphous and crystalline materials is vague. Very small crystals can appear amorphous to many tests of crystallinity. Such materials have been termed **cryptocrystalline**.

Allophane is a general name for amorphous aluminosilicate gels that have short-range order (poorly crystalline). Allophanes often convert ("ripen") to more crystalline clay minerals. The composition of allophane varies widely and includes Al_2O_3, Fe_2O_3, and SiO_2. Only minor amounts of Mg^{2+}, Ca^{2+}, K^+, and Na^+ are generally present. The Al:Si ratio is usually between 1 and 2. A general formula for allophane is $xSiO_2 \cdot Al_2O_3 \cdot yH_2O$ ($x = 0.8-2.5$, $y > 2.5$). Whether allophane is a mixture of individual silicon and aluminum oxide gels, or whether it is an amorphous hydrous aluminosilicate in which oxygen anions are shared *between* silicon and aluminum ions is unclear. In any case, silicon occurs in tetrahedral coordination and aluminum in octahedral coordination.

Allophane is commonly observed in soils that have significant volcanic ash input and forms as a result of rapid weathering of volcanic tephra. Allophane minerals are also common in soils as an intermediate product of weathering, especially as coatings of primary minerals.

The morphology of allophane is referred to as nanospherical (or nanoballs) because of the nanometer-sized spherical nature of the particles (**Figure 7.19**). Allophane contains a significant amount of associated water, and dehydration in the laboratory to measure properties such as surface area and cation exchange capacity may irreversibly alter its structural properties. Allophane can have a high cation exchange capacity at neutral to mildly alkaline pH; perhaps as large as 1500 mmol(+) kg^{-1}, but the CEC is highly dependent on pH and degree of hydration. Values reported in the literature range from 100 to 1500 mmol(+) kg^{-1}. The CEC measurement of allophane is indefinite because exchangeable cations are loosely adsorbed and hydrolyze extensively during washing used in CEC determination. Due to the small size and solution connectivity to both the interior and exterior surfaces, allophane has a high specific surface area (70–300 m^2 g^{-1}) that varies widely with degree of crystallinity and pH.

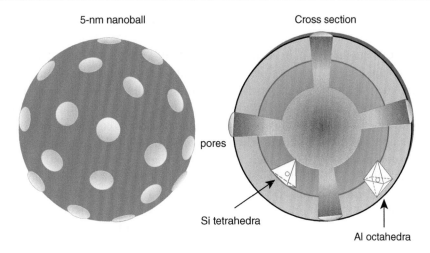

Figure 7.19 Idealized model of allophane nanoball illustrating the curved/spherical morphology and the pores in the sphere. The cross section shows the aluminum octahedral and silicon tetrahedral curved sheets. The model suggests that allophane is a symmetric and perfect crystal, but the structure is actually poorly crystalline and the morphology, while spherical, is irregular with little symmetry. Adapted from Iyoda et al. (2012). Reproduced with permission of Elsevier.

Imogolite $(Al_2SiO_3(OH)_4)$ was first described in Japanese soils derived from volcanic ash. Although not highly crystalline, imogolite can be recognized by X-ray diffraction. The morphology of imogolite is referred to as paracrystalline, which means the mineral has long-range crystallinity that extends in one direction, but shorter-range order in other directions. Imogolite exists as small threads or tubes that are 1–50 nm in diameter.

7.6   Zeolite

Zeolites are aluminosilicates (tectosilicates), analogous to feldspars, in which Si and Al cations in tetrahedral coordination are linked by sharing their O ligands with each other, as depicted in Figure 6.9. The tetrahedra are linked into 4-, 6-, 8-, and 12-membered rings, joined together less compactly than in the feldspars. Analcime $(NaAlSi_2O_6 \cdot H_2O)$ is a species of zeolite.

The open framework of zeolites leaves channels of different sizes that run in several directions through the crystal. The channels contain loosely held water molecules and charge-balancing cations that are freely exchangeable. The channels frequently interconnect and are usually larger than the diameters of the adsorbed cations, so both water and the charge-balancing cations diffuse readily through the crystals. Some zeolites have smaller-sized channels that effectively

prevent movement of large molecules, leading to the use of zeolites as molecular sieves. Recent work suggests that the electrostatic attraction between the surface charge of zeolites and hydrated cations is strong enough to dehydrate some cations, allowing them to migrate into some of the very small channels that they would otherwise be too large to fit when fully hydrated.

7.7   Oxide minerals

In soils, the most common **oxides, hydroxides,** and **oxyhydroxides** are minerals with iron, aluminum, or manganese cations coordinated with oxygen or hydroxide ions. Oxygen is also the major anion in silicates, but they are classed separately from oxides. In this text, the term *oxide* refers to minerals composed of oxygen or hydroxide. Oxides in soils occur with a range of crystallinities and with some isomorphic substitution.

A distinguishing characteristic of oxides is that the charge on the surface is mostly a result of unsatisfied bonds on the edges. This topic is covered in more detail in Chapter 9. An important concept in edge charge of oxides, however, is that it can vary from positive, zero, to negative, depending on the pH of the solution. The pH when the *net* charge on the mineral surface is zero is called the **point of zero charge** (**PZC**). At pH below the PZC, the surface is net positively charged due to H^+ adsorption. At pH above the PZC, the surface is net negatively charged due to H^+ loss (or its equivalent OH^- adsorption). In addition to pH, the magnitude of the charge depends on the mineral characteristics, such as particle size, surface area, and major cation composition of the oxide. **Variable charge minerals** are noted for their capacity to adsorb anions, such as phosphate and sulfate when the surface is positively charged (pH < PZC), and cations when the pH is above the PZC. Soils that contain large amounts of variable charge minerals are called **variable charge soils**. **Table 7.7** lists properties of oxides that occur in soils.

7.7.1 Aluminum oxides

Gibbsite ($Al(OH)_3$) is the stable low-temperature form of aluminum hydroxide, and is the most abundant aluminum hydroxide in soils. Gibbsite is particularly common in highly weathered soils; 20–30% gibbsite by mass, or more, is common. The crystallinity of aluminum hydroxides in soil varies from short-range ordered, to long-range ordered. Crystalline gibbsite consists of sheets, as described for the dioctahedral sheets of clay minerals (**Figure 7.7**), held together by hydrogen and van der Waals bonds between sheets. Some isomorphic substitution of Fe^{3+} in gibbsite likely occurs.

Table 7.7 Properties and occurrence of soil oxides, allophane, and imogolite. PZC is the point of zero charge (the pH at which the net surface charge is zero). At pH < PZC, the net surface charge is positive, and pH > PZC, the net surface charge is negative.

Mineral	Mineral Formula	PZC	Color	Occurrence
Gibbsite	$Al(OH)_3$	9.8	white	Common; forms in well drained, highly weathered soils
Allophane	$(SiO_2)_xAl_2O_3 \cdot yH_2O$ ($x = 0.8 - 2$, $y \geq 2.5$)	6–10	white	Common; forms in soils that have rapid weathering parent materials, such as Andisols
Imogolite	$SiO_2Al_2O_3 \cdot 2.5H_2O$	8.5–9	white	Common; forms in soils that have rapid weathering parent materials, such as Andisols
Goethite	$FeOOH$[a]	7.5–9	yellow	Common; forms in cool to moderate temperature climates
Hematite	Fe_2O_3	8.5	red	Common; forms in highly weathered soils of warmer climates
Ferrihydrite	$(Fe_5HO_8 \cdot 4H_2O)$[b]	6–9	dark reddish brown	Common; forms in soils with rapid weathering and oxidation
Magnetite	$Fe^{2+}Fe^{3+}_2O_4$	8.5	black	Trace to moderate; primary mineral inherited form parent material
Maghemite	Fe_2O_3	6.6	red to brown	Moderate in well weathered soils in warm climates; likely forms from oxidation of magnetite
Lepidocrocite	$FeOOH$[a]	6.7–7.5	reddish yellow	Trace to moderate; forms in cool climates in soils subject to repeated wetting and drying
Akaganeite	$FeOOH$[a]	9.6–10.3	brownish yellow	Trace to moderate in acid sulfate soils in coastal wetlands
Schwertmannite	$Fe_8O_8(OH)_6SO_4$	4	reddish yellow	Rare
Todorokite (tectomanganate)	$(Mg_{0.77}Na_{0.03})(Mg_{0.18}Mn^{2+}_{0.6}Mn^{4+}_{5.22})$ $O_{12} \cdot 3.07H_2O$	3.5	brown to black	Occurs in nodules in soils
Pyrolusite (tectomanganate)	MnO_2	4.4	brown to black	Rare
Birnessite (phylomanganate)	(Na,Ca,Mn^{2+}) $(Mn^{3+}Mn^{4+})_7O_{14} \cdot 2.8H_2O$	1.5–2.8	brown to black	Occurs in soils subject to alternating redox conditions
Lithiophorite (phylomanganate)	$LiAl_2Mn^{4+}Mn^{3+}O_6(OH)_6$	6.9	brown to black	Occurs in nodules in soils

[a] Polymorphs.
[b] Variable.

In acid soils, gibbsite and iron oxides react strongly with phosphate, and are responsible for making phosphate unavailable to plants, and for adsorbing sulfate and reducing its availability to plants. Gibbsite and the iron oxides are responsible for much of the pH-dependent charge in soils. Typical specific surface areas of gibbsite range from 10 to 100 $m^2\ g^{-1}$.

Boehmite (AlOOH) is a polymorph of gibbsite, and occurs in intensively leached, highly weathered soils. It also can be formed from gibbsite by heating to about 130 °C. Corundum (Al_2O_3) is a high-temperature aluminum oxide rarely found in soils. Corundum is sometimes studied to characterize the solid-solution interface reactions of aluminum hydroxides.

7.7.2 Iron oxides

Hematite (Fe_2O_3) and **goethite** (FeOOH) are the most common iron oxides found in soils. Hematite occurs in highly weathered soils and is pink to bright red. Goethite is also characteristic of strongly weathered soils and is yellow to dark yellowish-brown.

The structure of hematite is stacked sheets of dioctahedral iron oxide bonded together by face sharing of the oxygen atoms between sheets (**Figure 7.20**). Goethite forms pairs of Fe-octahedra that run in channels (**Figure 7.21**). The resulting goethite structure has open channels with hydrogen bonds between the oxygen ligands of chains above and below. Warmer climates favor hematite formation over goethite. Both hematite and goethite occur as coatings on soil particles, imparting the red, orange, and yellow colors characteristic of well-weathered soils. Such coatings are often poorly crystalline, and may be **ferrihydrite**. As with gibbsite, the soil hematite or goethite content in highly weathered soils ranges from 20% to 30%, but can be much greater in some soils.

Ferrihydrite is an iron oxide with short-range order and structural water molecules. Ferrihydrite is also known as hydrous ferric oxide (HFO); HFO is considered the iron oxide phase with the least amount of atomic order, and the most water integrated into its structure. Special Topic Box 7.5 discusses the current state of knowledge of the structure of ferrihydrite. Ferrihydrite is common in soils that undergo rapid weathering (such as in volcanic-ash influenced soils), in soils where iron is subjected to repeated reductive dissolution and oxidation cycles (such as in wetlands), and in soils where complexation of iron with soil organic matter inhibits formation of more crystalline oxides. **Figure 7.22** illustrates the influence of soil organic matter and iron supply rate on formation of various iron oxides. In soils with lots of SOM, ferrihydrite or SOM-Fe^{3+} complexes are favored.

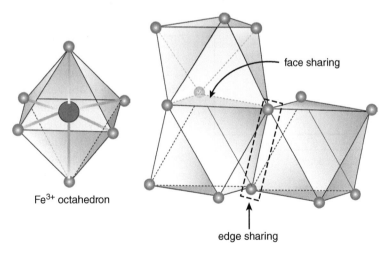

Fe^{3+} octahedron

face sharing

edge sharing

Figure 7.20 The hematite (Fe_2O_3) structure consists of sheets of edge sharing iron octahedron that are attached to sheets above and below through face sharing of three oxygen atoms per iron. The diagram shows only three of the octahedra to illustrate the edge and face sharing structure. The full unit cell for hematite contains 24-unique iron atomic positions, which are all coordinated to each other through edge or face-sharing polyhedra.

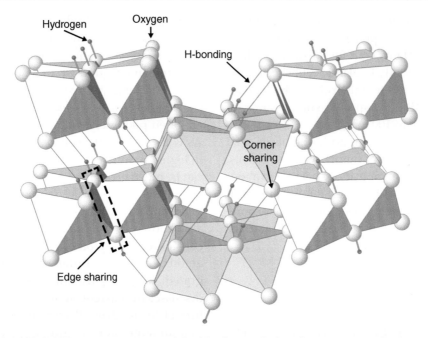

Figure 7.21 The goethite (FeOOH) structure consists of double chains of edge-sharing iron octahedra linked to other chains by corner-sharing oxygen ligands. The octahedral arrangement leaves channels in the structure where hydrogen bonds occur.

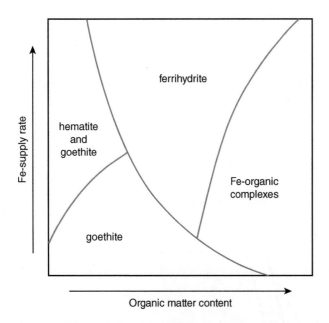

Figure 7.22 Representative stability diagram for iron oxides and soil organic matter Fe-complexes as a function of the rate of iron supply and amount of organic matter present. High rates of iron supply, and high amounts of soil organic matter cause ferrihydrite to be favored over the more crystalline goethite and hematite. At the highest organic matter levels, iron oxide precipitation is completely inhibited. Source: Cornell and Schwertmann (1996). Reproduced with permission of Wiley.

In warm environments, ferrihydrite transforms to hematite, and in cooler climates it dissolves and precipitates as goethite. The crystallization of amorphous iron oxides is responsible for the irreversible hardening upon drying of **plinthite** in tropical soils into stone-like materials.

Magnetite (Fe_3O_4) is a magnetic ferric and ferrous iron oxide inherited from igneous rocks or formed in reduced soils. It usually occurs as sand-sized grains of high specific gravity. Magnetite oxidizes to maghemite (Fe_2O_3), which is also magnetic.

The surfaces of iron oxides behave similarly to gibbsite, as described above. Most iron oxides have specific surface areas on the order of 20 to 100 m^2 g^{-1}. Some of the less well-crystallized phases have small particle sizes, and specific surface areas as high as 200 m^2 g^{-1}. Ferrihydrite particles are often of nanoparticle size, and can have specific surface areas as large as 400 m^2 g^{-1}, or more.

7.7.3 Titanium oxides

The titanium oxides commonly found in soils and clay sediments are rutile and anatase – polymorphs with mineral formulas of TiO_2. Both are inherited from the

Special Topic Box 7.5   The elusive ferrihydrite structure

Ferrihydrite is a common iron hydroxide mineral in soils, and is sometimes referred to as hydrous ferric oxide (HFO). It occurs in many soil environments, and is especially common in soils subject to rapid changes in redox conditions and weathering, such as wetlands and soils with volcanic ash inputs. Ferrihydrite gives soils a reddish brown color when present (see Color Plate 1.14, for example). Ferrihydrite has three important characteristics: (i) poor crystallinity; (ii) nanoparticulate in size (1–3 nm); and (iii) metastability. The last point means that given time, ferrihydrite will ripen to more stable iron oxides, such as hematite or goethite.

Ferrihydrite is difficult to detect using XRD because it lacks long-range atomic order. As such, the mineral properties of ferrihydrite are not well known. The most poorly crystalline phase of ferrihydrite is categorized as two-line ferrihydrite due to the two broad peaks that occur in the diffractogram (**Figure 7.23**). The more crystalline

ferrihydrite is known as six-line because of the six relatively sharp peaks that occur in the diffractogram.

The mineral formula of ferrihydrite is often listed as $Fe(OH)_3$. However, this formula is not correct. The mineral formula for six-line ferrihydrite is reported to be $Fe_5HO_8 \cdot 4H_2O$; but the number of structural waters is not well known and other mineral formulas have been proposed.

Modern tools have provided some insights into ferrihydrite's structure. In 1993 Drits et al. published a likely structure for six-line ferrihydrite, which consisted of Fe^{3+} octahedra linked to each other in various configurations. In 2007, Michel et al. published new work on the ferrihydrite structure based on results from a spectroscopic tool capable of discerning atomic coordinates in short-range order minerals. They proposed that Fe^{3+} in six-line ferrihydrite exists in both tetrahedral and octahedral coordination, which contrasts with the Drits et al. model. Since the

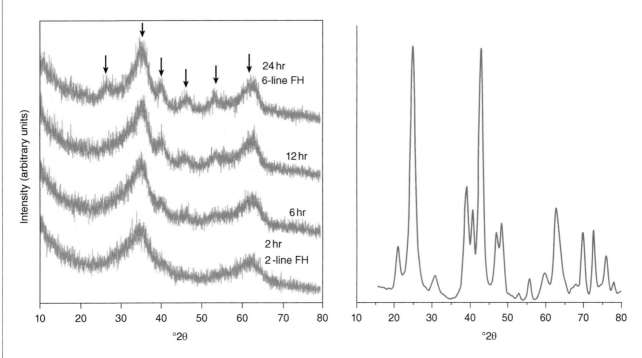

Figure 7.23 XRD of ferrihydrite (left panel) and goethite (right panel). The ferrihydrite XRD shows the progression of ferrihydrite (FH) from two-line to six-line as it develops more crystallinity. The peaks (arrows) in the FH XRD are low intensity and broad due to the short-range order of the mineral. The goethite XRD shows sharper peaks than ferrihydrite because it is more crystalline, and has more uniform diffraction from the planes of well-ordered atoms. Ferrihydrite XRD is from Majzlan et al. (2004); and Goethite XRD are from Anschutz and Penn (2005). Majzlan et al. (2004) reproduced with permission of Elsevier. Anschutz and Penn (2005). With permission of Springer.

2007 paper, a lively debate amongst scientists has ensued on whether ferrihydrite has tetrahedral Fe^{3+}, and what its structure is. The debate has resulted in dozens of new papers, and several corresponding letters published to critique the published papers. Such scientific debate puts scientific discovery in *hyper-mode*, and makes for exciting times for those within the discipline.

In soils, ferrihydrite minerals are generally considered to be a type of amorphous iron oxide mineral with significant amounts of structural water. Because they are metastable minerals, ferrihydrite persists in soils only when the rate of formation is equal to or greater than the rate of disappearance. Most soil ferrihydrite has Al^{3+} and other metal cations isomorphically substituted into the structure. Formation of well-ordered (i.e., six-line) crystals of ferrihydrite are inhibited by adsorption of ions such as silicate and phosphate, and association with soil organic matter (see illustration in **Figure 7.1** and stability trends in **Figure 7.22**). *Pedogenic*-ferrihydrite is typically far from the purity and crystallinity of the *ideal* six-line ferrihydrite and does not likely conform to the proposed structures. Rather, ferrihydrite in soils is a poorly crystalline to amorphous phase of predominantly iron octahedron with possibly some tetrahedral iron. The Fe-oxygen units are likely coordinated to each other with somewhat random edge- and corner-sharing bonds.

References

Drits et al. (1993)

Michel et al. (2007)

---

parent rock. Because titanium oxides resist weathering so well, they are often used as indicators of the original amount of parent material from which a soil has formed. Because they are present in soils much less than other oxides, they do not impart significant physical or chemical characteristics to the soil.

7.7.4 Manganese oxides

Manganese oxides in soils typically occur as poorly crystalline mixtures of Mn(III) and Mn(IV). In soils with fluctuating water tables, manganese oxides form concretions and nodules. Most soils likely have some manganese oxides that occur as coatings or small poorly crystalline nanoparticles that may be associated with iron oxides.

Based on their general structural arrangement, manganese oxides can be categorized into two types:

1 **Tectomanganates** have manganese octahedra corner sharing with each other (as in the *tectosilicates* with corner sharing tetrahedra described in Chapter 6).
2 **Phyllomanganates** have edge-sharing manganese octahedra arranged in sheets (**Table 7.7**).

Soils contain on average 0.1% manganese oxides by weight; minimal compared to the amounts of aluminum and iron oxides. Although they are less prominent than iron and aluminum oxides, manganese oxides can greatly influence the chemical characteristics of soils, especially redox reactions and sorption of cations.

Many transition metal ions have the same size as the manganese ions, so isomorphic substitution is common. Manganese oxides have small particle sizes and large specific surface areas (100 to 200 $m^2 g^{-1}$) that can adsorb cations and anions. **Figure 7.24** shows the distribution of Pb, Fe, Mn, Si, and P in a mine-waste impacted soil. Lead is predominantly associated with Fe and Mn, showing the importance of these oxides in contaminant speciation in soils.

7.8 Summary of soil clays

This chapter described characteristics of ideal soil minerals. The most common secondary soil minerals are phyllosilicates and iron and aluminum oxides. Allophane and manganese oxides are also important soil minerals; the degree to which depends on the soil's parent materials and amount of weathering. While it is instructive to study individual soil mineral properties, it must be emphasized that soil minerals are not typically ideal, but instead have a range of compositions and mineralogical characteristics that span from highly crystalline to amorphous and have numerous isomorphic substitutions.

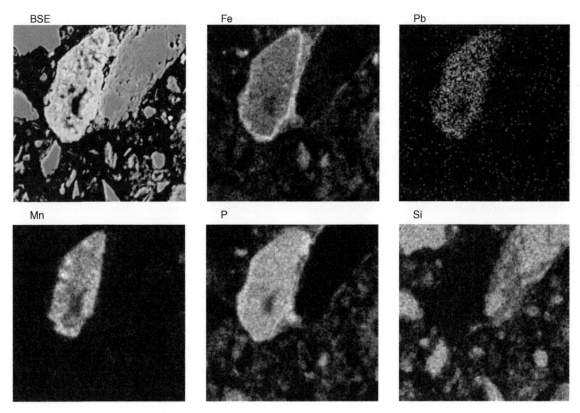

Figure 7.24 Distribution of elements in a 256 μm × 256 μm electron microprobe image from a thin section of a mine-waste contaminated wetland soil. The backscatter electron image (BSE) shows the distribution of soil grains in the sample. Intensity of white indicates the grains with the highest concentration of elements. Pb, P, Fe, and Mn are closely associated suggesting that the Pb and PO_4 ions are either precipitated with iron and manganese oxides, or adsorbed on the mineral surfaces. In contrast, the Si-rich mineral grains are depleted in Pb and P. Source: Strawn et al. (2007) reproduced with permission of Springer.

Soil minerals interact with each other and soil organic matter to create a solid mixture with properties different from their separate parts (e.g., **Figure 7.1**). Despite this complexity, understanding basic mineral structures provides a basis to better understand the properties of the soil minerals, allows detailed knowledge of soil characteristics, and allows for chemical processes to be predicted.

Questions

1　Why is the term *soil clay* ambiguous?
2　Distinguish between 1:1, 2:1, and 2:1:1 layer silicates by drawing a ball-and-stick diagram of their structures. Explain how layer charge influences the following layer silicate properties:
　a　Interlayer bonds
　b　Crystal size
　c　Swelling
　d　Surface area
　e　c-spacing
　f　Exchangeability of adsorbed cations
3　How do soil minerals differ from pure minerals? How are soil properties, such as cation exchange capacity, surface area, and swelling, expected to differ from the properties of pure minerals?
4　Write the half-unit cell formula for a montmorillonite mineral with a layer charge of 0.3, with 90% of the substitution in the octahedral layer (Mg^{2+} for Al^{3+}) and 10% of the substitution in the tetrahedral layer (Al^{3+} for Si^{4+}). The saturating cation is Na^+.
5　Calculate the cation exchange capacity of the mineral in Problem 4 and express the result in millimoles(+) per kilogram. Compare this value to the value normally given for montmorillonite.
6　What is meant by the point of zero charge?

7 Explain why oxide minerals are variable charge.

8 Calculate the proportions of mass and volume occupied by oxygen in kaolinite and goethite.

9 A given mass of clay mineral is digested in acid and analyzed for the elemental composition. The number of atoms per unit cell is determined based on the total number of atoms and charges in an ideal clay mineral formula, that is, $Si_4Al_2O_{10}(OH)_2$, or $Si_4Mg_3O_{10}(OH)_2$. Below are the results. Identify the clay mineral and calculate the molecular weight and CEC. Write the half-cell formula. Hint: Fill the tetrahedral and octahedral sheets; cations left make up permanent layer charge; in the octahedral sheet put Al, Fe, and Mg, in that order.

Atom	Atoms per half cell
Ca	0.125
Mg	0.435
Si	3.79
Al	1.86
Na	0.09
Fe	0.025
O	10
OH	2

10 What are the likely dominant sources of pH-dependent charge in the following?
   a A highly weathered mineral soil low in organic matter
   b A slightly weathered, montmorillonitic soil low in organic matter
   c A slightly acid forest soil
   d A volcanically derived soil low in organic matter

11 Draw a ball-and-stick two-dimensional representation of a dioctahedral 2:1 layer silicate. Label basal plane, octahedral, and tetrahedral sheets, interlayer, apical oxygens, and ditrigonal cavity. Calculate the charge distribution on an apical oxygen atom if a magnesium atom isomorphically substitutes for a aluminum atom in the octahedral sheet.

12 Why is potassium more easily weathered from biotite than muscovite?

13 Describe some important characteristics of poorly crystalline aluminosilicates.

14 Compare and contrast ferrihydrite and goethite (structure, formation, and environments found in).

15 Compare the mineral surface charge of a Vertisol to an Oxisol.

16 Compare factors that create surface charge on the edges of kaolinite and goethite.

17 Why does ferrihydrite typically have a much greater surface area than goethite?

18 What role does SOM have in soil mineral formation?

19 Why does rapid oxidation and high amounts of SOM cause ferrihydrite instead of more crystalline iron oxides?

20 What would the c-spacing of sodium-saturated montmorillonite be if there were three layers of interlayer water?

21 Based on the arrangement of atoms in crystals, explain why hematite has a greater density (5.26 g m^{-3}) than goethite (4.26 g m^{-3}). Which would have more adsorption capacity? Why? Hint: review the structural coordination figures in the chapter.

22 What is a major difference between montmorillonite and vermiculite with respect to shrink and swell behavior?

23 What are some important considerations in allophane mineral formation?

24 What are the major differences between beidellite and vermiculite?

25 Which clay minerals in soils would limit ammonium availability to plants? Why?

26 What are some important considerations for comparing soil minerals to ideal minerals?

Bibliography

Anschutz, A.J., and R.L. Penn. 2005. Reduction of crystalline iron(III) oxyhydroxides using hydroquinone: Influence of phase and particle size. Geochemical Transactions 6:60–66.

Cornell, R.M., and U. Schwertmann. 1996. The Iron Oxides: Structure, Properties, Reactions, Occurrence, and Uses. VCH, Weinheim; New York.

Drits, V.A., B.A. Sakharov, A.L. Salyn, and A. Manceau. 1993. Structural model for ferrihydrite. Clay Minerals 28:185–207.

Fuller, A.J., S. Shaw, M.B. Ward, S.J. Haigh, J.F.W. Mosselmans, C.L. Peacock, S. Stackhouse, A.J. Dent, D. Trivedi, and I.T. Burke. 2015. Caesium incorporation and retention in illite interlayers. Applied Clay Science 108:128–134.

Iyoda, F., S. Hayashi, S. Arakawa, B. John, M. Okamoto, H. Hayashi, and G.D. Yuan. 2012. Synthesis and adsorption characteristics of hollow spherical allophane nano-particles. Applied Clay Science 56:77–83.

Majzlan, J., A. Navrotsky, and U. Schwertmann. 2004. Thermodynamics of iron oxides: Part III. Enthalpies of formation and stability of ferrihydrite (similar to Fe(OH)(3)), schwertmannite (similar to FeO(OH)(3/4)(SO4)(1/8)), and epsilon-Fe2O3. Geochimica Et Cosmochimica Acta 68:1049–1059.

McBride, M.B. 1994. Environmental Chemistry of Soils. Oxford University Press, New York.

Meleshyn, A., and C. Bunnenberg. 2005. The gap between crystalline and osmotic swelling of Na-montmorillonite: A Monte Carlo study. Journal of Chemical Physics 122.

Michel, F.M., L. Ehm, S.M. Antao, P.L. Lee, P.J. Chupas, G. Liu, D.R. Strongin, M.A.A. Schoonen, B.L. Phillips, and J.B. Parise. 2007. The structure of ferrihydrite, a nanocrystalline material. Science 316:1726–1729.

Post, J.L., B.L. Cupp, and F.T. Madsen. 1997. Beidellite and associated clays from the DeLamar mine and Florida Mountain area, Idaho. Clays and Clay Minerals 45:240–250.

Strawn, D.G., P. Hickey, A. Knudsen, and L. Baker. 2007. Geochemistry of lead contaminated wetland soils amended with phosphorus. Environmental Geology 52:109–122.

Wilson, M.J., L. Wilson, and I. Patey. 2014. The influence of individual clay minerals on formation damage of reservoir sandstones: a critical review with some new insights. Clay Minerals 49:147–164.

8 PRODUCTION AND CHEMISTRY OF SOIL ORGANIC MATTER

8.1 Introduction

Soil organic matter (**SOM**) has long been recognized as one of the most important components of soils because it increases cation exchange capacity, soil aggregation, nutrient availability, pesticide retention, and overall soil health. Although SOM has been a topic of scientific research for more than a century, its molecular properties are not consistently predictable because it is a heterogenous material, and there are as many different types of SOM as there are environments.

Soil organic matter is an intermediate phase of carbon between photosynthetically fixed carbon in plants, animals, and microbes, and completely oxidized carbon (CO_2). **Table 8.1** shows the different components comprising SOM, and their categorization into labile, stable, and recalcitrant pools. Much of the interest in SOM is in the colloidal-sized, brown-to-black organic residue that is associated with soil minerals called **humus**. The carbon compounds in SOM are partially degraded plant, animal, and microbial molecules, and burned carbon that are integrated throughout the soil fabric. Isolating soil carbon as a type of compound outside of soil is not possible because the soil organic carbon is a milieu of carbon at various stages of degradation in association with soil minerals and microbes, as illustrated in Figure 7.1. The focus of this chapter is on the general properties of SOM and how soil properties affect SOM formation.

Soil organic matter is approximately 50% by mass carbon. The remaining is mostly oxygen, with lesser amounts of nitrogen, sulfur, phosphorus, and other atoms. As a carbon reservoir, SOM is the largest terrestrial pool of carbon on Earth – four times larger than the terrestrial biosphere (plants, microbes, etc.), and twice as large as the atmosphere and biosphere combined (Tables 1.2 and 3.8). Thus, understanding SOM characteristics and how environmental factors affect carbon fluxes from soils is important for understanding atmospheric carbon.

Soil Chemistry, Fifth Edition. Daniel G. Strawn, Hinrich L. Bohn, and George A. O'Connor.
© 2020 John Wiley & Sons Ltd. Published 2020 by John Wiley & Sons Ltd.

Table 8.1 Forms of soil organic matter with descriptions of the composition and relative degradability of the fractions. Source: Stockmann et al. (2010). Reproduced with permission of Elsevier.

Form	Composition	Relative turnover time of SOM
Surface plant residue	Plant material residing on the surface of the soil, including leaf litter and crop/pasture material	*Fast (or labile) pool* Decomposition occurs at a timescale of days to years.
Buried plant residue	Plant material greater than 2 mm in size residing within the soil	*Fast (or labile) pool* Decomposition occurs at a timescale of days to years.
Particulate organic matter (POM)[1]	Semi-decomposed plant residue smaller than 2 mm and greater than 50 μm in size, and not complexed to soil minerals	*Fast (or labile) pool* Decomposition occurs at a timescale of days to years.
Humus	Well-decomposed plant and microbial material smaller than 50 μm in size that is associated with soil mineral particles	*Slow (or stable) pool* Decomposition occurs at a timescale of years up to a century or more.
Resistant organic carbon (ROC)	Charcoal or charred materials that results from the burning of organic matter	*Passive (or recalcitrant) pool* Decomposition occurs at a timescale of decades to thousands of years.

[1] See Gregorich et al. (2006).

Soils vary greatly in organic matter content. Surface horizons of prairie grassland soils (e.g., Mollisols) may contain 1–6% organic carbon by mass, whereas a sandy desert soil (Aridisols) may contain as little as 0.1% organic carbon. Poorly drained soils (e.g., Aquepts) often have organic carbon contents greater than 5%, and peat soils (Histosols) may have more than 50% organic carbon. The greatest global mass of soil carbon is stored in Histosols, followed by Inceptisols; the latter are on average only 1.9% (1.1% median) organic carbon, but are broadly distributed.

The typical organic carbon content of A horizons of most mineral soils ranges from 0.5% (Alfisols) to 6% (Spodosols and Andisols). By mass, SOM is typically a minor fraction of soil solids; however, even in small amounts it strongly affects soil physical and chemical properties, including nutrient availability, cation exchange capacity (CEC), soil pH, and structure and aggregation that facilitates air–water exchanges in soils.

8.1.1 Components in SOM

SOM is composed of plant residues, fauna, bacteria, fungi, actinomycetes, **charcoal**, and degraded **biomolecules**. A biomolecule is a chemical compound (e.g., protein) synthesized by an organism. Biomolecules in soils exist within cells, and extracellularly when cells are broken apart or release exudates. Less than 4% total

soil carbon and 7% of the mass of SOM occurs in living organisms (**Figure 8.2** and **Table 8.2**). With the exception of **O horizons** and Histosols, less than 20% of SOM is detritus or **particulate organic matter** (POM). Particulate organic matter primarily consists of fragments of plants that still have some intact structure at

Table 8.2 Global estimates of terrestrial carbon, including biotic, soil organic matter, and soil inorganic carbon.

Terrestrial C Pool	Carbon (Pg)	Includes
SOC[1] 0–0.3 m	738	SOM, soil biota, permafrost, and peatland
SOC[1] 0–1 m	1486	
SOC[1] 0–2 m	2273	
SIC[2] 0–1 m	721	Ca and Mg carbonate C
Above-ground plant biomass[3]	320	Leaves, wood, flowers, fruit
Below-ground plant biomass[3]	130	Roots
Soil microbial biomass[3]	20	Fungi, bacteria, and archaea
Earthworms[3]	0.2	
Soil arthropods[3]	0.2	
Soil protist[3]	1.5	

[1] Jackson et al. (2017). Estimates from the literature in the past three decades vary from 991 Pg C in SOM to 2200 Pg C; the variation is due to spatial extrapolation, depth of measurement, bulk density assumptions, and inclusion of bogs, tundra, and permafrost. See Scharlemann et al. (2014).
[2] Batjes (1996).
[3] Bar-On et al. (2018).

the cellular level of the original plant leaf, root, or shoot. The smallest fraction of POM is the light fraction; further breakdown of light fraction components of POM creates biomolecules that are incorporated into humus. POM is an important component of available carbon, is enriched in microbes, and is degraded much quicker than other SOM components.

The bulk of SOM is the *residue* left after microorganisms consume *available* carbon and nutrients. Invertebrates (e.g., worms) also play an important role in SOM processes. The SOM residue consists of colloidal carbon molecules mixed with mineral colloids. There are three noteworthy aspects of SOM:

1 Residual SOM compounds persist in soils until environmental conditions are favorable for an organism to metabolize the carbon and nitrogen compounds.
2 In most cases, the organic molecules in SOM are not newly synthesized compounds, but are biomolecules occurring in various stages of degradation and oxidation.
3 SOM is not solely organic carbon compounds but consists of organic molecules *associated* with mineral colloids through bonding of SOM-functional groups to mineral surfaces, hydrophobic interactions, and

ion-facilitated bridging bonds between molecular carbon functional groups and mineral surfaces.
4 The terms *SOM* and *humus* are often used synonymously.

8.1.2 Studying SOM

Knowledge of SOM properties and behavior are critical for understanding global carbon cycling, as well as optimizing crop growth and making ecosystems sustainable. Because of its importance, scientists have long studied the formation and characteristics of SOM. Studies traditionally separated SOM from the soil using alkali extractions to isolate **humic substances** (**Figure 8.1**). However, current thinking is that the process of separating organic carbon compounds from soils alters its molecular structure, giving the humic substances distinct properties compared to SOM present *in* the soil. Thus, knowledge about molecular structure of SOM gleaned from extracted humic substances should be interpreted with caution.

Despite the complexity of identifying the molecular structure of SOM, there is some consensus that it can be categorized based on size of the organic particles and the degree of degradation (**Figure 8.1**).

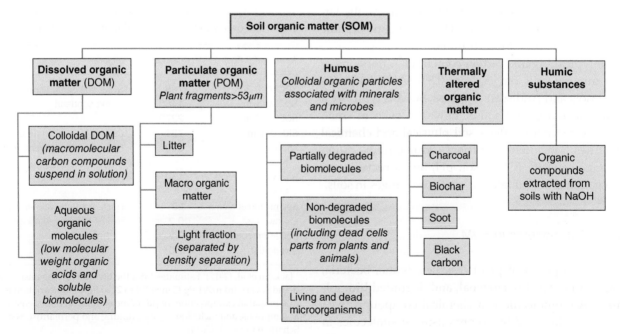

Figure 8.1 Overview of SOM fractions based on size or degree of degradation. Humic substances are separated from soils with strong bases and acids that cause changes to the molecular structure of the compounds.

In recent decades, advances in technology have allowed in-situ measurement of SOM properties, and results are providing new insights into structure-function relationships of soil organic compounds and their molecular interactions with each other and soil mineral particles.

8.2 Ecosystem carbon storage and fluxes

Figure 8.2 shows estimates of the distribution of global soil carbon between the different reservoirs. Living organisms in the soil represent only 4% of the total soil organic carbon (**Table 8.2**). Organic carbon in SOM is more than four times greater than the above-ground global terrestrial carbon pool. Within the soil, ~1/3 of the SOM carbon occurs in the surface horizons (0–30 cm), and rest occurs in the subsoil horizons (30–200 cm). Although A horizon carbon concentrations are typically much greater than subsurface horizons, the greater depth of the subsurface horizons creates an overall greater total mass of soil organic carbon. Carbonate minerals comprise about 37% of the global soil carbon.

At the plot and catchment scales, sizes of carbon reservoirs can be accurately estimated. However, as the scale increases to regional, biome, or global, estimation of carbon reservoir sizes becomes less accurate due to incorrect model assumptions and lack of detailed information on the soil and ecosystem processes. A survey of the literature reveals that estimates of the total global soil organic carbon reservoir ranges from 1500 to 3500 Pg C. Differences in global SOM mass are due to depth of the soil used in calculation, errors from extrapolation of soil carbon data across a given landscape, paucity of measurements for a region, and whether permafrost is included in the model. As ecosystem carbon storage models and data improve, carbon reservoir size and flux predictions will become more accurate.

Fluxes of organic carbon from soils are important sources of atmospheric CO_2. SOM is a sink of carbon to the terrestrial carbon reservoir. Scientists propose that enhanced sequestration of soil carbon may be a potential long-term sink for offsetting human inputs to the atmosphere. How to manage agricultural and forest systems to realize a positive carbon offset is a question of current debate and research. To predict

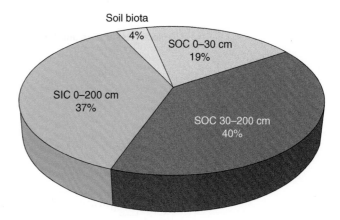

Figure 8.2 Global distribution of soil carbon: soil organic carbon (SOC), soil inorganic carbon (SIC) (carbonates), and carbon in living organisms (flora and fauna). (See **Table 8.2** for data sources.)

fluxes of carbon in and out of soil ecosystems, complex biogeochemical carbon models are used. **Figure 8.3** shows the fundamental approach to ecosystem carbon modeling. Modeling carbon fluxes involves predicting the size of vegetative, atmospheric, and soil carbon reservoirs, and the rate at which carbon moves between them. Models must include ecosystem processes that affect carbon fixation and loss of carbon to the atmosphere.

The movement of soil carbon can be very fast in geologic time. In the formerly glaciated areas of Canada, Europe, and Siberia, peat has been accumulating in lowlands since the glaciers retreated, 10 000 to 15 000 years ago (see points A and B in **Figure 8.4**). As a result, the amount of carbon in the northern peatlands and frozen soils and sediments represent nearly one third of the total soil organic carbon on Earth.

Increased atmospheric CO_2 may cause increased primary production, which fixes carbon in plants and increases litter for carbon storage in soils. In addition, increasing atmospheric CO_2 and other greenhouse gasses causes climate change, which will affect plant productivity and SOM degradation rates. Changes in plant productivity and climate will also affect soil nutrient availability and soil microbial processes. The feedback loop in carbon biogeochemical cycling suggests that increasing atmospheric CO_2 will drive SOM to a new **steady-state** (i.e., increase or decrease in total soil carbon). A goal of current research is to understand where that new steady-state will be.

Model Reservoirs and Fluxes

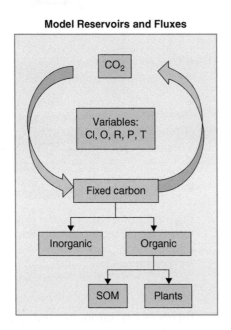

Ecosystem
properties input:

Vegetation type, SOM,
Soil clay, Temperature,
Nutrients, Soil texture,
Landscape, Microbial activity

Ecosystem
processes input:

• Degradation rates
• Nutrient cycles
• Gas exchange
• Vegetation processes
• SOM stabilization
• Temperature effects
• Microbe transformations
• Moisture effects
• Invertebrate activity
• Animal influences
• Fire
• Human influences
• Erosion and deposition
• Inorganic reactions
• Leaching

Figure 8.3 Modeling scheme for terrestrial carbon reservoirs and fluxes (Cl = climate, O = organisms, R = relief, P = parent materials, T = time).

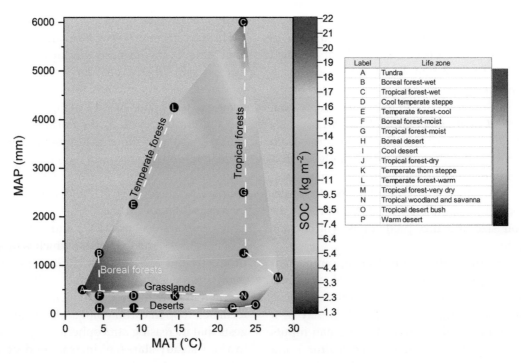

Figure 8.4 Global distribution of soil organic carbon (SOC) by biome (Life zone) as a function of mean annual precipitation (MAP) and mean annual temperature (MAT). Lines represent major climate gradients within ecosystem types. Data are from Post et al. (1982) who summarized soil data from 2700 soil profiles. See **Figure 8.4** in color plate section for color version of this graph.

Special Topic Box 8.1   Measurement of SOC and SOM content in soils

Total SOM and soil organic carbon (SOC) are basic soil characterization parameters used to assess suitability of soils for plant growth, overall soil quality, and to calculate soil carbon stocks. SOM can be measured directly, or indirectly by measurement of total SOC and conversion to total SOM. There are several artifacts that can make measuring or calculating SOM values inaccurate. Measuring total SOC is more accurate, although it too can be prone to inaccuracies if care is not taken.

A factor in measuring total SOC or SOM is that sample preparation can introduce some bias because of the size separation of the soil. Typically, soils are sieved to less than 2-mm particle size prior to SOM or SOC measurement. A drawback of sieving is that it removes coarser plant debris, such as dead root fragments and leaf litter. In the A horizon, the plant debris can be a significant amount of the SOM (greater than 20% in some soils). Thus, eliminating this from total SOM underestimates total soil carbon in surface horizons, and can lead to errors in estimations of SOC stocks.

The most common total SOC measurement method oxidizes SOM to carbon dioxide using either wet or dry methods; i.e., chemical or heat oxidation. Mass loss on ignition (LOI) can be a good estimate of SOM, but requires careful weighing, and may suffer from loss of some mineral structural water causing the LOI masses to be too high. Automated carbon analyzer instruments can measure carbon concentration evolved when a sample is heated and provide total SOC. However, soils with pH > 7 require pretreatment to eliminate carbonates that would otherwise release carbon at the high temperatures of the analysis

and would bias the total SOC value. Leaching a soil with a dilute acid is sufficient to remove the carbonate minerals from soils prior to SOC analysis.

SOC concentration can be converted to SOM concentration using an empirical relationship between the two measures; a common conversion is SOM = a × SOC, where a is an empirically fit parameter that typically ranges from 1.7 to 2.2 (Pribyl, 2010). For accurate measurement, the value of the conversion parameter a must be determined individually for each type of soil, after which it can be applied to other samples that are of the same soil type.

Chemical treatment of soils to measure total SOM relies on oxidation. A common chemical oxidizing method utilizes dichromate and sulfuric acid to oxidize the SOM, and the amount of carbon released is calculated by measuring the amount of dichromate left after the carbon oxidation. This method is called Walkley Black after the scientists that first reported the method in literature in 1946. Although it is labor intensive and produces toxic chromium solution that must be disposed of as hazardous waste, it is accurate when a correction factor for SOM that does not get completely oxidized is applied.

Accurately measuring total SOC and SOM for monitoring soil carbon processes requires understanding operational factors that affect the measurements. Given the access and ease of automated carbon analyzers, total SOC is becoming a more commonly reported measure than is SOM. The recent research and public policy focus on carbon fluxes from soils to and from the atmosphere and how management can affect them has cast an increased emphasis on direct SOC measures as opposed to SOM.

8.3   Soil organic matter formation factors

The same state-factors that affect soil development (climate, organisms, relief, parent material, time) also affect SOM formation. Understanding SOM formation processes and characteristics requires knowledge of the effects of environmental factors. For example, soil carbon turnover is as much a function of environmental factors as it is the type of organic molecules in the SOM.

8.3.1   Residence time of SOM

SOM **residence time** is the average time a carbon molecule spends in the soil. **Carbon turnover** time is the *rate* at which SOM molecules move through the soil; from entry to complete decay and loss as CO_2. The residence times for SOM molecules are highly variable (**Figure 8.5**). In temperate soils, the average residence time is 50 years, but in tropical and boreal forests, average SOM residence times are greater than 1000 years. Over the

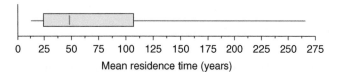

0   25   50   75   100   125   150   175   200   225   250   275

Mean residence time (years)

Figure 8.5 Mean residence times of SOM in temperate soils from 20 long-term research sites (the ends of the horizontal lines are 10th and 90th percentiles, the ends of the rectangle are 25th and 75th percentiles, and vertical bar is median). The mean residence time is the average time a SOM carbon atom will reside in the soil. Adapted from Amelung et al. (2008). Reproduced with permission of Elsevier.

past century, dominant thinking on the reason that organic matter persists in soils for long times has been that semi-stable organic compounds are created in the soil. In nature, however, organic compound persistence varies with water content, temperature, and mineral composition of the soil. Even complex biomolecules traditionally thought to slowly degrade, such as cellulose or lignin, readily degrade in some soils. Contrastingly, simple organic compounds that should be quickly degraded may persist for decades in some soils because of climate and soil properties. Thus, persistence of soil organic carbon is an ecosystem property, as well as a function of the type of organic carbon compounds in the soil. In well-drained soils, climate and vegetation are two of the most important factors that influence the amount of SOM. Topography, mineralogy, and age of the landscape are also important factors. Details of the effects of soil-forming factors on SOM are discussed below.

8.3.2   Climate effects on SOM

Climate affects the array of plant species, the quantity of plant material produced, and the intensity of soil microbial activity that degrades SOM. Vegetation and topographic effects are difficult to separate from climatic effects because the factors are related. Forest and grassland soils generally have more SOM than other well-aerated soils, whereas desert and semi-desert soils have very little SOM. North-facing slopes (in the Northern Hemisphere) are cooler and moister, and organic matter content is greater than in soils on south-facing slopes.

Low temperatures enhance SOM accumulation by slowing microbial activity. For example, major areas of peat are found in central Canada, which has cool and

moist climate. Smaller but substantial areas of Histosols occur on every continent where water saturation prevents oxidation of plant debris. Increased temperature causes increases in organic carbon additions as well as degradation rates.

The relation between SOC, mean annual temperature (MAT) and mean annual precipitation (MAP) for several biomes is shown in **Figure 8.4**. In desert environments, SOC decreases as MAT increases because the warmer temperatures promote more microbial activity that degrades the small amount of carbon input (desert biomass is relatively small). In tropical forests, TOC is directly correlated to MAP because high moisture increases plant growth, providing greater rates of carbon input into soil. To allow SOM levels to build up, degradation rates of SOM must be slower than input rates. In tropical soils, interaction of organic matter with iron and aluminum oxides and poorly crystalline minerals stabilizes the SOM against microbial decay, and thus decreases degradation rates. Tundra has much greater TOC than warmer grasslands because the low temperatures inhibit microbial degradation of SOM. The overall trend in **Figure 8.4** is that warm and dry environments create the least amount of SOM, while the most SOM is in either warm wet climates, or cool climates with moderate to high precipitation, such as boreal forests.

8.3.3   SOM in wetlands

In poorly drained lowland soils, such as wetlands and peat bogs, water inundation of plant debris causes low concentrations of oxygen, which prevents organisms from carrying out aerobic decomposition. In the absence of oxygen, fermentation can transform SOM, but the decay of cellulosic plant matter by fermentation is slow and insignificant with respect to the total amount of carbon present in most oxygen-limited environments (e.g., wetlands). Methane production from soil fermentation, however, is a concern because fluxes to the atmosphere exacerbate greenhouse gas effects.

Peat accumulation depends on the rate of organic addition versus the rate of oxidation. In the Sacramento–San Joaquin Delta of California, peat deposits were up to 30 m thick before those areas were drained and the oxygenated peat began degrading. The oxidation rate is a function of the rate of oxygen supply and temperature. The oxygen

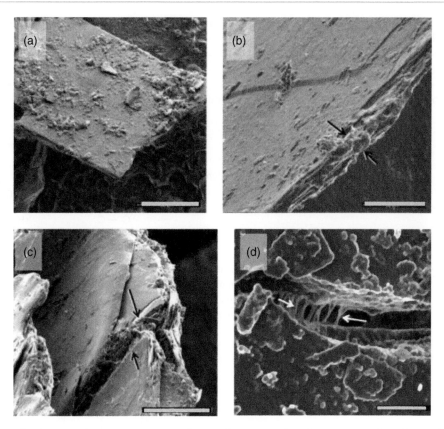

Figure 8.6 Ion microprobe images of organic matter residues on biotite incubated in a pine sapling rhizosphere. (a) Biotite with microbial organic material on its surface; (b) microbial exudate (extracellular polymeric substances, EPS) covering a biotite edge (indicated with arrows); (c,d) intercalation of microbial organic material in mineral pores and crevices. Arrows in (d) point to EPS strands. Scale bars: (a) 50 μm, (b) 20 μm, (c) 10 μm, (d) 1 μm. Images are from Dohnalkova et al. (2017), with permission.

supply, in turn depends on soil water content. When drained for cultivation, peat oxidizes as it contacts oxygen and shrinks as it dries. For example, shrinkage and oxidation has lowered the level of cultivated peatland in California and Florida, by as much as 25 to 50 mm y^{-1}. The rate has slowed in recent years because peatlands are better managed and because of the stabilization of the remaining organic matter by association with minerals. Wind erosion of dry surfaces and accidental burning of peat also contribute to the loss of these extremely productive soils.

8.3.4 Soil mineral effects on SOM

Soil minerals influence SOM by affecting soil texture and through surface reactions. In an area of similar climate and topography, SOM content tends to increase with soil clay content. The minerals physically isolate the SOM from microbes, and inhibit degradation. Aggregation and occlusion of organic molecules between colloids is an important mechanism of SOM protection (**Figure 8.6**), as are small (mostly water-filled) pores that limit oxygen diffusion needed for degradation.

SOM degradation is also limited by metals such as aluminum that complex to the SOM functional groups. 2:1 clays, like montmorillonite, have high adsorptive capacities for organic molecules, and may protect nitrogenous constituents of SOM from attack by microorganisms, particularly when the nitrogen exists as cationic amino groups that are attracted to negative layer charge of the clay minerals.

8.4 Organic chemistry of SOM

Understanding SOM molecular properties requires knowledge of the types of organic chemicals comprising the SOM. **Table 8.3** summarizes some of the basic organic chemicals that occur in biomolecules.

Table 8.3 Common organic functional groups and compounds in plant litter and SOM. **R** is the main carbon structure (backbone) of the molecule.

Molecule	Structure	Example compounds	Basic structure
Alkane/aliphatic	Carbon chains	Parts of fats, waxes, cutin on plant leaves	
Cycloalkane	3–6 carbon atom rings often branched, substituted, and polymerized	Basic structural unit of pheromones, steroids, and hormones	
Carboxyl	Carbon with double-bonded O and single-bonded OH	Acidic functional group, common on many biomolecules and their degradation products	
Arene (aromatic)	Benzene ring-six C atoms with hybridized bonds	Lignin, tannin	
Alcohol	OH group attached directly to a saturated carbon	Cellulose, hemicellulose, tannins, lignin, lipids	
Phenol	Aromatic ring with a hydroxyl	Lignin, tannin	
Monosaccharide (carbohydrate)	Chains or single rings of C atoms with O and/or H bonding to each C atom	sugars-glucose, fructose, galactose	
Amine	Nitrogen with lone pair of electrons bonded to 1–3 carbons and 0–2 H	proteins	

Table 8.3 (Continued)

Molecule	Structure	Example compounds	Basic structure
Polysaccharide	>10 rings of monosaccharides bonded together	cellulose, chitin, starch	
Lipid/fatty acids	Polar group attached to a nonpolar chain or ring structure	Common in plant leaves and needles	
Amino acid	Carbon structures with amine and carboxyl functional groups	proteins	
Thiol	Reduced sulfur (HS) bonded to carbon	proteins and enzymes	R-S-H
Phytic acid (inositol phosphate)	Six ortho-phosphate groups attached to a carbon ring	P storage and structural component	

Organic molecules in plants and soils are composed of conglomerates of organic functional groups or moieties linked to each other and attached to larger carbon molecules. **Lignin**, for example, is composed of three types of aromatic (benzene ring) alcohols bonded to each other through various linkages of propanol groups (3-carbon alkyl-OH groups) to create a larger macromolecule or polymer (**Figure 8.7**). The monomers of lignin are phenylpropanoids that are released in various forms upon degradation.

8.5 Plant and microbial compounds input into soil

Plant litter and roots are major sources of organic carbon input into soils. Root debris and root exudates are an important source of soil carbon in the lower profile. In forest soils, plant roots are responsible for up to about half of the total plant-derived carbon input to soils, and about one-third in grasslands. The annual input rate of plant litter varies widely between ecosystem types. In humid tropical forests, litter input rates are large, and in deserts litter input rates are minimal. Types of organic compounds input to soil are also diverse, even for a single ecosystem or organism. Despite the variability in rate and type of litter input, the bulk of SOM is mostly derived from a few types of natural organic chemicals input into soils in large amounts, such as **cellulose** (**Table 8.4**).

Degradation of plant matter releases carbohydrates, proteins, fats, lignins, tannins, and other biomolecules. **Table 8.5** shows plant biomolecules that eventually comprise litter and SOM. Most of the plant cell contents, including the cytoplasm, nucleus, and organelles, are quickly degraded in soils by microbes. Cell wall components and woody tissue are composed mostly of cellulose, hemicellulose, and lignin. These

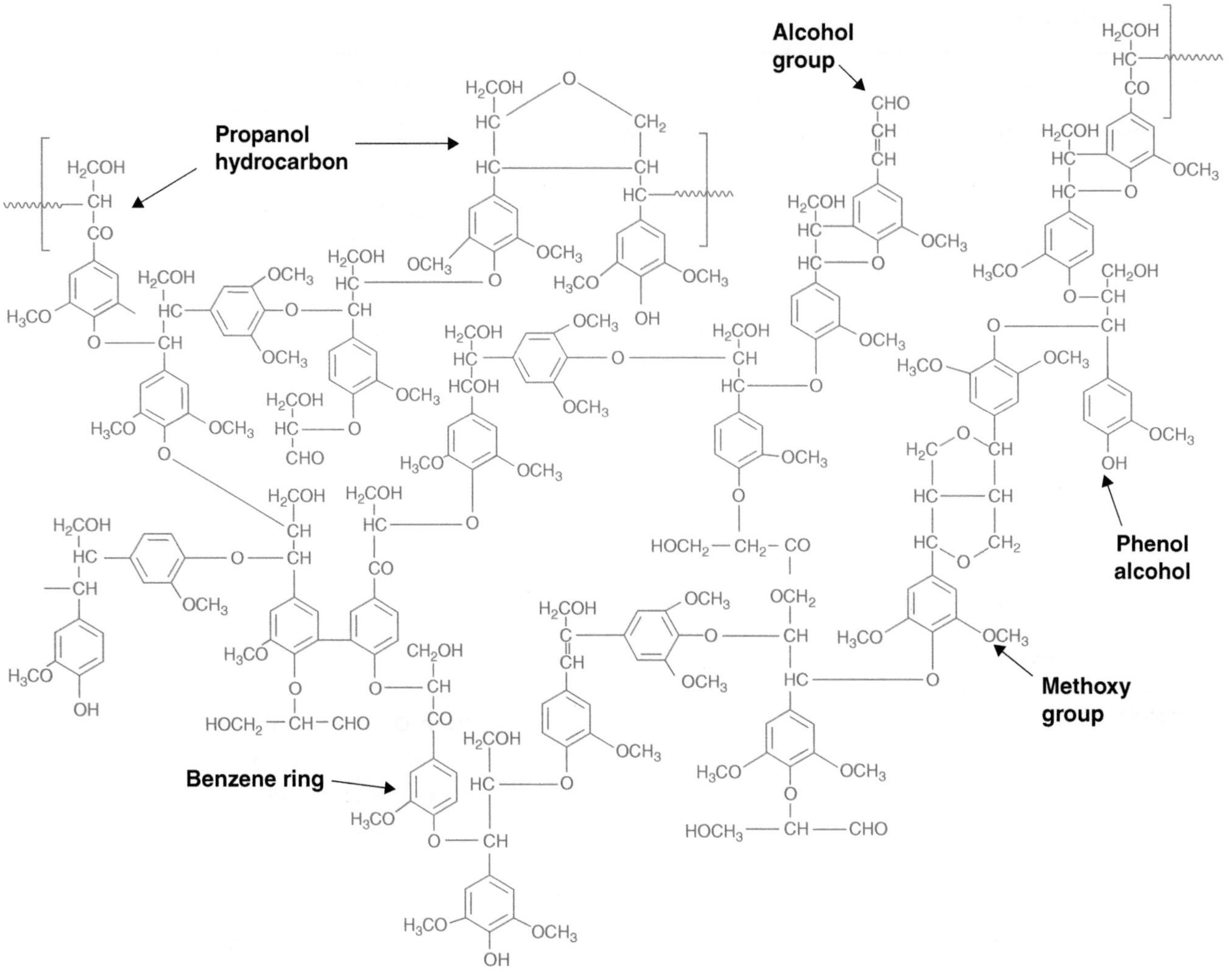

Figure 8.7 Proposed molecular structure of softwood lignin. Lignin is a macromolecule composed of three types of phenol-propanoids linked together. Phenolpropanoids are aromatic rings with propanol and various methoxy groups attached. The lack of acidic or basic functional groups renders lignin hydrophobic. Lignin and its partially degraded compounds are important components in formation of SOM. Adapted from Mohan et al. (2006). Reproduced with permission of Elsevier.

molecules, and their derivatives, are slower to degrade and dominate the plant material that becomes SOM. Amounts of plant-derived organic compounds input into soil depend on the biota, landscape, and ecosystem. For example, hardwood forest litter contains as much as 30% (by mass) lignin, while grassland litters typically have less than 10% lignin (see plant content ranges in **Table 8.4**).

Root exudates, such as chelates, are another important class of organic compounds input to soils. Chelates affect the soil chemistry in the rhizosphere, and dramatically alter nutrient availability. Plants actively input root exudates into soils to facilitate their biochemical and physiological processes. A **siderophore** is a commonly produced chelate that enhances the mobility of iron in soils. One type of siderophore is deoxymugineic acid ($C_{12}H_{20}N_2O_7$).

The carboxyl and amine functional groups of siderophores form multidentate bonds that are highly selective for Fe^{3+} (e.g., see Figure 4.17). The resulting Fe(III) complexes make ferric iron more soluble and available for plant uptake. Microbes also release siderophores, as well as several other extracellular enzymes that are important components of SOM and control nutrient availability in the soil.

Microbes are important contributors to SOM: One gram of soil can contain 10^6–10^8 bacteria cells and several km of fungal hyphae. Microbes mineralize 80–95%

of plant matter, and as much as 50% of SOM compounds are derived from microbial biomass. Many of the microbial inputs into the SOM pool are similar to plants, but some are unique. Murein, for example, is a protein and carbohydrate substance making up the cell walls of some bacteria and is more quickly degraded than cellulose and lignin molecules. Condensed lipids produced by bacteria have moderate degradation resistance. Chitin is a structural component of fungi (and insects), and has some resistance to degradation in soils. Many microbes exude a polymeric slime (e.g., exopolymeric saccharide (EPS)) that consists of cellulosic molecules that are only slowly to moderately degradable. Because of the diversity of soil microbes,

the types of compounds released by microbes to soil are highly varied; however, a few are the greatest contributors to what becomes soil organic matter. These include polysaccharides, proteins, and fats (**Table 8.4**).

8.6   SOM decay processes

Organic materials input into soil undergo microbial, enzymatic, and chemical reactions and form SOM. SOM, or humus, was historically classified as a *recalcitrant* (decay-resistant) organic material in soils. All soil organic compounds are, however, energetically unstable, and will degrade. **Table 8.1** proposes a relative order of SOM fractions based on degradation or recalcitrance. In this table, degradability does not describe the characteristics of molecular compounds in SOM, but instead describes the general time-dependent transformation of organic molecules to CO_2. The rate of degradation is a function of soil, climate, and organic matter properties. **Table 8.5** lists the relative degradability of biomolecules. Soil processes that influence susceptibility of SOM to degradation are:

1 Protection by adsorption on minerals.
2 Occlusion in micropores.

Table 8.4 Ranges of percent mass of cellulose, hemicellulose, and lignin in plant tissues.[1]

Plant material	Cellulose	Hemicellulose	Lignin
Wood	40–50	24–40	18–35
Grasses	25–40	35–50	10–30
Leaves/needles/litter	15–30	20–80	7–32

[1] Data summarized from:
Johansson (1995).
Rahman et al. (2013).
Prasad et al. (2007).

Table 8.5 Composition and relative degradability of compounds in typical plant litter. Degradability of organic compounds in soils is highly variable, depending on plant species and environment. Degradability of compounds are indicated using a relative plus-symbol scale – one plus indicates the most resistant compound.

Plant molecule type	Compounds	Plant use	Types of molecules	Degradability in soil
Water-soluble compounds	Simple sugars	Storage	Monosaccharides	++++
	Proteins	Enzymes, storage	Amino acids	
	Organic acids	Metabolism, defense	Low molecular weight organic acids (e.g., citric acid, malic acid)	
Organic solvent extractable compounds	Lipids, fats, oils, waxes	Cell walls, storage	Fatty acid and long chain aliphatic	++++
Hemicellulose	Branched polysaccharides	Cell walls	Multiple types of sugars linked and branched	+++
Cellulose	Polysaccharides	Cell walls	>10 000 linked glucose molecules	++
Lignin	Macromolecule of phenylpropanoids	Cell walls (especially wood)	Phenylpropanoids (phenol with 3-C chain aldehyde attached)	+
Other	Cutin	Cuticle	Fatty acid	++
	Tannin	Bark/woody tissue	Polyphenols	++
	DNA/RNA	Nucleus, mitochondria	Nucleic acids	++++
	Starch	Storage	Glucose polymers	+++

3 Specialization of decomposers toward different substrates.

4 Heterogeneous microhabitats in which different types of microorganisms are present.

5 Self-aggregation of organic molecules that protects parts of the molecules.

6 Adsorption of metals that makes the organic matter *unavailable* for degradation by microbes.

The decay of litter and biomass in soils involves oxidation of organic carbon by catabolic organisms that utilize the acquired electrons to sustain their life. The initial breakup of large plant materials is carried out by animals, insects such as termites, and earthworms. Saprophytic plants such as mushrooms and snowflowers also obtain energy from plant debris that is partially decomposed. New plant detritus is an excellent source of food for soil microorganisms. Decay proceeds as long as the oxygen, water, temperature, and nutrient levels are adequate for the decomposing organisms. For example, in the desert, the absence of water greatly hinders the oxidation rate of organic material at the soil surface, but deeper in the soil, where the moisture content is greater, decomposition can be faster.

Microbes initially degrade simple monomers of sugars, amino acids, and fats. The polymers, starch, and proteins follow close behind. Microorganisms are unable to transport large molecules through their cell walls, so larger substrates must first be broken down outside the cell by exoenzymes or physically by other organisms (e.g., earthworms). **Figure 8.8** is an example of the degradation rates of biomolecular compounds present in pine needle litter. Initially, the soluble and low molecular weight compounds in pine litter were degraded, followed by hemicellulose. Cellulose and lignin degradation followed (cellulose decayed sooner than lignin). Lignin is the most resistant to decay because it does not have hydrolysable bonds. Basidiomycetous fungi (white-rot fungi), and a few other fungi and bacteria, however, have biocatalytic enzymes that break apart lignin molecules and degrade them to CO_2. After five years, the pine needle litter lost 75% of its mass, and lignin was the predominant compound (48%) in the remaining SOM.

Degradability is not only a function of molecular size and chemical composition, but also the accessibility of the molecule to an organism. The latter may be particularly important because plant tissues, such as lignocellulose, glycoproteins, and cutin hinder accessibility of

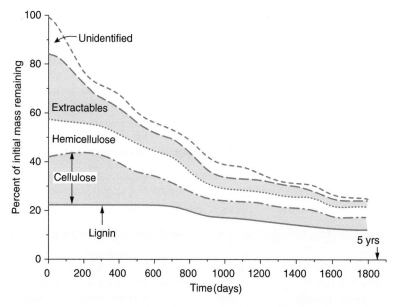

Figure 8.8 Degradation of biomolecules from pine needle litter in a spodic forest soil as a function of time. Extractables are low molecular weight compounds in needles that readily dissolve, including simple sugars, amino acids, phenolics, and hydrocarbons. Lignin is initially ~22% of the fresh litter, however due to relatively slow degradation compared to the other organic compounds in pine needles, after five years lignin is ~48% of the final litter mass. Adapted from Berg et al. (1982). Reproduced with permission of NRC Research Press.

biomolecules to microbes, and render them slowly degradable. For example, wood sawdust will decompose much more quickly than an equal mass of a wood log because the greater surface area in sawdust allows solutions and microbes such as ligninolytic fungi greater access to the cellulose and lignin.

Some plant materials (e.g., pine needles and oak leaves) contain inhibitory compounds such as tannins that deter microbial attack of the plant matter, except by specialized microbial species. Microbes, however, are opportunists, and in most soils there are organisms capable of degrading any natural organic molecule if oxygen, moisture, pH, nitrogen availability, etc., are sufficient for the microbes to live and gain energy. Under favorable conditions, even lignin will degrade relatively rapidly.

Microbial distribution and soil habitat are important determinants in the degradation rate of organic matter. Within the soil, microbial species establish themselves in niches or regions that are favorable for growth and survival. The distribution of microbes in the soil can be heterogeneous on many different scales, facilitating different SOM degradation conditions within a given soil. For example, microbial species on aggregate exteriors may be different than aggregate interiors. Such diversity also exists in different soil horizons and in different landscape positions.

Figure 8.9 shows a typical carbon degradation pathway in soils that leads to the development of SOM. The initial phase of microbial attack causes rapid loss of readily available (labile) organic material (**Table 8.5**) that have high energy yields and/or are physically accessible. Molds and spore-forming bacteria are especially active in consuming proteins, starches, and cellulose. Major degradation products are CO_2 and H_2O, plus new microbial tissue. Byproducts, especially in partially anaerobic conditions, include small amounts of NH_3, H_2S, organic acids, plus some incompletely oxidized substances. In subsequent phases of decomposition, when aerobic conditions are present, carbon compounds that are degraded at intermediate to slow rates are decomposed, producing new biomass and more CO_2. Once the chemically and physically available carbon is utilized, the nondegraded material and microbial cells left behind aggregate and assemble through hydrophobic and molecular forces with ions and minerals to form the dark colloidal SOM that is referred to as humus.

The final stage of litter decomposition is the gradual loss of plant compounds that are less available or larger,

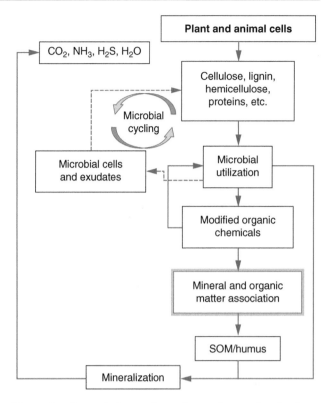

Figure 8.9 Degradation and reaction pathways that lead to formation of SOM. The release of biomolecules from microbes and plant roots is an important source of SOM compounds. The aggregation process (blue outlined box) describes likely polar and nonpolar interactions of organic compounds with each other and mineral surfaces.

such as lignin. The variable stages of plant compound degradation has led some investigators to conclude that SOM compounds exist in three major fractions:

1 Plant residues and the associated biomass that turn over every few years
2 Microbial metabolites and cell wall constituents that have a half-life of 5–25 years
3 More resistant organic molecules that evade degradation by association with other organic molecules or minerals, and range in age from 25 to 2500 years or more

When the age of soil organic carbon is measured by ^{14}C dating, a small but extremely old fraction raises the average age of carbon in SOM to about 1000 years in surface soils, and to several thousand years in subsoils. Under the right conditions, and over time, SOM will be completely oxidized, but because in most ecosystems carbon is continuously being added, the SOM content of soils remains more or less constant (i.e., it achieves a steady-state) for the particular environment in which the soil exists.

Figure 8.10 is an idealized graph of crop-residue decay rates in temperate region soils. Initially, plant residues are quickly degraded, but the rate of decay soon slows. Considerable plant carbon remains in the soil at this point, but part of the residual carbon occurs as microbial byproducts and part as the more resistant plant residues. While most of the biological input will eventually be converted into CO_2, fragments remain behind that *associate* with each other and soil minerals to form SOM, which resists degradation because of physical and chemical protection from microbial attack.

Figure 8.11 shows carbon losses over time from ryegrass residue added to soils in temperate (England, top *x-axis*) and humid tropical (Nigeria, bottom *x-axis*) climates. The shapes of the curves are similar, but the time scale is four times faster in the tropical climate. The approximate half-life of *fresh* organic matter (e.g., crop residue) in temperate region soils is about three to four months. In humid tropical soils, fresh organic matter degrades in as little as three to four weeks. The flatter portion of the curve represents the second stage of organic matter decomposition, which begins after

1.6 years in the humid tropics and 6.2 years in temperate regions (compare with **Figure 8.5**).

Fifty to 80% of freshly added organic matter is lost from most temperate soils during the first year. The smaller the particle size of the plant litter, the faster it degrades. The decomposition rate of organic materials in many soils increases in proportion to the addition rate – the more organic matter added, the more quickly it oxidizes. Thus, for a given soil and ecosystem, a steady-state SOM level exists, and increasing the total organic carbon in a soil is difficult. The steady state, however, is not fixed, but changes with management practices or natural processes that change the plant community or soil conditions.

In agricultural soils, continuous addition of plant or animal residues is required to maintain the favorable effects of organic matter on soil properties. An experiment in Rothamsted, England, has been adding about 30 Mg ha^{-1} of organic manures to a cultivated soil annually since 1843 without bringing the SOM content back to uncultivated levels. The inability to manage the requisite organic-mineral interactions in the soil to prevent organic matter decay may be the reason for the lack of increase in SOM in the Rothamsted soils. Knowledge of the critical factors that prevent organic matter degradation will allow new management practices (e.g., reduced tillage) to promote SOM buildup.

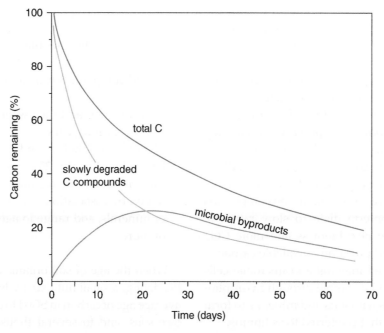

Figure 8.10 Idealized diagram for the decay of crop residues in soil under conditions that are optimal for microbial activity. Adapted from Stevenson (1972). Reproduced with permission of Wiley.

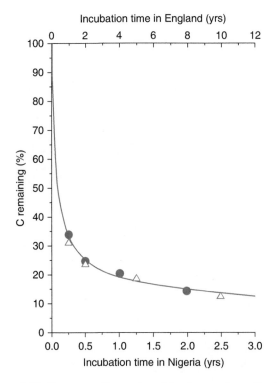

Incubation time in England (yrs)

Figure 8.11 Decomposition rates of fresh organic matter in soils in England and Nigeria. Triangles are soils in England, and circles are soils in Nigeria. Adapted from Jenkinson and Ayanaba (1977). Reproduced with permission of ACSESS.

8.7   SOM composition and structure

The organic compounds in SOM are complex mixtures of brown to almost black, amorphous colloidal substances derived from plant tissues and soil organisms containing primarily C, H, and O, with lesser amounts of N, P, and S, and minor amounts of other elements. The carbon-to-oxygen ratio of organic chemicals generally decreases with degradation because of increased oxygenation of functional groups. Oxygen in SOM carbon compounds exists in alcohol and acidic functional groups (e.g., carboxyl and phenols) that have lone pairs of electrons that attract and bond with cations or mineral surfaces through electrostatic forces and ionic or covalent bonding. Thus, the oxygen functional groups in SOM create substantial cation adsorption capacity.

Current theory proposes that SOM is largely non-degraded, or partially degraded plant and microbial biomolecules *associated* with each other and soil minerals. The relic biomolecules may be small molecular weight compounds, or large macromolecular compounds that resemble lignin or cellulose parent molecules. Typical *SOM-molecules* likely contain the following types of organic groups: di- or trihydroxyl phenols linked by bonds between O, N, H, and S; alcohol and carboxylic functional groups; heterocyclic and quinone (partially oxidized benzene) groups; aliphatic alkyl chains; and fatty acids. Within the soil, degraded organic compounds undergo various associations and reorganizations with each other and soil minerals via hydrophilic and hydrophobic interactions. SOM **humification,** or abiotic polymerization into new covalently bonded carbon compounds in soils is controversial, as there is little direct evidence that organic molecules polymerize to new compounds within the soil. However, this matter is still being vigorously debated.

The molecular characteristics of SOM are dictated by four factors:

1   Chemical properties of biomolecules
2   Hydrophobic character of the organic chemicals that causes them to organize their structures
3   Chemical interactions of organic functional groups with soil mineral surfaces and ions
4   Physical processes at solid-solution interfaces that enclose or isolate organic chemicals

The relative contribution of each of these factors to SOM composition and structure is controlled by physical and chemical interactions between organic chemicals, ions, and minerals. The organic molecules in soil react and associate to achieve the lowest *total* free energy (ΔG°_{rxn}). The SOM-mineral mixture is a *state* of partially degraded biomolecules, ions and minerals in the soil.

Chemical analysis of SOM has identified many aromatic carbon compounds. Given the aromatic character of lignin (**Figure 8.7**) and its relatively slow degradation, some SOM formation theories have proposed that lignin is the primary contributor to SOM. A lignin-only origin for SOM, however, would not account for the nitrogen content because lignin is nitrogen-free. Lignin would also not account for the large amount of alkanes (paraffins) observed in SOM. Black carbon, or charcoal, from vegetative fires is also an important contributor to phenolic compounds in SOM (see Special Topic Box 8.2).

Special Topic Box 8.2   Charcoal in soil

Charcoal is a thermally produced carbon material that results from incomplete combustion of organic material. The terms black carbon, pyrogenic carbon, and char are synonymous with charcoal. Charcoal produced intentionally from vegetative materials is called biochar. Charcoal is naturally occurring in soils, especially forest and grassland soils that undergo periodic wildfires. Charcoal is estimated to comprise 5–40% of the total soil organic carbon, depending on the ecosystem. Grasslands, for example, may contain as much as 35% of TOC as charcoal due to regular inputs from historical fires. Although often considered resistant to decay, charcoal does degrade in soils, and also dissolves and is transported out of the soil as DOC. Without degradation or export, many soils would be predominantly composed of charcoal. A metanalysis of published charcoal degradation research papers found that the average charcoal turnover time is 88 years, with a range from 1 to 750 years (Singh 2012).

Charcoal is mostly composed of carbon, up to 80%, arranged as macromolecules of bonded benzene rings that can have numerous functional groups. Charcoal often has a morphology inherited from the vegetative structure (**Figure 8.12**). Due to its high porosity, charcoal has a high specific surface area (10–500 m^2 g^{-1} or more, depending on biomass and burning temperature), allowing for numerous solid-solution interaction surfaces.

The hydrophobic and hydrophilic characteristics of charcoal vary, depending on the source of the vegetative material, the burning temperature, age, and weathering of the charcoal. As charcoal surfaces become oxidized, carboxylic acid and phenolic acid functional groups are created, which are weak acids, and the surface becomes overall negatively charged (**Figure 8.13**). The negatively charged functional groups attract cations and interact with mineral surfaces. The nonfunctionalized regions of the charcoal molecules are hydrophobic and thus attract nonpolar organic molecules and uncharged mineral surfaces.

Charcoal has significant impacts on soil properties such as cation exchange capacity, porosity, gas exchange, water-holding capacity, nutrient availability, and microbial activity. Even small amounts of charcoal added to soil can impart significant impacts to the chemical and physical characteristics of the soil. Negative surface charge of

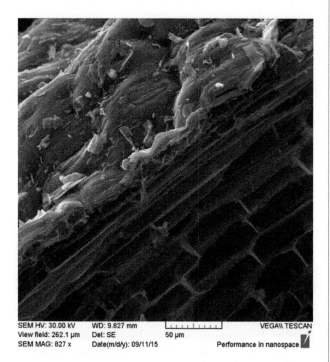

SEM HV: 30.00 kV   WD: 9.827 mm   VEGA\\ TESCAN
View field: 262.1 µm   Det: SE   50 µm
SEM MAG: 827 x   Date(m/d/y): 09/11/15   Performance in nanospace

Figure 8.12 Microscopic image (scanning electron microscopy (SEM)) of charcoal made from wheat straw. Image shows that the cell wall structure of the plant remains intact through pyrolysis, and the high internal pore structure of the charcoal. Image from Pranagal et al. (2017); with permission.

biochar can be as high as several thousand mmol kg^{-1}, especially after the biochar has aged for a long time, because surface oxidation creates weak acid functional groups that deprotonate as pH increases (**Figure 8.13**). Because of its impacts on chemical and physical properties of soil, charcoal can increase plant health and growth. In addition, charcoal is a significant source of organic carbon in soils that will persist for a long time and thus is a carbon sink when it is integrated into soils, potentially creating offsets to atmospheric carbon increases from fossil-fuel burning.

Charcoal produced by heating biomass in absence of oxygen (pyrolysis) is called biochar, especially when it is intended to be added to soils.

Hundreds of papers have been published investigating the benefits and limitations of adding biochar to soils. Consequently, scientists are gaining new understanding of

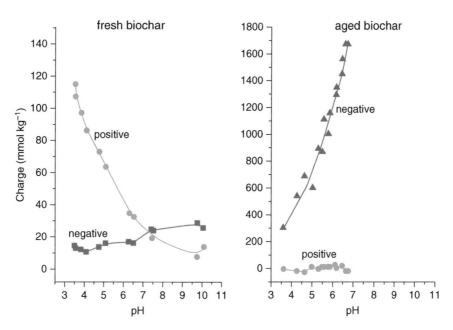

Figure 8.13 Surface charge of a fresh biochar made from wood and an aged biochar made in a kiln 135 years ago and allowed to age in the environment. Note the different y-axis scales. Adapted from Cheng et al. (2008); with permission.

the best environments for beneficial biochar amendment and are gaining a better understanding of how biochar addition changes soil properties. Undoubtedly, adding biochar to soils increases SOC and will offset fossil-fuel emissions. To what extent it will be utilized in agronomic and environmental ecosystems depends on the cost–benefit ratio and public policy.

Recent theories propose that SOM is composed of amphiphilic molecules that have a hydrophilic part and a hydrophobic part. **Amphiphiles** can contain either alkanes or aromatic rings. Amphiphilic compounds may be particularly important in long-term stabilization of organic molecules in soils on mineral surfaces (**Figure 8.14**), where the compounds *arrange* to create hydrophilic and hydrophobic regions within the SOM aggregates. The degree of hydrophobicity is also important in SOM formation. Hydrophobicity of nonpolar and amphiphilic organic compounds causes organic chemicals to associate with each other, creating **supramolecule** colloids, which are assemblies of molecules arranged according to hydrophobic and hydrophilic interactions, as well as molecular sizes. Supramolecules are similar to micelles, and in soils include mineral colloids. **Figure 8.14** is an illustration of a supramolecular association of SOM and mineral surfaces.

Sizes of organic compounds and soil minerals are important factors in SOM-mineral aggregation or association. **Figure 8.15** shows the sizes of SOM living and nonliving particles relative to soil minerals, aggregates, and pores. The interlayers and micropores of minerals are small enough to fit biomolecules, but too small to fit most bacteria. Thus, biomolecules that are small enough to mix with the smallest minerals may become physically isolated from microbes. Furthermore, the biomolecules can reside in micropores that stay hydrated even in dry conditions, causing them to arrange their hydrophobic and hydrophilic regions to create an organic-mineral matrix that further isolates the organic matter from microbes. The presence of water also limits gas exchange, slowing oxidative degradation.

8.8   NaOH extraction of SOM

A common method used to study SOM is to extract it from the soil using 0.1 to 0.5 M NaOH. This separation of organic matter from the inorganic matrix of sand, silt, and clay may alter the organic matter through hydrolysis and oxidation. The effects of base extraction

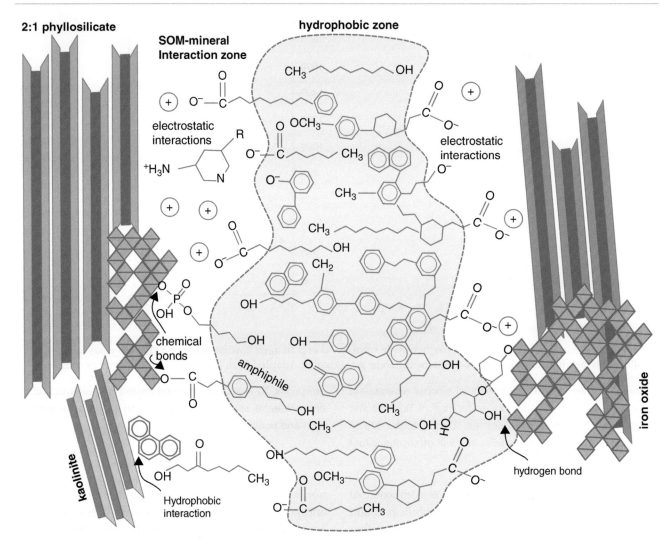

Figure 8.14 Two-dimensional conceptual diagram of molecular interactions of SOM with soil mineral surfaces to create an aggregated supramolecule-colloid. Water molecules are not shown but exert predominant influence on SOM interactions through hydrogen bonding and dipolar charge. The zones of interaction proposed are: 1) a contact zone where organic molecules interact with the mineral surface; 2) a zone of hydrophobic interactions where amphiphiles associate to decrease interactions with charged surfaces and water; and 3) a kinetic zone, where organic chemicals exist in soil solution and interact with each other and ions (not shown) would extend outside of the aggregate particle. The kinetic zone is a proposed intermediate between organic matter directly associated with a mineral surface and organic matter in bulk solution. Adapted from (Kleber and Johnson, 2010).

on the chemical properties of extracted SOM is topic of current debate. Separating SOM from soil creates distinct physical and chemical characteristics compared to when it is in soils because when separated it lacks the mineral and ion interactions that define its properties. Despite this caveat, useful information is gained by studying the isolated organic material; not unlike studying clay mineral properties when separated from soil.

Traditionally, organic components isolated from the soil using the base extracts are subsequently fractionated by precipitation with acids or metal salts, or by taking advantage of solubility differences in organic solvents. **Figure 8.16** shows the classical procedure for fractionation of soil organic matter using base and acid extractants. There are several refinements and purifications to the fractionation procedure. The extracted SOM is separated into a trio of organic compounds collectively called humic substances – and individually known as **humic acid, fulvic acid**, and **humin**. Humic acids are SOM compounds extracted in dilute alkali or neutral salt that coagulate

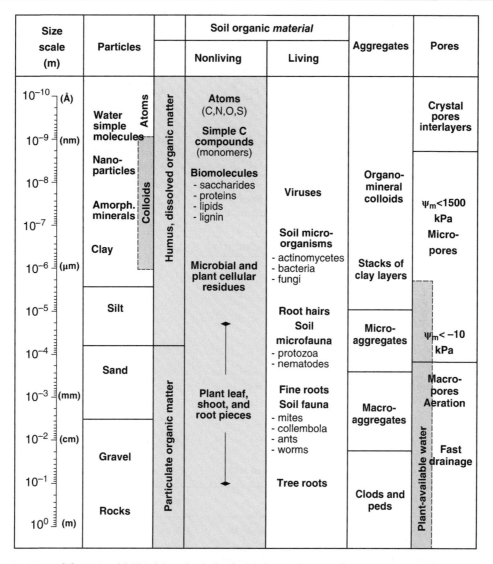

Figure 8.15 Comparison of the size of SOM (blue shaded columns) to soil minerals, aggregates, and soil pores. Adapted from Huang et al. (2005). Reproduced with permission of Elsevier.

when acidified. Fulvic acids are the material left in aqueous solution after acidification of the alkaline extract. Due to the size and greater amount of acidic functional groups, fulvic acids do not coagulate upon acidification like humic acids do. Humin is the SOM residue left in soils not extracted by dilute base, neutral salts, or acids.

Fulvic acids have lower molecular weights and higher oxygen-to-carbon ratios than humic acids. Fulvic acids may be similar to small SOM molecules that are soluble and can leach from soils into surface waters creating the brownish-yellow color of many natural waters. An extreme example is the Rio Negro ("black river") in South America.

8.9 Function of organic matter in soil

SOM contributes to plant growth by altering the chemical, biological, and physical properties of soil. SOM also supplies nitrogen, phosphorus, and sulfur for plant growth, serves as an energy source for soil microorganisms, and promotes good soil structure. Uptake of micronutrient and contaminant metals by plants is affected by interactions of the ions with SOM. The performance (availability) of herbicides and other agricultural chemicals is also affected by SOM. **Table 8.6** summarizes the properties of SOM and associated effects on soil.

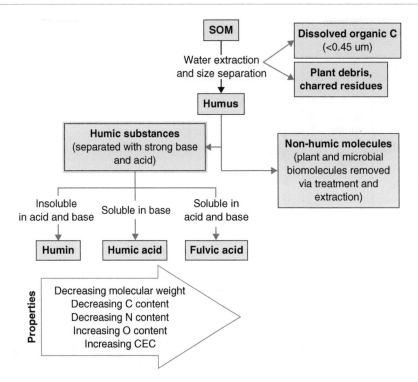

Figure 8.16 Categories of humic substances extracted from soil.

Table 8.6 General properties of SOM and associated effects in soil.

Property	Remarks	Effect on Soil
Color	The typical dark color of many soils is caused by organic matter.	Effects warming in spring and cooling in fall.
Water retention	Organic matter can hold up to 20 times its weight in water.	Helps prevent drying and shrinking; improves moisture retention in sandy soils.
Combination with clay minerals	Joins soil particles into structural units called aggregates.	Permits gas exchange; stabilizes structure; increases permeability.
Metal complexation	Forms stable complexes with Fe, Cu, Mn, Zn, and other polyvalent cations.	Buffers the availability of trace elements to higher plants.
Solubility in water	Insolubility of organic matter results partially from its association with clay; small organic matter fragments may be partly soluble in water as dissolved organic matter.	Little organic matter is lost by leaching.
pH relations	Organic matter buffers soil pH in the slightly acid, neutral, and alkaline ranges.	Helps to maintain a uniform pH in the soil.
Cation exchange	Total acidities (source of CEC) of extracted fractions of SOM (humic substances) range from 3000 to 14000 mmol+ kg^{-1}.	Increases the cation exchange capacity (CEC) of the soil; pH dependent.
Mineralization	Decomposition of organic matter yields CO_2, NH_4^+, NO_3^-, PO_4^{3-}, and SO_4^{2-}.	A source of nutrient elements for plant growth.
Combination with organic molecules	Affects bioactivity, persistence, and biodegradability of pesticides.	Modifies the application rate of pesticides for effective control.

8.9.1   Organic nitrogen, sulfur, and phosphorus

In soils formed in humid regions, carbon, nitrogen, and sulfur are found predominantly in SOM. With increasing aridity, organic matter content of soils decreases, and the fractions of the oxidized inorganic forms of carbon, sulfur, and nitrogen (i.e., carbonate, sulfates, and nitrate) increases. The mass ratio of C/N/S in SOM of temperate region soils is roughly 100/10/1. Carbon supplies the energy for nitrogen and sulfur reduction (fixation), as well as the matrix for compounds that reduced nitrogen and sulfur are incorporated into. Nitrogen and sulfur, in turn, are amongst the elements that govern the rate of plant growth and photosynthesis. The interdependency of these nutrients causes SOM to have a relatively constant C/N/S ratio under natural conditions. Soil amendments that alter the C/N ratio from a natural balance can promote or inhibit microbial decomposition of SOM.

Plant debris contains varying amounts of organic nitrogen compounds, ranging from 2–15% of the plant dry weight, but typically less than 6%. This nitrogen is input into the soil where it is degraded or is retained by association with other organic and mineral particles (**Figure 8.17**); in some cases organic nitrogen can persist for centuries. Microbes recycle much of the soil nitrogen, but some gets completely mineralized and is lost by gas, taken up by plants, or leached from the soil.

The average nitrogen content of SOM is about 3% by mass. More than 95% of organic nitrogen in surface soils is **amino acids** or amino sugars derived from proteins. A small proportion of nitrogen comes from nucleic acids in DNA and RNA.

Amino acids consist of an amine group (**Table 8.3**) attached to a carbon molecule that has a carboxylic acid functional group. Amine functional groups have a lone pair of electrons that protonate, making them positively charged when pH is less than pK_a:

$$R\text{-}NH_2 + H^+ = R\text{-}NH_3^+ \tag{8.1}$$

The positively charged protonated functional group attracts anions from the solution or can be attracted to negatively charged regions of clay minerals. Amines are the most important component of positive charge in SOM; however, the total positive charge created by amine groups is small compared to the amount of negatively charged carboxylic and phenolic acid functional groups. Thus, the net surface charge of SOM particles is always negative in soils.

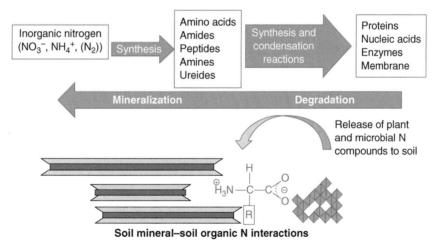

Soil mineral–soil organic N interactions

Figure 8.17 Synthesis of nitrogen from soil solution in plants and microbes creates organic nitrogen compounds like proteins. Input of organic nitrogen compounds into soil either from plant debris, root exudates, microbial cells, or microbial exudates is an important nutrient source. The organic nitrogen tissue and compounds are degraded and some of the degraded nitrogen compounds complex with minerals by either electrostatic interactions, such as cation association with clay minerals, or complexation with mineral surface functional groups. Association with minerals can protect the organic nitrogen compounds from further degradation and mineralization. The organic nitrogen accessible to microbes will be completely mineralized to create ammonium, nitrate, or nitrogen gas, which is taken up anew by plants and microorganisms, or lost from the soil.

The acidity of the amino acids varies according to the attachment of the carboxylic acid and amine functional groups to the carbon *backbone* of the compound, which may have several other different functional groups associated with it. However, between pH 3 and 10, most amine functional groups are protonated, and the carboxyl groups are deprotonated, making the molecule both positive and negatively charged (called a zwitterion). Positively charged functional groups are attracted to negatively charged surfaces, such as on clay minerals, and are an important mechanism for SOM-mineral bonding (**Figure 8.17**). Association with minerals creates aggregates that protects the organic nitrogen from further degradation, leading to its slow degradation in soils.

The average sulfur content of SOM is about 1% by mass. In soils developed in humid regions, more than 90% of the total soil sulfur in the surface horizon is associated with SOM. Sulfur functional groups principally exist as reduced sulfur bonded to carbons in the form of *R-O-S*, *R-N-S-R*, or *R-S-S-R* on proteins such as cysteine, cystine, and methionine. **Thiol** functional groups (*R-SH*) also exist that behave as weak Lewis acids, which are important sorption sites for metals such as Hg^{2+} and Cd^{2+}.

SOM consist of 1–3% phosphorus by mass. In surface soils, organic phosphorus can be as much as 60% of the total soil P, but is typically much less. Organic phosphorus has been identified as an important phase in forest soils and animal-manure amended soils. Many forms of organic phosphorus in soils are derived from plants and microbes, including, **inositol phosphates**, nucleic acid P (diesters: $R-PO_4-R$), and phospholipids. Insositol phosphate, also called phytic acid, is an orthophosphate monoester ($R-PO_4$), consisting of six orthophosphate groups attached to a hexane sugar ring (see **Table 8.3**). Inositol phosphate is synthesized by organisms as a phosphorus storage compound, occurs in high concentrations in seeds, and is one of the most stable organic P compounds in soils. Orthophosphate groups in inositol phosphate strongly adsorb on oxide mineral surfaces.

8.9.2 SOM influences on chemical processes

Between pH~4 and 7, the weak-acid carboxylic acids functional groups in SOM buffer soil pH. Buffering contributes significantly to the lime requirement of acid soils and the acidity of forest soils (Chapter 12). Additional buffering occurs beyond pH ~8 by phenolic groups that protonate and deprotonate at high pH. The negative charge of the deprotonated functional groups can be a major portion of a soil's CEC.

The cation exchange capacity of SOM ranges from 600 mmol(+) kg^{-1} at low pH, to as high as 3500 mmol(+) kg^{-1} at high pH (see Figure 9.12). Charged **carboxyl groups** (R-COO$^-$) enable SOM to retain cations in nonleachable, but exchangeable forms that are available to plants. For base cations, bonding to negative functional groups on SOM is primarily Coulombic or electrostatic (e.g., R-COO$^-$ -- K$^+$). For many transition metals, and Al^{3+}, Pb^{2+}, and Hg^{2+}, bonding to carboxyl functional groups is covalent, creating strong complexes (e.g., (R-COO$^-$)$_2$Pb^{2+}). Adsorption on solid phase SOM immobilizes many cations, especially metals. Surface charge properties of SOM are discussed in more detail in Chapter 9.

Dissolved organic matter, such as **low molecular weight organic acids**, form stable complexes (chelates) with Fe^{3+}, Cu^{2+}, Zn^{2+}, Al^{3+} Cd^{2+}, Hg^{2+}, and other polyvalent cations, greatly increasing metal solubility and mobility. For example, iron and aluminum oxides in Spodosols are leached out by soluble organic chelates (see Figure 4.19), creating an E horizon. Larger organic molecules may also have a high affinity for polyvalent cations, but are less water soluble, and can immobilize the cations. **Figure 8.18** shows an example reaction between a metal and the functional groups of an organic carbon compound. Complexation of aluminum by organic compounds in soil plays a significant role in reduction of aluminum toxicity to plants, and facilitates mobilization in the soil profile.

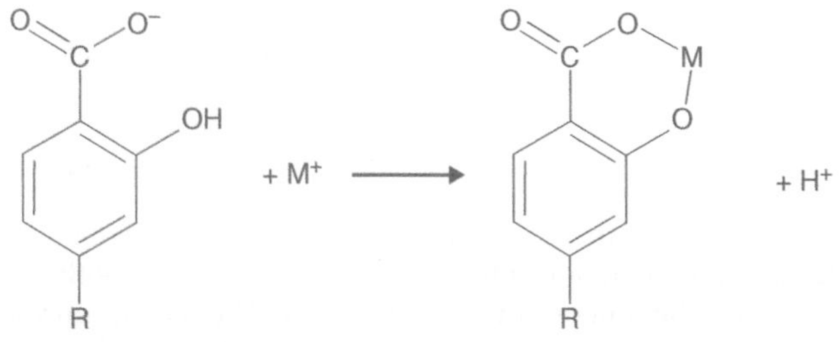

Figure 8.18 Possible reaction between a metal ion (M$^+$) and acidic functional groups on SOM. The metal is complexed via a bidentate bond with carboxylic acid and phenol functional groups.

8.9.3 SOM influences on physical properties

One of the most distinctive characteristics that SOM imparts to soil is color. Dark brown to black O and A horizons are rich in SOM; as are some B horizons in Spodosols. Color is an important factor in absorption, retention, and conduction of heat.

SOM affects soil structure, which is important for soil tilth, aeration, and moisture retention. The deterioration of structure that accompanies intensive tillage is usually less severe in soils that have a significant amount of organic matter. Soils that lose organic matter tend to become hard, compact, and cloddy.

SOM improves aeration, water-holding capacity, and permeability in soils. The frequent addition of organic residues leads to the formation of SOM that binds soil particles into aggregates. The intimate association of clay-sized phyllosilicates with SOM via cation bridges facilitated by Ca^{2+}, Mg^{2+}, Al^{3+}, Fe^{3+} cations promotes aggregation. Heavy (clayey) soils, in particular, benefit from organic matter additions that promote particle aggregation. Aggregation yields a loose, open, granular structure, promoting good water and air permeability. Bacterial polymeric slimes (biofilms) and fungal hyphae are also important contributors of soil aggregation.

SOM absorbs large quantities of water; as much as 80–90% water by weight. Additionally, micropores within organic-mineral aggregates hold water, increasing the plant-available soil water. Water retention is one of the major benefits of organic matter additions to soils.

8.9.4 Organic chemical partitioning

SOM is highly porous, allowing pesticides and other organic compounds added to soils to be *ab*sorbed into pores. Many pesticides are nonpolar, or have hydrophobic regions, and will adsorb to the hydrophobic regions of SOM. Hydrophobic regions of SOM compounds are rich in saturated carbon compounds, such as aliphatic chains or benzene rings. At low pH, many of the acidic functional groups on SOM protonate and these regions are also uncharged and have some degree of hydrophobicity.

In addition to hydrophobic regions, high surface area and high surface acidity makes SOM reactive for bonding organic chemicals via hydrogen bonds, cation

bridging, and ionic or covalent bonds. Thus, pesticides interact with SOM by several mechanisms. The surface charge properties of organic matter are discussed in more detail in Chapter 9.

Adsorption of pesticides by SOM strongly influences their behavior in soils, including effectiveness against target species, phytotoxicity to subsequent crops, leachability, volatility, and biodegradability. For soils high in SOM, herbicide application rates are often greater than application rates on soils with less SOM in order to compensate for increased adsorption by SOM. The behavior of organic chemicals like pesticides in soils is an active area of soil chemistry research. Pesticide and organic pollutant adsorption behaviors are discussed in Chapter 10.

8.10 Summary of SOM

SOM is a mixture of microbial and plant residue associated with soil minerals (**Figure 8.1** and **Figure 8.14**). It is very reactive and is an important reservoir of carbon storage (more than twice the size of atmospheric and plant carbon). Thus, characteristics of SOM and environmental factors that affect fluxes of carbon into and out of the soil must be understood to manage soil and ecosystem health and productivity. Important factors that affect SOM formation are type and amount of organic material input into soils, microbial activity, climate, and presence of minerals that interact with SOM molecules.

SOM has high specific surface area and variable (but high) negative charge and is thus very reactive with cations and minerals. Hydrophobic regions of SOM attract hydrophobic chemicals such as some pesticides and other trace organics, causing them to be adsorbed and retained in the soil. In addition, SOM is an important source of nitrogen, sulfur, and phosphorus storage in soils. Managing nutrient, contaminant, and pesticide availability for transport within and out of soils requires understanding the amount and characteristics of SOM. Details of surface properties and reactivity of SOM will be further discussed in the next three chapters.

Given the large global SOC reservoir, there is great interest in how best to manage environmental factors that change either emissions or sequestration of carbon into soil. Considering human-caused climate change, opportunities to manage soils for increased carbon storage are attractive and have been proposed. The late US President John F. Kennedy said in a 1963 speech at The American University, "*Our problems are man-made, therefore they may be solved by man. And man can be as big as he wants. No problem of human destiny is beyond human beings.*" It is with this hopeful spirit that scientists are busily researching ways to manage soils and the environment to increase SOM and rebalance atmospheric carbon.

Questions

1 Distinguish between soil organic matter, humus, and soil biomass.

2 Explain why the SOC content of soils in a given climatic zone tends to be higher in fine-textured soils than in coarse-textured soils.

3 How does SOM contribute to the chemical, physical, and biological properties of soil as a medium for plant growth?

4 How does SOM alter micronutrient and trace metal availability in soils?

5 How does SOM affect pesticide recommendations, and why?

6 Explain how only a few percent organic matter can exert a profound influence on soil properties.

7 How does SOM charge change with pH? Explain.

8 What are the primary functional groups responsible for charge development in SOM?

9 What are the molecular forms of nitrogen in soil organic matter? What are its *parent* sources?

10 Describe how a negatively charged soil organic matter functional group can be attached to an iron oxide surface at an alkaline pH.

11 Describe the role of water in SOM structure composition in soils. Hint: Consider hydrophobic and hydrophilic interactions.

12 What management factors are important to consider for long-term carbon storage in soils?

13 What role does charcoal have in SOM?

14 What are the species and properties of organic sulfur in SOM?

15 Why are lignin compounds more recalcitrant than carbohydrates in soil? What soil properties would cause lignin to be rapidly decomposed?

16 How might a soil clay impact the degradation rate of a SOM compound?

17 How do polar organic molecules influence the structure and properties of organic matter in soils?

18 How is particle size of organic fragments or compounds important for the formation and structure of SOM?

19 Describe the role of SOM in Spodosol soil formation and properties.

20 How does soil aggregation affect SOM degradation?

21 What is the difference between total SOM and SOC?

22 Why are organic phosphate compounds in soils important?

23 Explain how the amphoteric charge on soil organic nitrogen compounds originates and its importance in aggregation of SOM

Bibliography

Amelung, W., S. Brodowski, A. Sandhage-Hofmann, and R. Bo. 2008. Combining biomarker with stable isotope analyses for assessing the transformation and turnover of soil organic matter, pp. 155–250. *In* D. L. Sparks, (ed.) Advances in Agronomy. Soil Science Society of America.

Bar-On, Y.M., R. Phillips, and R. Milo. 2018. The biomass distribution on Earth. Proceedings of the National Academy of Sciences of the United States of America 115:6506–6511.

Batjes, N.H. 1996. Total carbon and nitrogen in the soils of the world. European Journal of Soil Science 47:151–163.

Berg, B., K. Hannus, T. Popoff, and O. Theander. 1982. Changes in organic-chemical components of needle litter during decomposition – long-term decomposition in a Scots pine forest.1. Canadian Journal of Botany-Revue Canadienne De Botanique 60:1310–1319.

Cheng, C.H., J. Lehmann, and M.H. Engelhard. 2008. Natural oxidation of black carbon in soils: Changes in molecular form and surface charge along a climosequence. Geochimica Et Cosmochimica Acta 72:1598–1610.

Dohnalkova, A.C., M.M. Tfaily, A.P. Smith, R.K. Chu, A.R. Crump, C.J. Brislawn, T. Varga, Z. Shi, L.S. Thomashow, J.B. Harsh, and C.K. Keller. 2017. Molecular and microscopic insights into the formation of soil organic matter in a red pine rhizosphere. Soils 1:4.

Gregorich, E.G., M.H. Beare, U.F. Mckim, and J.O. Skjemstad. 2006. Chemical and biological characteristics of physically uncomplexed organic matter. Soil Science Society of America Journal 70:975–985.

Huang, P.-M., M.-K. Wang, and C.-Y. Chiu. 2005. Soil mineral–organic matter–microbe interactions: Impacts on biogeochemical processes and biodiversity in soils. Pedobiologia 49:609–635.

Jackson, R.B., K. Lajtha, S.E. Crow, G. Hugelius, M.G. Kramer, and G. Pineiro. 2017. The ecology of soil carbon: pools, vulnerabilities, and biotic and abiotic controls. Annual Review of Ecology, Evolution, and Systematics 48:419–445.

Jenkinson, D.S., and A. Ayanaba. 1977. Decomposition of C-14-labeled plant material under tropical conditions. Soil Science Society of America Journal 41:912–915.

Johansson, M.B. 1995. The chemical-composition of needle and leaf-litter from Scots pine, Norway spruce and white birch in Scandinavian forests. Forestry 68:49–62.

Kleber, M., and M.G. Johnson. 2010. Chapter 3. Advances in understanding the molecular structure of soil organic matter: implications for interactions in the environment, p. 77–142. In L. S. Donald (ed.) Advances in Agronomy. San Diego, CA: Academic Press.

Mohan, D., C.U. Pittman, and P.H. Steele. 2006. Single, binary, and multi-component adsorption of copper and cadmium from aqueous solutions on Kraft lignin – a biosorbent. Journal of Colloid and Interface Science 297:489–504.

Post, W.M., W.R. Emanuel, P.J. Zinke, and A.G. Stangenberger. 1982. Soil carbon pools and world life zones. Nature 298:156–159.

Pranagal, J., P. Oleszczuk, D. Tomaszewska-Krojanska, P. Kraska, and K. Rozylo. 2017. Effect of biochar application on the physical properties of Haplic Podzol. Soil & Tillage Research 174:92–103.

Prasad, S., A. Singh, and H.C. Joshi. 2007. Ethanol as an alternative fuel from agricultural, industrial and urban residues. Resources Conservation and Recycling 50:1–39.

Pribyl, D.W. 2010. A critical review of the conventional SOC to SOM conversion factor. Geoderma 156:75–83.

Rahman, M.M., J. Tsukamoto, M.M. Rahman, A. Yoneyama, and K.M. Mostafa. 2013. Lignin and its effects on litter decomposition in forest ecosystems. Chemistry and Ecology 29:540–553.

Scharlemann, J.P.W., E.V.J. Tanner, R. Hiederer, and V. Kapos. 2014. Global soil carbon: understanding and managing the largest terrestrial carbon pool. Carbon Management 5:81–91.

Singh, N., S. Abiven, M.S. Torn, and M.W.I. Schmidt. 2012. Fire-derived organic carbon in soil turns over on a centennial scale. Biogeosciences 9:2847–2857.

Stevenson, F.J. 1972. Organic matter reaction involving herbicides in soil. J Environ Qual 1:333–343.

Stockmann, U., M. Adams, J. Crawford, D. Field, N. Henakaarchchi, M. Jenkins, A. McBratney, D.R.D. Courcelles, K. Singh, and I. Wheeler. 2010. Managing the soil plant system to mitigate atmospheric CO_2. The University of Sydney, Faculty of Agriculture, Food and Natural Resources; The United States Studies Centre at the University of Sydney, Sydney, Australia.

9 SURFACE PROPERTIES OF SOIL COLLOIDS

9.1 Introduction

The charge properties of the **solid–solution interfaces** of minerals and organic matter particles have important impacts on soil chemical processes. The solid–solution interface is where adsorption and desorption reactions occur, which are controlled by surface area and surface charge. For hydrophobic surface reactions on SOM and some mineral surfaces, lack of charge is an important surface characteristic.

Specific surface area is a direct result of particle size and shape (e.g., see Figure 1.9), so most of the total specific surface area of a soil is due to clay-sized particles and organic matter. Clay minerals have specific surface areas that range from 5 $m^2\,g^{-1}$ for kaolinite to 800 $m^2\,g^{-1}$ for montmorillonite. Iron and aluminum oxides can have a similar range, although the upper end would mainly occur for short range ordered minerals such as ferrihydrite and allophane. Larger surface area facilitates more solid-solution interface for surface reactions to occur. Fine-textured soils, therefore, have more surface reactivity than coarser-textured soils.

Minerals have two types of surface charge: (1) **permanent charge** originating from isomorphic substitution; and (2) **variable charge** that occurs from incomplete coordination on mineral edges. The sign and magnitude of charge on variable surface charge mineral surfaces changes as a function of pH. Mineral surfaces that change from negatively charged to positively charged are amphoteric. Soil organic matter also has pH-dependent charge because weak-acid functional groups protonate and deprotonate, causing them to be either uncharged or negatively charged.

9.2 Permanent charge

Isomorphic substitution is the replacement of a structural ion for another of similar size within a crystal lattice (see Chapters 6 and 7). The substituting ion may

Soil Chemistry, Fifth Edition. Daniel G. Strawn, Hinrich L. Bohn, and George A. O'Connor.
© 2020 John Wiley & Sons Ltd. Published 2020 by John Wiley & Sons Ltd.

have a greater, equal, or lower charge than the ion for which it substitutes. Isomorphic substitution occurs during crystallization of layer silicate minerals in magmas and in soils. If the primary cation is unavailable as the mineral forms, another cation of similar size – but not necessary the same charge – can take its place. Substitution of cations that have a different charge than the *parent* cation creates **permanent charge** that is independent of the soil-solution pH.

For 2:1 phyllosilicates, isomorphic substitution in tetrahedral or the octahedral sheets is the principal cause of negative **layer charge** (Figures 7.10 and 7.11). When a cation of lower valence substitutes for one of higher valence, such as Mg^{2+} for Al^{3+}, or Al^{3+} for Si^{4+}, the negative charges of O^{2-} and OH^- ions in the mineral structures are left unsatisfied, yielding a net-negative permanent charge. Permanent charge originating in the octahedral layer is further from the interlayer than that in the tetrahedral layer, and is thus a weaker cation attractive force (**Figure 9.1**).

In soils where expandable phyllosilicates are the predominant clay-size minerals, permanently charged minerals are the primary source of cation exchange capacity. Isomorphic substitution can also result in positively charged regions in phyllosilicates. For example, Al^{3+} substituting for Mg^{2+} in the brucite interlayer of chlorite or in the trioctahedral sheet of vermiculite creates positively charged regions in the octahedral sheets. However, positive permanent charge in clay minerals is typically small compared to negative charge from isomorphic substitution, so the net permanent charge is negative.

Cation exchange reactions with permanent charge in minerals occur through the hydrated interlayer between the mineral sheets. Chlorite and illite interlayers are unhydrated, thus solid–solution reactions with the layer charge in these minerals are negligible.

9.3 pH-dependent charge

Total charge of soil particles often varies with pH. **Figure 9.2** illustrates **pH-dependent charge** in a soil, where some portion of the soil changes from positive charge at low pH, to negative charge at higher pH. The soil's total charge is the sum of its negative and positive charges. The relative contribution of permanent and pH-dependent charge within a soil depends on the amount and type of soil minerals. Relatively young and weakly weathered soils characteristic of Europe and North America have a net negative charge because of their abundant 2:1 phyllosilicate mineral content with relatively high permanent negative charge. Highly weathered soils and volcanic soils, on the other hand, have high amounts of iron and aluminum oxides, and allophane, which, at pH < 7, have a net neutral to positive charge because of the high positive pH-dependent charge of the mineral edges. The permanent negative charge contribution of soils that have mostly hydroxides and allophane minerals would be much less than the contribution shown in **Figure 9.2**. As pH of well-weathered soils increases, the pH-dependent negative charge contribution increases and the positive

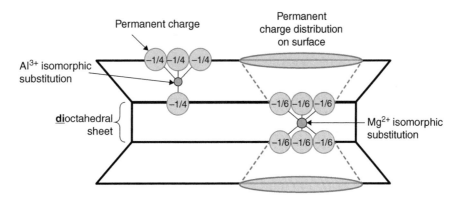

Figure 9.1 Permanent charge distribution on the basal plane of a 2:1 d̲ioctahedral phyllosilicate due to isomorphic substitution in the tetrahedral and octahedral sheets. The greater distance of the negative charge on the oxygen atoms created by octahedral isomorphic substitution of Mg^{2+} for Al^{3+} causes more dispersed charge on the basal plane than when Al^{3+} isomorphically substitutes for a Si^{4+} in the tetrahedral sheet.

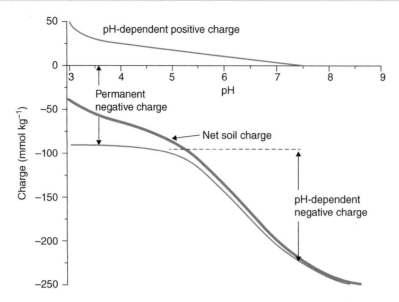

Figure 9.2 Example of positive and negative charges on soil particles as a function of pH for a typical soil containing 2:1 permanent charge minerals and some variable charge minerals. The sum of the two curves is the net surface charge on the particle.

charge contribution decreases, causing the net charge to be overall negative.

The primary source of pH-dependent charge in soils is adsorption and desorption of protons by inorganic solids and organic functional groups of SOM. The oxygen functional groups on the edges of soil minerals are capable of developing pH-dependent positive or negative charge, and are thus **amphoteric**. These minerals include phyllosilicates, allophane, and oxides.

pH-variable surface charge arises from basic mineral properties. At edges of the mineral, or where the mineral structure has holes, steps or kinks, the crystal lattice ends, leaving some of the structural cations and O^{2-} ions incompletely coordinated. Electrical neutrality on the mineral surface is necessary, and is achieved by adsorption of H^+, OH^-, water, and cations or anions from the soil solution.

The **net surface charge** (σ_p) on a particle is described by the following charge-balance equation:

$$\sigma_p = \sigma_o + \sigma_H + \sigma_{IS} + \sigma_{OS} \qquad (9.1)$$

The magnitude of the charge is measured in units of moles of charge per kg of solid (a commonly used unit is mmol($\pm$) kg^{-1}, where the charge can be positive or negative). The terms of this equation are described in **Table 9.1** and illustrated in **Figure 9.3**. The charge

Table 9.1 Surface charge components of minerals (see **Figure 9.3**).

Surface charge term	Description
σ_p	Total charge on the surface of a particle, not including the ions in the diffuse double layer.
σ_o	Permanent surface charge of the particle due to isomorphic substitution.
σ_H	Charge on the surface of a particle due to protonation and deprotonation of oxygen functional groups on the edge of the mineral.
σ_{IS}	Charge on the surface caused by adsorption of ions via inner-sphere adsorption.
σ_{OS}	Charge on the surface caused by adsorption of ions via outer-sphere adsorption; includes only hydrated ions directly next to surface charge, not diffuse ions.
σ_d	Total charge of ions in the diffuse double layer.

originating from the mineral structural properties (σ_o and σ_H) is called **intrinsic charge** because these charge components are inherent to the mineral regardless of solution conditions. The charge originating from adsorption of ions (excluding protons and hydroxides) as **inner-sphere** or **outer-sphere** ions, σ_{IS} or σ_{OS}, is called **Stern-layer charge**. Ions in the Stern layer are adsorbed on the surface of the mineral as either inner- or

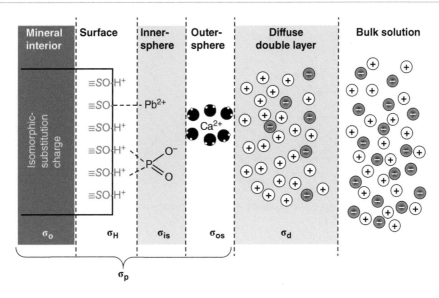

Figure 9.3 Model of surface charge components (see **Table 9.1**) on a mineral surface with negative permanent charge and pH-variable edge charge. The water molecules that exist on the ions in the double layer and bulk solution are not shown.

outer-sphere complexes. For variable charge surfaces, σ_o is small compared to the variable charge.

The magnitude of the surface charge can also be expressed in Coulombs per square meter (C m^{-2}; C is the unit Coulombs) by using the net surface charge in moles of charge per kg and specific surface area of the mineral:

$$\sigma_p\left(\text{C m}^{-2}\right) = \frac{\sigma_p\left(\text{mol kg}^{-1}\right) \times \text{F}}{\text{SA}} \tag{9.2}$$

where F is Faradays constant (96 485 C mol^{-1}) and SA is the specific surface area in m^2 kg^{-1}.

9.3.1 Balancing surface charge

The net surface charge (σ_p) in **Eq. 9.1** is balanced by **counter ions** adsorbed in a **diffuse swarm** adjacent to the mineral's surface called a diffuse layer (diffuse double layer is discussed in Chapter 10). The effective charge of the diffuse layer (σ_d) is of equal magnitude and opposite sign as the total particle surface charge (σ_p)

$$\sigma_p = -\sigma_d \tag{9.3}$$

Equations 9.1 and **9.3** show that the *net charge* of the diffuse layer is directly proportional to the intrinsic- and Stern-layer charges. These charge

balance expressions provide a conceptual model for understanding how pH and ion adsorption affects particle charge. For example, in soils with high concentrations of minerals with permanent charge (σ_o), and low concentrations of minerals with pH-variable charge (σ_H), adsorption capacity of ions varies little as pH changes. The charge balance expressions are also used to quantitatively model surface charge, which is needed to predict adsorption behavior of ions (discussed in Chapters 10 and 11).

9.3.2 Variable charge on phyllosilicates

On phyllosilicates, pH-dependent charge develops on the edges of the clay platelets (**Figure 9.4**). The Si-OH and Al-OH groups are called **silanol** and **aluminol** functional groups, respectively. The charge can be either positive, zero, or negative, depending on the pH of the soil solution. Deprotonation of Si-OH groups occurs only at high pH, and protonation occurs only at very low pH, thus the pH-dependent charge of layer silicates is most likely due to reversible protonation and deprotonation of Al-OH or (Si, Al)>OH functional groups (**Figure 9.4**).

The contribution of surface functional groups to pH-dependent charge on phyllosilicates is related to the acidity of the aluminol and silanol groups, and to the surface area of edges on the particles.

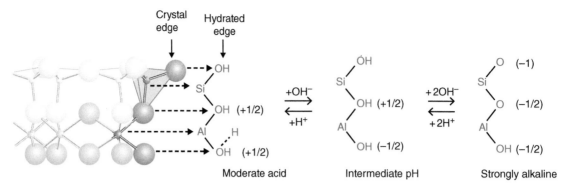

Figure 9.4 Protonation and deprotonation of two types of functional groups (aluminol and silanol) on the edges of kaolinite. Adapted from Schofield and Samson (1953).

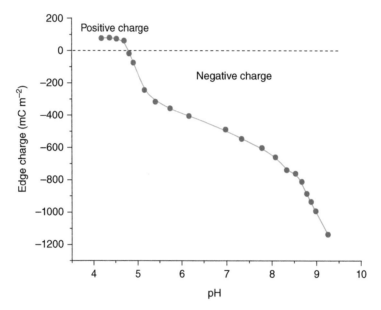

Figure 9.5 Edge charge, in milli-coulombs per unit area, of kaolinite sample from Georgia, USA. The pH where the curve crosses zero is where the number of adsorbed protons and hydroxides are equal, and is equivalent to the point of zero net proton charge (PZNPC). Adapted from Zhou and Gunter (1992).

For 2:1 minerals such as montmorillonite, the amount of edge surface is small relative to the basal (planar) surface. Kaolinite, on the other hand, has no interlayer access for solution and no permanent layer charge, so the edge area is much greater than the exposed basal plane area and pH-dependent charge is more important for kaolinite than for smectites or vermiculites. Approximately 5–10% of the surface charge on 2:1 layer silicates is pH dependent, whereas most of the charge developed on 1:1 kaolinite is pH dependent. However, as the particle size of 2:1 layer silicates gets smaller, the edge site to basal plane surface ratio increases, making edge site charge more important. Montmorillonite, for example, often occurs in soils as small colloids, and thus has relatively high edge-site charge.

Figure 9.5 shows the pH-dependent net surface charge (σ_H) on kaolinite. In solutions less than pH 5, the net charge is positive due to protonation of aluminol functional groups (e.g, $\equiv AlOH_2^{1/2+}$). In solutions greater than pH 5, the net surface charge changes to negative due to deprotonation of aluminol and silanol functional groups (see **Figure 9.4**). At the pH where the magnitude of positive and negative surface charges are equal, the net charge on the kaolinite particles is zero (~ pH 5). Edges of other phyllosilicates have a similar functional group speciation and pH dependency.

9.3.3 pH-dependent charge on iron and aluminum oxides

In highly weathered soils, iron and aluminum oxides that have considerable pH-dependent charge are abundant. Variable edge charge on manganese oxides is similar to iron and aluminum oxides. Edge charge originates from incompletely coordinated metal cations. For example, Fe^{3+} cations in goethite are coordinated by six O^{2-} atoms; the Fe-O bond in an iron octahedron supplies +0.5 charges to the coordinating O^{2-} ions (**Figure 9.6**). The remaining –1.5 charge of each coordinating oxygen atom in goethite is satisfied from either bonds with adjacent structural Fe^{3+} or bonds with H^+ (e.g., see hydrogen shown in Figure 7.21). On the edges of the goethite crystals, however, Fe^{3+} atoms are incompletely coordinated. Non-iron coordinated oxygen ligands on goethite's edges satisfy their charge by complexing H^+ ions from the aqueous solution (**Figure 9.6**). The result is a mineral with a layer of H^+ or OH^- ions on its surface. **Figure 9.7** shows the different types of functional groups on the edges of goethite. Surface functional groups in other oxides, such as ferrihydrite, hematite, and gibbsite,

are similar to goethite, although the number of ligand-metal linkages exposed on the surface of the mineral may vary.

Iron and aluminum oxides have minimal permanent charge, so their charge depends primarily on the pH of the soil solution. The generic reaction for protonation and deprotonation of mineral surface functional groups is

$$\equiv SOH = \equiv SO^- + H^+ \tag{9.4}$$

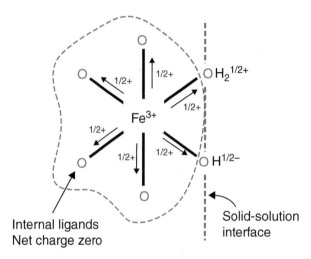

Figure 9.6 Examples of functional groups on a hydrated iron oxide, and the distribution of bond valence charge (see **Table 9.2**).

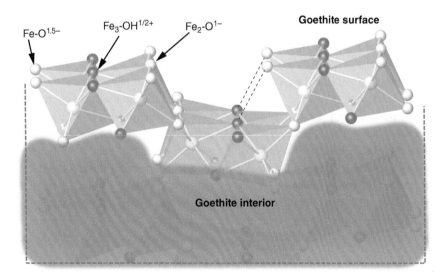

Figure 9.7 Surface functional groups on goethite (FeOOH) edges. See Figure 7.21 for complete structure. Charges are the bond valence charge on a nonhydrated surface. The reactivity (acidity) of the groups is different because they have different bond coordination to structural iron atoms. Different mineral faces expose different surface functional groups.

Table 9.2 Possible silicon and iron surface functional groups and their protonation and deprotonation reactions that cause variable surface charge. Aluminum surface functional groups are similar to Fe^{3+}. Protonation and deprotonation reactions depend on mineral properties. For some minerals, the functional groups are unreactive because their pK_a values are too high or low.

Functional group	Reaction in acidic solution	Reaction in basic solution
$\equiv SiOH$	does not protonate	$\equiv SiOH \leftrightarrow \equiv SiO^- + H^+$
$\equiv FeOH^{1/2-}$	$\equiv FeOH^{1/2-} + H^+ \leftrightarrow \equiv FeOH_2^{1/2+}$	$\equiv FeOH^{1/2-} \leftrightarrow \equiv FeO^{1.5-} + H^+$
$\equiv Fe_2OH$	$\equiv Fe_2OH + H^+ \leftrightarrow \equiv Fe_2OH_2^{1+}$	$\equiv Fe_2OH \leftrightarrow \equiv Fe_2O^{1-} + H^+$
$\equiv Fe_3OH^{1/2+}$	$\equiv Fe_3OH^{1/2+} + H^+ \leftrightarrow \equiv Fe_3OH_2^{1.5+}$	$\equiv Fe_3OH^{1/2+} \leftrightarrow = Fe_3O^{1/2-} + H^+$

where $\equiv SOH$ is a reactive surface functional group; the S indicates a structural cation (Al^{3+} or Fe^{3+}) and the O is a partially coordinated oxygen on the mineral surface. The protonation and deprotonation behavior of $\equiv SOH$ depends on soil pH and the acid-base properties of the mineral. **Table 9.2** lists examples of protonation/deprotonation reactions and charges for surface functional groups. Given the various configurations of surface functional groups on iron and aluminum oxide surfaces, the charge of $\equiv SOH$ ranges from –1.5 to +1.5, with half-unit increments[1].

Allophane, an amorphous hydrous oxide that has high specific surface area, also develops pH-dependent charge. Because allophane's specific surface area is greater than other more crystalline soil minerals, its total surface charge per mass is greater than most oxides. Ferrihydrite is also a high surface area mineral with a large pH-dependent surface charge. Because of the pH-dependent surface charge of soil oxides, soils containing large amounts of aluminum and iron oxides have a highly variable cation or anion exchange capacity, and are thus often referred to as **variable charge soils**.

Special Topic Box 9.1   Nanoparticles in soils

Nanoparticles are inorganic or organic particles with at least one dimension less than 100 nm and are distinguished because they often have unique surface characteristics compared to larger particles of the same composition and molecular or atomic structure. In soils, nanoparticles include minerals such as nanoball allophane (see Figure 7.19), and organic macromolecules such as root and microbial exudates that may become SOM.

Nanoparticles are clay *particles*, but not all clay particles are nanoparticles. Many clay *minerals* naturally occurring in nature exist as nanoparticles, e.g., montmorillonite. Nanoparticles have larger molecular masses than simple molecules or monomers, but are small enough to easily suspend and move in solutions. For this reason, nanoparticles are sometimes referred to as colloids, however, the two terms are not interchangeable; a nanoparticle is a particle less than 100 nm in one dimension, and a colloid is a small particle usually <1 micron (1000 nm) in effective diameter that is homogenously *suspended* in a *mixture*.

Nanoparticles occur naturally in nature, including soils, rivers and lakes, the atmosphere, and oceans. Very small particles have unique reactive surface properties compared to larger particles of the same composition. To take advantage of these properties, manufactured (engineered) nanoparticles are used in cosmetics, water treatment, electronics, and pharmaceuticals. As their use in products increases, the risks and impacts of engineered nanoparticles on the environment is a concern. For example, researchers are investigating if nanoparticles taken up and distributed to plant tissues or absorbed by animals can negatively impact the organism.

---

[1] The half unit increments are because the +½ bond valence distribution that is calculated by dividing the cation charge of Fe^{3+} and Al^{3+} by the coordination number, which is six for octahedrally coordinated cations.

Nanoparticles have huge specific surface areas, making their solid-solution interface very large. Allophane, for example, is formed in soils where dissolution of silicon and aluminum bearing minerals is rapid (e.g., volcanic ash-influenced soils) and is a spheroid nanoparticle with diameters of 3.5–5 nm (a nanoball) and a specific surface area of as much as 900 $m^2\ g^{-1}$. Functional groups at the edges of nanoparticles react with solutions, affecting redox reactions, acidity reactions, chemical adsorption reactions, interactions with microbes, and aggregation. Nanoparticle minerals (nanominerals) have more functional groups on their exterior per mass or volume than larger minerals of the same composition, making them very reactive (**Figure 9.8**).

One of the most interesting aspects of nanoparticles is that, in many instances, their surfaces do not react the same as the larger particles of the same type; even when normalized for surface area. The reactive groups on nanoparticle surfaces may have distinct molecular and quantum chemistries that changes their reactivity. The unique chemical processes occurring on nanoparticle surfaces include faster or slower reactions, different adsorption reactions, different protonation and deprotonation reactions (acidity), varying tendency to accept or donate electrons, and distinct mineral solubilities. The unique reactivity originates from the different molecular energies of the surface molecules created by the high surface functional group density compared to the bulk mineral atoms

(**Figure 9.8**), and the structural strain and disorder created at the surfaces of the nanoparticles.

In soils, many nanoparticles aggregate or coat larger minerals. For example, sand and silt grains are often coated by iron oxide nanoparticles. Given their unique reactivity and abundance, including engineered nanoparticles, a better appreciation of nanoparticle behavior is critical to understanding and predicting soil and ecosystem processes, and is an area of active research.

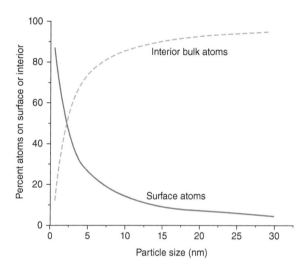

Figure 9.8 Calculated surface atom to bulk atom ratios for particles as a function of particle size. Adapted from Stark et al. (1996); with permission.

9.4 Point of zero charge of variable charged surfaces

The acid base properties of surface functional groups are predicted using surface-pK_a values called **intrinsic surface constants**. However, unlike aqueous acids and bases, surface-pK_a values are difficult to determine accurately, which makes using them to predict surface charge as a function of pH more complex than acidity constants. Instead, researchers experimentally measure how surface charge changes with pH. In this measurement, the pH value where the net charge on a mineral is zero ($\sigma_p = 0$) is referred to as the **point of zero charge (PZC)**. When the pH is above the PZC, the surface charge is negative, and when the pH is below the PZC, the surface charge is positive. In this way,

PZC is similar to a pK_a because it describes the protonation and deprotonation of the acid functional groups on the mineral surface. PZC values of common soil oxides are listed in Table 7.7.

One method to measure PZC is to place an aqueous suspension of particles between two oppositely charged electrodes and track particle movement (electrophoretic mobility) as a function of suspension pH (**Figure 9.9a**). The pH where the particles do not move (are not attracted to either the cathode or anode) is the **isoelectric point** (IEP), and is equivalent to the PZC of the mineral.

Another method to measure PZC of a mineral surface is to measure anion and cation adsorption as a function of pH (**Figure 9.9b**). The cations and anions used for charge measurements are typically

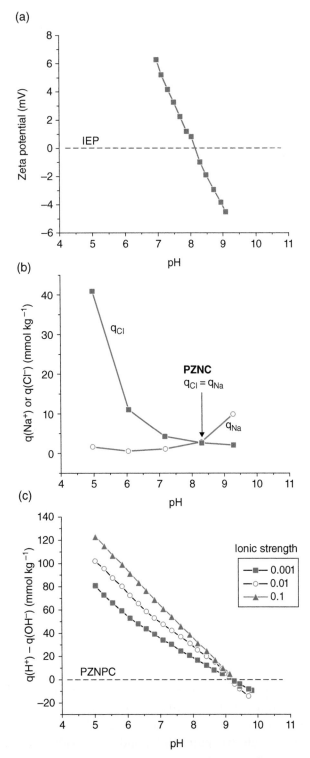

outer-sphere adsorbed ions such as Na^+ and Cl^-, called **index ions**. The point where the adsorbed cation charge equals the adsorbed anion charge is the point of zero net charge (**PZNC**). At the PZNC, cation exchange capacity equals the anion exchange capacity.

Instead of measuring adsorption of index cations and anions as a function of pH, H^+, and OH^- adsorption can be measured in a titration experiment to determine the PZC. The point where the number of protons adsorbed equals the number of hydroxides adsorbed is the point of zero net proton charge (**PZNPC**) (**Figure 9.9c**).

Charge properties (electrical potential) of variable charge surfaces change depending on the ionic strength of the solution used in the experiment, and thus the measured PZNPC varies. To account for the effects of solution ions on surface charge, PZNPC is measured at several ionic strengths. The point where the titration curves cross is the point of zero salt effect (**PZSE**), or the common intersection point (**Figure 9.9c**). In most cases, PZNPC is close to the PZSE (e.g., within 0.2–0.5 pH units for the three curves in **Figure 9.9c**).

Although each pH-dependent surface charge measurement uses different methods (e.g., PZNC, PZNPC, IEP), all of the measurements are *typically* within one pH unit of each other, and describe operationally defined PZC values for mineral surfaces. Values reported in the literature for a type of mineral, however, are variable because minerals used by different researchers are prepared differently. Different synthesis methods cause different amounts and types of surface functional groups on the mineral's surfaces. Also, whether the minerals are natural or synthetic influences surface charge characteristics. Use of different aqueous ion types and solid-solution ratios in the surface charge characterization experiments also causes variation. The reported PZCs of goethite, for example, range from 7.2 to 9.4. Such variability is a result of varying goethite sample

Figure 9.9 PZC of goethite as measured by electrophoretic mobility (IEP) (a); Na^+ and Cl^- adsorption (PZNC) (b); and proton or hydroxide adsorption (PZNPC) (c). The proton and hydroxide adsorption amounts (panel c) were measured at three ionic strengths to account for the effects of ionic strength on surface potential. Panels (a) and (b) adapted from Appel et al. (2003); reproduced with permission of Elsevier. Panel (c) adapted from Zeltner and Anderson (1988). Reproduced with permission of American Chemical Society.

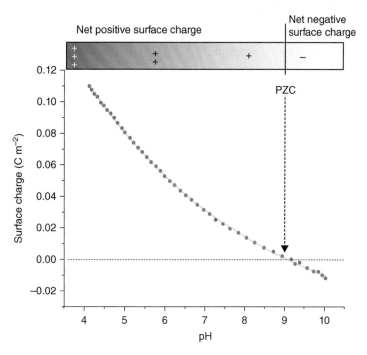

Figure 9.10 Net surface charge of goethite as a function of pH. The bar at the top shows the sign of the charge and the relative magnitude of the charge. pH further from the vertical line (PZC) causes greater surface charge. Surface charge data adapted from Gaboriaud and Ehrhardt (2003), with permission.

properties or the use of different measurement methods. Most reported PZC values for goethite are within the range of pH 8 to 9.

Mineral PZC values are reference points that show net charge behaviors of surfaces as a function of solution pH, and are useful for understanding colloid and adsorption phenomena. **Figure 9.10** shows a bar that spans the range of pH and indicates the surface charge as a function of pH. The vertical line in the bar represents the PZC. The further away from the PZC within the bar, the greater the net positive or negative charge. As pH of a solution in contact with a mineral surface decreases below the PZC, the surface becomes more positively charged (**Figure 9.10**). Conversely, as pH increases above the PZC, the surface becomes more negatively charged. At some pH above and below the PZC, the surface has maximum negative and positive surface charge, which is equivalent to the number of charged functional groups or surface sites on the mineral. This graphical concept is used to discuss pH-dependent adsorption processes in Chapter 10.

Electrophoretic mobilities and IEPs for several minerals are shown in **Figure 9.11**. The curves show

the surface charge characteristics for specific minerals and experimental conditions. Variances in mineral properties and experimental conditions can cause dramatic shifts in the curves, up to a pH unit or more. Additionally, the curves represent pure mineral phases that do not necessarily capture the actual behavior of soil minerals. Soil minerals are often impure, coated with other minerals and SOM, and have other ions adsorbed; factors that change PZC and surface charge characteristics. Despite these caveats, the data in **Figure 9.11** are useful to understand the relative charges of different soil minerals as a function of pH. For example, at pH values relevant to soils, the net charge on kaolinite is always negative. As shown in **Figure 9.4**, the edge charge on kaolinite is due to the silanol and aluminol functional groups that protonate or deprotonate as pH increases or decreases. Edge charge of other phyllosilicate minerals would be similar to kaolinite edges. Gibbsite, on the other hand, is positively charged at most soil pH values (PZC ~ 9). The weak acid character of the aluminol functional groups on gibbsite cause them to remain protonated until very high pH. Allophane and ferrihydrite surface charges can be positive or negative in typical soil pH

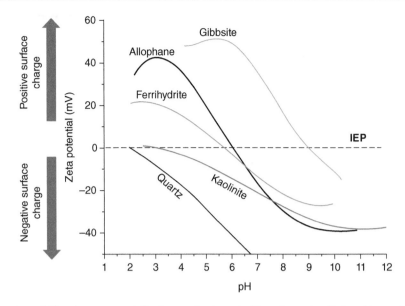

Figure 9.11 Electrophoretic mobility (zeta potential) of several minerals. Data from: kaolinite- (Au and Leong, 2016); gibbsite- (Adekola et al., 2011); quartz- (Fuerstenau and Jia, 2004); allophane- (Toyota et al., 2017); and ferrihydrite- (Li et al., 2015).

conditions, depending on whether the soil is acidic, neutral, or alkaline.

9.5   pH-dependent charge of SOM

SOM has higher specific surface area and adsorptive capacity per unit mass than phyllosilicate minerals. The reported specific surface of SOM varies, depending on how it is measured, but may be as high as 900×10^3 m^2 kg^{-1}. Cation exchange capacity of SOM ranges from 600 to 3500 mmol(+) kg^{-1}, depending on pH. At neutral to alkaline pHs, CEC of SOM is much greater than high charge clay minerals (vermiculite and smectite). **Figure 9.12** shows the pH-dependent CEC of SOM and clays in soils (SOM was not extracted). The slope of the SOM-CEC (506 mmol(+) kg^{-1} per pH unit) is approximately 10 times greater than the clay mineral CEC change (40.2 mmol(+) kg^{-1} per pH unit), indicating that the SOM has a much greater pH-dependent charge. Because of this behavior, SOM is a significant contributor to CEC in soils with pH greater than 6–7, even when much less SOM is present than clay minerals.

The negative charge (and hence the CEC) of SOM is due to the dissociation of H$^+$ from functional groups,

particularly weak acid **carboxylic** and **phenolic** groups. The acidic behavior of these two functional groups on SOM has been traditionally studied on humic and fulvic acids. Although, humic substances are operationally defined materials, their acidic behaviors are good models for the acidic functional groups in SOM. **Figure 9.13** shows a typical titration curve for soil humic acid. Fulvic acid has a similar shape. The slopes of humic and fulvic acid titration curves do not change as sharply as do the titration curves of monomeric acids because the organic acid functional groups on SOM have a range of molecular structures, and thus varying pK$_a$ values.

The *average* apparent **conditional acid dissociation constants** (pK$'_1$ and pK$'_2$) for humic and fulvic acids are listed in **Table 9.3**. They are apparent because they are the observed pK$_a$s for macromolecular species with various types of acid functional groups, and they are conditional because they are uncorrected for ionic strength effects. However, the reported standard deviations of several independent measurements at different ionic strengths suggest that ionic strength changes the constants by less than a third of a log unit. Weak acid dissociation behavior and mathematical relationships of pK$_a$ are discussed in Chapter 4 (e.g., Figure 4.5 and 4.6).

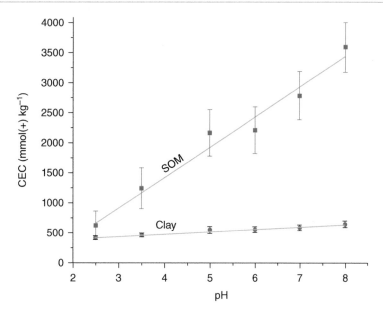

Figure 9.12 Mean cation exchange capacity (CEC) as a function of pH for SOM and soil clays. Relationships were derived from multiple regression of CEC as a function of soil clay and SOM content from 60 Wisconsin soils at six pH values. Error bars are standard errors of mean predicted CEC. Slope of regressed lines represents changes in CEC per unit pH: slope(clay) = 40.2 ± 2.4; slope(SOM) = 506 ± 36. Data from Helling et al. (1964). Reproduced with permission of ACSESS.

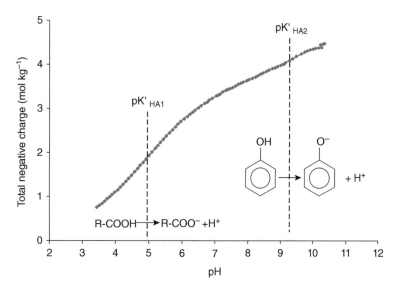

Figure 9.13 Titration of a purified humic acid. pK'_{HA1} represents the apparent acidity of the carboxylic acid functional groups on the humic acid, and pK'_{HA2} represents the apparent acidity of the phenolic acid functional groups on the humic acid. Adapted from Avena et al. (1999). Reproduced with permission of Elsevier.

The dissociation of carboxyl and phenol groups yields about 85 to 90% of the negative charge of SOM. Many carboxylic groups are sufficiently acid to dissociate at pH values below 6:

$$R-COOH = R-COO^- + H^+ \tag{9.5}$$

where R represents any number of organic species whose differing electronegativity alters the tendency for H^+ to dissociate, and thus changes the pK_a. As pH of a system increases beyond pH ~4, weaker acid carboxylic groups dissociate. At pH greater than 8, proton dissociation of phenolic and other very weak acids

Table 9.3 Average *apparent* pK$_a$s for proton dissociation from humic substances calculated from titration of 47 separate humic or fulvic acids. From Lenoir and Manceau (2010).

Humic substance	pK$'_1$	pK$'_2$
Humic acid	4.49 ± 0.18	9.29 ± 0.38
Fulvic acid	4.18 ± 0.21	9.29 ± 0.33

occurs. At very high solution pH, dissociation of H$^+$ from enolic (R-C-OH) and possibly other groups also contribute to the negative charge of SOM.

At normal soil pH values, SOM fractions are unlikely to possess a *net* positive charge. Weak base groups such as R-OH$_2^+$ and R-NH$_3^+$ can have positive charges, but there are much fewer of these weak bases than the weak acids, and thus the overall charge on SOM is negative.

9.6 Hydrophobic regions of soil organic matter

Soil organic matter has large regions of uncharged and nonpolar surfaces that are **hydrophobic**, and are water repellent. The degree of hydrophobicity varies as a function of the element content and structural arrangement of the atoms. Elements such as oxygen, sulfur, and nitrogen in SOM molecules may have nonionic molecular coordination, but have some polar charge because of differences in electronegativity from the carbon atoms. Uncharged mineral surfaces, such as the basal planes of kaolinite also have some hydrophobic character, but much less than SOM.

The most hydrophobic regions of SOM are the saturated aliphatic hydrocarbons and polyaromatic regions (**Figure 9.14**). To lower the free energy of the system, hydrophobic regions of the SOM molecules associate with each other in supramolecules, or pseudo-micelles, making large three-dimensional hydrophobic and hydrophilic regions. Figure 8.14 shows a two-dimensional model of the hydrophobic and hydrophilic regions in SOM.

Many pesticides and other organic chemicals in soils are hydrophobic, and nonpolar chemical sorption is greatly influenced by the amount of SOM present. Adsorption of hydrophobic chemicals to SOM occurs

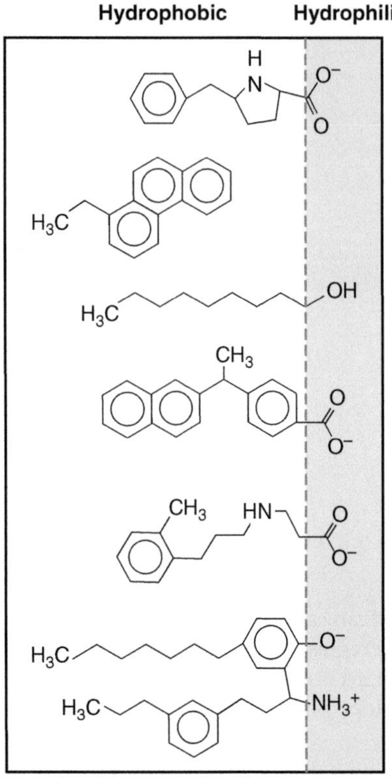

Figure 9.14 Examples of organic molecules that have hydrophobic and hydrophilic parts.

via van der Waal bonds and some weak dipole bonds (Figure 2.5). Although weak compared to covalent and ionic bonds, the multiplicity of the bonds between a hydrophobic chemical and hydrophobic SOM surfaces make the combined overall bond strength large. Pesticide adsorption is discussed in Chapter 10.

9.7 Summary of important concepts in soil surface charge

One of the most important characteristics of soils is the interactions between chemicals in solution and mineral and organic matter surfaces. Some minerals have high surface charge, which is dependent on mineral properties and can be positive or negative, and either permanent or pH-variable. Many clay minerals have large permanent negative charges, and thus have high CEC. Oxide minerals in soils have pH-variable charge surfaces, and adsorb anions at low pH and cations at high pH. The pH-dependent charge characteristics of mineral surfaces is characterized by measuring the point of zero charge (PZC). Soil organic matter also has variable-charged surfaces, but overall is negatively charged due to an abundance of carboxylic and phenolic acid functional groups on the macromolecular SOM structure that deprotonate as pH exceeds the pK_as of the functional groups. Uncharged SOM regions and mineral surfaces are hydrophobic and attract nonpolar molecules from solution.

A basic categorization of soils is based on soil charge characteristics. Permanent-charge soils have an abundance of high CEC clay minerals in them and occur in moderately weathered soil environments. Variable-charged soils are dominated by iron and aluminum oxides or kaolinite and occur in environments that have a high amount of weathering. The charge on mineral surfaces affects adsorption of chemicals, aggregation and dispersion, and shrink and swell behavior of soils. These soil properties are intrinsic to soil health because they impact nutrient and pesticide availability, contaminant mobility, soil salinity, soil acidity, water-holding capacity, gas permeability, and soil erosion. Thus, there is great interest in understanding and predicting soil charge characteristics. These topics are discussed further in Chapters 10 and 11.

Questions

1. Assuming that layer silicates and organic matter exist independently in a soil, calculate a reasonable cation exchange capacity of a soil containing 40% montmorillonite and 3% organic matter (assume neutral pH). *Hint:* Use estimated CEC values for SOM from **Figure 9.12** and mineral CEC values reported in Chapter 7. Repeat the calculation for a soil that contains 40% kaolinite and 3% organic matter.

2. Describe an experiment to measure PZC on the mineral hematite. What does this mean? Would you expect a cation such as Zn^{2+} to adsorb at pH 5?

3. Write a total charge expression (σ_p) that includes the relevant surface charge components for the following systems:
   a Na adsorbed on montmorillonite
   b Na adsorbed on goethite
   c Pb adsorbed on montmorillonite
   d Pb adsorbed on goethite
   Describe the relative magnitude of the different charge components in the four scenarios listed above.

4. Describe why protonation behaviors of variable charge mineral surfaces, such as goethite, cannot be theoretically described using a single acidity constant.

5. The *average* acidity constant of carboxylic acid functional groups on humic acids is reported as 4.49. What does this mean? Why is the acidity constant an *average?*

6. Why might PZNPC and PZSE be different?

7. If a soil is composed of 18% clay-sized minerals (of which 4% are oxides) and 4% SOM, draw the estimated surface charge as a function of pH (make y-axis values relative quantities).

8. What are the molecular forces involved in partitioning of chemicals to hydrophobic regions of SOM?

9. Why is it important to list the pH of a soil when describing CEC?

10. What soil conditions promote large AEC? Describe the soil components involved in AEC.

11. How can an anion be adsorbed onto montmorillonite?

12. At low pH, why don't amine functional groups on SOM cause the net surface charge on SOM to be positive?

13. When pH equals PZC, what is the charge on surface functional groups on mineral surfaces?

14. What role does charge separation distance have on adsorption (consider Coulomb's law)? Discuss relevance to soil minerals (see Figure 9.1).

15. Draw a graph similar to the one in **Figure 9.2** for an Oxisol (do not put numbers on y-axis, just relative magnitudes).

Bibliography

Adekola, F., M. Fedoroff, H. Geckeis, T. Kupcik, G. Lefevre, J. Lutzenkirchen, M. Plaschke, T. Preocanin, T. Rabung, and D. Schild. 2011. Characterization of acid-base properties of two gibbsite samples in the context of literature results. Journal of Colloid and Interface Science 354:306–317.

Appel, C., L.Q. Ma, R.D. Rhue, and E. Kennelley. 2003. Point of zero charge determination in soils and minerals via traditional methods and detection of electroacoustic mobility. Geoderma 113:77–93.

Au, P.I., and Y.K. Leong. 2016. Surface chemistry and rheology of slurries of kaolinite and montmorillonite from different sources. Kona Powder and Particle Journal:17–32.

Avena, M.J., L.K. Koopal, and W.H. van Riemsdijk. 1999. Proton binding to humic acids: electrostatic and intrinsic interactions. Journal of Colloid and Interface Science 217:37–48.

Fuerstenau, D.W., and R.H. Jia. 2004. The adsorption of alkylpyridinium chlorides and their effect on the interfacial behavior of quartz. Colloids and Surfaces a-Physicochemical and Engineering Aspects 250:223–231.

Gaboriaud, F., and J. Ehrhardt. 2003. Effects of different crystal faces on the surface charge of colloidal goethite (alpha-FeOOH) particles: An experimental and modeling study. Geochimica Et Cosmochimica Acta 67:967–983.

Helling, C.S., G. Chesters, and R.B. Corey. 1964. Contribution of organic matter and clay to soil cation-exchange capacity as affected by the pH of the saturating solution. Soil Science Society of America Journal 28:517–520.

Lenoir, T., and A. Manceau. 2010. Number of independent parameters in the potentiometric titration of humic substances. Langmuir 26:3998–4003.

Li, F., D. Geng, and Q. Cao. 2015. Adsorption of As(V) on aluminum-, iron-, and manganese-(oxyhydr)oxides: equilibrium and kinetics. Desalination and Water Treatment 56:1829–1838.

Schofield, R.K., and H.R. Samson. 1953. The deflocculation of kaolinite suspensions and the accompanying change-over from positive to negative chloride adsorption. Clay Minerals Bulletin 2:45–49.

Stark, J.V., D.G. Park, I. Lagadic, and K.J. Klabunde. 1996. Nanoscale metal oxide particles/clusters as chemical reagents. Unique surface chemistry on magnesium oxide as shown by enhanced adsorption of acid gases (sulfur dioxide and carbon dioxide) and pressure dependence. Chemistry of Materials 8:1904–1912.

Toyota, Y., M. Okamoto, and S. Arakawa. 2017. New opportunities for drug delivery carrier of natural allophane nanoparticles on human lung cancer A549 cells. Applied Clay Science 143:422–429.

Zeltner, W.A., and M.A. Anderson. 1988. Surface-charge development at the goethite aqueous-solution interface – effects of CO_2 adsorption. Langmuir 4:469–474.

Zhou, Z.H., and W.D. Gunter. 1992. The nature of the surface-charge of kaolinite. Clays and Clay Minerals 40:365–368.

10 ADSORPTION PROCESSES IN SOILS

10.1 Introduction

Sorption and **desorption** of chemicals are important soil processes describing the interaction of chemicals with soil solids. Sorption is removal of a chemical from aqueous solution by partitioning onto mineral or soil organic matter surfaces. Desorption is the reverse of sorption– the release of a sorbed chemical from the solid to the solution. By maintaining a distribution between the solid surface and solution, soil particles aid in the regulation of chemical concentrations in the soil pore water, facilitating chemical availability for plants, microbes, and animals.

There are several modes of sorption: **adsorption** of an ion to a charged surface by electrostatic attraction; adsorption by formation of bonds between a chemical and surface functional groups; formation of **multinuclear surface precipitates** distinct from the bulk mineral composition and structure; **hydrophobic partitioning** of an organic chemical to mineral and

SOM surfaces and in pore spaces; and physical entrapment (or <u>ab</u>sorption) of a chemical in pores of minerals or aggregates. Adsorption reactions are the processes most often associated with uptake of chemicals in soils. The two main types of adsorption reactions are:

1 *Outer-sphere adsorption.* At least one water molecule remains between the chemical and the surface.
2 *Inner-sphere adsorption.* The chemical makes a direct covalent or ionic bond with the mineral surface.

These two adsorption mechanisms are analogous to outer-sphere and inner-sphere *aqueous* ion complexation concepts presented in Chapter 4. Numerous variations of inner- and outer-sphere adsorption exist. The specific type of reaction mechanism depends on chemical and surface properties. **Figure 10.1** illustrates the different types of cation adsorption that can occur on clay mineral surfaces. Most of the adsorption on expandable clays occurs via outer-sphere

Soil Chemistry, Fifth Edition. Daniel G. Strawn, Hinrich L. Bohn, and George A. O'Connor.
© 2020 John Wiley & Sons Ltd. Published 2020 by John Wiley & Sons Ltd.

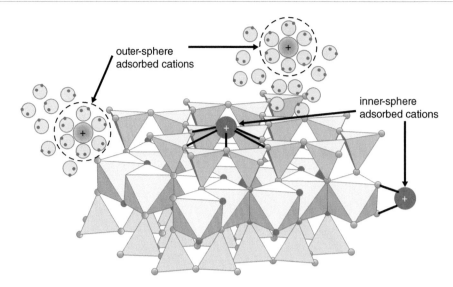

Figure 10.1 Adsorption mechanisms of cations on clay minerals. On the basal plane, some surface cations are adsorbed as inner-sphere complexes when they dehydrate. Most cations adsorb on clay minerals as outer-sphere complexes with their hydration sphere intact.

complexation in the interlayers; except for K^+, NH_4^+, and Cs^+ adsorption on illite and vermiculite surfaces, which form inner-sphere bonds to the clay layers because these ions have weak hydration spheres.

10.1.1 Outer-sphere adsorption

Outer-sphere adsorbed ions are located on the mineral surfaces, but they do not form direct chemical bonds with the surface functional groups. Instead, they maintain their hydration spheres. Electrostatically held outer-sphere adsorbed ions are the major reservoir of many plant nutrients, including Ca^{2+}, Mg^{2+}, Na^+, NO_3^-, K^+, and SO_4^{2-} (SO_4^{2-} is intermediate, and may form inner-sphere bonds on some soil mineral surfaces). Cations are electrostatically adsorbed on negatively charged surfaces, especially clay minerals, while anions are primarily adsorbed on positively charged oxide mineral surfaces. Factors that affect the outer-sphere electrostatic adsorption reactions are: type of surface charge, the valence of the ion, ion hydration properties (ion size), and type and concentration of ions in soil solution. Electrostatic adsorption reactions are fast and reversible, occurring on time scales of seconds to minutes, or faster.

On oxide surfaces and SOM, outer-sphere adsorbed ions exist as hydrated ions attracted to the charge of the **surface functional groups** (oxide shown in **Figure 10.2**). However, in this case, the charge of the surface functional groups on oxides and SOM is not strong enough to dehydrate outer-sphere adsorbed ions.

On minerals with a *delocalized* surface charge, such as phyllosilicate interlayers, adsorbed cations maintain their hydration spheres (**Figure 10.1**), and are mobile within the surface water film. Together, the surface charge (σ_p) and the charge from the swarm of ions in the film of surface water (σ_d) make up the **diffuse double layer** (DDL) (**Figure 10.3**). Ions adsorbed via outer-sphere mechanisms *directly on* surfaces with their hydration spheres intact are *assigned* a distinct surface charge (σ_o) to distinguish them from ions adsorbed in the diffuse double layer (σ_d, e.g., see Figure 9.3); however, at the molecular scale these two adsorption modes are not that different because they both have at least one water molecule between the surface and the ion.

Cation exchange is primarily an outer-sphere adsorption reaction and predominates in soils that have high clay mineral content. The **cation exchange capacity (CEC)** is the amount of cations a soil can adsorb, which is equal to the sum of negative charges on the soil particle surfaces. Soil clays and SOM provide most of a soil's CEC. Values range from less than 10 mmol(+) kg^{-1} up to 500 mmol(+) kg^{-1} and depend on the amount and type of clays and SOM.

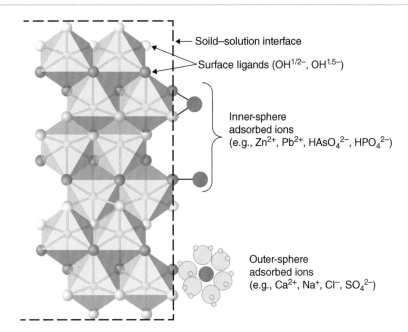

Figure 10.2 General adsorption mechanisms of ions on the surface of an oxide mineral.

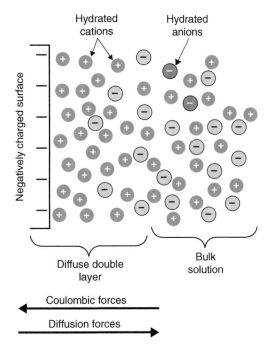

Figure 10.3 A diffuse double layer on a negatively charged mineral surface.

Anion exchange capacity (AEC) primarily describes outer-sphere adsorbed anions, although it is not as exclusive as CEC. AEC mostly comes from variable charged mineral surfaces that are positively charged when pH is less than PZC. AEC values of soils vary with pH, but are generally 5–50 mmol(–) kg^{-1}.

10.1.2   Inner-sphere adsorption

Inner-sphere adsorption reactions involve formation of covalent bonds between a chemical and a mineral or SOM surface. Surface complexation, inner-sphere adsorption, **chemisorption**, and **specific adsorption**, all refer to the same type of reaction in which a solution phase chemical adsorbs by forming bonds with surface functional groups. The transition metal cations, and Al^{3+}, Be^{2+}, Pb^{2+}, lanthanides, actinides, and oxyanions borate, phosphate, arsenate, and selenite form inner-sphere bonds with the functional groups that exist on aluminosilicate edges and oxide and soil organic matter surfaces.

2:1 phyllosilicates have minimal inner-sphere adsorption reactions in soils because most of their surface area consist of basal plane oxygen atoms in the interlayers that have permanent charge, but do not form chemical bonds with chemicals. Kaolinite is an exception because it lacks interlayer access and thus ions can only adsorb on edges where unsatisfied structural bonds create silanol and aluminol functional groups that can form inner-sphere bonds.

Figure 10.2 shows examples of mineral functional groups that participate in inner-sphere adsorption reactions (see also Figures 9.4 and 9.6). Amorphous materials have more surface functional groups than crystalline materials because they have greater specific surface areas, and more edge sites and unsatisfied bonds that create inner-sphere adsorption capacity. For example, allophane and ferrihydrite have much greater adsorption capacity than kaolinite and goethite.

Inner-sphere adsorption reactions can be predicted from the chemical properties of the adsorbing ions, such as water solubility, ionization energy, and ability to form covalent bonds. Mineral surface properties, such as charge, surface area, and ligand types are also important determinants in inner-sphere bond formation. Because functional group charge varies with pH, inner-sphere adsorption reactions are pH-dependent. Adsorption via inner-sphere reactions are rapid at first but continue at ever-slower rates for long periods. The slow continuing adsorption may be due to diffusion into the interior of the particles through small pores, formation of multi-dentate bonds, or formation of multi-nuclear sorption complexes (e.g., surface precipitates).

Chemicals may also adsorb on surfaces through an ion bridge linking the ion and a surface functional group. For example, a phosphate ion adsorbed on a mineral surface can complex with a Pb^{2+} ion, creating a **ternary adsorption complex** (**Figure 10.4**). Ternary surface complexes facilitate cation adsorption to positively charged surfaces; or the opposite, anion adsorption to negatively charged surfaces.

10.1.3 Adsorption of non-charged chemicals to soil particles

Many organic chemicals such as pesticides and other chemicals of emerging concern are nonionic (nonpolar), and thus hydrophobic. Nonionic chemicals interact with soil minerals via hydrogen bonds and van der Waal forces. Interactions of nonionic organic chemicals are similar to those illustrated in Figure 8.14. Soil organic matter is the most reactive soil component for adsorption of hydrophobic chemicals because it contains hydrophobic regions. Some soil mineral surfaces are also uncharged (e.g., the basal plane of kaolinite),

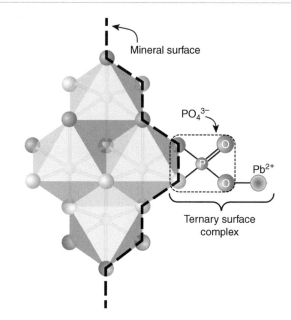

Figure 10.4 Ternary sorption complex showing Pb^{2+} and phosphate adsorption on an oxide mineral surface.

and facilitate some degree of hydrophobic chemical partitioning.

Inorganic molecules can also be uncharged because of protonation or hydrolysis. For example, $Pb(OH)_2^0$, H_4SiO_4, and $B(OH)_3$ are nonionic inorganic chemicals that occur in soil solutions. Although some hydrophobic partitioning may occur for such species, it is not a major adsorption mechanism. The lack of electrostatic interactions between nonionic inorganic chemicals and charged surfaces does not mean the uncharged species will not adsorb. An important distinction between inner-sphere and outer-sphere adsorption mechanisms is that inner-sphere bonds are not solely dependent on electrostatic interactions. Thus, inorganic nonionic molecules can adsorb to minerals and SOM by forming covalent bonds with surface functional groups.

10.1.4 Desorption

Much of the emphasis in soil chemistry is on adsorption of ions onto soil surfaces. **Desorption** reactions are the reverse of adsorption reactions and are equally important to the availability of chemicals. Desorption reactions release chemicals that are bonded on the surfaces of solid-phases to the soil solution. In this respect, desorption reactions are similar to dissolution reactions. Desorption reactions, however, are typically much faster than dissolution reactions, and thus, in many cases, are the reaction processes responsible for maintaining chemical concentrations in soil solution. For example, when a plant root *ab*sorbs ions from the soil solution, or ions are leached out of the soil profile, adsorbed ions will desorb to maintain equilibrium between the solid and solution. Plants release protons and hydroxides that promote desorption via exchange reactions.

Cation exchange reactions involve adsorption and desorption of cations on charged surfaces. The exchange is stoichiometric with respect to the amount of charge adsorbed and desorbed. Cation exchange is reversible, so adsorbed cations are readily desorbed to reestablish equilibrium when a cation comes into solution that preferentially adsorbs on the exchange sites.

Inner-sphere adsorbed ions are often strongly adsorbed, and thus may not readily desorb. This may cause the inner-sphere adsorption reactions to be **irreversible** or **hysteretic**. Irreversibility implies the reaction does not occur in the opposite direction, and hysteresis implies the reaction goes in the reverse direction, but does not reach the same equilibrium condition, or does not have the same rate of reaction. Causes of irreversibility or hysteresis are slow reactions (e.g., slow diffusion), existence of activation energy (see Figure 1.13), or occurrence of secondary reactions (e.g., precipitation).

When ions that have different adsorption mechanisms exchange on a surface, adsorption/desorption hysteresis commonly occurs. For example, Pb^{2+}-Ca^{2+} exchange on an oxide or SOM surface may be hysteretic because inner-sphere complexed Pb^{2+} cations are more strongly held than the outer-sphere complexed Ca^{2+} cations. Thus, the Pb^{2+} can displace the adsorbed Ca^{2+}, but the reaction is not reversible; that is, increasing Ca^{2+} concentration in solution may not displace the adsorbed Pb^{2+}.

Some cation exchange reactions on phyllosilicates are irreversible because interlayer collapse restricts accessibility to the adsorption sites. In this case, ions adsorbed in the interlayers are physically isolated from the bulk solution, and thus adsorbed irreversibly. For example, K^+ and NH^+ ions in the interlayer of vermiculite cause interlayer collapse, and create physical irreversibility, which is called **K^+- or NH_4^+-fixation**.

Another form of irreversible sorption occurs when surface precipitation occurs and isolates metal ions from the solution phase. Surface precipitation is a common adsorption mechanism for transition metal ions (further discussed in Chapter 11). Formation of hydrophobic micelles or supramolecules may also isolate chemicals from bulk solution and greatly restrict desorption or dissolution.

Data generated in adsorption/desorption experiments to measure the reversibility of boron adsorption on a soil are shown in **Figure 10.5**. Following adsorption *equilibrium,* the soil suspensions were diluted with boron-free solution. If boron adsorption/desorption was reversible, upon dilution, boron would desorb and the amount on the solid would decrease – the solid–solution trend would follow the adsorption line in decreasing concentration. However, the data in **Figure 10.5** show that the amount of boron remaining adsorbed following dilution is greater for a given solution concentration than occurred in the adsorption experiment (amount on solid is greater, and amount in solution is less). Thus, the reaction is hysteretic. The reasons for the boron hysteresis are unclear, and it appears to be a soil-specific phenomenon; that is, for

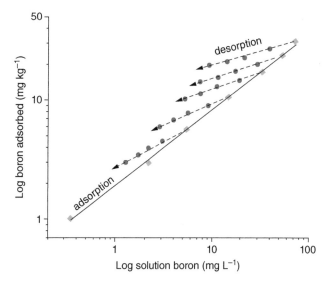

Figure 10.5 Boron adsorption and desorption in soil. Desorption was initiated by dilution of the adsorbed soil sample. Reversible boron reactions would retrace the adsorption trend when the sample is diluted. The desorption data are hysteretic because they do not fall on the adsorption curve. Samples were equilibrated for 23 hours (separate experiments showed that the reaction reached equilibrium within 12 hours). Adapted from Elrashidi and O'Connor (1982). Reproduced with permission of ACSESS.

some soils, boron adsorption is reversible. Because boron is a micronutrient, and is toxic at concentrations that only slightly exceed plant requirements, understanding the reversibility of boron adsorption in soils has major implications on plant nutrient management.

When evaluating the fate of chemicals in soils, both adsorption and desorption properties must be considered. Historically, adsorption reactions were studied more often than desorption reactions. In some cases, adsorption and desorption reactions are reversible, but in many others, assuming chemical availability based on adsorption reaction processes leads to errors because the reactions are not reversible, or are hysteretic. Reversibility depends on properties of the chemical and soil, and must be determined on a case-by-case basis.

10.2 Physical model of charged soil particle surfaces

The charge on a soil particle surface creates electrostatic forces that extend into the soil solution. The two-phase system is referred to as the solid–solution interface. Early descriptions of the solid–solution interface referred to it as a double layer because there are two sets of charges: surface charge and an oppositely charged layer of ions on the surface. The properties of the solid–solution interface affect ion adsorption and particle–particle interactions (i.e., flocculation and dispersion) – two very important processes in soils and several other disciplines, including water treatment, mining, and nanoparticle engineering, to name a few. Thus, there has been great interest in developing mathematical descriptions of the physical and chemical forces occurring at the solid–solution interface. While progress has been made, the assumptions about, and complexities of, the solid–solution interface render most models only approximate representations of true physical and chemical processes occurring in the system. However, understanding the surface charge models and the variables affecting them allows for insight and predicative ability of soil particle behavior.

10.2.1 Force of ion attraction to charged surfaces

In 1856, German physicist Hermon von Helmholtz developed a model of cation association with negatively charged surfaces on soil clays. He described a

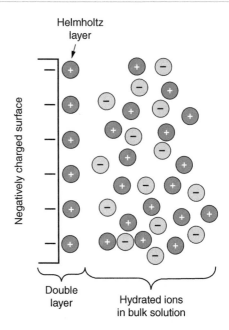

Figure 10.6 Distribution of cations and anions on a charged surface according to the Helmholtz model.

layer of exchangeable cations directly on the surface resulting in two slightly separated but oppositely charged layers, called a **Helmholtz double layer** (**Figure 10.6**). The model uses Coulomb's law and ion hydration properties to predict cation attraction behavior. The electrostatic attractive or repulsive force between an ion and a charged surface is described by **Coulomb's law**:

$$F = k\frac{qq'}{r^2} \tag{10.1}$$

where F is the force of attraction or repulsion in Newtons between two charges, such as an ion and a charged surface; q and q are the electrical charges in units of Coulombs; k is Coulomb's constant; and r is the separation distance of the charges. Coulomb's law predicts that the force of ion attraction to a charged surface increases with increasing ion charge, with increasing surface charge, and with decreasing distance between the surface charge and ion.

Ions that have small ionic radii have a greater density of charge per unit volume, which is the ionic potential (Figure 2.4). Such ions attract waters of hydration strongly and have large *hydrated radii* (Figure 2.8), causing them to be held less tightly than cations of smaller hydrated radii because the distance (r in **Eq. 10.1**) between the ion and the surface charge is greater, reducing the force of attraction. Ions with smaller hydrated radii, in contrast, approach surfaces more closely and have smaller separation distances between charges, resulting in greater attractive Coulombic force. Three relevant applications of Coulomb's law to cation adsorption are:

1 The force (F) between cations with higher valence and a charged surface is greater than cations with lower valence (e.g., $F(Ca^{2+}) > F(Na^+)$).
2 The force of adsorption between an ion and the permanent charge originating in the tetrahedral sheet of a clay mineral is greater than the force between an ion and the permanent charge originating in the octahedral sheet (i.e., $F_{tet} > F_{oct}$) because charge-separation distance (r) for the tetrahedral sheet is less than the octahedral sheet (see Figure 9.1).
3 The force between ions of the same charge and the surface is greatest for ions with the smallest hydrated radii because the separation distances (r) are less. Hydrated radius is inversely related to ionic radius (Figure 2.8). For example, the force of adsorption of Ca^{2+} is greater than that for Mg^{2+} because the hydrated radius of Ca^{2+} is less than that for Mg^{2+}.

10.2.2 The diffuse double layer

Georges Gouy and David Leonard Chapman independently advanced the double layer model in the early 1900s by introducing additional ion–ion and ion–surface forces into the physical description of the solid–solution interface. They proposed that Coulombic forces between the cations and the surfaces are opposed by diffusion forces that favor random distribution of ions. Thus, diffusion forces cause the cations attracted to negatively charged surfaces to disperse toward the bulk solution (**Figure 10.3**). The result of the electrostatic cation attraction and ion diffusion on negatively charged surfaces creates a *swarm* of cations in the aqueous phase near the charged layer called the **diffuse double layer** (DDL) (also called the electric double layer). The ions in the DDL are distinct from the **bulk solution** (the solution outside the DDL). The thickness of the DDL is equal to the distance over

which the cation concentration in solution exceeds the anion concentration. A DDL of anions is created for positively charged surfaces, however, in most discussions (especially for soils), the DDL is described in context of cation adsorption on negatively charged surfaces.

If the DDL contained only the amount of cations necessary to neutralize the mineral-associated charge (σ_p), the anion concentration within the DDL would be zero. Because diffusion continually drives anions from bulk solution toward the colloid surface, however, the total negative charge within the DDL is that of the anions plus the surface charge. DDL cations must equal the total negative charge density in the DDL (see Eq. 9.3). Cations in the DDL that neutralize the surface charge exchange with each other between the DDL and the bulk solution.

Guoy-Chapman theory makes three assumptions:

1 Exchangeable cations exist as point charges.
2 Colloid surfaces are planar and essentially infinite in extent.
3 Surface charge is distributed uniformly over the entire mineral surface.

These assumptions inaccurately describe actual systems, but the theory allows a semi-quantitative description of the DDL on particle surfaces that predicts the behavior of soil particles. DDL calculations can estimate the effects of ion charge and concentration on colloid mobility and flocculation, as well as forming a basis for estimating surface potentials needed to predict ion adsorption.

The Gouy-Chapman theory predicts the ion concentration at a specified distance (x) from a charged surface in the DDL as a function of **electric potential** (ψ_x), valence charge of the ion (z), and temperature. Electric potential is the energy in volts of a unit charge in an electric field. The electric potential of an ion near a charged surface varies with distance from the charged surface and the sign and magnitude of ion and surface charges, and cannot be directly measured. However, the total electric potential of ions next to a charged surface is directly related to the total **surface potential**, which is the total electric potential of charges if they are *directly* on the surface ($\psi_{x=0} \equiv \psi_0$). Surface potential is nonlinearly correlated to surface charge (**Figure 10.7**) and can be calculated from surface charge.

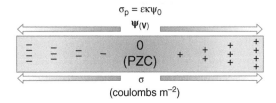

$$\sigma_p = \varepsilon \kappa \psi_0$$

$$\Psi_{(V)}$$

0
(PZC)

$$\sigma$$

(coulombs m^{-2})

Figure 10.7 Surface potential and surface charge are correlated. The equation shows the approximate relation between surface potential and surface charge that is valid for small potentials (parameters are defined in text).

An important outcome of the Gouy-Chapman theory is an equation that describes ion distribution in the DDL:

$$C_x = C_0 \exp\left(\frac{-ze\psi_0 \exp(-\kappa x)}{RT} \right) \tag{10.2}$$

where C_x is the concentration of an ion at a specified distance x from the charged surface, C_0 is the concentration of the ion in the bulk solution, z is the valence of the ion, e is the unit of electronic charge, R is the universal gas constant, T is the absolute temperature, and κ (the Greek letter *kappa*) represents the electrical properties of the ion solution. κ is calculated from the ionic strength (I) and temperature:

$$\kappa = \left(\frac{F^2 2000I}{\varepsilon RT} \right)^{1/2} \tag{10.3}$$

where ε is the relative dielectric constant for water (6.95×10^{-10}) and F is the Faraday constant. Units of κ are m^{-1}. **Equations 10.2** and **10.3** are mechanistic descriptions of ion distributions in the DDL. All of the parameters except surface potential can be measured or are known. Surface potential can be calculated from surface charge density (σ_p = charged sites per area (see Eq. 9.1), which can be measured. Using a relationship between surface potential and surface charge density (see, e.g., **Figure 10.7**), **Eq. 10.2** can be used to calculate ion distribution near a charged surface (C_x). The general relationship is

$$C_x = f\left(z, \sigma_p, T, C_0 \right) \tag{10.4}$$

Equation 10.4 indicates that the concentration of ions at a specified distance from the surface is a function of four measurable parameters: ion valence, surface charge, temperature, and bulk solution

concentration. The value of this relationship may not be readily apparent, however, the resulting quantitative description of ion distribution and surface potential is used to develop mechanistic adsorption models and to predict interactions of DDL from adjacent particles. Interactions of diffuse double layers control dispersion and flocculation, which are important factors for soil aggregation, erosion, and gas permeability.

Figure 10.8 shows the calculated concentrations of ions in the DDL next to a negatively charged particle for two bulk solution concentrations. The solution near the surface has an excess of cations and a deficit of anions; the y-axis is a log scale, thus anion concentration at the surface is 100 to 10000 times less than cation concentration at the two bulk solution concentrations. The two DDL concentration profiles in **Figure 10.8** show that increasing the bulk solution ion concentration reduces the tendency for diffusion away from the surface, and thus reduces (*shrinks*) the DDL ($C_1 > C_2$ and $C_3 > C_4$). The total ion charge *potential* (ψ_x) for both bulk solution concentrations shown in **Figure 10.8** is equal, and is related to the area under the curves, and is opposite the surface charge potential (ψ_0). For positively charged colloids, the behavior of cations and anions in the DDL and bulk solution are reversed; anions are attracted to the surfaces and cations are repelled.

The center of the charge distribution in the DDL, or *apparent* thickness of the DDL, is calculated using the inverse of κ (**Eq. 10.3**). For qualitative purposes, the **DDL thickness** in terms of the three measurable independent variables is reformulated as:

$$\text{DDL thickness} \propto \frac{T}{\sqrt{zC_0}} \tag{10.5}$$

where C_0 is the concentration of the ion in the bulk solution, z is the valence of the ion, and T is temperature of the solution. This form of the equation allows predicting how changing valence, concentration, and temperature affects the DDL thickness. **Table 10.1** gives double-layer thicknesses estimates for monovalent and divalent cations at three bulk solution salt concentrations on a constant charge flat surface. The values can be obtained by substituting ionic strength into **Eq. 10.3**. Two observations from the data in **Table 10.1** are: (1) the DDL thickness increases as electrolyte concentration decreases, and (2) the DDL thickness decreases by a factor 0.5 for valence increases

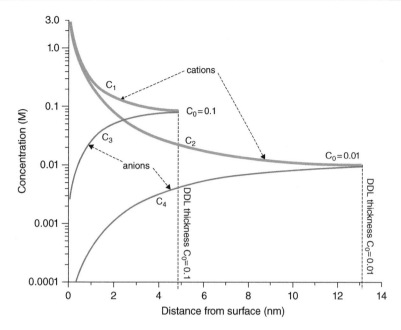

Figure 10.8 Distribution of monovalent cations and anions in a diffuse double layer near a negatively charged surface. Adapted from Boersma et al. (1972). Reproduced with permission of ACSESS.

Table 10.1 Thickness of DDL on a flat permanently charged particle for NaCl and $CaCl_2$ bulk solution concentrations at T = 298 K. Adapted from H. van Olphen (1964), with permission.

Bulk solution cations concentration $(mol(+) L^{-1})$	DDL thickness (nm)	
	Monovalent cations	Divalent cations
10^{-5}	100	50
10^{-3}	10	5
10^{-1}	1	0.5

Table 10.2 Effects of bulk solution changes on the thickness of DDL.

System change	Effect on DDL	Physical explanation
Increased valence (z)	DDL shrinks	Fewer ions are necessary for charge neutralization causing lower diffusion forces; more highly charged cations are attracted more strongly to the surface, causing DDL to contract.
Increasing bulk solution concentration (C_0) (or ionic strength)	DDL shrinks	The force of ions to diffuse away from a more concentrated DDL is lower because the bulk solution concentration is greater; thus, DDL contracts.
Increasing temperature (T)	DDL increases	Aqueous ions are moving faster and diffusion forces are greater; thus, DDL expands.

from +1 to +2. The effect of valence on the DDL thickness is called the **Schulze-Hardy Rule** after the nineteenth century scientist H. Schulze who discovered effects of valence on negatively charged particle flocculation in 1882, and W.B. Hardy who discovered the same effect on positively charged particles in 1900.

The DDL thicknesses on soil colloids are small compared to the diameters of soil pores, which are on the order of 1000 to 50000 nm. But in relatively dry soils, the DDL sizes are of the same order of magnitude as the much smaller water-film thicknesses. The effects of system changes on DDL thickness are summarized in **Table 10.2**. It is instructive to review these effects with the variables in **Eq. 10.5**.

10.2.3 Surface potential on variable charged surfaces

The Gouy-Chapman ion distribution equations and DDL calculations assume a constant-charge surface, such as the permanent charge on phyllosilicates clays ($\sigma_O = -\sigma_d$). Surface potential of variable charged minerals is also balanced by a DDL: $\sigma_p = \sigma_O + \boldsymbol{\sigma_H} + \boldsymbol{\sigma_{IS}} + \sigma_{OS} = -\sigma_d$, where bolded surface charges are pH-dependent charges. For variable charged surfaces, the relationship between surface potential and surface charge is derived using the Nernst potential equation, which calculates electrical potential as a function of magnitude of charge:

$$\psi_0 = \frac{2.3RT}{F}\left(PZC - pH\right) \tag{10.6}$$

The term (PZC – pH) represents the magnitude of pH-dependent charge, which is equivalent to, but not equal to σ_H (surface charge has units of mmol(±) kg^{-1} or Coulombs m^{-2}). There is a direct relation of surface charge to surface potential (**Figure 10.7**). For example, when soil solution pH = pH$_{PZC}$, the net surface charge is zero and the surface potential is zero. When surface potential is negative, surface charge is negative, and vice versa. Thus, **Eq. 10.6** shows that as pH goes above or below PZC, the sign of the surface potential changes. The *negative* surface charge in several soils as a function of PZC-pH is shown in **Figure 10.9**. As pH increases, the soils negative surface charge increases (compare with Figure 9.2).

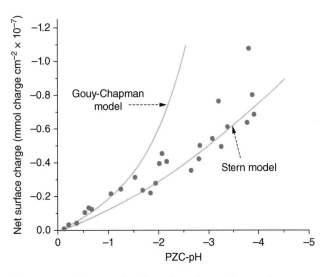

Figure 10.9 Measured and predicted negative surface charge determined by potentiometric titration of six different tropical soils as a function of how far the suspension pH is from the PZC (PZC-pH). The theoretical surface charge (solid lines) was calculated from the Gouy-Chapman and Stern models (see Section 10.2.4 for Stern model discussion). Adapted from Vanraij and Peech (1972).

To model surface charge (σ_O) from PZC-pH, surface-charge models are necessary. **Figure 10.9** shows the predicted surface charge for the Gouy-Chapman model and the Stern model. In the Gouy-Chapman model, **Eq. 10.2** is modified using electrochemical relationships not covered in this text; the result is a theoretical equation to predict surface charge from PZC-pH as included in **Eq. 10.6**. The model considers that the surface charge is all located in the diffuse layer, which is not a good model for the soils shown in **Figure 10.8**.

10.2.4   Stern modification of the Gouy-Chapman DDL theory

In 1924 Otto Stern, a twentieth-century Nobel laureate in physics, sought to improve the Gouy-Chapman DDL theory with a model that partitioned some of the ions directly on the surface. His model was similar to the Helmholtz layer depicted in **Figure 10.6** with the remaining surface potential satisfied by a diffuse layer. A satisfactory approximation of the Stern model can be made by assuming that the charge of the adsorbed ions (σ_{IS} and σ_{OS}) reduces the surface charge density of the mineral. The diffuse portion of the double layer thus forms to balance the surface potential extending from the Stern layer. **Figure 10.9** shows the calculations of surface charge as a function of PZC-pH on six tropical soils using the Stern model. Recall that PZC-pH implies the magnitude and sign of variable surface charge as used in **Eq. 10.6**. The Stern model more accurately predicts surface charge, supporting the idea that ions adsorbing directly on the mineral surface modify the surface potential that is then balanced by the potential of the diffuse layer.

Partitioning of ions in the Stern layer depends on their hydration energy. Ions with high hydration energy and large hydration spheres, such as Li$^+$, Na$^+$, and Mg^{2+} are not as likely to adsorb in the Stern layer as ions with lower hydration energies such as K$^+$, Cs$^+$, and Ca^{2+}.

10.2.5   Interacting diffuse double layers from adjacent particles

Fully expanded double layers are rare in field soils. Double-layer expansion normally is restricted by the limited thickness of water films on colloid surfaces or

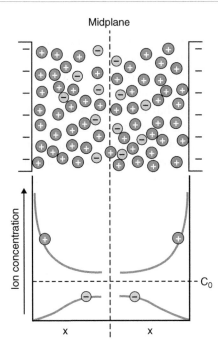

Figure 10.10 Illustration of ion distribution in two DDLs on interacting negatively charged surfaces. Compared to the bulk solution concentration (C_0), cation concentrations in the interlayer are elevated and anion concentrations are depleted.

by interactions with double layers on adjacent soil particles within aggregates. **Figure 10.10** represents such a restricted double layer, where the DDL of the two particles overlap. The distributions of cations, anions, and electrical potential in **Figure 10.10** are assumed to be symmetrical between two vertical charged surfaces. The cation concentration decreases with distance from the charged surface. At the midplane, the cation concentration remains greater than the bulk solution, causing an osmotic gradient. This, in turn, causes water imbibition, or swelling, of the colloid. Water imbibition continues until the tendency to swell is *balanced* by bonding forces between the two layers. The bonding forces are electrostatic interactions between the negatively charged surface, positively charged cations, and the adjacent negatively charged surface.

For most soils, swelling of the entire matrix is uncommon, but for montmorillonitic soils, such as Vertisols, it is significant. The swelling force for a single clay particle is relatively small, but, considering the number of clay colloids in a pedon, the cumulative force is large enough to move buildings and fracture foundations.

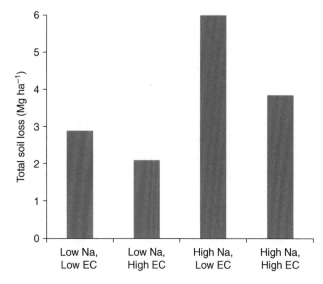

Figure 10.11 Effect of water quality on furrow soil loss during irrigation. Low Na refers to irrigation waters with relatively low Na^+ concentrations compared to Ca^{2+} and Mg^{2+}, and high Na refers to waters with greater Na^+ concentrations. EC is the electrical conductivity of the irrigation water (low = 0.5–0.7 dS m^{-1}; high = 1.7–2.1 dS m^{-1}). Data from Lentz et al. (1996).

Particle **flocculation** and **dispersion** can be predicted using DDL models of the electric fields of particle interactions, which is useful for predicting implications for the effects of irrigation water on erosion and runoff turbidity. Irrigation waters that have low concentrations of total dissolved solids (low ionic strength), or that are dominated by Na^+ ions, cause dispersion, making particles more susceptible to erosion. **Figure 10.11** shows erosion from a furrow irrigation system at high and low ionic strength (measured by electrical conductivity (EC)). The effects of low vs. high concentrations of Na^+, Ca^{2+}, and Mg^{2+} in the irrigation water are also shown. Erosion from the different irrigation waters corresponds to predicted DDL size, as interpreted in **Table 10.2**. Although Na^+ causes dispersion and erosion, differences in erosion between the Na^+-dominated high EC vs. low EC solutions are observed because of the effects of ionic strength on the thickness of the DDL (**Eq. 10.5**). The least amount of erosion occurs when Na^+ is low relative to Ca^{2+} and Mg^{2+}, and EC is high; this condition corresponds to irrigation water with high concentrations of Ca^{2+} and Mg^{2+}, such as hard water pumped from a calcareous aquifer.

10.3 Cation exchange on soils

Soil chemistry has historically stressed cation adsorption over anion adsorption. This bias occurred because most of the early soil chemistry research was conducted on soils of Europe and North America, which have more negatively charged minerals than positively charged minerals. Had early soil chemists studied soils of volcanic parent material or in highly weathered tropical soils, the bias might be toward anion exchange.

Cation exchange reactions are an important type of adsorption in soils. **Exchangeable cations** are those ions replaced by salt solutions flowing through soils (salt solutions are solutions containing dissolved ions). Salt solutions also exchange anions with soils, but in permanent charge soils, minerals are mostly negatively charged, and generally more cations exchange than anions. The major exchangeable cations are, in order of decreasing amounts, $Ca^{2+} > Mg^{2+} > K^+ > Na^+$. A typical agricultural loam soil contains about 20 000–30 000 kg ha^{-1} of exchangeable cations in its root zone (0.5 m depth). Roughly, 80% is Ca^{2+}, 15% is Mg^{2+}, 4% is K^+, and 1% is Na^+. As soil acidity increases, Al^{3+} and H^+ also occupy cation exchange sites on mineral surfaces. In some irrigated soils, exchangeable sodium increases, limiting the soil's use for crop productivity. Exchangeable cations and soil solution salts are very important to plant productivity, and can be manipulated by liming, irrigation, leaching, and fertilization. Hence, cation exchange has long been an important part of soil chemistry research.

In 1850, Thompson and Way conducted the first recorded studies of cation exchange in Rothamsted, England (see Figure 1.2). They showed that passing an

ammonium sulfate solution through soil columns leached calcium sulfate out of the soil. The predominant cation in the aqueous solution had changed from NH_4^+ to Ca^{2+} because of cation exchange. Thompson and Way showed that cation exchange was fast and reversible, and that the molar amount of ammonium ion retained was twice the amount of calcium released. Subsequent work has refined and supported Thomson and Way's findings, and has measured the cation exchange capacities of soil and soil components, the relative affinities of soils and their components for various cations, and the effects of changing soil pH on exchange reactions.

Because cations vary in charge, size, hydration energy, and polarizability, their adsorption behavior varies. Metal cations, such as Zn^{2+}, Pb^{2+}, and Al^{3+} tend to adsorb in soils by forming strong chemical bonds (inner-sphere) with surface functional groups on soil oxides and SOM surfaces. On expandable 2:1 clay minerals, metal cations will adsorb via outer-sphere mechanisms in the interlayer, and are exchangeable. Cations with large hydrated radii, and low charge, including Ca^{2+}, Mg^{2+}, and Na^+, adsorb via outer-sphere bonds on most minerals, and are typically the exchangeable cations measured in cation exchange.

Table 10.3 shows examples of exchangeable cations found in a wide variety of soils. Despite the wide range of soil-forming factors influencing the soils in **Table 10.3**, Ca^{2+} is the predominant cation in all but the extremely sodic (and barren) Merced soil. The Merced soil has high exchangeable Na^+ because of poor drainage and an arid climate where evapotranspiration leaves behind sodium salts. In the Netherland's soil, exchangeable Na^+ exceeds K^+ because of atmospheric inputs of NaCl from the nearby ocean. The high Mg^{2+}

Table 10.3 CEC values and major exchangeable cations on selected soils (Bear, 1964).

Soils	pH	CEC (mmol(+) kg^{-1})	Ca^{2+}	Mg^{2+}	K$^+$	Na$^+$	Al^{3+} (H$^+$)
			Exchangeable cations (% of total)				
Avg. Netherlands Ag. soils	7.0	383	79.0	13.0	2.0	6.0	-
Avg. California Ag. soils	7.0	203	65.6	26.3	5.5	2.6	-
Chernozem or Mollisol (Russia)	7.0	561	84.3	11.0	1.6	3.0	-
Sodic Merced soil (California)	10.0	189	0.0	0.0	5.0	95	-
Lanna soil, unlimed (Sweden)	4.6	173	48.0	15.7	1.8	0.9	33.6
Lanna soil, limed (Sweden)	5.9	200	69.6	11.1	1.5	0.5	7.3

content of the California soils may reflect the high-Mg content of rocks found in the region. Exchangeable Al^{3+} is present in appreciable quantities in acid soils (pH < 5.5), such as the Lanna soil in Sweden. This soil formed from granitic rocks in dense forest under conditions of high rainfall, good drainage, and the presence of organic acids from organic matter decomposition. The strongly acid Lanna soil contains considerable exchangeable Al^{3+} and some exchangeable H^+.

In most soils, the sum of exchangeable Ca^{2+}, Mg^{2+}, K^+, and Na^+ generally equals, for practical purposes, the soil's cation exchange capacity (CEC). In acid soils, exchangeable Al^{3+} is also a significant part of CEC. CEC in soils varies from 10 mmol(+) kg^{-1} for coarse-textured soils to 500 to 600 mmol(+) kg^{-1} for fine-textured soils containing large amounts of 2:1 layer silicate minerals and organic matter.

The original cation exchange research by Thompson and Way found cation exchange to be reversible, stoichiometric (the amount released, as moles of ion charge, equals the amount retained), and rapid. Since then, some refinements have been made, and some exceptions have been found, but the results of Thompson and Way are still generally valid. Because cation exchange reactions are stoichiometric with respect to charge, the sum of all exchangeable cation charges present varies little or not at all with cation species adsorbed to the mineral. For example, consider the exchange reaction

$$CaX_2 + 2NH_4^+ = 2(NH_4)X + Ca^{2+} \tag{10.7}$$

where X designates a cation exchanger (mineral or SOM surface). Two NH_4^+ ions replace one Ca^{2+} ion to preserve the stoichiometry of the reaction. Exchangeable cation composition and CEC values are normally expressed as millimoles of ion charge per kg of soil or clay (e.g., mmol(+) kg^{-1}). Other units used include cmol(+) kg^{-1} and milliequivalents $(100\ g)^{-1}$ (an equivalent is a mole of charge). One cmol(+) kg^{-1} is equal to one milliequivalents $(100\ g)^{-1}$ – a common unit for CEC in the older literature.

Exchange reactions are virtually instantaneous, but ion diffusion to or from the mineral surface is often a rate-limiting step. This is particularly true under field conditions, where ions have to move through tortuous pores, or through relatively stagnant water films on soil colloid surfaces to reach an exchange site.

Slow diffusion can produce hysteresis for some ion exchange reactions. To measure cation exchange behavior of soils and minerals, samples are normally shaken during exchange reaction experiments to speed ion movement through the water layers on soil particle surfaces.

Because of their reversibility, cation exchange reactions can be driven forward or reverse by manipulating the relative concentrations of reactants and products. In the laboratory, common techniques for removing all cations from a surface are to use high concentrations (≥1 M) of exchanging cations, and to maintain low concentrations of product cations by leaching or repeated washings. For example, to displace Ca^{2+} on an exchange site with Na^+, the soil colloids are suspended in a solution of 1M NaCl. The exchange reaction is

$$CaX_2 + 2Na^+ = 2NaX + Ca^{2+} \tag{10.8}$$

This exchange reaction has a reaction constant

$$K_{ex} = \frac{(NaX)^2 (Ca^{2+})}{(CaX_2)(Na^+)^2} \tag{10.9}$$

where K_{ex} is an **exchange constant**, and parenthesis are activities. Rearranging **Eq. 10.9** gives the relationship between the solid phase cation activities (left hand side of **Eq. 10.10**), and the aqueous cation activities (right hand side of **Eq. 10.10**):

$$\frac{(NaX)^2}{(CaX_2)} = K_{ex}\frac{(Na^+)^2}{(Ca^{2+})} \tag{10.10}$$

Equation 10.10 shows that if (Na^+) is increased, the system will reestablish equilibrium by adsorbing Na^+ (increasing (NaX)) and desorbing Ca^{2+} to the bulk solution (decreasing (CaX_2)).

10.3.1 Cation exchange selectivity

The magnitude of Columbic attraction between an ion and a charged surface describes the *strength* of electrostatic adsorption. Strongly adsorbed ions are preferentially adsorbed over less strongly adsorbed ions. This phenomenon creates **cation selectivity**.

Cation selectivity occurs because of the varying hydrated radii of ions, which is described by hydration energy of the ions (refer to Figure 2.8). Ions with more negative **hydration enthalpies** have stronger hydration spheres and larger hydrated radii, thus, they prefer to associate with the bulk solution more than ions with more positive hydration enthalpies, which prefer to associate with mineral surfaces. This exchange reaction concept is analogous to hydrophobic chemicals that selectively associate with each other to minimize their interactions with water molecules. Na^+, for example, is a more strongly hydrated cation (has a more negative hydration enthalpy) than Ca^{2+}, and thus in a Na^+-Ca^{2+} exchange system when solution concentrations are equal, Ca^{2+} selectively partitions to the mineral surface over Na^+.

Table 10.4 shows results of a cation exchange selectivity experiment. Data were generated by saturating montmorillonite with a given ion and then measuring the amount of ion released when an amount of NH_4^+ or K^+ equal to montmorillonite's cation exchange capacity was added. The cation exchange behavior of the ions in **Table 10.4** shows that ion valence is an important factor in determining ion exchange selectivity. Divalent cations, in general, are retained more preferentially (stronger) than monovalent cations, trivalent cations are retained even more strongly, and quadrivalent cations, such as thorium Th^{4+}, are essentially unreplaced by an equivalent amount of K^+.

The relative order of a group of cation's exchangeability, or ease of removal from clay minerals, is called a **lyotropic series**. The lyotropic series for cation exchange on montmorillonite (**Table 10.4**) is:

$$Li^+ \approx Na^+ > K^+ \approx NH^+ > Rb^+ > Cs^+ \approx Mg^{2+} > Ca^{2+}$$
$$> Sr^{2+} \approx Ba^{2+} > La^{2+} \approx Al^{3+} > Th^{4+}$$

This cation selectivity order is the reverse of the order of increasing strength of retention. Thus, Li^+ is the easiest to exchange and Th^{4+} is the most difficult to exchange. The order of the lyotropic series is explainable by Coulombic interactions that consider valence and hydrated radius of the adsorbed ions. However, many cations have other adsorption mechanisms than simply Coulombic interactions. If cations adsorbed as inner-sphere complexes make bonds with surface functional groups, electrostatics only partially accounts for the adsorption process, and the lyotropic series will change.

Valence and ionic radius describe the charge density (**ionic potential**) of the ions, which directly relates to hydrated radius and hydration enthalpy (Figure 2.9). **Figure 10.12** shows the relationship between selectivity and hydration energy for the cations listed in **Table 10.4**. Within a given valence series, the degree

Table 10.4 Relation of ion charge and size to ion retention as determined by either K^+ or NH_4^+ exchange on montmorillonite. Exchange data from (Jenny and Reitemeier, 1934). See Table 2.3 for ionic radii references.

Ion	Ionic radius (nm)	Hydrated radius (nm)	% Exchanged
Li^+	0.090	0.382	68
Na^+	0.116	0.358	67
K^+	0.165	0.331	49
NH_4^+	0.132	0.331	50
Rb^+	0.175	0.329	37
Cs^+	0.195	0.329	31
Mg^{2+}	0.086	0.428	31
Ca^{2+}	0.126	0.412	29
Sr^{2+}	0.140	0.412	26
Ba^{2+}	0.156	0.404	27
Al^{3+}	0.068	0.475	15
La^{3+}	0.117	0.452	14
Th^{4+}	0.108	0.450	2

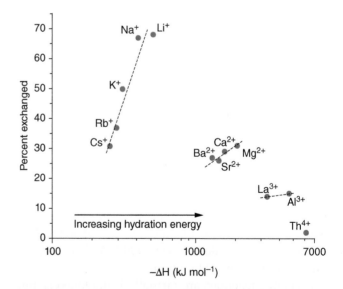

Figure 10.12 Relation of cation exchangeability with K^+ or NH_4^+ to hydration enthalpy. Within a given valence, cations with the greatest hydration energy (more negative enthalpy) are the most exchangeable. As valence increases, exchangeability decreases. Data from source listed in **Table 10.4**.

of exchangeability of an ion decreases as its hydration enthalpy decreases (becomes more positive). This effect occurs because of the increased Coulombic attraction for cations that dehydrate easier and because of the lower free energy of the system when the cations with the most negative hydration enthalpy are in solution.

Soil minerals with high **surface charge density** (high charge or CEC per unit surface area) generally have the greatest preference for highly charged cations. For example, vermiculite normally retains more Ca^{2+} than montmorillonite from a given Na^+-Ca^{2+} solution. In addition to Coulombic preferences related to hydration energy, certain minerals exhibit unusually high preferences for some cations. An example is the exchangeable Mg^{2+} content of vermiculite (**Figure 10.13**); hydrated Mg^{2+} apparently fits so well into the water network between partially expanded sheets of vermiculite that it is the preferred cation in the interlayer, even when ratios of other cations in solution, such as Ca^{2+}, are greater. The *dashed line* in **Figure 10.13** is the theoretical nonpreference line for Ca^{2+} or Mg^{2+} by the clay. The

non-preference line is indicative of a direct (1:1) relationship between exchange cation composition and solution composition. It can be derived from the homovalent exchange equilibrium reaction:

$$MgX_2 + Ca^{2+} = CaX_2 + Mg^{2+} \tag{10.11}$$

which has the following exchange equilibrium constant:

$$K_{ex} = \frac{(CaX_2)(Mg^{2+})}{(MgX_2)(Ca^{2+})} \tag{10.12}$$

If there is no preference for adsorption of either Ca^{2+} or Mg^{2+}, then the ratio of cations adsorbed on the surface is the same as their ratio in solution. In this case, K_{ex} equals one and **Eq. 10.12** transforms to

$$\frac{(MgX_2)}{(CaX_2)} = \frac{(Mg^{2+})}{(Ca^{2+})} \tag{10.13}$$

This equation needs to be rearranged and solved for the mole fraction of exchangeable magnesium (MgX_2/ ($CaX_2 + MgX_2$) as a function of the mole fraction of magnesium in solution (Mg^{2+}/($Mg^{2+} + Ca^{2+}$). The first step is to divide each side by the sum of cations on the exchanger phase or solution phase:

$$\frac{\dfrac{(MgX_2)}{(CaX_2 + MgX_2)}}{\dfrac{(CaX_2)}{(CaX_2 + MgX_2)}} = \frac{\dfrac{(Mg^{2+})}{(Ca^{2+} + Mg^{2+})}}{\dfrac{(Ca^{2+})}{(Ca^{2+} + Mg^{2+})}} \tag{10.14}$$

Next, solve for the exchanger phase magnesium mole fraction:

$$\frac{(MgX_2)}{(CaX_2 + MgX_2)} = \frac{(CaX_2)}{(CaX_2 + MgX_2)} \times \frac{\dfrac{(Mg^{2+})}{(Ca^{2+} + Mg^{2+})}}{\dfrac{(Ca^{2+})}{(Ca^{2+} + Mg^{2+})}} \tag{10.15}$$

Next, using the relation that the sum of the fractions of calcium species equals one:

$$\frac{(CaX_2)}{(CaX_2 + MgX_2)} + \frac{(Ca^{2+})}{(Ca^{2+} + Mg^{2+})} = 1 \tag{10.16}$$

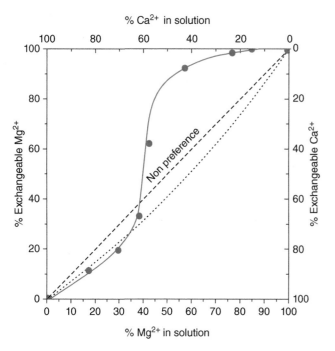

Figure 10.13 Ca^{2+}-Mg^{2+} exchange in vermiculite suspension. The dashed line shows an exchange constant of one (no preference); dotted line represents a theoretical Ca^{2+}-Mg^{2+} exchange constant of 1.5 (Ca^{2+} is preferentially adsorbed). Source: Peterson et al. (1965). Reproduced with permission of ACSESS.

which can be rearranged:

$$\frac{(CaX_2)}{(CaX_2 + MgX_2)} = -\frac{(Ca^{2+})}{(Ca^{2+} + Mg^{2+})} \qquad (10.17)$$

and substituted into **Eq. 10.15**:

$$\frac{(MgX_2)}{(CaX_2 + MgX_2)} = -\frac{(Ca^{2+})}{(Ca^{2+} + Mg^{2+})} \times \frac{\dfrac{(Mg^{2+})}{(Ca^{2+} + Mg^{2+})}}{\dfrac{(Ca^{2+})}{(Ca^{2+} + Mg^{2+})}} \qquad (10.18)$$

This simplifies to

$$\frac{(MgX_2)}{(CaX_2 + MgX_2)} = \frac{(Mg^{2+})}{(Ca^{2+} + Mg^{2+})} \qquad (10.19)$$

This is the *theoretical* nonpreference isotherm line plotted in in **Figure 10.13** (multiply by 100 for percent values on *y*- and *x*-axis). This line indicates no preference for Ca^{2+} or Mg^{2+} adsorption (defined in **Eq. 10.13** by setting $K_{ex}=1$).

The dotted line in **Figure 10.13** shows a more typical theoretical case of Ca^{2+}–Mg^{2+} exchange in which Ca^{2+} preferentially adsorbs ($K_{ex} = 1.5$). Ca^{2+}–Mg^{2+} exchange data on vermiculite (circles) show that at low Mg^{2+} concentrations, vermiculite prefers Ca^{2+} over Mg^{2+} because the hydrated Mg^{2+} is larger than hydrated Ca^{2+}, which is in agreement with Coulombic attraction forces and selectivity. As soon as enough Mg^{2+} is present in solution to exert a significant effect on the inter-lattice water network (>40%), the curve shifts to a pronounced preference for Mg^{2+}; likely due to unique interlayer forces of vermiculite and the greater entropy associated with the Mg^{2+}-water solution in the interlayer. Although normal soil solutions have relatively high Ca^{2+}/Mg^{2+} ratios, the crossover in **Figure 10.13** occurs at Ca^{2+}/Mg^{2+} ratios that are attainable under some natural conditions, such as when calcium carbonate is precipitating and removing Ca^{2+}, when former marine sediments are contributing Mg-enriched soluble salts, or when high Mg^{2+} micas (biotite) are weathering to vermiculite. Under the high Ca^{2+}/Mg ratios, the preferential adsorption of magnesium ions on vermiculite causes it to become nearly saturated with Mg^{2+}, even when exposed to appreciable Ca^{2+} concentrations or monovalent cations. To replace interlayer Mg^{2+} in vermiculite, it must be repeatedly leached with high concentrations of a replacing ion. This procedure lowers the relative Mg^{2+} concentration below the point of preferential adsorption.

Another case of ion selectivity by a mineral is the preference of vermiculite and illite for K^+, NH_4^+, Cs^+, and Rb^+. Mica weathers to illite or *vermiculite-like* minerals, with a decrease in layer charge accompanying the weathering process. The preferential adsorption of large ionic radii cations like K^+ (small hydrated radii) by such minerals occurs because of the excellent fit of the ions in the hexagonal or ditrigonal holes on vermiculite surfaces. Cations with small hydration radii have small hydration energies and are thus easily dehydrated when adjacent silicate sheets approach one another. The hydration energy theory explains why Ba^{2+}, which has essentially the same ionic radius as NH_4^+, is not fixed by trioctahedral micas or vermiculite – the divalent Ba^{2+} ions have greater hydration energy and are not as easily dehydrated, and thus cannot be entrapped by *collapse* of adjacent mineral lattices.

Preferential retention of K^+ and NH_4^+ by vermiculite is of such importance that a sizable literature has accumulated on this fixation reaction. Drying of soils also accentuates cation fixation because it removes additional water from the clay interlayers. Fixation generally decreases with soil acidification. This occurs because at low pH, aluminum and iron hydroxide interlayers form between vermiculite layer lattices creating hydroxyl-interlayered clays (like chlorite), preventing ion entry and preventing the lattices from collapsing completely.

10.3.2   Cation exchange equations

Cation exchange models are useful for predicting plant nutrient availability, availability of toxic metal cations for leaching and attenuation, and effects of irrigation, liming, weathering, fertilization, and acid rain on soil chemical properties. Soil physical properties such as dispersion and erosion are also predicted by cation exchange modeling.

There are several different models for predicting cation exchange processes, and each has its own set of parameters. Major deterrents to adoption of a more widespread and uniform approach to modeling cation exchange are: (1) exchange coefficients are typically only applicable to the soil they were derived from; and (2) lack of an absolute method for quantifying the adsorbed ion (**exchanger phase**) activity. For these reasons, exchange *constants* are typically not constant, and are better referred to as **exchange coefficients** or **selectivity coefficients**. *Selectivity* coefficient refers to the selectivity of a surface for one cation over another. Three other assumptions are required for using cation exchange models for predicting cation partitioning in soils:

1 The mineral surface is assumed to possess constant exchange capacity.
2 Ion exchange is assumed to be stoichiometric (1 to 1) with respect to ion charge.
3 Ion exchange is assumed to be completely reversible.

Despite limitations of these assumptions, careful characterization of a soil system and cation exchange reactions allows for modeling to predict partitioning of cations between the solid and solution phases. Utilization of selectivity coefficients across different soil types is done with some success for prediction of the relative changes of cation composition on soil surfaces under varying solution cation compositions (see, e.g., the use of cation exchange selectivity coefficients to predict irrigation water effects on soil salinity in Chapter 13).

Cation exchange reactions in which the exchanging cations are the same valence are called **homovalent exchange**. For example, the exchange of aqueous Na^+ for adsorbed K^+ (Na^+–K^+ exchange) on clay minerals is described by the cation exchange reaction:

$$KX + Na^+ = NaX + K^+ \tag{10.20}$$

where X indicates the negatively charged exchange site on the mineral surface. The exchange coefficient for the homovalent Na^+–K^+ reaction in **Eq. 10.20** is

$$K_{ex} = \frac{(NaX)(K^+)}{(KX)(Na^+)} \tag{10.21}$$

where parentheses denote activities. Mg^{2+}–Ca^{2+} exchange is a divalent homovalent exchange reaction.

For ions of different valences, the reaction is a **heterovalent exchange** reaction. For example, a heterovalent Ca^{2+}–Na^+ exchange reaction is:

$$2NaX + Ca^{2+} = CaX_2 + 2Na^+ \tag{10.22}$$

The exchange coefficient for the heterovalent Ca^{2+}–Na^+ reaction in **Eq. 10.22** is

$$K_{ex} = \frac{(CaX_2)(Na^+)^2}{(NaX)^2(Ca^{2+})} \tag{10.23}$$

Since activity is a function of concentration, exchange coefficients can be expanded to use activity coefficients (γ), for example, **Eq. 10.21** can be written as:

$$K_{ex} = \frac{\gamma_{NaX}[NaX]\gamma_{K^+}[K^+]}{\gamma_{KX}[KX]\gamma_{Na^+}[Na^+]}$$

$$K_{ex} = \frac{\gamma_{NaX}\gamma_{K^+}}{\gamma_{KX}\gamma_{Na^+}}\frac{[NaX][K^+]}{[KX][Na^+]} \tag{10.24}$$

where brackets are concentrations. The equation for heterovalent exchange reactions is similar to **Eq. 10.24**. Writing exchange coefficients in terms of concentrations allows for the cation concentrations adsorbed or in solution to be predicted, but this requires the activity coefficients to be calculated. The Debye-Hückel or Davies equations (Eqs. 4.16 and 4.17) are used to calculate the activity coefficients for ions in aqueous solution. However, models for predicting **exchanger phase activities** are complex and less reliable.

If aqueous and exchanger phase activities are assumed to be ideal, then concentration can be used in place of activities ($\gamma_i = 1$). This is the approach of the **Kerr-exchange coefficient** (K_{Kerr}). For the Na^+–K^+ exchange reaction in **Eq. 10.20**, the Kerr exchange coefficient is

$$K_{Kerr} = \frac{[NaX][K^+]}{[KX][Na^+]} \tag{10.25}$$

where brackets denote concentrations. Likewise, the K_{Kerr} for the heterovalent exchange reaction in **Eq. 10.22** is

$$K_{Kerr} = \frac{[CaX_2][Na^+]^2}{[NaX]^2[Ca^{2+}]} \tag{10.26}$$

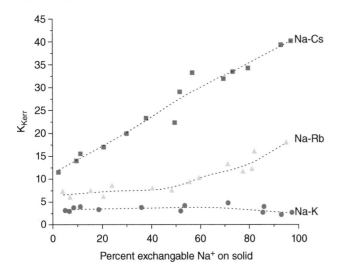

Figure 10.14 Exchange coefficient for Na$^+$ exchange on K$^+$-, Cs$^+$- and Rb$^+$-saturated montmorillonite clay as a function of percent Na$^+$ on solid (Gast, 1972).

The Kerr modeling approach assumes that concentrations and activities are the same, which is valid if activity coefficients are close to one, or if activity coefficients in the denominator and numerator cancel (i.e., when activity coefficients are equal). Unit or equal activity assumptions works for some homovalent exchange reactions, such as for Ca^{2+}–Mg^{2+} or Na$^+$–K$^+$ exchange; particularly over narrow concentration ranges and at low ionic strength.

Figure 10.14 plots Kerr equation exchange coefficients for a homovalent clay mineral system. For Na$^+$–K$^+$ exchange, the exchange coefficient is constant over most of the range of percent adsorbed Na$^+$. For Na$^+$–Rb$^+$ and Na$^+$–Cs$^+$ exchange, however, K$_{Kerr}$ is not constant. An exchange coefficient is most useful if it is constant over the range of exchange compositions for which it is used. The variability of K$_{Kerr}$ for Na$^+$–Rb$^+$ and Na$^+$–Cs$^+$ exchange on montmorillonite, as shown in **Figure 10.14**, limits the practical usefulness of the Kerr exchange coefficient to small ranges of exchanger phase composition for these cations.

The challenge to derive accurate cation exchange constants to predict heterovalent exchange reactions over a range of concentrations has been the subject of numerous studies, resulting in many different exchange coefficient models. Two of the most often used are the Gapon and the Vanselow equations. The Gapon equation, proposed by Russian scientist Y.N. Gapon in 1933, normalizes the exchange reaction

to unit charge for all reactants by using ½ as the stoichiometric coefficient for the divalent cation. For example, the Ca^{2+}–Na$^+$ exchange reaction in **Eq. 10.22** becomes

$$NaX + \tfrac{1}{2}Ca^{2+} = Ca_{\frac{1}{2}}X + Na^+ \tag{10.27}$$

The ½ stoichiometric coefficient on Ca^{2+} makes the exchanged charge on the exchanger site the same for both Ca^{2+} and Na$^+$. The Gapon exchange coefficient for this reaction is

$$K_G = \frac{[Ca_{\frac{1}{2}}X][Na^+]}{[NaX][Ca^{2+}]^{\frac{1}{2}}} \tag{10.28}$$

where the concentrations of the exchanger phase cations are in *moles of charge* (**equivalents**) per kilogram, and soluble-cation concentrations are in moles per liter. For homovalent exchange, using aqueous concentration instead of activity is adequate over narrow, though important, ranges of soluble-ion composition. In the Gapon equation, aqueous activity is often used in place of concentration to account for nonideality in the solution phase, which may extend the range of its accuracy.

Ca^{2+}–Na$^+$ exchange is important in irrigated regions because the exchange of Na$^+$ and Ca^{2+} ions between soils and solution causes dispersion and physical deterioration of many soils when exchangeable Na$^+$ becomes too high. Although the Gapon equation is unsatisfactory over the entire range of Ca^{2+}–Na$^+$ compositions, in many irrigated soils of the western United States it works fairly well over the range from 0 to 40% exchangeable Na$^+$.

The Vanselow equation assumes that activities of exchangeable ions are proportional to their mole fractions. This is equivalent to assuming that ions on soil colloid surfaces behave as ideal solutions. The mole fraction of NaX (N$_{Na}$) in the Ca^{2+}–Na$^+$ exchange reaction described in **Eq. 10.22** is

$$N_{Na} = \frac{[NaX]}{[CaX]+[NaX]} \tag{10.29}$$

where brackets are the number of moles per unit mass. The mole fraction for CaX (N$_{Ca}$) is

$$N_{Ca} = \frac{[CaX]}{[CaX]+[NaX]} \tag{10.30}$$

Substituting the mole fractions into **Eq. 10.23** for the exchanger phase activity yields the **Vanselow exchange coefficient** (K_V) for the Ca^{2+}–Na^+ exchange reaction:

$$K_V = \frac{N_{Ca}\left(Na^+\right)^2}{N_{Na}^2\left(Ca^{2+}\right)} \tag{10.31}$$

Many studies have shown that the Vanselow equation successfully predicts cation exchange on simple, relatively uniform exchangers. For soils, however, the Vanselow exchange selectivity constant is not always accurate because soils have mixed mineralogy and ion compositions, and the assumptions that the adsorption mechanisms occurring within the soil are only cation exchange are inaccurate. Despite these limitations, the Vanselow equation is useful in many instances where the Gapon equation fails to model cation exchange on soils. Special Topic Box 10.1 shows an example of calculating and interpreting Ca^{2+}–Na^+ exchange coefficients on a soil.

Other cation-exchange equations that have a more theoretical basis than the Gapon and Vanselow equations have been developed, but the Gapon, and to some extent the Vanselow, are simpler to apply. Many workers are willing to sacrifice a little theoretical rigor in a cation-exchange equation to gain simplicity; especially if the model is to be implemented in a reactive transport model where the number of parameters needs to be minimized.

10.3.3   Measuring CEC

There are two common methods used to measure CEC (Sparks, 1997):

1 Use a high concentration of displacing cation solution to exchange off all the exchangeable base cations (Ca^{2+}, Na^+, K^+, Mg^{2+}) and perhaps NH_4^+ or Al^{3+}, and sum them up for the total CEC.
2 Saturate the soil with an **index cation**, such as Na^+ or Ca^{2+}, which is then displaced with another saturating cation such as K^+ or NH_4^+. The total concentration of the index cation displaced is summed to calculate the CEC.

CEC measurement is affected by the presence of variable charge, exchangeable acidity, and soluble salts and carbonate minerals. Thus, the CEC method used must be carefully chosen so that an accurate measure of the soil's exchange capacity is obtained. In soils with variable charge, the pH of the CEC measurement is critical. In soils with exchangeable Al^{3+} or H^+, special care to displace and account for the Al^{3+} is needed. In soils with soluble salts, their removal during the CEC measurement must be accounted for so that they do not artificially elevate the CEC estimation. Given the sensitivity of CEC measurement to methodology, it is important to select the correct CEC measurement method and note the conditions of measurement when reporting CEC.

10.4   Inner-sphere adsorbed cations

The metal cations Cu^{2+}, Cd^{2+}, Co^{2+}, Fe^{3+}, Al^{3+}, Hg^{2+}, Mn^{2+}, Ni^{2+}, Pb^{2+}, Be^{2+}, and Zn^{2+} are essential micronutrients, or occur as potentially toxic ions in the environment. These metal cations can form **inner-sphere complexes** on mineral surfaces because their electron orbital energies favor formation of covalent bonds. The soil minerals reactive for inner-sphere adsorption are edges of oxides and allophane, clay mineral edges, and acidic functional groups on SOM. Inner-sphere complexed cations have variable availability for cation exchange; some inner-sphere adsorbed cations are mostly exchangeable, while others are not exchangeable. The exchangeability depends on the covalent bond strength between the cation and mineral surface functional group, and the exchanging ion in solution.

With the exception of iron and aluminum, metal concentrations in soils are typically low (see Table 3.6). In soils contaminated by municipal and industrial wastes, contaminated water, or fly ash, concentrations can be elevated. Even when concentrations are elevated, however, plant uptake is typically minor due to the strong adsorption of the metal cations by oxide minerals and soil organic matter, or because the cations precipitate out of solution as metal oxides. For example, the root zone of a typical agricultural soil might contain as much as $300\,000$ kg ha^{-1} of total Fe^{3+} and Al^{3+}, but their plant availability is only a few kg ha^{-1} because the metal cations exist as oxides with low solubility and desorption and dissolution of the metals from surfaces is minimal. This section discusses the strong, inner-sphere adsorption mechanisms of metal cations on soils.

Special Topic Box 10.1   Example of calculation of exchange coefficients

Table 10.5 shows data from a Ca^{2+}–Na^+ exchange experiment in vadose zone sediments, which has the following reaction:

$$2NaX + Ca^{2+} = CaX_2 + 2Na^+ \tag{10.32}$$

The Vanselow exchange coefficients for the Ca^{2+}–Na^+ exchange data are calculated using **Eq. 10.31**, and values for the mole fraction of surface cation compositions are calculated from **Eqs. 10.29** and **10.30**. For example, for the first row of **Table 10.5**, the mole fraction of Na^+ is

$$N_{Na} = \frac{[NaX]}{[CaX] + [NaX]} = \frac{1.48}{71.9 + 1.48} = 0.0202 \tag{10.33}$$

The mole fraction value is unitless. N_{Ca} for the exchange data in the first row of **Table 10.5** is

$$N_{Ca} = \frac{[CaX]}{[CaX] + [NaX]} = \frac{71.9}{71.9 + 1.48} = 0.980 \tag{10.34}$$

Substituting the mole fraction values and the aqueous activities into **Eq. 10.31** yields

$$K_V = \frac{N_{Ca} \left(Na^+\right)^2}{N_{Na}^2 \left(Ca^{2+}\right)} = \frac{0.980 \left(0.00469\,mol L^{-1}\right)^2}{\left(0.0202\right)^2 \left(0.0115\,mol L^{-1}\right)} = 4.59\,mol L^{-1} \tag{10.35}$$

K_V values for all of the experimental exchange date in **Table 10.5** are shown in **Table 10.6**.

Table 10.5 Exchange equilibrium data for Ca^{2+}–Na^+ exchange reactions on a vadose zone sediment. Data from Baker et al. (2010).

Percent Na+ on solid	Aqueous activity		Cations on exchange sites (mmol kg⁻¹)	
	Ca²⁺	Na⁺	Ca²⁺	Na⁺
2.02	0.0115	0.00469	71.9	1.48
3.94	0.0110	0.00983	64.6	2.65
10.7	0.00895	0.0218	67.6	8.08
20.7	0.00679	0.0372	60.2	15.7
33.2	0.00408	0.0548	52.8	26.2
63.5	0.00122	0.0727	36.1	62.8

The Gapon exchange coefficients for data in **Table 10.5** are calculated using **Eq. 10.28**, with the amount of Ca^{2+} and Na^+ converted to equivalents of charge adsorbed. **Table 10.6** lists K_G values.

The exchange coefficients are plotted in **Figure 10.15** as a function of percent Na^+ on the exchanger phase. K_G decreases as percent exchangeable Na^+ increases, suggesting it is not constant. K_V values also decrease, but have a smaller range as exchanger phase composition changes. Because K_V is less variable, it is the better exchange coefficient for modeling the Ca^{2+}–Na^+ exchange reactions on this soil.

Table 10.6 Mole fractions (N_{Ca} and N_{Na}) and Gapon and Vanselow exchange coefficients for exchange equilibrium data from **Table 10.5**.

Percent Na+ on solid	N_{Ca} solid	N_{Na} solid	K_G	K_V
2.02	0.980	0.0201	4.25	4.59
3.94	0.960	0.0394	4.57	5.43
10.7	0.893	0.107	3.86	4.16
20.7	0.793	0.207	3.46	3.77
33.2	0.668	0.332	3.46	4.47
63.5	0.365	0.635	2.39	3.92

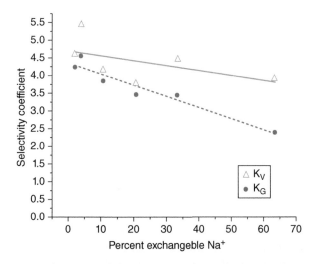

Figure 10.15 Gapon and Vanselow exchange coefficients as a function of percent Na^+ on exchanger phase calculated from data shown in **Table 10.5**.

Special Topic Box 10.2   <u>Ab</u>sorption of metal cations from soils by plants

Plant uptake of ions from soil is an important process required for assimilation of essential elements. Accumulation of metal cations that exceed plant nutrient requirements, or uptake of metal cations that are nonessential can harm plants as well as animals that consume the plants. The *bioavailability* of metal cations from soils varies with concentration, metal speciation, plant species, and soil properties. Important soil properties include pH, salinity, amount of organic matter, type of minerals, and presence of interacting cations or anions. Of these properties, soil pH is one of the most important factors influencing metal bioavailability (see Figure 3.1).

Many contaminated-site assessments base risks on total concentration, which is a reasonable assumption for screening purposes (see Figure 1.7), but is a poor indicator of actual risks at a site. Because of limited resources, remediation decisions need to be based on accurate assessments of metal bioavailability, not simply total concentration. There are several empirical models that use soil properties to predict metal bioavailability. In addition, soil extraction tests have been devised to simulate metal bioavailability. Such extractions are called *bioaccessibility* tests because they do not directly measure bioavailability. A ligand solution such as diethylenetriaminepentaacetic acid (DTPA) is a common extractant used to estimate plant bioavailability of metal cations in soils. The strong ligand complex is designed to simulate the metal release in the rhizosphere, but it may over- or underestimate actual plant uptake.

Measurement of metal uptake in plant tissue provides direct evaluation of metal bioavailability to plants. To compare the *relative* uptake of metal cations across species and soils, plant concentration (C_{plant}) is normalized by the total metal concentration in the soils (C_{soil}). This ratio is called the bioaccumulation factor:

$$BAF = \frac{C_{plant}}{C_{soil}} \qquad (10.36)$$

where concentration units of the plants and soils are dry weights.

Figure 10.16 shows Pb^{2+}, Cd^{2+}, Zn^{2+}, and Cu^{2+} bioaccumulation factors for various plant tissues from several soils that were contaminated from a nearby mine. Metal ion uptake by the plants was highly variable among the different plants and tissues. The most bioavailable metal was Cd^{2+}, followed by Zn^{2+}, Cu^{2+}, and Pb^{2+}. This order is the reverse of the adsorption affinity series predicted for metal adsorption in soils (see Section 10.4.2), suggesting that adsorption and desorption reactions are controlling the metal bioavailability in the soils.

Bioaccumulation factors for metal cations are usually small. For example, BAFs range from 0.001 to 1 for the metal cations in **Figure 10.16**. Given the great mass of soil compared to the plant biomass, the fraction of soil-metal cations absorbed by the plants is very small. The potential toxicity of the plant-absorbed metal cations to organisms that consume the plants depends on the amount consumed, health and diet of the organism consuming the plant, and biochemistry of the metal cations within the plant.

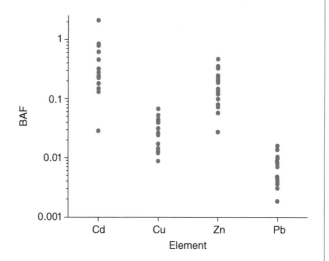

Figure 10.16 Bioaccumulation factors for uptake of metal cations by food plants grown in four different soils contaminated from mining activities. The points represent the ranges in concentrations for leafy vegetables, fruit vegetables, root vegetables, and rice. Data are from Zhuang et al. (2009).

10.4.1 Inner-sphere adsorption of cations on minerals

Inner-sphere metal cation adsorption occurs on clay edges (e.g., **Figure 10.1**) and surfaces of soil oxides (**Figure 10.2**), where unsatisfied bonds create reactive surface functional groups or ligands. The charge on the surface ligands is amphoteric, so metal cation sorption is pH-dependent (**Figure 10.17** and **Figure 10.18**; also see Figure 3.19).

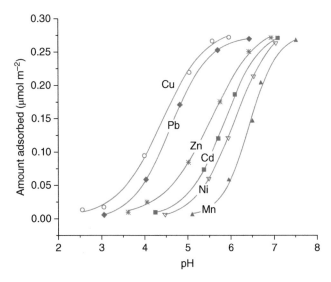

Figure 10.17 pH-dependent adsorption edge for metal cation adsorption on goethite. Source: Cornell and Schwertmann (1996). Reproduced with permission of Wiley.

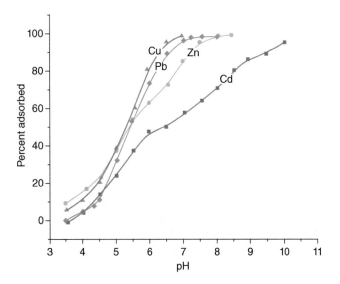

Figure 10.18 pH-dependent adsorption edge for metal cation adsorption on kaolinite. Adapted from Srivastava et al. (2005).

A *generic reaction* for deprotonation of a surface functional group and adsorption of a metal (M^{2+}) as an inner-sphere complex is

$$\equiv SOH + M^{2+} = \equiv SOM^+ + H^+ \tag{10.37}$$

where $\equiv SOH$ is a reactive surface functional group on the edge of a mineral or SOM. The protonation and deprotonation behavior of $\equiv SOH$ depends on soil pH and the PZC of the soil mineral (or pK_a of the SOM functional group). The actual charge of $\equiv SOH$ or $\equiv SOM^+$ depends on the surface ligand coordination (see Table 9.2 and Figure 7.21) and the valence and number of surface bonds of the adsorbed metal cation.

To study pH-dependent adsorption, the amount of an ion adsorbed in a soil suspension at different pH values is measured from low to high pH. At low pH, few cations are adsorbed, as pH increases, cation adsorption increases, dramatically increasing until the adsorption capacity is reached, after which adsorption levels off. The pH-dependent behavior is called an **adsorption edge** (**Figure 10.17** and **Figure 10.18**).

The ability of metal cations to form inner-sphere adsorption complexes depends on both the electrostatic interaction of ions with surfaces and the bonding characteristics of the metal cations. The sign and magnitude of surface charge controls electrostatic interactions of ions. At low pH, most of the surface functional groups are positively charged because pH is below the PZC (e.g., goethite in Figure 9.9) and the positively charged surface functional groups electrostatically repel metal cations. As pH increases, fewer positively charged, and more negatively charged functional groups exist, allowing metal cations to more readily approach the surface and form inner-sphere complexes. The charge behavior of SOM is similar, where the surface functional groups become negatively charged as pH exceeds the pK_a of the functional groups (e.g., carboxyl functional groups) on the SOM.

Net surface charge (σ_p) of minerals is predicted using their PZC values (Table 7.7). While, PZC describes the *overall* surface charge, it does not describe the charge of individual surface functional groups. When pH < PZC, the net surface charge is positive, but some surface functional groups are uncharged or negatively charged. Thus, when pH < PZC and the net surface charge is positive, negatively charged functional groups can still exist and cations can adsorb via inner-sphere complexation.

The bonding characteristics of cations can be predicted from the tendency of the cations to form aqueous hydrolysis complexes because cations that commonly exist in aqueous solution as metal-OH complexes ($MeOH_x^{z-x}$), also form inner-sphere surface complexes. Alkali and alkaline earth metal cations do not readily form hydrolysis complexes, and do not typically form inner-sphere surface complexes. The relationship between inner-sphere complexation and hydrolysis exists because surface complexation involves formation of metal-OH bonds, similar to hydrolysis.

The rate of inner-sphere adsorption in soils is typically initially rapid, but adsorption may continue over longer periods (days to months). The reversibility of metal cation adsorption tends to decrease over longer periods of incubation, possibly due to formation of more inner-sphere bonds, or formation of precipitates that dissolve slower than adsorption complexes. Slow adsorption/desorption reactions are also caused by slow diffusion of ions through small pores in mineral particles and aggregates.

10.4.2 Metal adsorption selectivity on minerals

pH-dependent adsorption behavior of metal cations is used to predict the relative preference for adsorption (**Figure 10.17** and **Figure 10.18**). Preference can also be derived from experimentally determined adsorption coefficients. The order of preference for metal adsorption varies with initial concentration, solid preparation method, and ionic strength. For adsorption of divalent cations on iron oxides, the *typical* selectivity order is

$$Pb^{2+} > Cu^{2+} > Zn^{2+} > Ni^{2+} > Cd^{2+} \geq Co^{2+} > Sr^{2+} > Mg^{2+}$$

Retention of the divalent cations by aluminum oxide is similar:

$$Cu^{2+} > Pb^{2+} > Zn^{2+} > Ni^{2+} > Co^{2+} \geq Cd^{2+} > Mg^{2+} > Sr^{2+}$$

Adsorption preference on edges of clay minerals is similar to the preferences listed for the metal oxides. The divalent cations Hg^{2+}, Fe^{2+}, and Mn^{2+} are not listed in the above series; Hg^{2+} is typically amongst the more preferred cations for adsorption, and Fe^{2+} and Mn^{2+} are typically less preferred than Cd^{2+} and Co^{2+}, but more preferred than Mg^{2+} and Sr^{2+}.

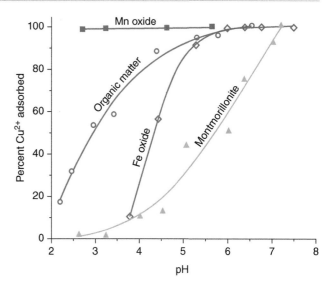

Figure 10.19 Copper adsorption as a function of pH on several different minerals. One-gram samples were equilibrated at different pHs with 200 mL of 5 mg L^{-1} Cu^{2+} and 0.05 M $CaCl_2$. Adapted from Mclaren and Crawford (1973). Reproduced with permission of Wiley.

Figure 10.19 shows the pH-dependent adsorption of Cu^{2+} on three common soil minerals and SOM. Manganese oxides have a low PZC and a very high affinity for Cu^{2+} adsorption; thus, nearly 100% of the Cu^{2+} is adsorbed at pH > 2.8. Montmorillonite exhibits the lowest Cu^{2+} adsorption affinity in **Figure 10.19**; maximum adsorption likely occurs when pH > 7. Adsorption of metals in the interlayers of expandable clay minerals is not pH dependent, but the edges of clay minerals have pH-dependent charge, just like kaolinite. The permanent charge of montmorillonite and vermiculite is much greater than edge charge, so pH-dependent metal adsorption on edge sites goes unnoticed. However, when the ratio of Na^+, K^+, Ca^{2+}, or Mg^{2+} are much greater than a metal in the solution (e.g., in saline soils), metal adsorption in the interlayer is blocked because the greater concentration of alkali and alkaline earth metal cations outcompete the metal cations for the outer-sphere adsorption sites. On the edge sites, metal cations form inner-sphere covalent bonds with functional groups and are preferentially adsorbed over alkali and alkaline earth cations. Thus, in solutions that have high concentrations of salts, metal adsorption on clay minerals is pH-dependent; for example, pH-edges like kaolinite in **Figure 10.18** and the copper adsorption edge for montmorillonite in

Figure 10.19. For the Cu^{2+} adsorption edge on montmorillonite experiment the solution had 0.05 M Ca^{2+}, which is much higher than the copper concentration, and thus the interlayers were saturated by Ca^{2+}, leaving only the edge sites for the Cu^{2+} to adsorb.

10.4.3   Inner-sphere metal adsorption on soil organic matter

Soil organic matter is one of the most reactive components for metal cation adsorption. For example, the adsorption affinity of Cu^{2+} for SOM is much greater than that for iron oxides and montmorillonite (**Figure 10.19**). Thus, despite the relatively small amount of SOM in many soils (typically less than a few percent), it often plays a predominant role in metal cation adsorption behavior.

Carboxylic and phenolic acid functional groups on SOM (Figure 3.15) are particularly reactive for metal cations. Electron-rich nitrogen-containing functional groups may also be active adsorption sites for metal cations, but they are present at only a fraction of the carboxyl and phenol functional groups, and thus are not as important for metal adsorption.

Cu^{2+}, Pb^{2+}, and Hg^{2+} have a particularly high affinity for reduced sulfur-containing functional groups of SOM (thiols (R-SH)), which are weak acids. Even though sulfur functional groups are present at a fraction of the carboxyl groups, there may be enough to adsorb most of the Hg^{2+} present in soils because it is typically present at relatively low concentrations (low relative concentrations of mercury are still potentially hazardous).

The average pK_a of carboxyl groups in SOM is ~4.5 (Table 9.3). At $pH < pK_a$, the carboxyl groups are protonated, and are less likely to adsorb ions. At $pH > pK_a$, the carboxyl groups deprotonate, become negatively charged, and the electrostatic attraction between metal cations and the SOM functional groups is favorable; facilitating adsorption via inner-sphere complexes. The order of adsorption affinity of metal cations on humic acid extracted from SOM is

$$Hg^{2+} = Fe^{3+} > Al^{3+} > Pb^{2+} > Cu^{2+} > Cr^{2+}$$
$$> Cd^{2+} = Zn^{2+} = Ni^{2+} > Co^{2+} = Mn^{2+}$$

which is similar to the order of adsorption preference of metal cations for mineral surfaces.

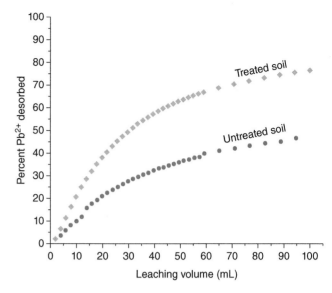

Figure 10.20 Percent of adsorbed Pb^{2+} desorbed from a Delaware, USA soil before and after treatment to remove organic matter. The x-axis is the volume of 0.05 M $NaNO_3$ leached through a stirred-flow reactor containing the Pb^{2+}-adsorbed soil. Adapted from Strawn and Sparks (2000), with permission.

Ions adsorbed on SOM are typically tightly held, and adsorption/desorption reactions are frequently hysteretic or irreversible (see Special Topic box 10.3). **Figure 10.20** shows Pb^{2+} desorption from a soil before and after removal of the SOM. In the *whole soil,* only 45% of the adsorbed Pb^{2+} desorbed. In the soil treated to remove the organic matter, greater than 75% of total adsorbed Pb^{2+} desorbed. Spectroscopic investigation of the whole soil confirmed that the Pb^{2+} was complexed to SOM, whereas in the treated soil Pb^{2+} was sorbed on minerals.

10.4.4   Inner-sphere metal adsorption in soils

The specific order of the metal cation adsorption affinities in whole soils varies with soil pH and type and amount of minerals and SOM in the soil. In general, amongst the divalent cations, Hg^{2+}, Cu^{2+}, and Pb^{2+} are more strongly adsorbed than Cd^{2+}, Zn^{2+}, Mn^{2+}, Fe^{2+}, and Ni^{2+}; which in turn are more strongly adsorbed than alkaline earth metal cations. Trivalent metal cations Fe^{3+}, Al^{3+}, and Cr^{3+} typically exist in soils as mineral phases that are not very soluble, but some fraction of these ions adsorbs to mineral and SOM surfaces. Adsorption of trivalent metal cations is stronger than most divalent cations on soil surfaces.

Special Topic Box 10.3   Monitored natural attenuation as a tool for remediation

Chemical contamination of soils at many sites often encompasses large areas, and total remediation is too expensive or not technologically feasible. Part of a remediation plan may include active remediation by treatment or removal of soils that pose the greatest risks. Soils posing less risk are possible candidates for *monitored natural attenuation* (MNA). MNA relies on soil processes to limit the mobility and bioavailability of the chemical contaminant within an acceptable time frame.

MNA is used for both organic and inorganic chemicals, although use for organic chemical remediation is more common. Important considerations for successful MNA are: (1) demonstrating contaminant sequestration, (2) estimating attenuation rates, (3) estimating the attenuation capacity, and (4) evaluating potential reversibility issues. Soil processes that attenuate contaminants are adsorption, precipitation, degradation, and redox. For organic chemicals, degradation decreases the concentration of the contaminant, but the MNA plan must consider risks from the degradation products. Inorganic chemicals cannot be removed from soils by degradation, unless they are volatile (e.g., hydrogen selenide), but they can be converted into species that have lower relative risks.

Natural attention relies on chemical reactions occurring in soils that decrease the availability of the contaminant to the soil solution. In natural settings, this may be rate-limited by either physical transport or chemical reactions. The time dependency of the attenuation process can range from months, decades, or even centuries. Several possible attenuation processes are:

1 When chemicals are first input to soil, they adsorb onto SOM and mineral surfaces.
2 Chemicals convert from one type of adsorption, such as outer-sphere (cation exchange), to another, such as inner-sphere.
3 Chemicals diffuse to the interior of particles and aggregates, becoming physically isolated.
4 Organics degrade.
5 Inorganics change oxidation state, affecting their solubility or adsorption behaviors.
6 Adsorbed chemicals form surface precipitates.
7 Solutions become oversaturated and minerals precipitate.
8 Surface precipitates and minerals ripen to less-soluble species.

Time-dependent attenuation processes occur because of initial disequilibrium of the contaminant chemical in the soil, and because fluxes are continuously changing the status of soil solutions, minerals, and organic matter (see Figure 1.1). Remediation managers often apply an in-situ treatment prior to MNA (e.g., adding a microbe, food source, or chemical reactant).

A key aspect of MNA is identification of the major processes controlling contaminant solubility or availability to organisms and identifying how the contaminant changes over time. Measuring such factors requires understanding of soil characteristics and processes, contaminant species within the soil, and accurate models for predicting changes over time.

Figure 10.21 shows the desorption of cadmium from a soil incubated for different lengths of time. As incubation time increased, the amount of cadmium desorbed using an EDTA chelate extractant decreased, suggesting that attenuation occurred. Studies have observed that metal cation adsorption occurs initially on low-affinity adsorption sites, and over time, ions transfer to high-affinity adsorption sites. The amount of metal cations desorbed from low-affinity sites is greater than high-affinity sites (**Figure 10.22**). Thus, the

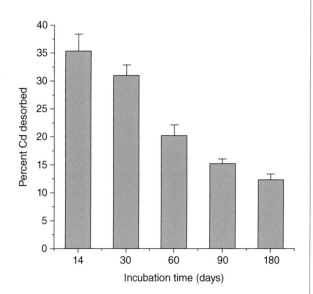

Figure 10.21 Amount of cadmium desorbed from a soil incubated for different lengths of time. Desorption was measured using EDTA extractant, which simulates a natural organic chelate such as a root exudate. Data are from Zhang et al. (2018).

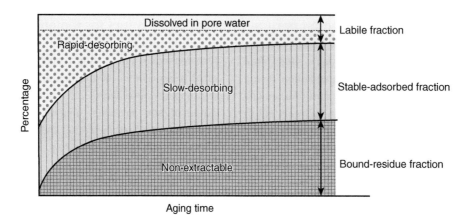

Figure 10.22 Relative availability of chemicals that show aging behavior in soils to soil solution as aging time increases. The chemical transforms from the labile fraction to the stable and bound fractions as aging time increases. Stable-adsorbed chemicals are slowly available to the soil solution controlled by diffusion or strong adsorption processes, while the bound-residue chemicals are entrapped within particles or irreversibly sorbed on particle surfaces. The allocation of chemicals in the different fractions is a function of soil and chemical properties, and the distribution into the different fractions with aging time is highly variable. Source: Cheng et al. (2019). Reproduced with permission.

conversion to high-affinity sites is an MNA process. Diffusion to the interior of aggregates or solids may also explain this behavior, especially for organic chemicals.

For chromium-, arsenic-, and selenium-contaminated soils, redox is a critical aspect of MNA because these elements readily change oxidation states when redox status of soils changes (see Table 5.1), which changes their adsorption and solubility behaviors. For example, Cr(III) is insoluble, while Cr(VI) is soluble; and As(III) and Se(VI) are not strongly adsorbed to soils, while As(V) and Se(IV) are strongly sorbed. In the case of MNA, redox stabilization can be an important aspect of a remediation plan, which would include managing oxygen availability and food source for microbes that create reducing conditions. The MNA management strategy may be to optimize reducing or oxidizing conditions, depending on the desired oxidation state of the target contaminant.

In reducing environments, some metals, such as Pb, form metal sulfide phases that are not very soluble. Thus, keeping a system reduced promotes metal sulfide mineral formation and may be a MNA strategy. This can be achieved by keeping a soil, such as a wetland, permanently flooded instead of allowing cyclic flooding that causes variable redox conditions.

Many rare earth elements (lanthanides) also form strong inner-sphere complexes in soils, as do actinides such as Pu^{4+} and U^{4+}, and precious metal cations such as Ag^+. Strong adsorption of these metal cations occurs because they are large ions with weak hydration spheres, have polarizable electron shells that are favorable for electron sharing with surface functional groups, and, in some cases, have high charge.

Variable charged oxygen surface ligands are prevalent in soils, and can adsorb large amounts of metal cations if pH is greater than approximately 6.5, which may cause plant deficiencies of metal micronutrients (Cu, Mn, Zn, etc.). Changing the soil pH by adding a soil acidifier (e.g., sulfur) is often more effective than adding these microelements as soil fertilizers. Plant deficiencies of essential micronutrients, such as Fe-chlorosis and Zn-deficiency may occur above soil pH 8. Deficiencies in fruit and nut trees and in some varieties of sorghum are common, especially in arid regions under irrigation, where fast plant growth exceeds soil metal release rates. Native plants growing in natural ecosystems do not typically show deficiency symptoms.

When the concentration of metal cations is too high, such as manifested by Al^{3+} toxicity in acid soils or Fe^{2+} and Mn^{2+} toxicity in rice paddies, raising the soil pH by liming is an effective immobilization strategy. Increasing soil pH can also be used in polluted soils to attenuate metal cations, which makes them less available for bio-uptake and less mobile to surface and groundwaters.

10.5 Anion adsorption

Anion adsorption behavior in soils depends on the characteristics of the anions, soil mineral composition, and soil pH. Anions adsorb by electrostatic adsorption on positively charged mineral surfaces, and by forming chemical bonds with surface functional groups. The anions of concern to agriculture include Cl^-, $H_3SiO_4^-$, HCO_3^-, NO_3^-, SO_4^{2-}, HPO_4^{2-}, $H_2PO_4^-$, $B(OH)_4^-$, OH^-, and F^-. In addition, some micronutrients (MoO_4^{2-}, and SeO_3^{2-}, SeO_4^{2-}) and contaminants (CrO_4^{2-}, AsO_3^{3-}, $H_2AsO_4^-$) exist as anions in soils, as do some pesticides such as the deprotonated phenoxyacetic acids (2,4,5-T and 2,4-D). Inorganic anions are either individual ions (halogens and sulfide), or **oxyanions**, such as HPO_4^{2-} and SO_4^{2-}. Many oxyanions are weak acids, and thus occur in various protonated forms in soil solutions (e.g., $H_2PO_4^-$, HPO_4^{2-}).

Soil-anion adsorption reactions vary from weak, outer-sphere adsorbed ions to very strong, inner-sphere adsorbed ions (**Table 10.7**). Inner-sphere anion adsorption typically occurs through a **ligand exchange** reaction, where the anion (A^-) displaces a surface OH^- ligand:

$$\equiv SOH + A^- = \equiv SA + OH^- \qquad (10.38)$$

Phosphate, for example, adsorbs to mineral surfaces via multiple strong bonds (**Figure 10.23**), where the O-ligand of the phosphate displaces the surface-O ligands. The strong bi-dentate inner-sphere complexation of phosphate causes it to be not readily released to the soil solution because of either irreversible or very slow desorption.

Weakly adsorbed anions that have only electrostatic interactions undergo **anion exchange** reactions. Anion exchange is rapid and reversible. All anions interact electrostatically with charged surfaces, but sulfate is the only macronutrient ion for plants that is retained to a significant extent as an exchangeable anion (phosphate is a strongly adsorbed anion not considered exchangeable). Each category in **Table 10.7** covers a range of retention. Among the *weakly retained anions*, sulfate and selenate are retained the strongest by soils because of their divalent charge. Nitrate, chloride, and perchlorate, are weakly adsorbed, and thus mobile in most soils. As in cation exchange, anion adsorption depends on the size and charge of the hydrated ion,

Table 10.7 Anions in soil and their adsorption characteristics.

Outer-sphere adsorbed Weak	Inner-sphere adsorbed Intermediate	Inner-sphere adsorbed Strong
Cl^-, HCO_3^-, NO_3^-, SO_4^{2-}, SeO_4^{2-}, AsO_3^{3-}, HCO_3^-, I^-, Br^-	$H_3SiO_4^-$, CrO_4^{2-}, VO_4^{2-}, $HSbO_4^{2-}$, $B(OH)_4^-$	HPO_4^{2-}, $H_2PO_4^-$, OH^-, MoO_4^{2-}, SeO_3^{2-}, $HAsO_4^{2-}$, HS^-, F^-,

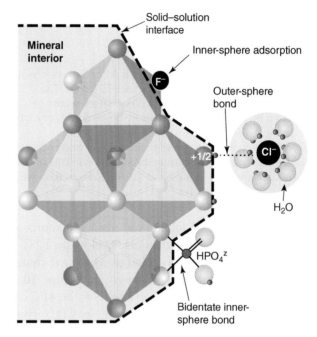

Figure 10.23 Adsorption of F^-, Cl^-, and HPO_4^{2-} on an oxide surface. The chloride maintains its hydration sphere and adsorbs as an outer-sphere complex. Fluoride adsorbs via ligand exchange with a surface hydroxyl group. Phosphate forms inner-sphere bonds via ligand exchange reactions with the two surface hydroxyl groups.

and on the ability of the ion to covalently bond with mineral-bound metals.

In well-weathered soils of low pH and low soil organic matter, and in soils derived from volcanic parent material, the **anion exchange capacity** (AEC) may be similar to or exceed the CEC (**Figure 10.24**). The predominance of AEC or CEC can change from one horizon to the next in the same soil as the pH and composition of the soil changes. Strongly weathered soils such as Oxisols and Ultisols contain an abundance of aluminum and iron oxides that have positive surface

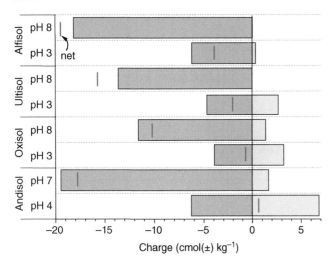

Figure 10.24 Charge characteristics of A horizons of several well-weathered soil orders at different pH values. The vertical bar shows the net charge. The charges were measured by cation and anion adsorption amounts at each pH. In the Alfisol and Ultisol soils, at pH 8, negative adsorption occurred, meaning no anion adsorption occurred and anions were desorbed from the soil. Oxisol, Ultisol, and Alfisol data are from Morais et al. (1976); Andisol data are from Auxtero et al. (2004).

charges. In Ultisols, kaolinite can also be a source of positive surface charges when pH is below the PZC.

SOM has mostly negative charge, so the net charge in the A horizon samples shown in **Figure 10.24** are highly negative, but in the B horizons, the AEC can be of similar or greater magnitude as CEC. Thus, variable charge soils often adsorb significant amounts of anions. Many of the soils that have high AEC are located in the Southern Hemisphere, but anion exchange is also significant in acidic and highly weathered soils of the southeastern United States, and in European forest soils. Soils, including Andisols, developed in regions with volcanic ash deposition in regions in New Zealand, Japan, northwestern United States, and Central America have abundant short-range ordered minerals such as allophane and ferrihydrite that have high AEC.

10.5.1   Outer-sphere adsorbed anions

Anions are attracted to positively charged sites on clay mineral edges, hydrous oxides, and allophane surfaces, and are repelled by negative charges on permanent charge clay mineral interlayers. Soils that contain high amounts of oxides or kaolinite can have an overall positive charge. Anions approaching positively charged sites are electrostatically attracted to the mineral surfaces. The effects of ion concentration and valence on the distribution of exchangeable anions are similar to the effects described for cations. **Figure 10.23** shows outer-sphere adsorption of Cl⁻ to a positively charged mineral surface site. The positive charge in this case is the result of surface protonation, which increases as soil pH decreases. Anion exchange occurs on positively charged surfaces. For example, a NO_3^-–Cl⁻ exchange reaction is:

$$XCl + NO_3^- = XNO_3 + Cl^- \tag{10.39}$$

where X is an **anion exchange site**, such as $\equiv FeOH_2^{1/2+}$ on an iron oxide surface. Exchange equations similar to those developed for cation exchange describe such reactions because outer-sphere adsorbed anions are in the solution adjacent to the solid surface, and are readily exchangeable.

Cl⁻, NO_3^-, and SO_4^{2-} are common anions in soils that are adsorbed as outer-sphere ions. At low pH, sulfate may form inner-sphere bonds on some mineral surfaces. Selenate (Se(VI), SeO_4^{2-}) and arsenite (As(III), AsO_3^{3-}) are also typically outer-sphere complexed anions. **Table 10.8** and **Figure 10.25** shows typical data for Cl⁻ and SO_4^{2-} adsorption by soils. The capacity of soils to adsorb anions increases with increasing acidity, and is much greater for the kaolinitic soil than the montmorillonitic soil (**Table 10.8**). At all pH values, more of the divalent SO_4^{2-} ion is adsorbed than the monovalent Cl⁻ ion, as would be expected on the basis of stronger electrostatic attraction forces of the divalent

Table 10.8 Adsorption of Cl⁻ and SO_4^{2-} by kaolinitic soil as a function of pH, and of Cl⁻ by a montmorillonitic soil (Bear, 1964).

	Kaolinitic soil			Montmorillonitic soil	
pH	Cl⁻ (mmol kg⁻¹)	SO_4^{2-} (mmol kg⁻¹)		pH	Cl⁻ (mmol kg⁻¹)
7.2	0	0		6.8	0
6.7	3	10		5.6	0
6.1	11	27		4.0	0.5
5.8	24	36		3.0	1
5.0	44	52		2.8	4
4.0	60	–			

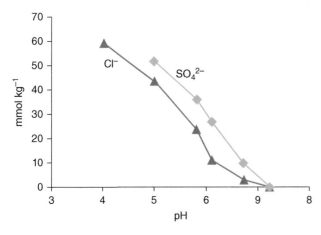

Figure 10.25 Sulfate and chloride adsorption as a function of pH on a kaolinitic soil. Data source provided in **Table 10.8**.

anion charge compared to the monovalent anion charge. For the montmorillonitic soil, where pH-dependent charge is much smaller than the permanent negative surface charge, Cl⁻ adsorption is minimal; especially at neutral to alkaline pH range. Anion adsorption trends shown in **Figure 10.25** are typical of outer-sphere adsorbed anions.

Chloride, nitrate, and sulfate are common and important anions in most soils. Because Cl⁻ mobility is similar to NO_3^-, it is often used as an indicator of NO_3^- mobility in soils. This avoids complications from biological transformations of nitrogen that complicate experiments and monitoring.

10.5.2   Inner-sphere adsorption of anions

Anions *strongly retained* as inner-sphere complexes by soils include the various protonated forms of PO_4^{3-}, AsO_4^{3-}, MoO_4^{2-}, CrO_4^{2-}, S^{2-}, SeO_3^{2-}, $B(OH)_4^-$, and F⁻. The state of these anions in the soil is a matter of great economic and environmental concern since they are added to soils in fertilizers, agricultural wastes, fly ash from coal combustion, and municipal and industrial wastes. Initially, addition of strongly retained anions to soils increases the soil solution concentrations and plant availabilities. With time, however, the anions adsorb to soil colloids, and their plant availability and leaching decrease.

Oxide minerals in soils are common sorbents for inner-sphere complexed anions. In addition to electrostatic interactions, oxide surfaces and edges of clay

minerals also form chemical bonds with anions (**Figure 10.23**). Iron oxides and other oxides scavenge (remove) arsenate, phosphate, molybdate and other anions from solution with high efficiency.

Phosphate and fluoride anions displace O^{2-} ligands coordinating mineral cations (Al^{3+} or Fe^{3+}) (called ligand exchange or anion penetration) (**Figure 10.23**). Ligand exchange takes place on the mineral surface of oxides, as described in the reaction in **Eq. 10.38**. Ligand exchange reactions do not directly depend on surface charge, but the *approach* of anions to the surface depends on surface charge. This contrasts with outer-sphere anion adsorption, which occurs only when the surface carries a net positive charge (when pH < PZC).

Although formation of inner-sphere complexes on mineral surfaces is not directly affected by electrostatic interactions, the anions must contact the mineral surfaces to adsorb. Thus, inner-sphere adsorption reactions depend on surface charge, and the charge of the anion. The sign and magnitude of the net surface charge can be predicted from PZC measurements. The charge of weak acids is predicted from pK_a values (e.g., Figures 4.5 and 4.6).

The bar characterizing surface charge in Figure 9.10 and a bar indicating the charge on the weak acid can be used together as a guide to evaluate electrostatic interactions between anions and surface charge. For example, the pK_as for phosphoric acid deprotonation are $pK_{a1} = 2.15$, $pK_{a2} = 7.20$, and $pK_{a3} = 12.35$ (see Figure 4.5). The bar showing predominant phosphate speciation as a function of pH is shown together with the surface charge and phosphate adsorption behavior on goethite in **Figure 10.26**. At low pH, the goethite surface has maximum net positive charge, and between pH 2.2 and 7.2 $H_2PO_4^-$ is the predominant species, with a minus one valence. The negative charge of the anion and the highly positive surface charge create an electrostatic attraction force, and the phosphate is attracted to the surface and adsorbs via a ligand exchange reaction, such as

$$\equiv FeOH_2^{1/2+} + H_2PO_4^- = \equiv FeOH_2PO_3^{1/2-} + H_2O \quad (10.40)$$

This inner-sphere adsorption reaction describes a single **monodentate complexation** reaction, where one of phosphate's oxygen ligands displaces a surface oxygen functional group (or a water molecule). Phosphate adsorption is actually more likely to

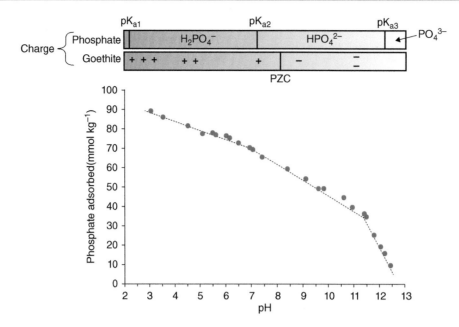

Figure 10.26 Phosphate adsorption as a function of pH on goethite. The bars at the top indicate the change in charge on the phosphate and goethite surface as pH changes. Data from Hingston et al. (1972), with permission of Wiley.

occur via a bidentate ligand exchange reaction, which is similar to the reaction in **Eq. 10.40**, but includes two surface $\equiv FeOH_2^{1/2+}$ ligands bonding to a single phosphate:

$$2 \equiv FeOH_2^{1/2+} + H_2PO_4^- = (\equiv FeO)_2 PO_2H^{1-} + 2H_2O + H^+$$
$$(10.41)$$

In this reaction, two oxygen ligands from phosphate displace two OH_2^- surface ligands. The two OH_2^- surface ligands can be coordinated to the same surface iron atom, or neighboring surface iron atoms. **Figure 10.23** depicts a **bidentate surface complex**.

As pH increases past the pK_{a1} of phosphate, the surface charge becomes less positive, and the pH-dependent phosphate-adsorption slope is the shallowest, suggesting the smallest amount of desorption with increasing pH (**Figure 10.26**). When the pH exceeds the pK_{a2} value for phosphate, the phosphate valence becomes minus two, and, even though the surface is still net positively charged, it is nearing the PZC and some of the functional groups deprotonate and become negatively charged, repelling HPO_4^{2-} anion. Thus, the slope of the adsorption-pH curve becomes steeper, indicating a greater amount of desorption as pH increases. At pH greater than the PZC, some phosphate is still adsorbed, even though the surface is net negatively charged, because phosphate adsorbs

through inner-sphere bonds that can overcome the electrostatic repulsion. In addition, there are some functional groups that are still positively charged (PZC describes the *net* charge). When pH > pK_{a3}, phosphate deprotonates to the minus-three species and the negative surface charge is large, causing the slope of the adsorption-pH curve in **Figure 10.26** to become steeper.

Phosphate is an important example of inner-sphere adsorbed anions. Many soils fix large quantities of phosphate by converting readily soluble phosphate to forms less available to plants. The long-term capacity of most soils to adsorb phosphate is orders of magnitude greater than the amounts of phosphate added as fertilizer. A phosphate adsorption mechanism is shown in **Figure 10.23**. Phosphate replaces OH groups forming very stable binuclear bonds with two Fe^{3+} or Al^{3+} cations on the mineral surface. In the laboratory, phosphate adsorption by layer silicates is rapid for a few hours, and then continues more slowly for weeks. The initial rapid reaction is a combination of inner-sphere adsorption and ligand exchange reactions on mineral edges. The slower reaction probably consists of a complex combination of mineral dissolution and precipitation of added phosphate with exchangeable cations, or cations within the lattices. Thus, in most soils, phosphate is retained by a multi-stage process, probably involving several mechanisms including

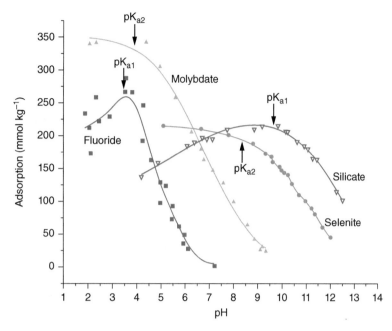

Figure 10.27 Adsorption of anions on goethite as a function of pH. Initial concentrations of anions were different, thus total adsorption amount is not comparable between anions. The pK_a values are acid dissociation constants of the weak acids. Adapted from Hingston et al. (1972). Reproduced with permission of Wiley.

inner- and outer-sphere adsorption as well as mineral precipitation.

Figure 10.27 shows the pH-dependent adsorption of several anions on goethite. At pH values below the first pK_a, adsorption is low because the weak acid anions are protonated, and thus there is no electrostatic attraction between the protonated acid and the positively charged surface. As pH increases toward the pK_{a1}, the adsorption–pH curve increases as the anion species concentration increases due to deprotonation of the weak acids. As pH increases past pK_{a1}, anion adsorption decreases because the surface and more of the anions are becoming negatively charged. This behavior is the same as described for phosphate adsorption in **Figure 10.26**. The shape of the anion adsorption curve is called an **adsorption envelope** because it looks like the top flap of an envelope.

High concentrations of F^- in soils cause high concentrations in plants that can be toxic to the plants and grazing animals. However, fluoride adsorption on soil particles is strong; thus, little fluoride is released to soil solution in typical soils, and toxicity is not widespread. Fluoride adsorption occurs by ligand exchange. In acid soils at equal anion concentrations, fluoride adsorption predominates over that of other common anions; probably because of the close similarity in

size of F^- and OH^- anions. The preferential fluoride adsorption over most other anions makes it an effective desorbing agent that can be used in lab assays for concentration of anions adsorbed on soil particles.

Boron is a micronutrient, but at high concentrations is toxic to plants. In aqueous environments, boron exists as the oxyanion borate ($B(OH)_3$ or $B(OH)_4^-$). In arid soils, relatively high concentrations of borate are common, thus monitoring and predicting soil pore-water concentrations in agricultural soils is required. Strong inner-sphere borate adsorption is an important soil chemical process that controls pore-water concentrations. Borate is a weak acid that hydrolyzes water ($pK_a = 9.24$), becoming negatively charged past pH 9:

$$B(OH)_3 + H_2O = B(OH)_4^- + H^+ \qquad (10.42)$$

Thus, borate adsorption decreases past pH 9 because it is repelled by the abundance of negatively charged surface functional groups. **Figure 10.28** shows a borate adsorption envelope for a soil. Modeling of the borate adsorption behavior predicts solid–solution partitioning of boron so that plant absorption as a micronutrient or toxicant can be estimated.

Selenite (SeO_3^{2-}), chromate (CrO_4^{2-}), molybdate (MoO_4^{2-}), vanadate (VO_4^{2-}), antimonate (SbO_4^{3-}), and

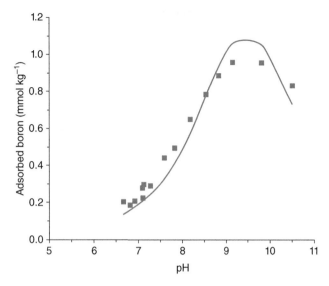

Figure 10.28 Boron adsorption by a California, USA, soil as a function of pH. The solid line is the data fit using a mechanistic model (see Chapter 11). Adapted from Goldberg (2005). Reproduced with permission of ACSESS.

either intentionally or inadvertently added to soils. Soil adsorption affects compound reactivity (including efficacy of pesticides), degradation rate, and mobility. Adsorption processes of some natural organic chemicals (e.g., dissolved organic matter), and chemicals of emerging concern are similar to pesticide adsorption processes. This section focuses on pesticides, but the principles apply to other anthropogenic organic chemicals in soils.

The reactions and fate of pesticides in soils depend on the partitioning between the solid, liquid, and gas phases in the soil, and the pesticide's degradation rate. These properties vary, depending on the soil properties. Given the number and variety of pesticides and widely different soil properties, reaction mechanisms of pesticides in soils will vary. Understanding the reactions, species, and fate in soils requires consideration of the pesticide's vapor pressure, solubility, degradation rate, degradation products, and soil properties, such as mineralogy, pH, soil porosity, and microbial activity.

arsenate (AsO_4^{3-}) occur at low concentrations in soils and are moderately to strongly adsorbed as inner-sphere oxyanions. When elevated concentrations of these oxyanions occur in soils, they are contaminants. The oxyanions are weak acids that remain protonated in the pH ranges of 2 to 9 and have pH-dependent adsorption because their charge changes as they protonate and deprotonate. Arsenate (As(V)) is chemically analogous to phosphate, and adsorption behavior of arsenate on soils mimics that of phosphate, forming strong bidentate bonds on oxide surfaces. Selenite (Se(IV)) is a strongly adsorbed, reduced form of selenium that readily oxidizes to the less strongly adsorbed (outer-sphere) selenate (SeO_4^{2-}) oxyanion. Chromate (Cr(VI)), vanadate (V(VI)), antimonate (Sb(V)), and molybdate (Mo(VI)) are oxyanions that have adsorption affinities somewhere between sulfate (lowest) and phosphate, which may vary, depending on the solution and solid conditions.

10.6 Adsorption of anthropogenic organic chemicals in soils

Industrial organic chemicals, organic chemicals of emerging concern (C_hEC, Table 3.4) and most pesticides are anthropogenic organic chemicals that are

10.6.1 Mechanisms of organic chemical retention

There are three mechanisms of organic chemical retention in soils:

1 Electrostatic attraction
2 Formation of inner-sphere bonds
3 van der Waal or hydrogen bonds involved in hydrophobic partitioning

The type of retention depends on the soil-particle properties and the properties of the organic chemical. Surface properties of soils that affect pesticide retention include surface area, surface charge and **hydrophobicity**. Soil organic matter is one of the most active components for pesticide retention. In addition to surface interactions of organic chemicals, hydrophobic organic chemicals also partition into pore spaces in soil particles and aggregates to lower their exposure to soil solution; driven by the decrease in entropy of the system. Similarly, hydrophobic organic chemicals will partition into regions of the supramolecule structures of SOM that are hydrophobic. The partitioning of hydrophobic chemicals involves van der Waal forces between the organic chemicals and the hydrophobic regions of the soil particles.

Properties of organic chemicals that control retention are molecular size, hydrophobicity, ionization, and covalent character of functional groups. The sign of charge, or lack of charge (**nonionic**), on an organic molecule is the most important property for predicting retention behavior. The charge characteristics of organic chemicals are:

1 **Positively charged organic chemicals** have **weak-base** amino functional groups that protonate when pH is less than pK_a (see Figure 3.15) or are cations such as quaternary nitrogen-containing groups (four carbons bonded to a nitrogen, e.g., R_4N^+).
2 **Nonionic organic chemicals** are uncharged and hydrophobic. The degree of hydrophobicity depends on the presence or absence of polar-covalent bonds in the molecule, such as alcoholic functional groups, and the occurrence of saturated hydrocarbons and aromatics.
3 **Negatively charged organic chemicals** have weak acidic functional groups, such as carboxyls or phenols.

Table 10.9 shows examples of pesticides from each of the above groups. Availability of pesticides for biological uptake or leaching to groundwater is directly linked to the charge character of the pesticide and the soil's surface charge characteristics. The adsorption mechanisms of the nonionic pesticides are either H-bonding or van der Waal forces. Weak base pesticides adsorb by cation exchange, and weak acids adsorb by **cation bridging** or anion exchange.

10.6.2 Adsorption of charged pesticides

Pesticides with nitrogen in an amino group bonded to three other carbon or hydrogen atoms are **weak bases** that have a lone-pair of electrons that can protonate in acid solutions:

$$R\text{-}NH_2 + H^+ = R\text{-}NH_3^+ \tag{10.43}$$

Protonation of the amine indirectly hydrolyzes a water molecule creating a OH^-, which is why they are called weak bases:

$$R\text{-}NH_2 + H_2O = R\text{-}NH_3^+ + OH^- \tag{10.44}$$

The tendency of an acid molecule to deprotonate is characterized by its pK_a. The tendency of a base molecule to accept protons is predicted by a basicity constant (pK_b). The relationship between acid and base constants is $pK_a + pK_b = 14$. Thus, for simplicity, base behavior can be predicted using acidity constants $(pK_a = 14\text{-}pK_b)$. For the protonation reaction in **Eq. 10.43** the equilibrium constant is

$$K_a = \frac{\left(R\text{-}NH_3^+\right)}{\left(H^+\right)\left(R\text{-}NH_2\right)} \tag{10.45}$$

When pH is less than the pK_a, weak bases are protonated and positively charged. Important weak base pesticides include triazine and triazole compounds (**Table 10.9**). Weak base cations are adsorbed on negatively charged soil solids such as clays and SOM. For example, the pesticide amitrole (**Table 10.9**) is a weak base that remains protonated up to pH ~ 4.5, and in acidic soils will adsorb on negatively charged clays.

Quaternary nitrogen organic compounds $(R_4\text{-}N^+)$, such as paraquat, are cationic, and adsorb to negatively charged minerals. Adsorption of cationic pesticides in soils is similar to cation exchange reactions on clay minerals, except that molecular size of the organic pesticides may limit compound access to surfaces inside small pores or interlayers.

Weak acid pesticides deprotonate when soil pH is greater than pK_a and are negatively charged. The most common weak-acid functional groups are carboxyls:

$$R\text{-}COOH = R\text{-}COO^- + H^+ \tag{10.46}$$

The pH-dependent speciation of carboxyl functional groups (R-COOH) on acetate is shown in Figure 4.4. When pH is less than pK_a, weak acid functional groups are protonated, and the pesticide is nonionic. Weak acid anion pesticides adsorb onto positively charged sites on iron and aluminum oxides or layer silicate edges. In soils with an abundance of permanently charged clay minerals, anionic pesticides are often highly mobile because there is low adsorption between the negatively charged pesticide and negatively charged clay colloids. Phenoxyacetic acids such as 2,4-D and 2,4,5-T are important weak acid pesticides that dissociate to anions. The low pK_a values for 2,4-D (**Table 10.9**) and 2,4,5-T mean that the pesticides usually exist as anions in most soils.

Table 10.9 Selected pesticides and their properties.

Pesticide	Use	Structure	pK_a	Notes
Weak bases				
Atrazine (triazine)	herbicide		1.68	Low-moderate water solubility; low adsorption
Amitrole (triazole)	herbicide		4.17	Moderate adsorption
Cationic				
Paraquat	herbicide			High water solubility; high adsorption
Chlormequat	plant growth regulator			Quaternary N compound, soluble, high adsorption, quickly degraded in soils
Nonionic				
Carbaryl (carbamates)	insecticide		none	Very low water solubility
DDT	insecticide		none	Very low water solubility
Methylisothiocyanate (MITC) (active ingredient in metam sodium)	universal pesticide		none	Volatile, polar molecule, low-moderate water solubility
Weak acids				
2,4-D	herbicide		2.80	Moderate water solubility, low adsorption in soils
Glyphosate	herbicide		$pK_{a2} = 2.30$ $pK_{a3} = 6$	Chelation with iron decreases mobility

Glyphosate (**Table 10.9**) is a commonly used herbicide that is negatively charged and strongly adsorbed as a chelate (multiple ligand-cation bonds) to iron and aluminum oxides through the phosphate functional groups. Its mobility in soils is minimal. Many naturally occurring low molecular weight organic compounds that occur in soil rhizospheres or released by microorganisms are weak acids and behave similar to anionic pesticides.

The degree of acidity or basicity (i.e., the pK_a) of a functional group is controlled by how electron rich the main organic molecule is, which is a function of electronegativity of the elements and bonds. For example, a carboxyl functional group attached to a benzene ring (**Figure 10.29**) has a lower pK_a (4.20) and is a stronger acid than a carboxyl group attached to a simple aliphatic carbon structure such as acetate ($pK_a = 4.76$). Benzene has greater electronegativity than an aliphatic carbon and withdraws more electrons from carboxyl functional groups, making the oxygen of the carboxyl less electron rich and less attractive to a proton; that is, more acidic.

In addition to creating charge on organic chemicals, functional groups can form covalent bonds with cations and mineral surfaces. Carboxyls are especially important for forming covalent bonds; organic chemicals that have carboxyl groups can adsorb onto mineral surfaces by direct bonding with the mineral's

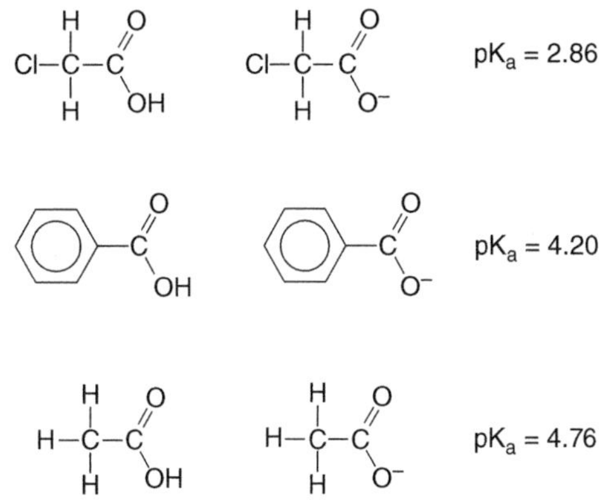

Figure 10.29 Effect of electronegativity on pK_a of carboxylic acid functional groups of chloroacetic acid, benzoic acid, and acetate. Chlorine is the most electronegative and draws the electron cloud away from the carboxyl-OH, making it more acidic (lower pK_a). The benzene ring is more electronegative than the saturated carbon in acetate, but not as electronegative as chlorine.

cations (similar to a ligand exchange reaction shown in **Eq. 10.38**), or through cation bridging where a cation such as Fe^{3+} or Ca^{2+} acts as a link between the negatively charged surface and the negatively charged pesticide.

10.6.3 Retention of nonionic organic chemicals

Many pesticides have uncharged regions, or are completely uncharged. Typically, uncharged molecular moieties are aromatic or aliphatic carbon molecules that lack acidic or basic functional groups (e.g., Figure 9.14). Alcohols are uncharged, but participate in H-bonding because of the polar oxygen bond. Hydrophobic chemicals adsorb on soil through van der Waals forces (Figure 2.5). Hydrogen bonding occurs between pesticides and nonionic oxygen ligands of minerals or SOM. Basal planes of kaolinite are examples of nonionic-mineral oxygen ligands. Individual hydrogen bonds are relatively weak, but many nonionic organic chemicals have numerous sites capable of hydrogen bonding with soils. For example, carbaryl is a nonionic insecticide (**Table 10.9**) that forms hydrogen bonds on SOM, is strongly retained in soil, and has low leachability.

Many organic molecules that are uncharged and have no apparent hydrogen bonding are nonetheless strongly retained by soils. Nonpolar pesticides prefer an environment less polar than that of the highly polar water existing in soil pores. If some other less-polar phase is present, such as SOM surfaces, the organic compounds partition from the aqueous phase to the hydrophobic regions of the SOM.

Adsorption of uncharged, nonpolar, molecules in soils occurs through van der Waals forces. Charge-induced dipole interactions and dipole-induced dipole interactions are the forces thought to be responsible for van der Waal bonding (Figure 2.5). The van der Waals attractions are weak and short-ranged, but they are additive, making the adsorption forces between hydrophobic chemicals and hydrophobic surfaces relatively strong. Hydrophobic regions of chemicals may orient themselves to form micelles or regions of hydrophobic chemical seclusion. Nonpolar molecules such as DDT or PCB, for example, associate with hydrophobic regions of SOM, and mobility is severely restricted.

Some organic chemicals that are accidently input into soils, such as trichloroethylene, are extremely hydrophobic and denser than water (dense nonaqueous phase liquids (DNAPL)). Because of their physical properties, they have little interactions with solid surfaces, and sorb very little in soils, especially subsoils. Due to their mobility, DNAPLs are a concern for groundwater contamination.

10.6.4 Predicting organic chemical retention in soil

Soils' adsorption capacity for organic molecules is continually renewed by degradation of the adsorbed molecules. Solutions containing organic molecules pass through an intricate network of soil pores, and the organic molecules adsorb (partition) out of the soil solution onto the solid phase. The ratio of a molecule's concentrations in the water and soil is the partition or **distribution coefficient** (K_d):

$$K_d = \frac{C_{soil}}{C_{solution}} \tag{10.47}$$

where C is the concentration of the organic chemical on the soil and in the soil solution. K_d is a measure of the relative solubility of a chemical in soil with units $L\,kg^{-1}$ soil. For substances that are only slightly soluble in water, the values of K_d are very large. Paraquat, a cationic pesticide, and DDT, a hydrophobic pesticide, adsorb strongly to soils and have K_d values greater than $10\,000\,L\,kg^{-1}$. The compounds are relatively immobile because they are sorbed to the soil particles instead of occurring in the soil pore water solution. Dichloropropene ($C_3H_4Cl_2$), on the other hand, a soil fumigant used to control pests, has a low K_d ($< 0.2\text{-}2\,L\,kg^{-1}$), meaning that little of the pesticide adsorbs, and it remains in the soil solution (it also has a low vapor pressure, causing it to volatilize). Atrazine, a weak base, is deprotonated at most soil pH values, and thus is nonionic. Atrazine partitions to SOM and minerals via weak van der Waal bonds, and its K_d typically ranges from $< 1\,L\,kg^{-1}$ to $50\,L\,kg^{-1}$, depending on SOM concentration (see Special Topic Box 10.4).

Soil organic matter content is often the single best characteristic for predicting pesticide adsorption in soils, especially hydrophobic pesticides. The distribution coefficient of the pesticide can be normalized for the organic matter content to a constant called K_{oc}:

$$K_{OC} = \frac{K_d}{f_{OC}} \tag{10.48}$$

where f_{OC} is the fraction of organic carbon in the soil:

$$f_{OC} = \frac{\text{Mass of organic C}}{\text{Total mass of soil}} \tag{10.49}$$

Units of K_{OC} are $L\,(kg\,OC)^{-1}$; i.e., the distribution coefficient is normalized to soil organic carbon content. K_{OC} is often relatively independent of soil type and approximates a pesticide's adsorption affinity in many different soils. Thus, by knowing the K_{OC} for an organic chemical, adsorption behavior can be predicted in several soils if their organic carbon contents are known. This is much more efficient than determining the K_d for every soil for a particular pesticide.

The relation between K_d and K_{oc} of hydrophobic organic chemicals varies with the hydrophobicity of the pesticide. Hydrophobicity is determined by measuring partitioning between a mixture of a hydrophobic and hydrophilic solution mixture, such as octanol and water. The distribution of an organic chemical between octanol and water is called **K_{OW}** (*OW* refers to the *octanol-water* partitioning):

$$K_{OW} = \frac{\text{Mass of pesticide in octanol}}{\text{Mass of pesticide in water}} \tag{10.50}$$

The correlation of K_{OW} to K_{OC} in many soils is

$$K_{OC} = 0.35\,K_{OW} \tag{10.51}$$

The variability in the K_{OC} predicted from K_{OW} in **Eq. 10.51** is a factor of ±2.5, which is a large degree of uncertainty. The uncertainty is because pesticide adsorption in soils is inherently more complex than octanol-water partitioning.

Measurement of K_{OW} is a simple laboratory test, and values for various pesticides are listed in reference and product information tables. These values and the amount of organic carbon in the soil can be used to predict K_{OC} (**Eq. 10.51**), which can be used to predict K_d; however, K_d values predicted from K_{OW} have three degrees of separation from actual chemical behavior in soils and may suffer from relatively large modeling errors. Nevertheless, K_{OW} values are useful for

Special Topic Box 10.4   Adsorption of atrazine by SOM

Atrazine is one of the most commonly used herbicides in crops and horticulture. It is an uncharged molecule in soil solutions at most pHs, and is thus hydrophobic. Atrazine, however, has electron-rich nitrogen and chlorine groups that make it polar, yielding moderate water solubility, which facilitates mobility in soils and natural waters. The hydrophobic character of atrazine causes it to preferentially partition to hydrophobic regions on SOM that can shield it from the soil solution in soils with abundant SOM.

Figure 10.30 shows the relationship between distribution coefficient of atrazine (K_d) and SOM content. Larger K_d values indicate that more atrazine is adsorbed. Most soils have SOM less than 10%; thus, the atrazine distribution coefficients are typically less than 6 ml g^{-1}, which helps explain the common occurrence of atrazine in groundwaters in agricultural areas.

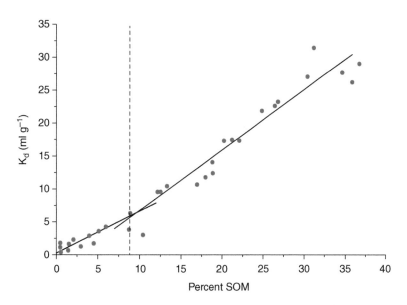

Figure 10.30 Atrazine adsorption distribution coefficients (K_d) on 32 soils as a function of soil organic matter content. Below about 8% SOM, adsorption of atrazine was a function of both organic matter content and clay mineral interactions (not shown). Adapted from Stevenson (1972), with permission.

approximations of pesticide adsorption affinities in soil and are useful for relative comparisons of pesticide mobility.

10.6.5   Aging effects on organic chemical adsorption

Adsorption of many organic chemicals is slow, and it continuously changes as the chemicals age in soils. Most of the slow adsorption process is due to diffusion of the organic chemical, either through small pores of soil aggregates (including organic matter) or within mineral particles, including clay mineral interlayers. The organic chemicals may become entrapped within organic matter structures that isolate the chemical from the bulk solution due to hydrophobic and hydrophilic interactions.

The slow, continuous adsorption processes cause an aging effect, limiting availability of the organic chemical to the soil solution (see **Figure 10.22**) and likely availability to organisms. This may be advantageous in the case of mobilization of an organic chemical out of a soil into groundwater or surface water, but negatively affects degradation of organic pollutants or the availability to work on the intended pest or plant.

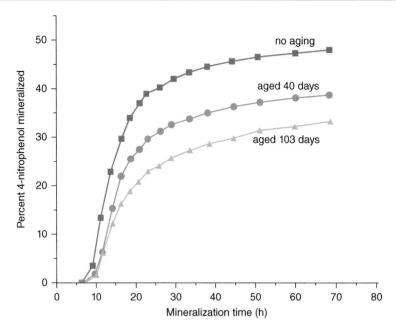

Figure 10.31 Degradation rate of 4-nitrophenol incubated in a soil for different lengths of time by an inoculated bacterium. Adapted from Hatzinger and Alexander (1995), with permission.

For example, **Figure 10.31** shows a decrease in degradation rate of 4-nitrophenol by a microorganisms by nearly 35% upon aging for 103 days. 4-nitrophenol is a nitrogen containing phenol compound that is a model compound for many pesticides and industrial chemicals input into soils.

10.7 Summary of important concepts for adsorption and desorption reactions in soils

Adsorption and desorption reactions on soil particle surfaces are amongst the most important soil processes. They are influenced by soil mineralogy, soil organic matter characteristics, specific surface area of solids, and soil pH. The most reactive soil solids, clay minerals, oxides, and SOM, have charged surfaces that create electrostatic surface potential between the surface and the ions in solution. The electrostatic potential in soils controls cation and anion adsorption reactions, especially outer-sphere cation and anion exchange reactions. Soils with high amounts of permanent charge clays have high cation exchange capacity (CEC). In soils containing oxide minerals, surface charge varies from positive at low pH, where anion adsorption is favored, to negative at high pH, where cation adsorption is favored; thus, adsorption and exchange capacity (CEC and AEC) vary with pH.

Some ions only interact with charged surfaces via electrostatic interactions, in which case, attractive and diffusive forces are the dominant factors affecting ion surface exchange. Adsorbed ions form an electrical potential on mineral surfaces that equals the surface potential. Models such as the Gouy-Chapman diffuse double layer model predict the surface potential characteristics, account for how ion size and valence affect adsorption selectivity and explain surface-to-surface interactions that cause flocculation and dispersion.

Ions that form covalent bonds with surface functional groups on minerals and organic matter are adsorbed as inner-sphere surface complexes. These reactions often displace protons or hydroxides and are pH-dependent. As pH increases, metal cation adsorption increases, creating pH-adsorption edges. At low pH, anion adsorption increases. Many anions (e.g., phosphate) are weak acids, so when soil pH is less than the weak acid pK_a, protonation decreases the negative charge and decreases the retention of anions on the soil surfaces. This type of anion adsorption creates a pH-dependent adsorption behavior called an adsorption envelope.

Adsorption of organic chemicals such as pesticides depends on their molecular characteristics – especially whether the chemical has ionizable functional groups or is uncharged. Cationic and anionic functional groups on organic chemicals promotes adsorption to mineral and SOM surfaces similar to inorganic cations and anions. Uncharged organic chemicals have hydrophobic characteristics and partition to hydrophobic regions of soils, especially SOM, which can be predicted using a distribution coefficient (K_d).

The solid–solution interfaces of soil minerals and soil organic matter are important soil properties that control adsorption of chemicals in the environment. Adsorption of chemicals removes them from solution, and their availability for plant or microbe uptake and leaching potential are decreased. Thus, predicting nutrient and pesticide availability in soils and contaminant fate relies on understanding details of chemical adsorption processes. Quantitative adsorption modeling is discussed in more detail in the next chapter.

Questions

1 The following distribution of cations and anions exists near a soil colloid surface:

Distance	4.0 nm	3.0 nm	2.0 nm	1.0 nm	0.5 nm	0.25 nm
Cation concentration (mol(+) L⁻¹)	0.10	0.12	0.17	0.35	1.0	2.0
Anion concentration (mol(−) L⁻¹)	0.10	0.08	0.06	0.04	0.01	0.00

Assume that the excess of cations at the midpoint of each increment represents the entire increment (e.g., that the cation concentration is 2.0 mol charge L^{-1} from the colloid surface to 0.375 nm from the surface, etc.). Estimate the CEC for a colloid having $800 \times 10^3 \, m^2 \, kg^{-1}$ of reactive surface (Ans. = 12.0 mmol charge kg^{-1}).

2 Generate a selectivity diagram like **Figure 10.13** for two monovalent cations (A and B) having Kerr-type coefficients of
 a 0.5
 b 1.0
 c 2.0

3 A surface soil with vermiculite in it can fix 25 mmol kg^{-1} of K^+ or NH_4^+. What rate of $(NH_4)_2SO_4$ or KCl fertilizer (in kg ha⁻¹) would be required to saturate this fixation capacity for a 30-cm depth of soil?

4 A saturation and displacement method is used to determine CEC of a soil. If 5 g of soil retains 3 g 0.1 M(+) index solution after centrifugation and decanting, and if the total index cation retained is subsequently determined to be 1.6 mmol(+), what is the CEC of the sample? Based on the anion distribution of Problem 1, what percentage error is contributed in this case by anion exclusion if the soil has a reactive surface area of $200 \times 10^3 \, m^2 \, kg^{-1}$?

5 What is the relative attraction for a dehydrated K^+ ion residing directly on the mineral surface of a tetrahedrally substituted 2:1 mineral when compared to an octahedrally substituted 2:1 mineral with the same layer charge?

6 Explain the role of *surface activity* in exchange equilibrium models and assumptions in the Kerr, Vanselow, and Gapon models.

7 For a CEC procedure that uses Na^+ as the index cation, H_2O/ethanol as the wash solvent, and Mg^{2+} as the displacing cation, discuss the effect of each of the following on CEC measurements.
 a Hydrolysis due to excess washing
 b Presence of large amounts of lime or gypsum in the soil
 c Incomplete index-cation saturation
 d Precipitation of an insoluble Na^+ salt in the ethanol
 e Incomplete removal of the index cation by Mg^{2+}

8 What are the forces acting on anions with distance from mineral surfaces? How will these forces vary from soil to soil? What is the dominant force acting on anions in most agricultural soils?

9 How do the following factors affect anion repulsion?
 a Anion charge
 b Anion concentration
 c Exchangeable cation
 d Soil pH
 e Other anions

10 Explain the different ways in which anions adsorb to surfaces in soils. Give examples of each reaction type.

11 What reactions are responsible for the fixation of phosphate in acid and basic soils?

12 Explain in your own words the peaks and inflection points of **Figure 10.27**.

13 Maximum phosphate availability in soils tends to occur around pH 6 to 6.5. Explain why in terms of PZC and pK$_a$ values for phosphate.

14 Draw a graph of As(V) (a triprotic oxyanion) adsorption on goethite as a function of pH. Hint: Draw a qualitative shape to show relative changes; the pK$_{a1}$, pK$_{a2}$, and pK$_{a3}$ for As(V) are 2.24, 6.74, and 11.6, respectively.

15 Determine which conditions in each of the following scenarios will create more dispersion or flocculation and give reasons.

   a High Na concentration vs. high Ca concentration

   b High Na concentration vs. low Mg concentration

   c High temperature vs. low temperature

16 Within the following pairs of cations, determine which will preferentially sorb into the interlayer of clays and explain why.

   a Na$^+$ or Cs$^+$

   b Al^{3+} or Ca^{2+}

   c Ca^{2+} or Mg^{2+}

17 Describe how a system containing goethite, copper, and sulfate would be expected to complex sulfate at pH > 7. Assuming copper is adsorbed as an inner-sphere complex and sulfate is adsorbed as outer-sphere complex, show which surface adsorption planes (see Figure 9.3) copper and sulfate exist on the goethite surface.

18 Why is desorption important?

19 Put the following in order of their relative availability for desorption: outer-sphere complex, surface precipitate, monodentate complex, bidentate complex.

20 Which soil components would you expect the following pesticides to adsorb to? Which would have the greatest K$_{OW}$?

   a DDT

   b Paraquat

   c 2,4-D

21 What is the difference between surface potential, surface charge, and the electric potential at a point within the DDL? How would inner sphere adsorption of a cation change surface potential?

22 What assumptions are required for the Vanselow selectivity coefficient to be valid?

23 Derive the K$_{Kerr}$ coefficients for the data in **Table 10.5**, and compare to the values in **Table 10.6**.

Bibliography

Auxtero, E., M. Madeira, and E. Sousa. 2004. Variable charge characteristics of selected Andisols from the Azores, Portugal. Catena 56:111–125.

Baker, L.L., D.G. Strawn, and R.W. Smith. 2010. Cation Exchange on Vadose Zone Research Park Subsurface Sediment, Idaho National Laboratory. Vadose Zone Journal 9:476–485.

Bear, F.E. 1964. Chemistry of the Soil. 2nd ed. Reinhold Pub. Corp., New York.

Boersma, L., J.W. Cary, D.D. Evans, A.H. Ferguson, W.H. Gardner, R.J. Hanks, R.D. Jackson, W.D. Kemper, D.E. Miller, D.R. Nielsen, and G. Uehara. 1972. Soil Water. Soil Science Society of America, Madison. WI.

Cornell, R.M., and U. Schwertmann. 1996. The Iron Oxides: Structure, Properties, Reactions, Occurrence, and Uses VCH, Weinheim; New York.

Elrashidi, M.A., and G.A. O'Connor. 1982. Boron Sorption and Desorption in Soils. Soil Science Society of America Journal 46:27–31.

Gast, R.G. 1972. Alkali-Metal Cation-Exchange on Chambers Montmorillonite. Soil Science Society of America Proceedings 36:14-.

Goldberg, S. 2005. Equations and models describing adsorption processes in soils, pp. 489–518, *In* M. A. Tabatabai and D. L. Sparks, (eds.) Chemical Processes in Soils. Soil Science Society of America, Madison, WI.

H. van Olphen. 1964. An Introduction to Clay Colloid Chemistry. Interscience Publishers, New York, NY.

Hatzinger, P.B., and M. Alexander. 1995. Effect of aging of chemicals in soil on their biodegradability and extractability. Environmental Science & Technology 29:537–545.

Hingston, F.J., J.P. Quirk, and A.M. Posner. 1972. Anion adsorption by goethite and gibbsite .1. Role of proton in determining adsorption envelopes. Journal of Soil Science 23:177–192.

Jenny, H., and R.F. Reitemeier. 1934. Ionic exchange in relation to the stability of colloidal systems. The Journal of Physical Chemistry 39:593–604.

Lentz, R.D., R.E. Sojka, and D.L. Carter. 1996. Furrow irrigation water-quality effects on soil loss and infiltration. Soil Science Society of America Journal 60:238–245.

Mclaren, R.G., and D.V. Crawford. 1973. Studies on soil copper 2. Specific adsorption of copper by soils. Journal of Soil Science 24:443–452.

Morais, F.I., A.L. Page, and L.J. Lund. 1976. The effect of pH, salt concentration, and nature of electrolytes on the charge characteristics of Brazilian tropical soils. Soil Sci. Soc. Am. J. 40:521–527.

Peterson, F.F., J. Rhoades, M. Arca, and N.T. Coleman. 1965. Selective adsorption of magnesium ions by vermiculite. Soil Science Society of America Journal 29:327–328.

Sparks, D.L. (ed.) 1997. Methods of Soil Chemical Analysis. Soil Science Society of America, Madison, WI.

Srivastava, P., B. Singh, and M. Angove. 2005. Competitive adsorption behavior of heavy metals on kaolinite. Journal of Colloid and Interface Science 290:28–38.

Stevenson, F.J. 1972. Organic matter reaction involving herbicides in soil. J Environ Qual 1:333–343.

Strawn, D.G., and D.L. Sparks. 2000. Effects of soil organic matter on the kinetics and mechanisms of Pb(II) sorption and desorption in soil. Soil Science Society of America Journal 64:144–56.

Vanraij, B., and M. Peech. 1972. Electrochemical properties of some oxisols and alfisols of tropics. Soil Science Society of America Proceedings 36:587.

Zhang, X., S.Q. Zeng, S.B. Chen, and Y.B. Ma. 2018. Change of the extractability of cadmium added to different soils: Aging effect and modeling. Sustainability 10:1–14.

Zhuang, P., M.B. McBride, H.P. Xia, N.Y. Li, and Z.A. Lia. 2009. Health risk from heavy metals via consumption of food crops in the vicinity of Dabaoshan mine, South China. Science of the Total Environment 407:1551–1561.

11 MEASURING AND PREDICTING SORPTION PROCESSES IN SOILS

11.1 Introduction

The Nobel Prize winning physicist Wolfgang Pauli (1900–1958) wrote that *God made the bulk; surfaces were invented by the Devil* – referring to the complexity of surficial processes and the difficulty in describing them using physics and mathematics. Reactions occurring on particle surfaces are indeed complex, yet the complexity that has perplexed modelers and theorists is one of nature's most important processes with respect to evolving and supporting life. This chapter presents information on quantitative modeling of adsorption reactions, **adsorption kinetics**, **surface precipitation**, and methods for measuring adsorption processes. These topics are fundamental to predicting the fate, transport, and bioavailability of chemicals in soils.

11.2 Adsorption experiments

The aim of an adsorption experiment is to measure the amount of chemical that attaches to a surface as a function of some variable. The three most common variables in adsorption experiments are: (1) equilibrium solution concentration of the chemical of interest; (2) reaction time; and (3) equilibrium solution pH (**Figure 11.1**). Other variables that may be altered are **solid-to-solution ratio**, temperature, ionic strength, and background electrolyte. In a well-designed adsorption experiment, only a single parameter is varied. This is sometimes difficult because equilibrium may be slow, pH can fluctuate, and secondary reactions such as precipitation may occur.

The basic setup of an adsorption experiment is to combine a soil or mineral with a chemical of interest,

Soil Chemistry, Fifth Edition. Daniel G. Strawn, Hinrich L. Bohn, and George A. O'Connor.
© 2020 John Wiley & Sons Ltd. Published 2020 by John Wiley & Sons Ltd.

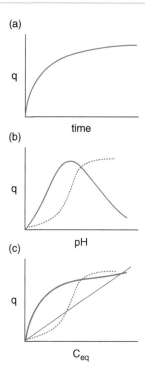

(a)

time

(b)

pH

(c)

C_{eq}

Figure 11.1 Amount of chemical sorbed (q) as a function of three variables: (a) time, showing an initial fast reaction rate; (b) pH, showing a weak acid adsorption-envelope and cation adsorption-edge; and (c) equilibrium concentration (C_e), showing a convex-adsorption curve, sigmoidal-adsorption curve, and linear adsorption.

and, after a given time, separate the soil solution from the solid. At the end of the incubation, the concentration of the chemical of interest left in solution is measured (C_{final}). If the adsorption reactions in the experiment are at equilibrium, C_{final} is the equilibrium concentration (C_{eq}). The amount of chemical retained by the solid (q) is calculated using the following mass balance equation:

$$q = \left(C_{initial} - C_{final} \right) \times \frac{V}{m} \quad (11.1)$$

where C is concentration in the *initial* and *final* solutions, V is the volume of solution in liters, and m is the mass of solid in kilograms. Units of concentration are typically either mmol L^{-1} or mg L^{-1}. Units of adsorbed chemical are then either mmol kg^{-1} or mg kg^{-1}, where the latter is equivalent to a part per million (ppm) (see Section 1.8.1).

A critical factor in an adsorption experiment is the mass of solid to the volume of solution, called the solid

to solution ratio (solid : solution), or suspension density. In nature, the soil's solid phase is much greater than volume of solution, even when the soil is saturated with water. To eliminate physical transport and diffusion factors from an adsorption experiment, lower solid to solution ratios are used, typically ranging from 1:2 to 1:200 (in units of g:ml or the equivalent kg:L). The lower solid-to-solution ratios used in laboratory experiments allow for uniform mixing of the soil suspensions, which eliminate time-dependent chemical transport processes so that only the soil chemical reaction processes are being studied. Separate experiments using soil columns or field studies are required to measure physical movement factors for solutions and chemicals. Since the amount of sorption is normalized by the solid to solution ratio in **Eq. 11.1** (V/m), the results of adsorption experiments at different solid to solution ratios should be the same, allowing transfer of the adsorption behavior to the higher solid to solution field conditions. Sometimes, however, varying solid to solution ratio causes unintended secondary reactions such as surface precipitation or changes to equilibrium pH, which will alter the adsorption behavior.

11.3 Predicting adsorption using empirical models

The amount of chemical adsorbed is commonly expressed as the **distribution coefficient** (K_d) (also called a *partition coefficient*). K_d is calculated by dividing the amount of chemical on the solid by the amount in solution (C_{eq}):

$$K_d = \frac{q}{C_{eq}} \quad (11.2)$$

Units of K_d are L kg^{-1}. High values of K_d represent a high affinity of the surface for the chemical, and low values indicate a low affinity.

Commonly, K_d is evaluated over a range of concentrations (e.g., **Figure 11.1c**), called an **adsorption isotherm**. An adsorption isotherm shows the amount of chemical sorbed as a function of its equilibrium solution concentration (C_{eq}). Thus, K_d is the slope of an adsorption isotherm for a linear adsorption isotherm, or a section along its curve in a nonlinear adsorption isotherm.

11.3.1 Linear adsorption isotherms

If the concentration on the solid increases in *direct* proportion to the concentration in solution, the isotherm is linear, and K_d is constant. The concentration on the solid can then be calculated for any solution concentration by

$$q = K_d \times C_{eq} \tag{11.3}$$

Linear isotherms occur for solids and chemicals in which the chemical partitions to the surface without any interactions from the changing surface properties, and there is no maximum adsorption capacity for the chemical. However, if chemical concentrations are increased beyond a normal range, partitioning may deviate from the linear trend because secondary reactions occur at higher concentrations.

Many organic chemicals partition to soils via linear isotherms. **Figure 11.2** shows the adsorption of atrazine, a nonionic solute, on soil as a function of concentration in solution.

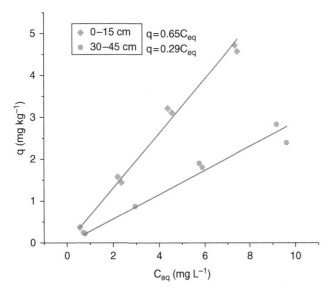

Figure 11.2 Atrazine adsorption on a surface and subsurface fine sand soil at pH ~6.5. The surface 0–15 cm soil had 6.7 g kg^{-1} of organic carbon and the subsurface 30–45 cm soil had 2.8 g kg^{-1} of organic carbon. The decreased organic carbon in the 30–45 cm soil resulted in a much-lower K_d (0.29 L kg^{-1}) as compared to the 0–15 cm soil sample ($K_d = 0.65$ L kg^{-1}). Adapted from Mbuya et al. (2001), with permission.

11.3.2 Nonlinear adsorption isotherms

For most inorganics and charged organic chemicals, adsorption is nonlinear because the affinity of the surface for the chemical changes as the concentration on the surface changes, and because the surface may have a finite (maximum) adsorption capacity. A variety of isotherm *shapes* are possible, depending on the affinity of the chemical for the solid surface and the surface properties. The two most common are a sigmoidal curve (**S-curve**) and a convex curve (**L-curve**) (**Figure 11.1c**).

The S-curve represents an adsorption process that has a low K_d slope at low solution concentrations, but as concentrations increase, the slope increases and then levels off at the highest concentrations. The cause of the increasing slope may be a secondary reaction occurring, such as co-ion interaction, surface precipitation, or formation of ion complexes.

The slope of the L-curve is high at low solution concentrations, but as solution concentrations increase, the slope of K_d decreases. There are several deviations of the L- and S-curves that affect the slopes and maximum adsorption values. L-curves are more common than S-curves for most adsorption processes. Two important types of L-curves differ based on whether there is a **maximum adsorption** amount or a gradual but continuous increase in adsorption amount as solution concentration increases.

To model L-curve adsorption behavior (q as a function of C_{eq}), the Langmuir and the Freundlich equations are commonly used. The **Langmuir equation** was initially derived for the adsorption of gases by solids by Nobel Prize winning chemist Irving Langmuir. The derivation was based on three assumptions: (1) a constant energy of adsorption that is independent of the extent of surface coverage (i.e., a homogeneous surface); (2) adsorption on specific sites, with no interaction between the chemicals once adsorbed; and (3) maximum adsorption is equal to a complete mono-molecular layer on all reactive adsorbent surfaces. The Langmuir equation is

$$q = \frac{K_L C_{eq} q_{max}}{1 + K_L C_{eq}} \tag{11.4}$$

where K_L is a constant and q_{max} is the maximum adsorption capacity for the soil or mineral (i.e., a theoretical mono-molecular layer). Rearranging **Eq. 11.4** yields a

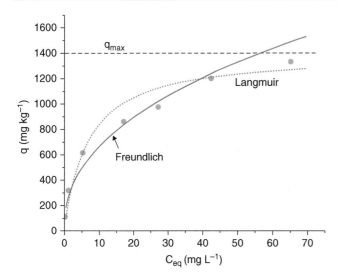

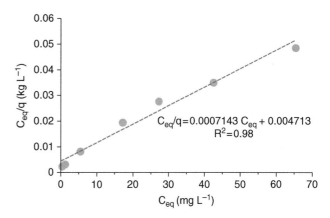

Figure 11.4 Isotherm data from **Figure 11.3** plotted according to linear form of the Langmuir equation. Best-fit linear line is shown.

Figure 11.3 Isotherm of phosphate adsorption on a soil (silty clay, pH = 4.6). The data were fit using Langmuir and Freundlich equations. The predicted isotherms are based on derived parameters in **Table 11.1**. Data from Osborne et al. (2015).

Table 11.1 Isotherm model fit parameters for phosphate adsorption data on soil shown in **Figure 11.3**.

	Langmuir equation	Freundlich equation
Linear equation	$C_{eq}/q = 1/q_{max} \times C_{eq} + 1/(K_L q_{max})$	$\log q = n \log C_{eq} + \log K_F$
Linear fit	$C_{eq}/q = 0.00071 C_{eq} + 0.0047$	$\log q = 0.43 C_{eq} + 2.4$
Slope	$1/q_{max} = 0.00071$	n = 0.4285
Intercept	$1/(K_L q_{max}) = 0.0047$	$\log K_F = 2.4$
K	0.15	250
n	–	0.43
q_{max} (mg kg^{-1})	1400	–

linear form that can be used to graphically fit data and derive the equation parameters K_L and q_{max}:

$$\frac{C_{eq}}{q} = \frac{1}{q_{max}K_L} + \frac{C_{eq}}{q_{max}} \tag{11.5}$$

It may not be readily apparent that this is a linear equation of form $y = mx + b$, but, in a plot of C_{eq}/q (*y-axis*) versus C_{eq} (*x-axis*), the slope (*m*) is $1/q_{max}$ and the intercept (*b*) is $1/(q_{max}K_L)$. If adsorption conforms to the Langmuir model, plotting C_{eq}/q versus C_{eq} yields a straight line, and the parameters in **Eq. 11.5** can be calculated. **Figure 11.3** shows an adsorption isotherm for phosphate on soil. The data are plotted in the linear form of the Langmuir equation in **Figure 11.4**. **Table 11.1** shows the results of the linear fit and the derived fitting parameters. The *modeled* adsorption isotherm based on the Langmuir equation is shown in **Figure 11.3**.

An advantage of using the Langmuir equation for describing adsorption is the identification of an adsorption maximum for the solid. This limit has been used to estimate the adsorption capacity of soils for metals, anions and various herbicides. When solute concentrations exceed the precipitation saturation point of a particular solid, the *measured q* can exceed q_{max}. Thus,

q_{max} is a maximum adsorption capacity only in the range of concentrations for which it is measured, and for the solution conditions (i.e., pH and ionic strength) of the suspension for which it was measured.

The **Freundlich equation** models L-curve data that do not have a clear adsorption maximum:

$$q = K_F \left(C_{eq}\right)^n \tag{11.6}$$

where K_F and n are empirical constants[1]. The equation implies that the adsorption affinity decreases logarithmically as the amount of covered surface increases. Like the Langmuir equation, the Freundlich equation has a theoretical underpinning. The assumptions of the

---

[1] The theoretical Freundlich model uses $1/n$ instead of n. For soils however, the exponential is an empirical fitting parameter, thus there is no benefit to using the fractional adjustable parameter over just n.

equations, however, are not valid for soil systems, and it is best to consider them as empirical curve-fitting of adsorption data rather than describing mechanisms of adsorption.

The linear form of the Freundlich equation is

$$\log q = \log K_F + n \log C_{eq} \qquad (11.7)$$

where the slope of *log q* versus *log C_{eq}* equals *n*, and the intercept equals *log K_F*. The adsorption data in **Figure 11.3** are plotted in the linear-Freundlich equation form in **Figure 11.5**. Fit parameters are listed in **Table 11.1**. The modeled adsorption isotherm based on the Freundlich equation is shown in **Figure 11.3**.

The frequent good fit of adsorption data to the Freundlich equation is influenced by the insensitivity of log-log plots and by the flexibility afforded curve fitting by the two empirical constants (K_F and n). This flexibility does not guarantee accuracy if the data are extrapolated beyond the experimental range. The Freundlich equation has the further limitation that it does not predict a maximum adsorption capacity, which is a useful parameter to use as an adsorption property of a soil or mineral (even though it may be a pseudo-maximum). The Freundlich equation is commonly included in chemical transport models for predicting adsorption behavior in soil because of its relatively simple mathematical form and ability to represent the commonly observed L-curve adsorption behavior.

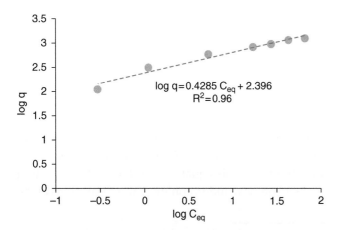

Figure 11.5 Isotherm data from **Figure 11.3** plotted according to linear form of the Freundlich equation. Best-fit linear line is shown.

The linearization of adsorption equations such as in **Eqs. 11.5** and **11.7** is useful for fitting data. However, modern fitting algorithms implemented in computers can fit nonlinear equations such as the Langmuir and Freundlich by minimizing residuals between the fit and the untransformed data. Nonlinear fitting models often result in slightly better fits to the data than the linear fits of the transformed equations.

A number of studies involving the adsorption of solutes from solution by mineral surfaces have suggested that adsorption isotherm behavior is a combination of multiple L-type adsorption isotherm shapes; especially as the range of C_e values increases. Multi-site adsorption models have been developed that postulate several different arrays of sites, each fulfilling the requirements of the Langmuir model. An S-curve is an example of an adsorption isotherm behavior that has a multi-site adsorption behavior. To fit multi-site adsorption isotherms, two-, three-, and four-site Langmuir models have been invoked that sum the adsorption in each section of the isotherm.

11.4 Predicting adsorption using mechanistic models

In aqueous solutions, thermodynamic models are used to predict chemical speciation of acids, complexes, and precipitation and dissolution of solids (Chapter 4). The theory of aqueous thermodynamic models for prediction of redox reactions is also well developed, although implementation to natural environments is limited.

Thermodynamic models are **mechanistic models** because they include the **chemical speciation** of the reactants and products. Mechanistic aqueous geochemical models are accurate if the species are correct and accurate thermodynamic data are available. For surface adsorption reactions, mechanistic models, while accurate in some cases, have not been as widely adopted as aqueous geochemical modeling because surface species and reactions are not as well known, and accurate thermodynamic data are not available.

Several mechanistic models of surface chemical reactions exist, and in many cases, prediction of *possible* reactants and products is reliable. However, because of the uncertainty in the surface reactions, there is not universal agreement that the model's adsorption mechanisms are correct. Furthermore, the **mechanistic adsorption models** make assumptions that may not be valid, and require complex mathematical modeling and parameter fitting. Because of these limitations, mechanistic adsorption models are often only applicable to systems for which they are tested and parameters optimized. Despite the limitations and challenges of surface complexation modeling, powerful computers and analytical tools capable of providing molecular details of adsorption reactions are allowing for progress in model development. Newer models are capable of reasonable accuracy in predicting adsorption when soil mineralogy, soil organic matter content, and soil physicochemical properties (e.g., pH and CEC) are included. This section presents the basics of mechanistic adsorption models.

Adsorption reactions are thermodynamically driven along a free energy gradient (Figure 1.13). Using Gibbs free energy terms to represent adsorption thermodynamics of chemical and electrostatic interactions yields

$$\Delta G^\circ_{ads} = \Delta G^\circ_{int} + \Delta G^\circ_{coul} \tag{11.8}$$

where ΔG°_{ads} is the overall adsorption energy of the reaction at equilibrium, ΔG°_{int} is the free energy involved in making and breaking of bonds on the surface (**intrinsic adsorption energy**), and ΔG°_{coul} is the Coulombic energy associated with the electric potential of the chemical in the electrostatic field of the surface charge. Recall from Chapter 10 that electric potential near a charged surface is calculated from **surface potential** (ψ_0), which is related to surface charge (σ_p). Using the relationship $\Delta G^\circ = -RT \ln K$ to transform free energy components in **Equation 11.8** to equilibrium constants yields

$$K_{ads} = K_{int} K_{coul} \tag{11.9}$$

where K_{ads} is a **conditional equilibrium constant** because it changes with surface properties, solution ionic strength, and pH. K_{int} is a constant that represents the *chemical interaction* of the surface and the adsorbing chemical – making and breaking bonds and dehydration. K_{coul} is a constant that accounts for the influence of electrostatic interactions of the surface upon adsorption. To model K_{coul}, the Coulombic free energy $\left(\Delta G^\circ_{coul}\right)$ is related to surface potential (ψ_0) by:

$$\Delta G^\circ_{coul} = F \Delta Z \psi_0 \tag{11.10}$$

where F is Faraday's constant and ΔZ is the change in the surface charge after adsorption of the chemical. K_{coul} is derived from **Eq. 11.10** using the relationship $\Delta G^\circ = -RT \ln K$:

$$K_{coul} = e^{\frac{-F\Delta Z\psi_0}{RT}} \tag{11.11}$$

Substituting **Eq. 11.11** in **Eq. 11.9** yields

$$K_{ads} = K_{int} e^{\frac{-F\Delta Z\psi_0}{RT}} \tag{11.12}$$

K_{ads} is related to surface concentrations, and K_{int} is the adsorption equilibrium constant that modelers need to predict adsorption equilibrium. Rearranging **Eq. 11.12** to solve for K_{int} yields

$$K_{int} = K_{ads} e^{\frac{F\Delta Z\psi_0}{RT}} \tag{11.13}$$

There are two aspects that make calculations with **Eq. 11.13** challenging: (1) Surface potential is not measured, but modeled; and (2) the change in charge of the surface (ΔZ) depends on the surface functional groups and the type of surface complex formed with adsorbing chemical.

Challenge 1   Modeling surface potential

Several models relate surface potential to surface charge ($\psi_0 = f(\sigma_p)$). The diffuse double layer model and the pH-variable model discussed in Chapter 10 are two such models. The Stern model is another. These models are amongst the simplest, but make some assumptions that are incorrect, such as the ions are point charges and uniformly distributed. Variants of these models have been developed to relate the different surface charge factors to surface potential by dividing the surface charge into individual surface-potential components. Recall that the surface charge components are related through the charge balance equation $\sigma_p = \sigma_o + \sigma_H + \sigma_{IS} + \sigma_{OS} = -\sigma_d$ (Table 9.1). Thus, each surface charge component has a distinct contribution to the surface potential. **Figure 11.6** shows examples of the surface-charge components and their relationships to surface potential. The challenge in modeling surface potential is using a mathematical model that

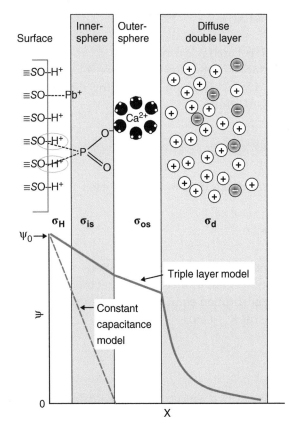

Figure 11.6 Model of adsorption mechanisms on a mineral surface and surface potential as a function of distance (x) from the surface. The surface charge is comprised of proton adsorption, Pb^{2+} and PO_4^{3-} inner-sphere adsorption, hydrated Ca^{2+} outer-sphere adsorption, and cation and anion adsorption in the diffuse double layer. The constant capacitance model includes surface charge from proton adsorption and inner-sphere adsorption ($\sigma_H + \sigma_{is}$). The triple-layer model includes surface charge created by proton adsorption, inner- and outer-sphere adsorption, and a diffuse layer.

accurately represents the relation of surface charge to surface potential.

Challenge 2   Determining change in surface charge

Inner-sphere adsorption of ions or protons ($\sigma_H + \sigma_{IS}$) changes the surface charge (ΔZ in **Eq. 11.13**), and depends on the surface functional groups and the number of bonds the adsorbing chemical makes with the functional groups. Modelers determine surface functional group structure based on theoretical descriptions of the mineral surfaces, such as in Table 9.2. In some cases, particularly for pure minerals, surface

functional group structure is determined experimentally; for example, using a technique such as FTIR spectroscopy. Thus, the molecular coordination of the adsorbed chemical is either theoretically modeled based on possible structures or experimentally measured. In either case, determining a molecular structure on a surface is challenging, especially in complex systems where multiple types of surfaces and surface complexes occur. Thus, predicting ΔZ and charge distribution is not a trivial matter.

A somewhat simple example of a mechanistic surface complexation model is the deprotonation reaction of a generic surface functional group, such as occurs on oxide mineral surfaces:

$$\equiv SOH = \equiv SO^- + H^+ \tag{11.14}$$

where $\equiv SO^-$ represents a surface functional group. The conditional adsorption equilibrium constant for this reaction (K_{ads}) is

$$K_{ads} = \frac{\left[\equiv SO^-\right]\left(H^+\right)}{\left[\equiv SOH\right]} \tag{11.15}$$

where brackets indicate concentrations, and activity coefficients of the surface species are assumed to be equal ($\gamma_{\equiv SO^-} = \gamma_{\equiv SOH}$), and thus cancel each other.

Substituting **Eq. 11.15** in **Eq. 11.13** yields

$$K_{int} = \frac{\left[\equiv SO^-\right]\left(H^+\right)}{\left[\equiv SOH\right]} e^{\frac{F\Delta Z\psi_0}{RT}} \tag{11.16}$$

Assuming there are only two species of surface sites, the total concentration of surface sites is

$$C_T = \left[\equiv SO^-\right] + \left[\equiv SOH\right] \tag{11.17}$$

Solving **Eq. 11.17** for concentration of protonated surface sites, substituting into **Eq. 11.16**, and rearranging yields

$$\left[\equiv SO^-\right] = \frac{C_T K_{int}}{\left(H^+\right)K_{coul} + K_{int}} \tag{11.18}$$

where K_{coul} is the Coulombic term defined in **Eq. 11.11**. Thus, **Eq. 11.18** shows that the concentration of the deprotonated surface sites is a function of the total surface sites (C_T), ψ_0, K_{int}, and (H^+). Total surface sites can be measured, and surface potential can be calculated using a model that relates surface potential to surface charge such as the diffuse double layer model. Thus, if K_{int} is known, the number of deprotonated sites can be calculated as a function of pH. K_{int} for many minerals are determined by fitting models to data acquired by pH titration of mineral suspensions. Calculating the protonation of surface functional groups is analogous to thermodynamic calculation of the distribution of an aqueous acid protonation and deprotonation reaction, as shown in Chapter 4 for phosphoric and carbonic acids.

The reaction in **Eq. 11.14** is a simple model for the protonation of a mineral surface functional group. Mineral surfaces have multiple types of surface functional groups, and ion adsorption makes surface complexation modeling even more intricate.

Figure 11.7 shows the result of surface protonation and deprotonation calculations on the surface charge

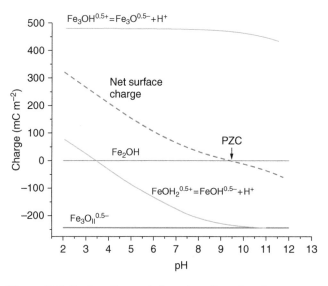

Figure 11.7 Protonation and deprotonation of surface functional groups on goethite modeled using theoretical surface protonation constants (K_{int}). Each line represents the sum of the protonated and deprotonated species in the reaction. This model predicts that the goethite surface has two types of $Fe_3OH_n^{n-0.5}$ functional groups (see Figure 9.7). K_{int} for one surface functional group is very low and it does not protonate. The net surface charge is the sum of the charge on the surface functional groups. When the sum of charges equals zero, the pH is equal to the mineral's PZC. Adapted from Hiemstra et al. (1996). Reproduced with permission of Elsevier.

of goethite, and the net surface charge reactions. The surface functional group speciation curves in **Figure 11.7** used protonation and deprotonation reactions for several different surface species, which were input into algorithms in a computer model that included a surface charge to surface potential model, a fixed number of reactive sites, and intrinsic equilibrium constants for the various protonation and deprotonation reactions on the goethite surface.

Information from surface protonation speciation of goethite can be used to predict surface reactivity for adsorption of cations and anions. In addition, if K_{int} values for the adsorption reactions are known, cation and anion adsorption can be quantitatively predicted using a similar modeling approach, as done for the protonation/deprotonation reaction predictions shown in **Figure 11.7**. K_{int} for adsorption of many cations and anions on pure minerals have been measured, and are reported in reference tables, which allows thermodynamic modeling of adsorption behavior.

A major advantage of mechanistic surface adsorption models is the ability to predict adsorption behavior across a wide range of conditions, such as pH and concentration. Thus, adsorption isotherms and pH-dependent adsorption can be predicted a-priori. This may be cheaper and faster than actually measuring the adsorption behavior for soil samples. In addition, output from mechanistic adsorption models can be included in reactive-transport models used to predict chemical fate and transport.

Results of a fit of a mechanistically based adsorption model for boron adsorption on a soil are shown in Figure 10.28. The mechanistic model is based on the boron adsorption reaction

$$\equiv SOH + H_3BO_3 = \equiv SH_3BO_4^- + H^+ \tag{11.19}$$

which has a constant of

$$K_{int} = \frac{\left[\equiv SH_3BO_4^-\right]\left(H^+\right)}{\left[\equiv SOH\right]\left(H_3BO_3\right)} \times e^{\frac{-F\psi_0}{RT}} \tag{11.20}$$

where the right-hand exponential is the K_{coul} (**Eq. 11.11**) with $\Delta Z = -1$ (see reaction in **Eq. 11.19**). Solving **Eq. 11.20** for [$\equiv SH_3BO_4^-$] requires knowing the value of the adsorption constant (K_{int}), and a model to predict overall surface charge balance and surface potential (i.e., a surface adsorption model (**Figure 11.6**)). For the fit curve in Figure 10.28, the surface potential was solved using a model that assigns a constant capacitance to the electric potential. The constant capacitance assumes a uniform (linear) surface potential related to net surface charge (σ) (**Figure 11.6**):

$$\sigma_p = \frac{CSa}{F}\psi_0 \qquad (11.21)$$

where C (coulombs V^{-1} m^{-2}) is a capacitance density parameter, S (m^2 kg^{-1}) is the specific surface area, a (kg m^{-3}) is the concentration of the solid in aqueous suspension, and F (96485 coulombs mole^{-1}) is the Faraday constant. Surface charge (σ_p) is determined as a function of pH and adsorption complex (described in Eq. 9.1):

$$\sigma_p = \sigma_o + \sigma_H + \sigma_{IS} + \sigma_{OS} \qquad (11.22)$$

where the adsorption charge is predicted from a reaction equation such as in **Eq. 11.19**. Combining the adsorption complex charge and surface charge from protonation and permanent charge, the net surface charge for boron adsorption (Goldberg et al. 2000) is

$$\sigma_p = \left[\equiv SOH_2^+\right] - \left[\equiv SO^-\right] - \left[\equiv SH_3BO_4^-\right] \qquad (11.23)$$

where brackets indicate concentration in mol L^{-1}. From this equation, surface potential can be modeled and used with the intrinsic adsorption complexation constant to calculate the amount of boron adsorbed at certain system conditions.

Using empirical relationships, it has been shown that K_{int} values can be satisfactorily predicted from soil properties such as soil CEC, carbonate content, iron oxide content, and organic matter composition. This makes predicting chemical adsorption possible without a-priori conducting adsorption experiments. **Figure 11.8** shows a boron adsorption isotherm for the same soil used for the boron adsorption data shown in Figure 10.28. The predicted isotherm was modeled using a mechanistically based adsorption model (**Eqs. 11.19–11.23**), with K_{int} derived from an empirical relationship of soil properties (CEC, SOM, etc.). Such prediction ability is a powerful tool to base best management strategies for optimizing availability of boron in soils for plant availability and leaching.

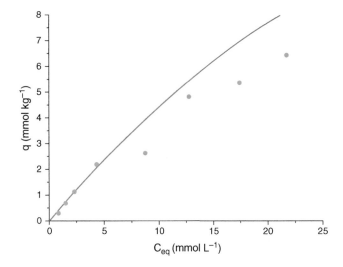

Figure 11.8 Boron adsorption isotherm on a California soil. The solid line is the predicted adsorption isotherm using a mechanistic adsorption model and an intrinsic equilibrium constant derived from soil properties (CEC, organic carbon, inorganic carbon, iron oxide content). Source: Goldberg et al. (2005). Reproduced with permission of ACSESS.

11.5 Rates of adsorption

Adsorption reactions are often slow and, thus, time-dependent (see Figure 4.20). If soil pore water is moving through the soil faster than a chemical adsorbs or desorbs, the surface reaction will not come to equilibrium. In such cases, predicting chemical fate and transport requires kinetic modeling approaches.

Adsorption reaction rates range from seconds to minutes, to hours and months. Cation exchange reactions are the fastest adsorption reactions. Reactions that take longer are caused by slow pore and film diffusion processes, formation of surface precipitates, or formation of inner-sphere complexes that require significant activation energy to make and break the associated bonds (see Figure 1.13). **Figure 11.9** shows arsenate and selenite adsorption as a function of time on a soil. At the first sampling point (<10 minutes), arsenic and selenium adsorption was ~65% of the amount adsorbed after 24 hours. Following the initial fast adsorption reaction, arsenic and selenium adsorption continued to increase rapidly until about 4 hours, after which the adsorption rate was slow, but continuous for the next 20 hours. The reaction did not level off, and it appears that longer incubation times would result in more adsorption.

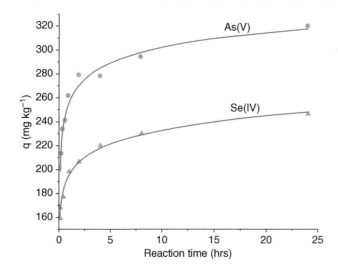

Figure 11.9 Adsorption of As(V) and Se(IV) on a tropical soil as a function of time by a gravelly clay loam, pH 4.6. Adapted from Goh and Lim (2004). Reproduced with permission of Elsevier.

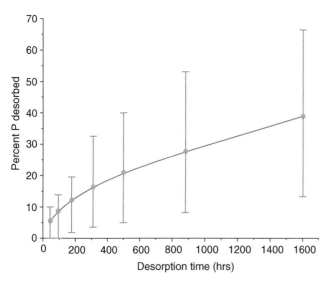

Figure 11.10 Average phosphorus desorbed from 44 soils as a function of desorption incubation time. Percent phosphorus desorbed is the percentage of total available phosphorus measured by oxalate extraction. Bars indicate the minimum and maximum of phosphorus desorbed from the soils. Adapted from Lookman et al. (1995). Reproduced with permission of American Chemical Society.

Desorption reactions are often much slower than adsorption reactions. **Figure 11.10** shows the desorption behavior of phosphate from several soils. Much of the phosphate in soil is apparently not desorbed, or slowly desorbed. This hysteresis or slow desorption is

commonly observed for strongly adsorbed cations and anions (e.g., see Figure 10.5).

11.5.1 Modeling adsorption kinetics

Kinetic theory of adsorption reactions follows the same principles described for kinetics of aqueous chemical reactions in Chapter 4 (Section 4.10). For a metal (M^{2+}) adsorption reaction:

$$\equiv S^- + M^{2+} = \equiv SM^+ \tag{11.24}$$

the rate of adsorption (r_{ads}) is written as the change of the aqueous ion (M^{2+}) concentration over a specified time ($d[M^{2+}]/dt$):

$$r_{ads} = \frac{d[M^{2+}]}{dt} = -k_f[M^{2+}][\equiv S^-] + k_r[\equiv SM^+] \tag{11.25}$$

where k_f is the forward reaction **rate constant,** and k_r is the reverse reaction rate constant. The differential form $d[M^{2+}]/dt$ indicates the change in concentration of the metal in solution over an infinitesimal change in time.

Equation 11.25 is an elementary reversible adsorption reaction that implies the rate of adsorption is directly dependent on the concentration of each of the species in the reaction (first order with respect to each species). Other ordered equations exist that predict reaction rates that are independent of the concentration (called zero ordered); or the concentration changes the reaction rate as an exponential factor (e.g., second order). Rate equations are often simplified by assuming only one reaction component is changing. For example, if a reaction is assumed to only go forward, and the concentration of surface sites ($\equiv S^-$) is assumed to be constant, then **Eq. 11.25** reduces to

$$\frac{d[M^{2+}]}{dt} = -k_f[M^{2+}] \tag{11.26}$$

Equation 11.26 is a *pseudo* **first-order adsorption equation**, and is much simpler to model and determine the reaction rate coefficient (k_f) than **Eq. 11.25**. Kinetic experiments are often designed to allow simplified rate equations to be used to model the adsorption or desorption data. This is done by controlling the reaction conditions so that the assumptions are true (e.g., reverse reactions are not occurring and there is an excess of surface sites).

The rate of adsorption predicted in **Eq. 11.25** or **Eq. 11.26** shows that time-dependent adsorption can be predicted from the concentrations of species in the reaction and the rate constants. Desorption reactions are modeled using the same type of rate reactions as adsorption models. Rate constants for kinetic reaction equations are obtained by fitting adsorption data as a function of time with an integrated form of a rate equation (fitting models derived from rate expressions are not covered in this text). If rate constants are known, then the concentration of M^{2+} in solution at any time can be solved. For adsorption reactions that are relatively slow, such as in the arsenic and selenium adsorption shown in **Figure 11.9**, rate equations are more accurate for predicting adsorption in fate and transport models than equilibrium models.

Many other types of rate equations exist other than the ordered expression shown in **Eq. 11.25**. Empirical rate expressions are often used to fit adsorption data for reactions that are complex, such as when multiple rate-controlling processes are occurring. Empirical equations typically include a rate constant and concentration of one or more of the species in the system. In soils, diffusion through micropores, and even diffusion through the film of solution on a particle, can be a rate-controlling step in adsorption and desorption reactions. In such cases, diffusion models are appropriate for describing the time-dependent adsorption process.

11.6 Reactive transport

In soil chemistry, much of the emphasis is on reactions occurring at the solid–solution interface. In soils, however, reactions and solute transport occur simultaneously, and prediction of the fate and bioavailability of chemicals requires that both processes be considered. Physical transport processes can be predicted using a model that accounts for fluxes of solutes and mixing due to diffusion and dispersion; called the **advective-dispersion equation**. Deviations of the solution transport model have been developed to account for saturated and unsaturated conditions, preferential flow, and soil heterogeneity. Soil physics and hydrology courses cover this topic. To predict chemical

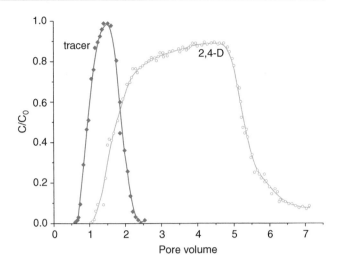

Figure 11.11 Pulse breakthrough curves for tritiated water, a nonreactive tracer, and 2,4-D (a common herbicide) leached through a soil column. The breakthrough curve of 2,4-D is delayed compared to the tracer because it is adsorbed to the soil. Adapted from Rao and Davidson (1979). Reproduced with permission of Elsevier.

movement in soils, *reaction* models are coupled with *transport* models in what is referred to as **reactive transport modeling**.

Reactive transport experiments are often done in water saturated columns, where the chemical solution is input to one end, and the solution flows through the column of soil and outflows the opposite end. **Figure 11.11** shows the **breakthrough** curves for a nonreactive tracer (tritiated water) and 2, 4-D, which is a weak acid and has a nonlinear adsorption behavior in soils. For continuous input, the breakthrough is the midpoint of the minimum and maximum concentrations that flow through the soil. For pulse input, the breakthrough is the peak.

When the soil adsorption sites are saturated with the chemical, maximum outflow concentrations occur (i.e., $C_{in} = C_{out}$). The breakthrough curve for the 2, 4-D in **Figure 11.11** is delayed compared to the nonreactive tritiated water because of adsorption reactions. Furthermore, the outflow of 2,4-D does not reach a maximum value of one in the experiment because adsorption decreases the solution concentration – to reach a maximum, the input of 2,4-D would have to continue beyond the 4.5 pore volumes prior to switching to the elution phase of the flow-through experiment. Predicting the delay using adsorption and transport modeling is the goal of reactive transport modeling.

Mathematical modeling of reactive transport uses a solute flux model that includes a reaction term. For saturated flow through a homogenous medium, a common solute transport model is the advective-dispersion equation, which predicts the change in concentration of the outflow with time:

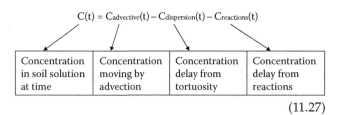

$$C(t) = C_{advective}(t) - C_{dispersion}(t) - C_{reactions}(t)$$

Concentration in soil solution at time	Concentration moving by advection	Concentration delay from tortuosity	Concentration delay from reactions

$$(11.27)$$

where $C(t)$ is the concentration at the outflow at a specific time and the subscript terms refer to the different chemical reaction and flow components:

- **Advective flow** describes the mass transport of the fluid through the porous media and is a function of flow rate and distance; for example, water moving through a straight pipe that is not interacting with the edges.
- **Dispersive flow** describes the hydrodynamic dispersion and diffusion through the tortuous soil paths, and is a function of the soil tortuosity, soil diffusion characteristics, distance, and concentration gradients.
- Reactions describe chemical processes that remove chemicals from the solution (*sink*) or add them to the solution (*source*) in the time the solute is in contact with the soil.

When a chemical is dissolving or desorbing, the reaction component is negative (i.e., the last term in **Eq. 11.27** becomes + $C_{reactions}$).

Soil chemistry is concerned with the reactive component in **Eq. 11.27**, which may be an equilibrium or time-dependent adsorption or desorption process, a dissolution or precipitation reaction, or a decay or degradation process. Application of the advective dispersion equation uses differential equations $(\partial C / \partial t)$ to model concentration as a function of time.

When a chemical moving through a porous medium reacts with the solid, the reaction retards the movement of the chemical with respect to the transport of the nonreactive solution (water). The simplest equation to model the chemical retardation ($C_{reactions}$) is a linear partition coefficient:

$$q = K_d \times C_{eq} \tag{11.28}$$

Since many adsorption reactions are nonlinear, this equation may not be a good model to use in reactive transport. However, in many instances, nonlinear adsorption behavior can be modeled using a linear equation over a limited range. The equilibrium concentration (C_{eq}) is equivalent to C(t) in **Eq. 11.27**, and $C_{reaction}(t)$ can be calculated from the adsorption equilibrium concentration (q):

$$C_{reaction}(t) = q \times \frac{\rho}{\theta} \tag{11.29}$$

where ρ is the soil bulk density (mass of solid per volume of bulk soil) and θ is the volumetric water content (volume water per volume bulk soil). Substituting **Eqs. 11.28** and **11.29** into **11.27**:

$$C_{eq}(t) = C_{advective}(t) - C_{dispersion}(t) - K_d C_{eq}(t) \times \frac{\rho}{\theta} \tag{11.30}$$

Isolating for C_{eq}:

$$C_{eq}(t) = \frac{\left(C_{advective}(t) - C_{dispersion}(t)\right)}{1 + \frac{\rho K_d}{\theta}} \tag{11.31}$$

The term in the denominator is a **retardation factor** that models the removal of chemical from solution onto the solid via a linear distribution reaction:

$$R = 1 + \frac{\rho K_d}{\theta} \tag{11.32}$$

The retardation factor can also be related to the relative velocity of a chemical through a soil profile, which is useful for comparing the absolute delay of chemical movement as related to the bulk solution movement (e.g., movement of *nonreactive* water). This approach is a simplistic solution of the advective dispersion equation using a tracer to define the boundary conditions of the advection and dispersion components in the equation. The ratio of the average velocity of nonreacting solution (v) to the reacting chemical solution (v_c) is equal to the retardation factor (R):

$$R = \frac{v}{v_c} \tag{11.33}$$

The retardation factor is greater than or equal to one ($R = 1$ means no reaction). At a given time (t), the distance a chemical travels (X_c) can be computed from flow-rate of the chemical (v_c) multiplied by time:

$$X_c = v_c \times t \tag{11.34}$$

Rearranging **Eq. 11.33** and substituting in **Eq. 11.34** yields the relative distance of the retarded chemical transport:

$$X_c = \frac{v}{R} \times t \tag{11.35}$$

$$X_c = \frac{X}{R}$$

where $v \times t$ is defined as X, which is the distance that the nonreactive solution (tracer) travels. Tracer travel distance can be measured experimentally or modeled using the advective-dispersion equation.

Equation 11.35 predicts the distance that the chemical will travel (X_c) relative to the nonreacting solution (X) in a given time. As derived above, for systems in which the chemical partitioning to the solid can be predicted with a linear adsorption equation, the retardation factor can be calculated (**Eq. 11.32**). Thus, if the linear partition coefficient is large ($q > C_{eq}$) the retardation factor is large, and the chemical breakthrough is significantly delayed ($X_c < X$).

An example of calculating chemical transport retardation through a soil profile is atrazine transport, for which the linear adsorption coefficient for the surface soil is 0.65 ml g^{-1} (**Figure 11.2**). In a saturated soil that has a bulk density of 1.6 g ml^{-1} and saturated water content of 0.4, the retardation factor predicted using **Eq. 11.32** would be

$$R = 1 + \frac{\rho K_d}{\theta} = 1 + \frac{1.6\,\text{g ml}^{-1} \times 0.65\,\text{ml g}^{-1}}{0.4} = 3.6 \tag{11.36}$$

Considering a pulse application of the pesticide, and using **Eq. 11.35**, the chemical will only move 0.28 times the distance of the nonreactive solution (wetting front):

$$X_c = \frac{X}{3.6} = 0.28 \times X \tag{11.37}$$

In the lower soil profile (30–45 cm), K_d of atrazine is 0.29 ml g^{-1} (**Figure 11.2**), and bulk density is 1.66 g ml^{-1}.

Thus, the retardation factor for atrazine in the subsurface is 2.3 and the atrazine will travel 0.43 times the distance of the nonreactive solution; much further than in the surface horizon. The decreased mobility in the surface horizon is due to the increased organic carbon that adsorbs more of the atrazine and retards its migration down the profile into the groundwater.

For nonlinear adsorption, such as the Freundlich equation (**Eq. 11.6**), the retardation factor is

$$R = 1 + \frac{\rho K n C^{n-1}}{\theta} \tag{11.38}$$

Nonlinear retardation factors are concentration dependent, as shown by the C term in **Eq. 11.38**. This means that for a flowing and reacting chemical in solution in which the concentration is continuously changing, the retardation factor is constantly changing. Thus, an analytical solution for nonlinear reactive transport is not possible, and it must be solved using numerical methods. Using a linear model for a small region of the nonlinear adsorption curve is a workaround that simplifies the transport model and allows for an analytical solution.

Equations 11.32 and **11.38** show how adsorption modeling is used to predict chemical transport in soils. The advection and dispersion components in **Eq. 11.27** use partial differential equations for modeling the physical transport processes across space as a function of time. The reaction component in **Eq. 11.27** creates more complexities to the transport model. Using a linear retardation factor is the simplest form of the equation to solve, and has an analytical solution. As adsorption model complexity increases, the number of parameters to be modeled also increases. Advanced mathematical predictions that incorporate nonlinear adsorption models (e.g., Freundlich or Langmuir isotherms), mechanistic models, or rate-limited reactions can be made using computers that are able to simultaneously solve for adsorption concentration and solution transport using numerical algorithms.

There remain many challenges to bridge the gaps between field transport and laboratory-based experiments. For example, test-tube sorption experiments lack the varying fluxes and heterogeneity that occur in the field. Also, the use of low solid to solution ratios in adsorption experiments may not be transferable to the obviously much higher solid to solution ratios that exist in soils (discussed in Section 11.2). Similar limitations occur when solute transport experiments are done in laboratory soil columns. Conducting reactive transport experiments at the field scale can overcome the limitations inherent in laboratory experiments, but field measurements are much more costly, require permissions from landowners and regulatory agencies, are more complicated because of heterogeneity and lack of control of variables, and are time consuming. Bridging the gap between laboratory and field experiments is an important area of research for many environmental scientists.

11.7 Surface precipitation

Molecular-level investigations have revealed that metals form **multinuclear precipitates** on mineral surfaces well below theoretical bulk solution saturation conditions. Multinuclear surface precipitates have been observed for Pb^{2+}, Cu^{2+}, Ni^{2+}, Co^{2+}, and Zn^{2+}. The surface precipitates can be small, such as dimers, or completely cover the surface. Such surface complexes may explain why soil pore water concentrations of metals are often observed to be less than predicted from solubility of the *pure* (hydr)oxide phases.

Surface precipitation can occur in systems when the bulk solution is undersaturated with respect to solid-phase precipitation; i.e., when the saturation index is less than zero. Surface conditions that promote precipitation are:

1 The surface acts as a nucleation site decreasing the activation energy for the onset of precipitation.
2 Differences in solution pH and ionic strength next to the surface that cause *local* oversaturation as compared to the bulk solution.
3 Contributions of co-ions from the mineral that make surface *co-precipitates* saturated at the mineral surface.

The metals Ni^{2+}, Co^{2+}, and Zn^{2+} often form surface co-precipitates with Al^{3+} and Fe^{3+} cations released from solid surfaces.

Formation of surface precipitates affects the release of the sorbed metals, and creates distinct surface conditions not predicted from the mineral's bulk

chemistry. Reaction models of surface precipitation are similar to kinetic and thermodynamic models of bulk solution precipitation reactions, but are less certain because the solid phase (surface precipitate) activity is not ideal (lacks the unit activity of pure minerals).

Surface precipitates, and especially mixed multinuclear precipitates, often have lower solubility than pure mineral phases. For example, the solubility products for Ni-Al precipitates that are observed on soil clay minerals are several orders of magnitude lower than pure $Ni(OH)_2$ (saturation index modeled in **Figure 11.12**).

Modeling of *pure* mineral solubility uses ion activities and K_{sp}. A similar approach can be used for multinuclear surface precipitates, however, the solubility product (K_{sp}) for the mixed-phase dissolution reaction is not the same as the pure solid solubility product, and needs to be adjusted for nonideality. This is done using an activity coefficient for the surface species (λ)

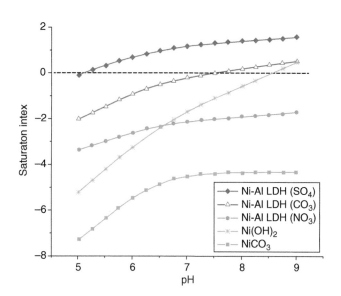

Figure 11.12 Saturation index (Eq. 4.61) for several Ni minerals in a simulated groundwater in the presence of montmorillonite. The Ni-layered double hydroxide (Ni-LDH) phase with a sulfate anion between the sheets is the least soluble nickel mineral in the groundwater. The graph shows that the mixed nickel-metal precipitates are less soluble than the pure nickel hydroxide and carbonate phases, which implies they control the aqueous nickel solubility in equilibrium systems. Adapted from Peltier et al. (2006). Figure 3. Reproduced with permission of The Clay Minerals Society. Publisher of Clays and Clay Minerals.

and scaling the K_{sp} by the fraction of the metal in the coprecipitate (X_M); thus, K_{sp} for a surface precipitate (MOH_n, where M indicates the metals in the coprecipitate) is

$$K_{sp\,surf\,precip} = \frac{\left(M^{n+}\right)\left(OH^-\right)^n}{\left(MOH_n\right)} = \lambda K_{sp} X_M \qquad (11.39)$$

where K_{sp} is the solubility product of the pure metal phase. The surface activity coefficient (λ) is obtained from modeling, and accounts for the nonideality of multinuclear precipitates. For a multinuclear surface precipitate, λ and X_M are less than 1, and thus **Eq. 11.39** predicts that the solubility of M^{n+} in a surface precipitate is less than the pure mineral phase.

For the theoretical groundwater modeled in **Figure 11.12**, the thermodynamic model predicts that a Ni-Al layered double hydroxide phase that incorporates a sulfate anion between the hydroxide sheets is supersaturated throughout the pH range, and is likely the phase controlling the solubility of Ni^{2+} in the system. Similar mixed metal-Al hydroxide phases have been observed for Zn^{2+}, Co^{2+} and Cr^{2+}. The decreased solubility of mixed metal phase surface precipitates is one reason that predictions based on thermodynamics of pure mineral phases often fail to model metal solubility in soils.

11.8 Analytical methods for determining adsorption mechanisms

Interest in the molecular mechanisms of surface sorption processes dates back to the first observation of cation exchange in the nineteenth century. Early experiments focused on understanding the properties of minerals and organic matter in soils. However, the complexity of surfaces and adsorption reactions cannot be predicted from the properties of the bulk soil solids alone because the surfaces have incompletely coordinated atoms. In recent decades, great strides have been made in understanding surface reaction mechanisms using advanced spectroscopic and microscopic tools that probe the nature of the surface chemical species at a molecular-level.

Many of the same tools used in analyses of soil mineralogy (Table 6.15) are used to probe surface

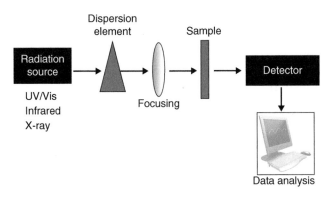

Figure 11.13 Components of a spectrometer. The radiation passes through a dispersion element to separate the energy into distinct wavelengths. The wavelength of interest is focused to maximize intensity and optimize size on the sample. The radiation impinges on the sample, and the outgoing radiation can be collected either from reflection or transmission detectors. The data are stored and analyzed using a computer.

chemistry. Amongst these, X-ray absorption spectroscopy and FTIR spectroscopy are responsible for much of our current knowledge and understanding of molecular species on surfaces. **Figure 11.13** shows the setup for a spectroscopic experiment.

X-ray absorption spectroscopy (**XAS**) is well suited to probing adsorbed chemicals because it can be focused to collect signal only from adsorbed species, which are typically present at a fraction the bulk solid atoms. For example, Zn^{2+} adsorbed on a montmorillonite surface may be present at a ratio of a few hundred atoms out of a million atoms in the solid phase. X-ray spectroscopy can be focused and tuned to collect data only from the Zn^{2+} atoms, thereby filtering out the signal from the much more prevalent bulk mineral atoms.

X-ray spectroscopic methods measure how the *absorption* of photons by the sample changes the outgoing signal that is either passed through or reflected from the solid. Two types of X-ray absorption spectroscopy commonly used are X-ray absorption near edge spectroscopy (**XANES**), and extended X-ray absorption fine structure (**EXAFS**) spectroscopy. Both types of spectra are measured in the same experiment, and are collectively referred to as X-ray absorption fine structure (**XAFS**) (**Figure 11.14**). XAFS experiments require very bright X-ray sources produced by large electron storage rings called synchrotrons.

XANES spectra are especially useful for measuring oxidation state, and can be used to infer molecular

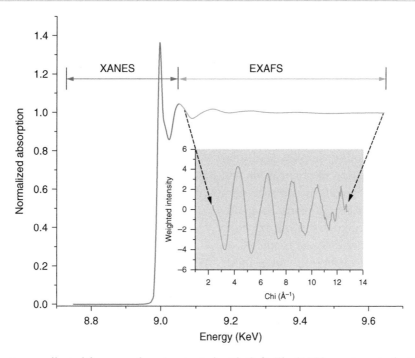

Figure 11.14 XAFS spectrum collected from a soil contaminated with Cu^{2+}. The XAFS spectrum is divided into the XANES and EXAFS energy regions. The inset shows the EXAFS region transformed to a chi spectrum, which better shows the EXAFS features and allows data analysis (*x-axis* is wavenumber energy of the incoming photoelectron). The amplitude, frequency, and wavelengths of the EXAFS spectrum indicate the molecular coordination of the copper in the soil. The XANES and the EXAFS spectra show that the Cu^{2+} is bonded to functional groups on organic matter in the soil. Adapted from Strawn and Baker (2008). Reproduced with permission of American Chemical Society.

structure by comparing the unknown adsorbed species spectra to the spectra of a known molecular coordination. An EXAFS spectrum represents the molecular structure surrounding a probed atom, and provides knowledge of the surrounding atoms' identification, bond distances, and coordination numbers. EXAFS spectra are fit using theoretical molecular structures, allowing for derivation of the molecular surface structures. **Figure 11.14** shows an example of the XAFS spectroscopic measurement of soil that had Cu^{2+} spilled onto it from a storage tank. Analysis of the EXAFS spectrum by fitting bond distances and atomic coordination for the first three atomic shells surrounding the soil copper showed that the molecular structure of the adsorbed Cu^{2+} consisted of copper bonded via an inner-sphere complex onto functional groups on SOM.

Fourier transform infrared (**FTIR**) spectroscopy uses infrared light to excite the molecular bonds in a molecule, causing them to bend, rotate, and vibrate. The distortions cause fluctuations in the outgoing signal. FTIR spectra are indicative of the molecular coordination within the sample. Comparing pre- and post-chemical adsorption shows the surface functional groups involved in the reaction. For molecules that have functional groups, such as oxyanions like phosphate, FTIR spectra can reveal which of these groups is reacting with the surface, and how.

11.9   Summary of important concepts for modeling surface reactions in soils

Quantitative predictions of sorption and desorption reactions in soils require mathematical models of the solid-solution distribution of a chemical (K_d). Empirical models, such as the linear, Langmuir, and Freundlich equations are blind to adsorption mechanisms, but often accurately describe adsorption behavior as a function of equilibrium concentration. However, the empirical models in most cases cannot account for variability in system properties, including soil pH and differences in mineral and organic matter composition, thus they are only accurate for the systems for which they were derived.

Mechanistic adsorption models use a thermodynamic approach (i.e., free energy relationships) to develop a system of equations that model reaction processes. Although more difficult to develop than empirical models, they have greater flexibility for predicting surface reactions as soil properties vary, and thus have more utility for use. However, the limited availability of accurate thermodynamic constants (K_{int}) to use for mechanistic modeling, and difficulty in describing the adsorption reaction mechanism on soil particle surfaces, prevents their widespread adoption. The constants can be calculated from many measurable soil properties, but this approach has not yet been universally adopted by soil scientists and modelers.

Fate and transport of chemicals in soils can be modeled by including a chemical reaction parameter into the advective-dispersion equation. The chemical reaction models can be empirical or mechanistic, and predict equilibrium or rate-limited kinetic processes. As models become more capable and more easily integrated into computer programs, they will become increasingly used to predict fate and transport of chemicals in soils. The most accurate models will be those that include soil properties and integrate reaction mechanisms (e.g., ion exchange, inner-sphere adsorption, or surface precipitation).

The next era of fate and transport modeling for soil chemicals will use advanced computer approaches such as artificial intelligence computing that make models even more accurate and allows for modeling of increasingly complex systems such as chemical transport through heterogenous and dynamic soil systems. With such advanced models, scientists will one day be able to select a location on a map, input a chemical, and get output predicting the concentration of the chemical available for transport to a plant root or groundwater under varying input concentrations and weather conditions… At the rate of technological advancement, this day may not be far off!

Questions

1 Given the data below for adsorption of 2,4,5-T to soil, determine if the adsorption behavior conforms to the Langmuir or the Freundlich models and determine the appropriate adsorption parameters for the best-fit model.

Initial solution concentration (mg L^{-1})	Final solution concentration (mg L^{-1})	Volume of solution (mL)	Weight of soil (g)
5	2	10	5
15	7	10	5
25	13	10	5
35	19	10	5
55	33	10	5
75	48	10	5
95	64	10	5

2 Describe an experiment to measure the time dependency of copper adsorption on a soil. What are the experimental factors and assumptions required to use a pseudo first-order kinetic equation to model the data? Write a copper adsorption reaction and show the first-order reaction model.

3 Write a reaction and derive an intrinsic equilibrium constant expression for zinc adsorption on an iron oxide surface. Assume the surface functional groups are either SOH or SOZn$^+$. What are the variables and constants in the expression, and how can they be modeled or obtained?

4 A tank is leaking a copper sulfate pesticide onto soil overlying a shallow groundwater. The range of adsorption over a selected concentration of Cu^{2+} is modeled using a linear partition coefficient with $K_d = 0.45$ ml g^{-1}. Calculate how far down the soil profile the Cu^{2+} will move compared to the wetting front (i.e., a nonreactive tracer). Soil bulk density is 1.54 g ml^{-1} and saturated water content is 0.43.

5 Explain why surface precipitation reactions cannot be modeled using equilibrium constants from pure mineral phases.

Bibliography

Goh, K.H., and T.T. Lim. 2004. Geochemistry of inorganic arsenic and selenium in a tropical soil: effect of reaction time, pH, and competitive anions on arsenic and selenium adsorption. Chemosphere 55:849–859.

Goldberg, S., S.M. Lesch, and D.L. Suarez. 2000. Predicting boron adsorption by soils using soil chemical parameters in the constant capacitance model. Soil Science Society of America Journal 64:1356–1363.

Goldberg, S., D.L. Corwin, P.J. Shouse, and D.L. Suarez. 2005. Prediction of boron adsorption by field samples of diverse textures. Soil Science Society of America Journal 69:1379–1388.

Hiemstra, T., P. Venema, and W.H. VanRiemsdijk. 1996. Intrinsic proton affinity of reactive surface groups of metal (hydr)oxides: The bond valence principle. Journal of Colloid and Interface Science 184:680–692.

Lookman, R., D. Freese, R. Merckx, K. Vlassak, and W.H. Van Riemsdijk. 1995. Long-term kinetics of phosphate release from soil. Environmental Science & Technology 29:1569–1575.

Mbuya, O.S., P. Nkedi-Kizza, and K.J. Boote. 2001. Fate of atrazine in sandy soil cropped with sorghum. Journal of Environment Quality 30:71–77.

Osborne, L.R., L.L. Baker, and D.G. Strawn. 2015. Lead Immobilization and Phosphorus Availability in phosphate-amended, mine-contaminated soils. J Environ Qual 44:183–190.

Peltier, E., R. Allada, A. Navrotsky, and D.L. Sparks. 2006. Nickel solubility and precipitation in soils: A thermodynamic study. Clays and Clay Minerals 54:153–164.

Rao, P.S.C., and J.M. Davidson. 1979. Adsorption and Movement of Selected Pesticides at High-Concentrations in Soils. Water Research 13:375–380.

Strawn, D.G., and L.L. Baker. 2008. Speciation of Cu in a contaminated agricultural soil measured by XAFS, μ-XAFS, and μ-XRF. Environmental Science & Technology 42:37–42.

12 SOIL ACIDITY

12.1 Introduction

pH, the negative logarithm of the *activity* of hydrogen cations in a solution, affects speciation and availability of many chemicals in soils (e.g., see nutrient availability in Figure 3.1) and is often considered the master variable for characterizing soil chemical behavior. **Figure 12.1** shows typical soil pH values, and pH influence on soil conditions and plant growth.

In many regions of the world, soil acidification is a concern for sustainable management of forest and agronomic systems. Approximately 30% of Earth's soils are acidic. Soil acidification can be detrimental for agriculture because it decreases availability of anionic nutrients (e.g., see Figure 10.27), causes cationic nutrients to be leached from the soil profile, and causes Al^{3+} and Mn^{2+} toxicity. Forest soils and wetland soils are often naturally acidic. This chapter presents information on processes that cause soil acidity, and how to neutralize soil acidity.

Aqueous solution pH measurements represent the activity of **hydronium ions** (H_3O^+), and not the **H^+ ion activity**. Protons are very reactive and have extremely low activity in aqueous solution. In this chapter, however, we use the term **proton** or H^+ instead of hydronium.

12.1.1 Measurement of soil acidity

H^+ activity of aqueous solutions is measured using either a pH probe or colorimetrically. In the laboratory, soil pH measurement methods use different ratios of soil and solution, and different soil-wetting solutions (salt or deionized water), all of which may affect the measured pH (see Special Topic Box 12.1). Thus, it is important to note the pH measurement method used when measuring soil pH.

Acidification and neutralization processes in soils are complex, and measurement of soil solution pH

Soil Chemistry, Fifth Edition. Daniel G. Strawn, Hinrich L. Bohn, and George A. O'Connor.
© 2020 John Wiley & Sons Ltd. Published 2020 by John Wiley & Sons Ltd.

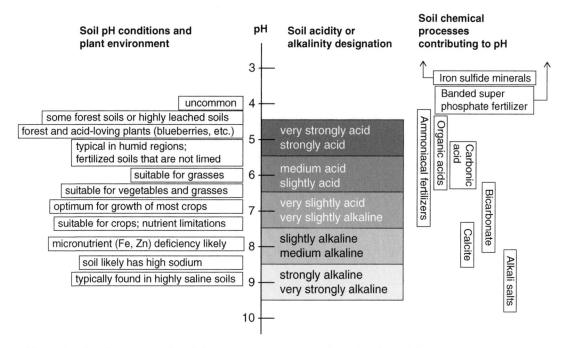

Figure 12.1 The soil-pH scale. Major soil-acidification processes are indicated. Adapted from Adriano (2001). Reproduced with permission of Springer.

Special Topic Box 12.1 Soil pH measurement method

Modern pH meters and electrodes make measurement of soil pH seemingly simple; however, if care is not taken, errors from artifacts occur.

A pH probe is an ion-selective electrode that relies on a membrane to develop a junction potential (separation of charge in a liquid) (**Figure 12.2**). In the case of pH, the junction potential is sensitive to the hydrogen ion activity in solution. Inside of the pH electrode, a silver wire coated with AgCl dips into an HCl solution. The HCl solution is separated from the test solution by a membrane of special glass. Differences in H^+ activity across this glass membrane create an electrical potential, which is measured by a sensitive potentiometer. A *reference electrode* is necessary to complete the electrical circuit. Modern electrodes include the reference electrode and hydrogen electrode in a single combination-pH probe.

The electrical current created by the junction potential is on the order of nano- or picoamperes, and diffusion of trace quantities of cations in the glass carries the current. The potential across the glass membrane can be closely

calibrated to the value of the H^+ activity using solutions with known pH values (i.e., buffers).

Reference electrodes use a KCl solution to carry the charge across the junction to the solution. In a colloidal suspension, the colloids cause variance in the K^+ and Cl^- diffusion, and spurious potentials. Attraction or repulsion by charged colloids causes ion separation at the junction between the electrode solution and the soil suspension. The separation produces a charge or electrical potential called the liquid–liquid junction potential. Such potentials are unpredictable and can cause errors in accuracy in pH measurement. Errors from spurious electrode potentials greater than 0.5 pH units are probably uncommon for soils. Measuring pH in salt solutions of 0.01 M $CaCl_2$ or greater concentration virtually eliminates this error.

Another simple way to minimize liquid-junction potential errors in soil pH measurements is to allow the tip of the reference electrode to contact only the supernatant solution above the colloidal phase. The rates of K^+ and Cl^- diffusion are then unaffected by the colloids. The glass

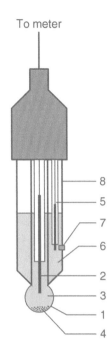

To meter

1. bulb made from a H$^+$-sensitive glass
2. internal electrode, usually silver chloride or calomel
3. internal solution, usually 1×10^{-7} mol/L HCl
4. AgCl may precipitate inside the glass electrode
5. reference electrode, usually the same type as 2
6. reference internal solution, usually 0.1 mol/L KCl
7. porous junction
8. body of electrode

8
5
7
6
2
3
1
4

Figure 12.2 pH electrode components. Adapted from https://commons.wikimedia.org/wiki/File:Glass_electrode_scheme_2.svg. January (2019).

hydrogen electrode, on the other hand, can be placed either in the supernatant solution or in the colloidal suspension. The H$^+$ activity is the same in both phases, and the glass electrode is unaffected by the presence of colloids.

The glass membrane of the pH electrode has proved to be by far the most successful ion-sensitive membrane. The glass has a uniform response to a wide range of H$^+$ activities, requires little maintenance, is resistant to contamination, is structurally strong, and is mostly insensitive to interfering ions. Other ion-sensitive electrodes are not nearly as effective in screening out interfering ions as the hydrogen electrode.

Experimental conditions of the soil suspension or paste during pH measurement are major sources of variance. Different volumes of soil-wetting solutions (mass solution to mass solid) and solution salt composition create differences in pH measurement. A common method is to wet the soil with an equivalent mass of solution as the mass of soil (1:1). However, some researchers use a saturated paste, which is less than 1:1, or ratios greater than 1:1

(e.g., 5:1 or 10:1). The differences in pH measurements from different solid to solution ratios are typically less than 0.5 pH units.

Another source of error in soil-pH measurement occurs when a soil sample is mixed under atmospheric conditions that are different than those that occur in the soil. In the rhizosphere, for example, CO_2 may be much greater than the partial pressure in the atmosphere and can have a significant effect on soil solution pH (see, e.g., Figure 4.9). Most soil pH values are recorded at atmospheric P_{CO2}, and no adjustment is necessary unless unusual P_{CO2} values are expected in the soil.

Artifacts are typically dealt with by noting the pH-measurement procedure used when reporting soil pH, and making sure that comparison of soil pH and pH-dependent processes is done using the same soil pH measurement methods. Digital pH meters commonly record pH values out to the thousands place. However, given the inherent variability in soil pH measurements and the spatial variability in soils (even at the millimeter scale), the precision is likely limited to the tenths place.

does not accurately represent the acid-base properties of soils. The complexity occurs because soil solids continuously add and subtract chemicals to and from the soil solution that change the soil solution's acid-base characteristics. To determine soil acid–base behavior, three different types of acidity are measured:

1 *Active soil pH*. pH of the soil solution measured in a soil wetted with deionized water or a 0.01 M CaCl$_2$ solution.
2 *Exchangeable acidity*. pH of the soil solution when all the Al^{3+} and H$^+$ are released from the cation exchange sites; typically measured in a soil wetted with 1 M KCl solution.
3 *Residual acidity*. pH of a soil solution when all the titratable acids and bases that would readily dissolve from the solid phase are neutralized; typically measured by first wetting and incubating a soil for several hours with a pH 8 buffer solution, and then titrating the buffer solution to pH 5. The amount of protons required to reach pH 5 is the residual acidity of the soil.

The most commonly reported soil pH is *active soil pH*, which is a useful measure of the relative acidity or basicity of the soil. For example, soil solutions with pH measurement less than 5.5 are considered strongly acidic. Measurement of *exchangeable acidity* accounts for a soil's immediate acid buffering ability, and is useful for determining the amount of lime needed to raise soil pH. *Residual acidity* correlates to the overall buffering capacity of the soil and is related to the amount of weathering that the soil has undergone. Residual acidity is not typically used for acid-soil management, but is useful for understanding the overall soil acid–base behavior, such as changes in soil acidity over time; and may be factored into lime-amendment rate calculations. Soils that have sulfide minerals present (e.g., wetlands and mine-impacted soils) need special consideration for residual acidity because sulfide oxidation produces acid.

12.2   History of soil acidity

For centuries, farmers have known that productivity of some soils is improved by adding powdered limestone or crushed sea shells to neutralize the soils' acidity.

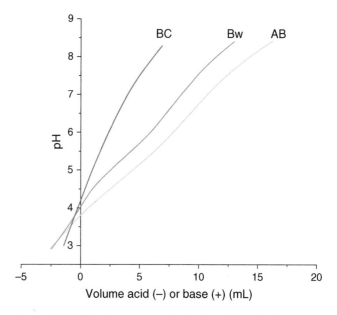

Figure 12.3 Titration curves of soils from three horizons from a Japanese Entisol. Adapted from Funakawa et al. (2008), with permission.

Around the turn of the twentieth century, experiments by F.B. Veitch of the US Bureau of Chemistry showed that accurate determination of the acidic behavior of a soil required suspending the soil in a cationic salt solution to exchange all of the soil's acidity. Studies throughout the 1930s, 1940s, and early 1950s reported the properties of hydrogen-saturated soils and clays. The aim of such studies was to predict the amount of lime needed to counteract soil acidity. A common approach was to titrate the soil using a strong base such as NaOH, and measure the change in pH as a function of the amount of hydroxide ions added (potentiometric titration). For example, **Figure 12.3** shows titration curves of three soil horizons from an Entisol from Japan. The AB horizon has the strongest buffering capacity because it has a greater amount of secondary minerals in it as well as some organic matter. The Bw horizon has an intermediate buffering capacity, and the BC has the least buffering capacity because it has the fewest weathered minerals.

Figure 12.4 shows typical curves for the potentiometric titration of acid montmorillonite suspensions. *Curve 1*, a freshly prepared acidic clay suspension, looks similar in shape to titration of strong acids such as HCl: The pH remains relatively constant until nearly all of the acid is neutralized, then the pH rises rapidly until it is determined by the concentration of the added

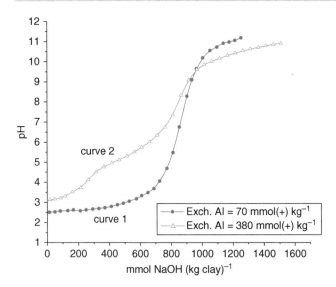

Figure 12.4 Titration of acidic montmorillonite clay suspensions with NaOH. The clays were prepared differently so that they have different exchangeable acidity (see text). Adapted from Aldrich and Buchanan (1958), with permission.

base. The clay used for *curve 1* has H^+ cations adsorbed that release as the clay is titrated with base.

Curve 2 in **Figure 12.4** shows NaOH titration of an acidic suspension of the same clay used in *curve 1*, but it was acidified more slowly than the *curve 1* clay; thus, *curve 2* clay has five times the exchangeable Al^{3+} as the acidified clay in *curve 1*. As base is added, the pH increases in *curve 2*, with an occasional plateau, which corresponds to pK_a values for Al^{3+} hydrolysis species ($AlOH^{2+}$ pK_{a1} = 4.95, $Al(OH)_2^+$ pK_{a2} = 9.3). Potentiometric titration curves of acid soils look similar to *curve 2*, suggesting that the soil acidity is exchangeable Al^{3+} and not exchangeable H^+.

In 1947, V.A. Chernov of the former Soviet Union summarized the results of many studies about the nature and properties of acid soils and clays. He recognized that hydrogen-saturated minerals were highly unstable and rapidly broke down to release Al^{3+}, Mg^{2+}, and Fe^{3+} from within clay lattices. He suggested that most acidic soils were in reality saturated primarily with Al^{3+} and Fe^{3+}. Chernov's work was supported by Jenny, Coleman, and others from the United States during the early 1950s. They confirmed that the weak-acid properties commonly attributed to hydrogen clays were the result of partial or complete saturation of exchange sites with weakly acidic Al^{3+} ions. For example, in *curve 2* in **Figure 12.2**, 55% of the cation

exchange capacity of the acid clay was occupied by exchangeable Al^{3+} as compared to only 8% in the acid clay used for *curve 1*. This suggests that the treatment used to prepare the acid clay used in *curve 2* dissolved the mineral cations and released Al^{3+} (proton-promoted dissolution). Aluminum is strongly adsorbed by exchange sites, and displaces Ca^{2+}, Mg^{2+}, K^+, and Na^+ cations on the clay minerals, which leach out of the profile. This process results in exchangeable acidity. Because of these early experiments, the focus of soil acidity is on both active acidity in the soil solution, and the exchangeable acidity that occurs from exchangeable H^+ and Al^{3+} ions on soil-mineral surfaces.

12.3   The role of aluminum in soil pH

Soil acidification implies that species exist that will contribute protons to the soil solution when it is titrated; i.e., the soil will neutralize bases added to soils. Thus, soil acidification includes processes that directly produce protons, and processes that create **exchangeable acidity**. An important process that creates exchangeable acidity is weathering of minerals and leaching of **base cations** (Ca^{2+}, Mg^{2+}, K^+, Na^+), which causes an enrichment of exchangeable Al^{3+} in the soil, leading to soil acidification.

12.3.1   Creation of exchangeable aluminum

Rainfall is slightly acidic because gaseous CO_2 forms H_2CO_3 (see Figure 4.9), and SO_2 and NO_x gasses from volcanos and fossil-fuel combustion form H_2SO_4 and HNO_3 (the components of acid rain). In addition, rainfall picks up organic acids from plants as it leaches through organic matter in the O and A horizons. Soil-weathering reactions from the infiltrating acidic rainfall dissolves primary and secondary minerals. Mineral dissolution reactions consume protons – for example, the dissolution of feldspar consumes eight protons from soil solution:

$$CaAl_2Si_2O_8 + 8H^+ = 2Al^{3+} + 2H_4SiO_4 + Ca^{2+} \qquad (12.1)$$

Subsequently, the products of feldspar dissolution either form new minerals, leach out of the weathering zone, or adsorb onto soil clays. Precipitation of new minerals releases protons. For example, the precipitation reaction for kaolinite is

$$2Al^{3+} + 2H_4SiO_4 + H_2O = Al_2Si_2O_5(OH)_4 + 6H^+ \quad (12.2)$$

Adding **Eqs. 12.1** and **12.2** yields a weathering reaction for dissolution of feldspar and formation of the mineral kaolinite:

$$CaAl_2Si_2O_8 + 2H^+ + H_2O = Al_2Si_2O_5(OH)_4 + Ca^{2+} \tag{12.3}$$

The overall weathering reaction **Eq. 12.3** consumes two protons from the soil solution, and is thus an alkalinity-producing reaction. Another weathering reaction example is the formation of gibbsite $(Al(OH)_3)$ from feldspar dissolution:

$$CaAl_2Si_2O_8 + 2H^+ + 6H_2O = 2Al(OH)_3 + 2H_4SiO_4 + Ca^{2+} \tag{12.4}$$

Silicic acid (H_2SiO_4) is a weak acid $(pK_{a1} = 9.8)$, and thus does not contribute protons to the soil solution. In both examples, weathering dissolves minerals, and protons are consumed. Thus, *mineral weathering reactions neutralize acids, meaning most minerals are bases, and their dissolution does not directly create soil acidity.*

Following mineral dissolution in the acid-neutralizing weathering reactions shown in **Eqs. 12.3** and **12.4**, Al^{3+} is adsorbed on clay minerals and organic matter and dissolved base cations are leached from the soil instead of adsorbed because the Al^{3+} outcompetes them for the exchange sites. This leaves the soil enriched in exchangeable Al^{3+}, creating soil acidity. The exchangeable Al^{3+} maintains equilibrium with aqueous Al^{3+} in the bulk soil solution. Aqueous Al^{3+} undergoes hydrolysis reactions that produce protons. The first hydrolysis is

$$Al^{3+} + H_2O = Al(OH)^{2+} + H^+ \tag{12.5}$$

Figure 4.14 shows the complete hydrolysis species for **Al-hydroxide monomers**. In addition to monomers, many aqueous **Al-hydroxide polymers** form that also contribute protons to the soil solution, for example:

$$13Al^{3+} + 28H_2O = Al_{13}O_4(OH)_{24}^{2+} + 32H^+ \tag{12.6}$$

Because of base cation leaching and exchangeable Al^{3+} enrichment, weathered soils become slightly to moderately acid over time. As weathering proceeds, acidic components are also leached from the soil. At this stage, the surface soil pH, and ultimately the pH of the entire profile, approaches neutrality, and iron and aluminum oxides become the predominant minerals.

12.4 Base cations in soil solutions

The cations Ca^{2+}, Mg^{2+}, K^+, and Na^+ are called base cations because they form strong base salts with hydroxide ions and do not readily hydrolyze when dissolved in solution. However, the name is misleading because in most cases base cations do not directly release or create hydroxides in soils. Base cations occur in alkaline soils as salts, carbonates, minerals, and exchangeable cations. In solutions, hydrated base cations do not hydrolyze water (cause water to release H^+) to any significant extent, and are electrically balanced by anions.

The concentration of base cations in soil solutions is an indirect measure of the ability of a solution to neutralize acid, which is equivalent to a solution's **alkalinity** or **acid neutralization capacity.** Soils with high concentrations of base cations are typically alkaline. Fluxes of base cations into a soil create alkalinity, and fluxes out of a soil are associated with soil acidification. The relationship between base cations and soil acidity and alkalinity is complex, and involves both solution and solid phase reactions.

12.4.1 Aqueous chemistry of base cations

Solution phase reactions of base cations are best understood by considering the electrical neutrality equation for soil waters (cation charge = anion charge):

$$\begin{aligned}
2 \times [Ca^{2+}] &+ [Na^+] + 2 \times [Mg^{2+}] + [K^+] + [H^+] \\
&+ 3 \times Al^{3+} + 2 \times AlOH^{2+} + Al(OH)_2^+ \\
&= [HCO_3^-] + 2 \times [CO_3^{2-}] + [OH^-] \\
&+ [org^-] + [Cl^-] + 2 \times [SO_4^{2-}]
\end{aligned} \tag{12.7}$$

where brackets indicate concentrations of the ions, and the concentration is multiplied by ion valence to account for the charge of the ion. Dissolved organic compounds (org⁻ in **Eq. 12.7**), such as carboxylic acid functional groups on SOM, may be a significant anion contributor in the soil solution. Nitrate and phosphate anions can be added to the anion tally on the right-hand side of **Eq. 12.7**, but are excluded for this exercise. **Eq. 12.7** can be rearranged to show species that are generally *fixed* on the left, and those that are *pH-dependent* on the right:

$$
\begin{aligned}
2\times\left[Ca^{2+}\right]&+\left[Na^+\right]+2\times\left[Mg^{2+}\right]+\left[K^+\right]-\left[Cl^-\right]\\
&-2\times\left[SO_4^{2-}\right]\\
&=\left[HCO_3^-\right]+2\times\left[CO_3^{2-}\right]+\left[OH^-\right]+\left[org^-\right]\\
&-\left[H^+\right]-3\times Al^{3+}-2\times AlOH^{2+}-Al(OH)_2^+
\end{aligned}
\tag{12.8}
$$

Species on the left-hand side of **Eq. 12.8** are fixed, or conservative, because they do not change concentrations with pH (unless they precipitate). Species on the right hand side are pH-dependent (nonconservative), because they change concentrations with pH – that is, they take on or release protons or hydroxides as pH changes, and thus *buffer* the soil solution from pH changes. **Eq. 12.8** can be simplified by assuming Cl⁻, SO₄²⁻, and CO₃²⁻ are negligible compared to the other species. The simplified relationship between base cations in solution and common pH-buffering ions is

Base cations	$2\times\left[Ca^{2+}\right]+\left[Na^+\right]+2\times\left[Mg^{2+}\right]+\left[K^+\right]\approx$
Basic species	$\left[HCO_3^-\right]+\left[OH^-\right]+\left[org^-\right]$
Acidic species	$-\left[H^+\right]-3\times Al^{3+}-2\times Al(OH)^{2+}-Al(OH)_2^+$

$$\tag{12.9}$$

This equation establishes an equality between base cations and basic and acidic species in solution. When base cation concentrations in soil solution are high, concentrations of protons or aqueous Al³⁺ species are low, and concentrations of bicarbonate, hydroxide, and deprotonated organic acids are high; thus, pH is alkaline. When base cation concentrations in soil solution are low, concentrations of H⁺ or aqueous Al³⁺ species are high, and concentrations of carbonate, OH⁻, and deprotonated organic acids are low, thus pH is acidic. Thus, the main influence of base cations on soil pH is

Base cations in solution do not directly create alkaline pH, but, the charge-balancing hydroxide, bicarbonate, and organic anions consume protons when acids are added to solutions; thus buffering the soil solution against acidification.

12.4.2   Exchangeable base cations

When base cations are removed from solution by precipitation, absorption by plants and microbes, or leaching, the lost base cations are replenished by desorption of base cations from the soil's exchangeable base cations. The presence or absence of base cations on exchange sites greatly influences exchangeable acidity. The percent of exchange sites occupied by exchangeable base cations is the percent **base-cation saturation** (%BS):

$$
\%BS = \frac{2\times\left[XCa^{2+}\right]+\left[XNa^+\right]+2\times\left[XMg^{2+}\right]+\left[XK^+\right]}{CEC}\times 100
\tag{12.10}
$$

where [XM⁺] indicates the concentration of base cation (M⁺) on the exchange site (X), and CEC is the cation exchange capacity. The pH used for CEC measurement must be specified because CEC increases with pH. A soil of pH 5, for example, may have 5 mmol(+) kg⁻¹ of exchangeable bases (Ca²⁺, Mg²⁺, K⁺, and Na⁺), and 1 mmol(+) kg⁻¹ of exchangeable acidity for a total CEC of 6 mmol(+). The %BS at pH 5 is

$$
\%BS(pH\,5) = \frac{5\,mmol(+)kg^{-1}}{6\,mmol(+)kg^{-1}}\times 100 = 83\%
\tag{12.11}
$$

At pH 7 the CEC is 8 mmol(+) kg⁻¹, and at pH 8.2 the CEC is 10 mmol(+) kg⁻¹. The %BS at pH 7 is

$$
\%BS(pH\,7) = \frac{5\,mmol(+)kg^{-1}}{8\,mmol(+)kg^{-1}}\times 100 = 63\%
\tag{12.12}
$$

Similarly, at pH 8.2, % BS = 50%.

In the early soil acidity literature, soils were characterized by their percent base saturation. Soils with low percent base saturation values were considered to be dominated by kaolinite and hydrous oxide minerals. Soils of high percent base saturation were considered to be dominated by 2:1-type minerals such as montmorillonite, vermiculite, chlorite, and the micas.

Base saturation is a criterion used to classify soil orders in the US soil classification scheme: Alfisols and Ultisols, are distinguished from one another on the basis of base saturation percentage – Alfisols have >35% BS, Ultisols have <35% BS. Percent base saturation is also used as a criterion for distinguishing Mollic epipedons (dark, organic matter-enriched surface horizons) from their umbric counterparts; Mollic epipedons have greater than 50% BS (based on soil CEC *at pH 7*) and umbric have less than 50% BS.

Unfortunately, %BS is as much a measure of the pH-dependent charge of soils as it is of the actual percentage of cation exchange sites occupied by exchangeable bases. The denominator includes any additional charge (CEC) generated by SOM and oxide-mineral complexes between the initial soil pH and the reference pH (7 or 8.2) that the CEC is measured. Since neither exchangeable Al^{3+} nor exchangeable H^+ is appreciable above pH 5.5, the CEC above this pH should be 100% base saturated. However, % BS in soils in the pH range 5.5 to 8.2 are often well below 100%. In alkaline soils dominated by pH-dependent charged minerals, low percent base saturation values are common, and most likely reflect the incorrect CEC value used in **Eq. 12.10**.

Although imprecise, %BS is useful for soil classification purposes, and for empirical liming recommendations. From the standpoint of soil chemical properties and reactions, %BS is considered an acidity index useful for relative comparisons of soil pH and buffering properties. To understand soil acidification processes, however, exchangeable acidity and pH-dependent charge on minerals and SOM are more meaningful than %BS; in some soils residual acidity may also be a significant contribution to titratable or exchangeable acidity.

12.4.3 Total exchangeable acidity

Total exchangeable acidity (TExA) is defined as the amount of exchangeable H^+ and Al^{3+}, measured in mmol(+) kg^{-1} soil:

$$TExA = \frac{3 \times \left[Al^{3+}\right] + \left[H^+\right]}{CEC} \tag{12.13}$$

$$TExA = CEC\left(1 - BS\right) \tag{12.14}$$

where *BS* is the fraction of base saturated cation on the ion exchange sites (%BS/100). For a given pH, CEC is fixed, thus base saturation and TExA are inversely related. The exchangeable acidity represents the resupply capacity of soil acidity. Soils with high %BS tend to have low TExA and are alkaline. Soils with low %BS tend to have high TExA and are acidic.

12.5 Soil acidification processes

Soil acidification occurs when acids are added to soils, or bases are lost. A common approach to understanding soil acidification processes is to compartmentalize the soil system into rhizosphere processes, bulk soil processes, and above-ground plant processes; and then attempt to identify reactions that produce acid and base in those compartments. While useful for studying soil acidification processes, this approach often fails to completely explain soil acidification because fluxes between the compartments are ongoing and complex, and no discreet boundaries exist. Instead, soil acidification is a continuum of *reactions* and *fluxes* that cause net changes in the active, exchangeable, and reserve acidity components. In addition, the vast array of proton producing and consuming microbial and plant biochemical reactions makes the reality of proton balancing in soils via compartmentalization increasingly complex.

Some reactions or processes in soils are main contributors to acidification. Such processes include natural fluxes (e.g., weathering followed by leaching), and human-caused fluxes to the soil (fertilization and harvest). Acid rain containing natural and anthropogenic H_2SO_4 and HNO_3 is also a common source of soil acidification.

In most cases, soil acidification includes fluxes that alter the biogeochemical cycling of carbon, nitrogen, or sulfur by changing either their concentrations or species. Proton producing and consuming reactions from carbon and nitrogen biogeochemical cycling in soils are shown in **Table 12.1**. Export or import of carbon and nitrogen compounds in open systems can cause proton buildup in the soil, decreasing the solution pH, and increasing the exchangeable acidity. For example, removal of plant material removes base cations from soils, and promotes acidification. Ammonium fertilizers cause acidification through nitrification.

Table 12.1 Common proton generation and consumption processes in carbon and nitrogen biogeochemical cycling. Adapted from Bolan and Hedley (2003). Reproduced with permission of Taylor & Francis.

Process	Net Reaction	H^+ produced $(mol_{H^+} mol_{acid}^{-1})$
Carbon cycle		
CO_2 dissolution	$CO_2 + H_2O \rightarrow H_2CO_3 \rightarrow$ $\mathbf{H^+} + HCO_3^-$	+1
Organic acid production (SOM)	Organic C $\rightarrow$ R-COOH $\rightarrow$ R-COO$^-$ + $\mathbf{H^+}$	+1
Degradation of organic acid (decarboxylation)	R-COO$^-$ + $\mathbf{H^+} \rightarrow$ RH + CO_2 (g)	−1
Nitrogen cycle		
N fixation	$N_2 + H_2O + 2R\text{-}OH \rightarrow$ $2R\text{-}NH_2 + 1.5\ O_2$	0
Organic N mineralization (ammonification)	$R\text{-}NH_2 + \mathbf{H^+} + H_2O \rightarrow$ $R\text{-}OH + NH_4^+$	−1
Ammonia volatilization	$NH_4^+ \rightarrow NH_3 + \mathbf{H^+}$	+1
Nitrification	$NH_4^+ + 2O_2 \rightarrow NO_3^- +$ $H_2O + 2\mathbf{H^+}$	+2
Denitrification	$2NO_3^- + 2\mathbf{H^+} \rightarrow N_2$ $+ 2.5O_2 + H_2O$	−2

Reactions important for soil acidification are classified as either net **proton sources** to soil solution or net **proton sinks** that remove protons from soil solution.

12.5.1 Organic matter influences on pH

Organic carbon cycling influences soil pH by producing CO_2, releasing base cations upon degradation, and producing aqueous organic acids. Research on the effects of organic matter input (e.g., crop residue, plant litter) or export (e.g., harvest) on soil pH has shown that pH changes are variable, and depend on soil properties (mineralogy, temperature, porosity, initial pH, etc.), plant species, and management practices (e.g., no-till, or burning). The complexity of the various soil organic matter interactions makes predicting soil responses to harvest and residue management difficult. Crop residue input generally increases soil pH, and harvest decreases soil pH. **Figure 12.5** shows the effects of crop residues on soil pH on a relatively short time scale. The variability in base cations in the crop residue has a large influence on the soil pH. For example, in **Figure 12.5** chickpea residues have the highest base cation content (1420 mmol(+) kg^{-1}) and the soils

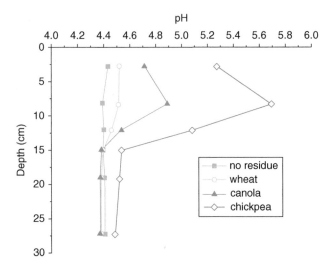

Figure 12.5 Effect of residue additions (14 t ha^{-1}) on soil pH in an Australian Podzol (Spodosol) after three months of incubation. The chickpea had a base cation charge content of 1420 mmol(+) kg^{-1}, the canola 1140 mmol(+) kg^{-1}, and the wheat 430 mmol(+) kg^{-1}. Plant residue decreased soil acidity. Adapted from Butterly et al. (2013). Reproduced with permission of Springer.

with the residue incorporated have the highest pH. In contrast, wheat residues have the least amount of base cations (430 mmol(+) kg^{-1}), and the soils have the lowest pH.

Agronomic-pH management has created guidelines for lime requirements to neutralize the alkalinity removed from soil by a given crop that are based on yield. For example, for a corn grain harvest of 10 Mg ha^{-1} (about 150 bushels acre^{-1}), it is recommended that 22 kg ha^{-1} of agricultural lime be added. Oats on the other hand, requires only about half the lime addition rate for an equivalent harvest yield. Such estimates are largely based on empirical observations, and the exact mechanisms and accurate management methods for pH changes based on agricultural practices of crop residue and harvest are not well known. This is because several carbon-related acid and base reactions are simultaneously occurring, as are fluxes of nitrogen and base cation compounds that also greatly influence soil pH. This makes construction of a proton balance for an open system, such as a field soil, difficult.

Dissolved organic acids produced by microbes and plants are typically weak acids that create soil-solution acidity. However, they can also complex Al^{3+}, removing it from solution, reducing acidity. Protonation and deprotonation of organic acids depends on their pK$_a$, which typically ranges from 2 to 7. Soil organic matter behaves similarly to dissolved organic acids. Weak acid functional groups donate and accept protons (Table 9.3) so SOM can be a proton source or sink. In forest soils, organic acids are present in high concentrations and are especially important in acidifying soil. In alkaline soils, organic matter functional groups are deprotonated, and thus when titrated to lower pH, they consume protons and buffer the soil pH. Protonation and deprotonation of organic acids can be modeled using acidity constants (e.g., pK_a), however, deriving an overall acidity constant for a heterogenous substance such as SOM is not straight forward.

Organic acids in forest soils mobilize Al^{3+} (and also Fe^{3+}), causing enrichment in subsurface horizons (e.g., see the Spodosol in Figure 4.19). Aluminum that dissociates from organic acids undergoes hydrolysis to produce protons:

$$Al\text{-}org + H_2O = Al(OH)^{2+} + H^+ + org^- \qquad (12.15)$$

where *Al-org* is an Al^{3+} cation complexed by functional groups of an organic acid (*org⁻*). The extent of the reaction in **Eq. 12.15** is predicted using metal complexation and Al^{3+} hydrolysis constants. Fluxes of other cations and organic matter degradation also contribute to the release of Al^{3+} from the organic ligands.

The complexity of acid-base processes affected by organic matter is apparent when considering organic matter degradation and organic acid reactions. Degradation of organic acids is a decarboxylation reaction that reduces acidity. Degradation of organic matter, however, also creates CO_2 gas that dissolves in water to create carbonic acid (**Table 12.1**). The first pK_a of carbonic acid is 6.35 (Figure 4.9), so carbonic acid creates only a slightly acidic soil solution by itself, but the lowering of pH causes the nondegraded organic acids (SOM) to further deprotonate and increase soil acidity. The effects of organic acid degradation on proton source-sink reactions are influenced by soil processes, such as microbial respiration rates, soil weathering, agricultural practices, and so on, as well as other proton source–sink reactions (e.g., nitrogen transformations), highlighting the complexity of predicting soil acidification from soil organic matter processes. Oxidation reactions (mineralization, ammonification, metabolism of various organics, etc.), in general, are usually acid-producing reactions.

Figure 12.6 Non-allophanic soil profile from Costa Rica showing deep accumulation of SOM. The melanic epipedon complexes Al^{3+} preventing formation of allophanic minerals. Photo courtesy of Paul McDaniel, University of Idaho. See Figure 12.6 in color plate section for color version.

In forest soils with an abundance of readily soluble minerals (e.g., volcanic ash in Andisols), and where precipitation and litter input is high, low pH leads to an abundance of SOM-Al^{3+} complexes. Aluminum complexation appears to slow SOM degradation. Some of the SOM-Al^{3+} complexes are soluble and mobile, and cause deep accumulation of SOM (**Figure 12.6**). In such soils, the preferential complexation of Al^{3+} with organic materials inhibits formation of allophane, and the soils are classified as non-allophanic Andisols. pH in these soils can be acidic because of the soluble organic acids and exchangeable aluminum.

12.5.2 Acidity from the nitrogen cycle

The nitrogen cycle involves multiple oxidation-state transformations (see Figures 3.20 and 5.20). It also is closely associated with the carbon cycle, and is greatly influenced by amount and type of biological processes. Thus, separating N-cycling acidification processes

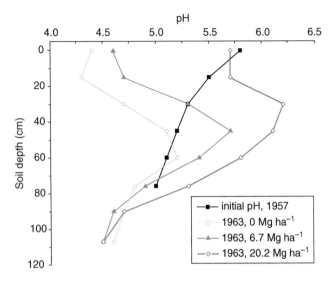

Figure 12.7 Influence of varying lime application rates on soil profile acidity following annual ammonium nitrate fertilization of Coastal Bermudagrass. Lime was applied following each fertilizer application (four per year). Soil is an Ultisol with sandy loam texture. Adapted from Adams et al. (1967), with permission.

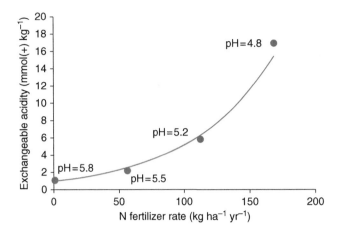

Figure 12.8 Effect of urea or NH_4NO_3 fertilizer amendment rates on exchangeable Al^{3+} in Wisconsin, USA, soils annually amended with fertilizers for 30 yrs. Final average pH is listed above each point. The loam soil originally was pH 6.5 to 7. Crops were tobacco, corn, and soybean. Each point is an average; standard deviations were less than 1 mmol(+) kg^{-1}. Adapted from Barak et al. (1997). Reproduced with permission of Springer.

from soil carbon cycling processes is difficult. One of the largest nitrogen fluxes causing wide-scale acidification of agricultural soils is nitrification of ammonia-based fertilizers (**Table 12.1**). Less acidity is generated from NH_4NO_3 per unit of nitrogen than from $(NH_4)_2SO_4$ because only half of the nitrogen in NH_4NO_3 is oxidized.

Figure 12.7 shows the pH of a soil profile in which NH_4NO_3 was added multiple times per year for Bermuda grass fertilization. After six years (from 1957 to 1963), soil pH of the top 20 cm of the profile had decreased by more than one pH unit. pH decreases in the subsoil were less dramatic (**Figure 12.7**).

Figure 12.8 shows soil pH and exchangeable acidity after 30 years of agricultural production under annual ammonium nitrate or urea fertilizer inputs of different amounts. At the highest fertilization rate, the soil pH decreased from an initial average value of 6.5–4.8 over 30 years, and exchangeable acidity showed a corresponding increase with increasing fertilizer application rates.

If all of the nitrogen input to soils was consumed by plants and microorganisms, then the input of nitrogen fertilizers would be acid-base neutral. For example, ammoniacal-N fertilization (anhydrous ammonia, ammonium nitrate, or urea) results in acid production

by nitrification (shown in **Table 12.1**); but subsequent nitrate uptake and assimilation by biota consumes protons, thereby neutralizing the acidity from the nitrification reactions. A summary nitrate assimilation reaction that occurs in a plant when nitrate is absorbed and fixes nitrate into a protein is

$$NO_3^- + ROH + H^+ + 2CH_2O = RNH_2 + 2CO_2 + 2O_2$$
(12.16)

where *ROH* and *RNH$_2$* are the biomolecules (e.g., proteins) that fix the nitrate in the plant or microbe. This reaction occurs within a plant and occurs in multiple reaction steps, including reduction of the nitrate to ammonia, followed by synthesis into a biomolecule. Consider urea application to a soil that undergoes nitrification within the soil (**Table 12.1**); if all the nitrate is absorbed and assimilated, as described in **Eq. 12.16**, the net reaction is proton-neutral:

$$CO(NH_2)_2 + 2ROH + 4O_2 + 4CH_2O$$
$$= 2RNH_2 + 5H_2O + 5CO_2$$
(12.17)

Similar reactions can be written for other ammoniacal fertilizers. The proton sink of nitrate assimilation, however, occurs in the plant, and does not directly offset the proton source of nitrification occurring in soils. **Eq. 12.17** represents a closed system; that is, the plant, soil, and nitrogen reactions are allowed to physically equilibrate, transferring acidity or alkalinity in the form of base cations, nitrogen compounds, organic acids, anions, or protons. When plant debris is not completely degraded, as is the case when it is stored as SOM or is removed from the system by harvest, the nitrogen assimilation alkalinity is not returned to the soil, and the nitrification-caused soil acidity remains.

Nitrogen-use efficiency by plants also affects soil acidification. If plants used 100% of the applied nitrogen, farmers could save money and apply less. But, because some nitrogen is not used by the plant, farmers apply nitrogen fertilizer at greater rates. The nitrate produced from nitrification of ammoniacal fertilizers that is not assimilated by plants leaches from the soil, leaving behind the acidity created in the nitrification reaction.

12.5.3 Phosphate and sulfate fertilizer additions to soil acidity

Phosphoric acid released by dissolving phosphate fertilizer granules can lead to pH values as low as 1.5 near the granule. For example, dissolution of monocalcium phosphate fertilizer (superphosphate) produces a proton, dicalcium phosphate, and a phosphate anion in the reaction

$$Ca(H_2PO_4)_2 = CaHPO_4 + H_2PO_4^- + H^+ \qquad (12.18)$$

Some of the phosphate acidity is neutralized by hydroxides released when the phosphate adsorbs on oxide mineral surfaces (e.g., ligand exchange (Eqs. 10.40 and 10.41). But some of the acidic reaction products may remain to influence soil properties.

Adding elemental sulfur fertilizer produces sulfuric acid when it oxidizes:

$$2S^0 + 2H_2O + 3O_2 = 2SO_4^{2-} + 4H^+ \qquad (12.19)$$

The oxidation is facilitated by chemolithoautrophic bacteria, and can lead to severe acidification if lime additions are not included in the land management practices. If the soil contains $CaCO_3$ minerals, additional elemental sulfur will be required to acidify the soil due to the neutralization reactions of the carbonate anions dissolved from the carbonate minerals ($CO_3^{2-} + H_2O = HCO_3^- + OH^-$).

12.5.4 Plant root influences on soil acidity

When roots take up cation and anion nutrients they maintain electrical neutrality in plant root cells by absorbing or releasing H^+ ions, organic acids, and OH^- ions. When plants take up more cations than anions, acidification occurs. This causes rhizospheres to be more acidic than the surrounding soils. Production of organic acids by plant roots and microbes, and respiration that increases the partial pressure of CO_2 in the rhizosphere also cause rhizosphere acidification. Nitrogen fixation by nodules releases protons, depending on how the nitrogen is assimilated; for example, production of amino acids from fixed nitrogen creates carboxylic acids. To maintain pH neutrality during nitrogen fixation, plants export protons and take up base cations. Legumes, for example, export 0.2 to 0.7 mole H^+ per mole of fixed nitrogen.

12.5.5 Protonation and deprotonation of mineral surfaces

The edges of clay minerals have weak acid Fe-O, Al-O, and Si-O functional groups that change the soil solution concentration of protons (e.g., Figure 9.7 and Table 9.2). Protonation and deprotonation of these weak acids are important sources and sinks of protons to the soil solution, and buffer changes in soil pH.

12.5.6 Pollution sources of soil acidity

Mine wastes containing iron pyrite (FeS_2) and other sulfides added to soils can create highly acidic soil conditions. The oxidation occurs via biotic and abiotic oxidation processes that produce sulfuric acid and iron oxide minerals (e.g., ferrihydrite), or iron sulfate minerals (e.g., jarosite or schwertmannite). The pyrite oxidation reaction is

$$2FeS_2 + 7O_2 + 2H_2O = 2Fe^{2+} + 4SO_4^{2-} + 4H^+ \qquad (12.20)$$

The oxidation is catalyzed by chemotrophic bacteria that gain energy from the electrons released from the sulfide (details are provided in Special Topic Box 5.1). The ferrous iron released is further oxidized by chemotrophs, or can be abiotically oxidized. The Fe^{3+} can subsequently be abiotically reduced by pyrite, producing sulfuric acid and creating an abiotic–biotic feedback loop for pyrite oxidation. Pyrite oxidation can result in soil pH values less than two. Over time, pyrite minerals are exhausted, the acidity is leached out of the soil, and the soil pH returns to near neutral due to alkalinity caused by mineral dissolution. Drainage of estuaries or wetlands containing high-sulfide sediments (acid-sulfate soils) also creates low-pH soils via acidification processes similar to pyrite oxidation.

Acid rain caused by industrial releases of SO_2 and NO_x gases is a widespread problem for natural landscapes. The gases emitted from coal combustion, for example, create nitric and sulfuric acid vapors that rain down and acidify soils. Nitric acid from NO_x emitted from automobile exhaust also creates soil acidity problems. Agriculture, for the most part can ameliorate the effects of acid rain by liming. But acidification of forests and prairies from acid rain is a major concern because of the detrimental effects on plant growth and leaching of aluminum to surface waters, where it poisons aquatic organisms. Sulfur and nitrogen deposition, on the other hand, is beneficial for soils low in these nutrients, improving plant growth. Modern coal combustion facilities and smelters have installed traps to reduce nitrogen and sulfur emissions, and automobiles and gasoline formulations do not emit as much NO_x. However, problems remain in regions where such reduction measures are not used.

12.5.7 Redox reaction influence on soil acidity

Reduction and oxidation reactions also cause soil pH changes. Electrons and protons follow each other in half reactions. However, redox requires a couple, thus for every reducing reaction that consumes a proton, there must be a simultaneous oxidation reaction that likely produces a proton. The net effect would be neutral, except for the fact that the products of the redox reaction may change the alkalinity of the soil solution by changing oxidation state (adding charge to the cation or anion charge balance in **Eq. 12.7**). Additionally, the concentrations of ions in solution may change because of phase changes (e.g., precipitation, adsorption, or gas fluxes). An increase in solubility of species upon redox also facilitates leaching fluxes of the chemicals out of the soil.

' Many of the reactions discussed in carbon and nitrogen transformations in this chapter are redox processes. In soils that have fluctuating oxygen concentrations, iron, manganese, and sulfur oxidation and reduction are common redox processes that can change acidity:

$$FeOOH + e^- + 3H^+ = Fe^{2+} + 2H_2O \qquad (12.21)$$

$$SO_4^{2-} + 8e^- + 8H^+ = S^{2-} + 4H_2O \qquad (12.22)$$

These are reducing half reactions that consume protons. The relative change in solution acidity, however, depends on the other *half* of the reactions. For example, if the electrons for the goethite reduction reaction are coming from reduced carbon, such as in carbohydrate (CH_2O):

$$CH_2O + H_2O = CO_2 + 4e^- + 4H^+ \qquad (12.23)$$

then the net reaction for iron reduction consumes protons, increasing soil pH:

$$4FeOOH + 8H^+ + CH_2O = 4Fe^{2+} + 7H_2O + CO_2 \quad (12.24)$$

Using the carbohydrate as the reductant (**Eq. 12.23**) to reduce the sulfate to sulfide (**Eq. 12.22**) would be a net-neutral reaction, with no net change in the protons in solution. In both the iron and sulfate reduction reactions, however, the CO_2 from carbohydrate oxidation will dissolve in the soil solution, creating some acidity or reducing the alkalinity of the reaction.

When wetlands become reduced, they initially become acidic because of the release of organic acids that are slowly degraded in anoxic conditions (i.e., decarboxylation is minimal). Under prolonged reducing conditions, however, soil pH increases because of volatilization of CO_2 and conversion of organic acids to methane. The amount of organic matter, types of minerals present, and input of ions from external sources have major influences on wetland soil pH. For example, the high sulfur content of brackish water wetlands causes low pH soils (acid-sulfate soils), because of the production of sulfuric acid upon oxidation of iron sulfide minerals (**Eq. 12.20**).

12.6 Aluminum and manganese toxicity

Many plants grow poorly in acid soils. Early workers supposed that this was a consequence of either hydrogen ion toxicity or Ca^{2+} and Mg^{2+} deficiencies. The soil pH must be less than about pH 3, however, before the H⁺ concentration is high enough to be toxic to most plant species. Although exchangeable Al^{3+} and H^+ are emphasized in acid soils, in soils with pH greater than about five, the major exchangeable cations are Ca^{2+}, Mg^{2+}, and to a lesser extent K^+. Thus, under these moderately acid conditions, the exchangeable Al^{3+} and H^+ do not limit availability of base cations for plant growth. Rather, in acid soils, the Al^{3+} appears to stunt plant growth. The low pH may also limit nitrogen or phosphate availability.

Plant growth problems associated with poor root penetration into acid subsoils are frequently associated with high plant availability of Al^{3+} or Mn^{2+}, which are toxic to most plants. Aluminum restricts or stops root growth at solution concentrations as low as 1 mg L^{-1}. Some plants, such as conifers, appear to be more tolerant of Al^{3+} toxicity; forest soils are typically acidic. Plants tolerate higher levels of soluble manganese, but reducing conditions in flooded or periodically inundated acid soils can result in soluble manganese concentrations as high as 100 mg L^{-1}, which can be toxic.

12.7 Plant nutrients in acid soils

The effects of low pH on plant growth may also arise from nutritional imbalances because the nutrient availability can increase or decrease as soil acidity changes (see Figure 3.1). The effects of pH on plant-nutrient levels in soils are complex, but some generalizations are possible.

Plants able to utilize ammonium forms of nitrogen have a considerable advantage in acid soils because below pH 5.5 nitrification is slow. In acid forest soils, ammonium ions may accumulate because the microbes that mineralize organic nitrogen to ammonia are less dependent on soil pH than the microorganisms that oxidize ammonia to nitrate.

Acid soils generally provide sufficient micronutrients to plants; sometimes even toxic amounts. Because of the small quantities of micronutrients required for plant growth, adequate amounts can be taken up from small portions of the root zone if such regions are sufficiently acidic. In basic soils, the acidity from fertilizers plus elemental sulfur or sulfuric acid added to a portion of the root zone may provide adequate micronutrients to plants.

An exception for micronutrient availability in acid soils is molybdenum, which exists as the molybdate oxyanion. Molybdate is less available to plants at low pH (more strongly adsorbed). Occasionally, the harmful effect of soil acidity on leguminous plants is caused by molybdenum deficiency rather than by aluminum toxicity. Molybdenum is required for nitrogen fixation by legumes, and in low-pH soils, crop yields of nitrogen-fixing plants have been shown to respond equally to liming or molybdenum fertilization. The acidification and leaching associated with weathering can also remove boron causing boron deficiency. Adding molybdenum or boron to soils is difficult because of the narrow window between deficiency and toxicity.

12.8 Managing acidic soils

A major challenge for managing acid soils is to estimate the quantity of lime required to raise the soil pH to a certain level. **Figure 12.7** shows the response of soil pH to liming rates. Field trials, such as those shown in **Figure 12.7**, are the best way to estimate liming rates. In most cases, however conducting field trials is impractical and time-consuming.

An economical way to estimate the lime requirement of acid soils is to measure the quantity of base required to raise soil pH to a specified level. To be realistic the titration must be slow enough for the added base to react completely with the soil. Both exchangeable and titratable acidity will be neutralized during the titration.

12.8.1 Predicting lime requirement

An important effect of lime is to provide hydroxyl ions that convert exchangeable Al^{3+} to $Al(OH)_3$ (gibbsite). Increased quantities of soluble and exchangeable Ca^{2+} and Mg^{2+} are byproducts of liming, which serve to displace exchangeable acidity, and may be beneficial to plants, such as legumes, having high calcium requirements.

The calculation of $CaCO_3$ lime required from a titration is based on production of two hydroxides ions per mole of $CaCO_3$:

$$CaCO_3 + H_2O = Ca^{2+} + CO_2(g) + 2OH^- \qquad (12.25)$$

For example, if a soil titration, like shown in **Figure 12.3**, indicates that 2.0 mmol of OH⁻ is consumed per 100 g of soil for each unit increase in pH, then 5 Mg CaCO₃ per ha (30 cm) is required for every unit increase in pH desired. The calculation is

$$\frac{2.0\,\text{mmol}\,\text{OH}^-}{100\,\text{g soil unit pH}} \times \frac{1\,\text{mmol}\,\text{CaCO}_3}{2\,\text{mmol}\,\text{OH}^-} \times \frac{100\,\text{mg}\,\text{CaCO}_3}{1\,\text{mmol}\,\text{CaCO}_3}$$
$$\times \frac{1\,\text{Mg}}{10^9\,\text{mg}} \times \frac{4.5 \times 10^9\,\text{g soil}}{\text{ha}\,(30\,\text{cm})} = 5.0 \frac{\text{Mg}\,\text{CaCO}_3}{\text{ha}\,(30\,\text{cm})\,\text{unit pH}}$$

(12.26)

This is the *apparent* lime-based buffer capacity (BC) of the soil, which is the amount of lime required to add to soil per pH-unit increase. The calculation requires the density of the soil (assumed 4.5×10^9 g soil per ha(30 cm) in this example), and assumes that pure CaCO₃ that provides 2 mmol OH⁻ per mmol CaCO₃ is used as the lime amendment (and that it completely dissolves). Corrections for actual density of the soil, and type of amendment are required for accurate predictions of buffer capacity of a given soil. The **lime requirement** (LR), in Mg ha⁻¹ is calculated using the soil buffer capacity:

$$\text{LR} = \left(\text{pH}_{\text{optimal}} - \text{pH}_{\text{current}}\right)\text{BC}$$

(12.27)

where pH_{optimal} is the soil pH required for the specified cropping system, pH_{current} is the pH of the topsoil, and BC is the calculated buffer capacity of the soil, such as in **Eq. 12.26**.

Acidity neutralization by field-liming is typically incomplete because of incomplete mixing and slow reaction times. The lime dissolution reaction rates vary inversely with pH, limestone particle size, and solubility of the liming agent. Hence, the laboratory-based lime requirement value is often further multiplied by a conversion factor to better estimate the amount of lime needed to achieve a given field pH. Such a conversion factor is regionally specific, and dependent on the type of lime.

The titration of individual soil samples is impractical for soil-testing purposes because of the time and experimental precision required. The usual procedure to estimate lime requirement is to add a pH buffer solution to the soil, measure the amount of buffer consumed or the resulting pH of the soil-buffer suspension, and calibrate results with field-lime requirements for similar soils from the same geographical area. For example, a pH change of 0.1 unit from the initial buffer pH might correspond to 1 Mg limestone ha⁻¹, which corresponds to a rate of 10 Mg limestone ha⁻¹ for a full unit pH change of the buffer (1 Mg ha⁻¹ × 10). If a soil has an initial pH of 5.5, the buffer solution has an initial pH of 6.8, and the final mixture has a pH of 6.3 (pH$_{\text{desired}}$ − pH$_{\text{buffer}}$ = 0.5), then the lime requirement for this soil would be

$$\frac{10\,\text{Mg limestone}}{\text{pH unit ha}} \times 0.5\,\text{pH units} = 5\frac{\text{Mg limestone}}{\text{ha}}$$

(12.28)

Using a calibration curve from the titration of one soil can be used to estimate the lime requirements of other soils from the same geographic region if soil texture and measurements of initial soil pH are incorporated in an empirical model, but the predicted lime rates will be less precise.

12.8.2 Optimal management of soil pH

To achieve maximum crop production in acid soils, soil pH must be raised to the optimum level for the crop in question. Little is gained by raising the pH to still higher levels. The growth increase from each successive increment of lime diminishes, but the cost of adding the increment remains the same. Additionally, raising pH beyond what is recommended for the plant may limit the availability of micronutrients, such as Fe^{3+}.

Plant growth in strongly weathered soils can be hampered by acidic subsoils. Surface application and mixing by plowing are ineffective in treating the subsoils. In one study, it was observed that even 10 years after application, lime diffused less than 10 cm. Adding gypsum and lime at the same amounts (ca. 5–10 Mg ha⁻¹ each) on the surface improves plant growth and considerably improves root growth and penetration of the treatment into the subsoil. Gypsum ($CaSO_4 \cdot 2H_2O$) is more soluble than CaCO₃. The sulfate anion is thought to penetrate to the subsoil, saturate the positively charged clays, raise pH by a small amount, and complex some of the soluble Al^{3+} making it less toxic for the plant roots. Lime products that have micron- and nanometer-sized lime particles are promoted as having better subsoil penetration, but they are more expensive than the traditional lime products.

Following lime amendment, attaining the desired soil pH may require 6–8 months, and the pH may change appreciably for as long as 18 months thereafter. Adequate soil water is necessary to permit hydroxyl and calcium ion diffusion, and to carry out the associated liming reactions. When slowly reacting dolomitic limestone is used as a lime source, soil pH may increase for as much as five years. In general, more finely ground-liming materials cost more but react faster and more thoroughly with the soil. Finer particles also disperse more evenly through the soil than smaller numbers of large particles.

Because of the challenges in lime amendment rate estimation, and the potential cost to farmers to apply lime, a proposed better soil acidity management uses a more holistic soil–plant system approach called **biological-lime requirement** (BLR). BLR is the amount of lime required to eliminate limitations for plant growth, and is based on soil pH and buffering capacity, soil fertility, plant species, and agronomic factors (tillage, crop residue management, etc.). Managing soil pH for all of these factors is more sustainable than basing lime-amendment rates on laboratory measurements alone. BLR determinations, however, require more initial investment into research for site-specific lime requirements. Field trials of lime-amendment rates are not difficult but take many years to conduct due to the slow reactivity of lime, and are regionally and crop-system specific.

12.9 Summary of important concepts in soil acidity

Soil pH is a major property of soils that determines the soil health because it affects most chemical processes, impacts microbial population and dynamics, and is a major determinant in plant growth. Some soils are naturally acidic (e.g., forest soils), while others are naturally alkaline (e.g., soils with calcite in them). Accurately characterizing soil pH requires assessment of the active and exchangeable pH components. Percent base saturation is used as an indicator of soil acidity. Exchangeable acidity (i.e., exchangeable

aluminum) is another important parameter for assessing soil acidity.

Soil acidification is a major threat to soil health that causes aluminum toxicity to plants and can leach aluminum into surface waters negatively impacting aquatic organisms. Acidification occurs from application of ammoniacal fertilizers, acid rain, and addition of pyrite-containing mine wastes to soils. To remediate soil acidification of agricultural fields, lime ($CaCO_3$) is amended to soils. Accurately calculating the amount of lime to amend to the soils is difficult because the integration of lime into soils and the weathering and chemical reactions in soils are variable. The best approach for estimating lime requirement, aside from actual field trials for each soil and plant type, is laboratory titrations matched with soil physicochemical characteristics and plant needs. However, some field testing should be done to validate the lab-based lime requirement accuracy, especially when soil and site properties in a region are variable.

Questions

1 Calculate the relative acidifying tendencies of 100 kg ha^{-1} of nitrogen applied as (NH_4)$_2SO_4$, NH_4NO_3, or NH_3. *Hint:* Upon nitrification, anhydrous ammonia releases one proton per mole of NH_3.

2 A soil of pH 5.5 retains 60 mmol(+) kg of exchangeable bases and has a CEC at pH 7 of 80 mmol(+) kg^{-1}. What is its percent base solution? What is its approximate CEC at pH 5.5?

3 A soil has a pH of 5.2, retains 70 mmol(+) kg^{-1} of exchangeable bases, 10 mmol(+) kg^{-1} of exchangeable Al^{3+}, and has CEC at pH 7 of 100 mmol(+). What is its percent base saturation? What is its percent exchangeable acidity? What is the approximate field lime requirement if the pH is to be raised to 6.5?

4 In a soil titration, 2.3 mmol of base is consumed per 100 g of soil for each unit increase in pH. What is the lime requirement for a soil to increase pH from 4.78 to 6.0? Assume bulk density of 4.5 × 10^9 g soil per ha(30 cm) and complete utilization of lime for neutralization.

5 To understand soil acidification processes, soil processes are often compartmentalized; for example, proton production and consumption processes of plants, soil residue and minerals are often studied

independently. What are advantages and limitations to this approach for studying and predicting soil acidification?

6 Describe how an Al^{3+} ion exchanged off of a soil particle causes the release of a proton.

7 Write reactions that describe why a soil will consume more limestone than active soil pH predicts. *Hint:* Consider exchangeable acidity.

8 How much limestone with a $CaCO_3$ equivalent of 87% would you need to apply to eliminate exchangeable Al^{3+} in a soil with CEC = 7.5 cmol(+) kg^{-1} and an aluminum saturation of 58%?

Bibliography

Adams, W.E., A.W. White, and R.N. Dawson. 1967. Influence of lime sources and rates on coastal bermudagrass production soil profile reaction exchangeable Ca and Mg. Agronomy Journal 59:147–149.

Adriano, D.C. 2001. Trace Elements in Terrestrial Environments: Biochemistry, Bioavailability, and Risks of Metals. 2nd ed. Springer.

Aldrich, D.G., and J.R. Buchanan. 1958. Anomalies in techniques for preparing H-bentonites. Soil Science Society of America Proceedings 22:281–286.

Barak, P., B.O. Jobe, A.R. Krueger, L.A. Peterson, and D.A. Laird. 1997. Effects of long-term soil acidification due to nitrogen fertilizer inputs in Wisconsin. Plant and Soil 197:61–69.

Bolan, N.S., and M.J. Hedley. 2003. Role of Carbon, Nitrogen, and Sulfur Cycles in Soil Acidification, *In* Z. Rengel, (ed.) Handbook of Soil Acidity. Marcel Dekker, Inc., New York.

Butterly, C.R., J.A. Baldock, and C. Tang. 2013. The contribution of crop residues to changes in soil pH under field conditions. Plant and Soil 366:185–198.

Funakawa, S., K. Hirooka, and K. Yonebayashi. 2008. Temporary storage of soil organic matter and acid neutralizing capacity during the process of pedogenetic acidification of forest soils in Kinki District, Japan. Soil Science and Plant Nutrition 54:434–448.

13 SALT-AFFECTED SOILS

13.1 Introduction

Salt-affected soils are common in arid and semiarid regions, where annual precipitation is insufficient to meet the evapotranspiration needs of plants and to prevent salt buildup (**salinization**). "Making the desert bloom" has been a dream for many generations. Salt problems, however, are not restricted to arid or semi-arid regions; under some conditions, soil salinity can develop in subhumid and humid regions. Salt accumulation in soils is detrimental to plant growth, and thus management of salt affected soils is necessary.

Sustainability in growing crops in arid regions requires that adequate drainage to remove accumulated salts is provided. Drainage is costly and requires water use in excess of what is required for the plant growth. Correctly managing soil salinity requires understanding soil chemical processes related to soil and irrigation water salinity so that the amount of water needed for drainage to prevent salinization can be minimized.

13.2 Distribution and origin of salt-affected soils

As much as 20% of all irrigated lands in the world (or approximately 45 million ha) may have irrigation-induced soil salinity problems. In addition, vast areas of land have native salinity and limited arability. Salt buildup in soils is a problem in much of Australia and Africa, western United States, Middle East, north and central Asia, and western South America. Wherever there is a low amount of rainfall and limited availability of water for leaching salts out of the soils, salinization is a risk. Salt problems associated with greenhouse crops, mine spoils, and waste disposal further increase the amount of land that has salt problems.

Soil Chemistry, Fifth Edition. Daniel G. Strawn, Hinrich L. Bohn, and George A. O'Connor.
© 2020 John Wiley & Sons Ltd. Published 2020 by John Wiley & Sons Ltd.

The three main natural sources of soil salinity are mineral weathering, atmospheric precipitation, and fossil salts (those remaining from former marine or lacustrine environments). The human activities that add salts to soil include irrigation and land application of saline industrial wastes. Seawater encroachment can also harm soils.

An example of human induced salinity problems is Australia's Murray-Darling Basin. In this region, salts released during weathering *naturally* accumulated below the roots of native vegetation at the weathering front, which represents the depth of water infiltration. Conversion of the native landscape to shallower rooting agricultural crops caused excess water to infiltrate the salt-rich subsurface, raising the water table. The higher water table created an upward flux of *historic* salts that existed below the native rooting zone, resulting in salinization of the surface horizons. When the occasional heavy rains occur in the Murray-Darling region, the salt in the upper horizons mobilizes and seeps from the landscape into streams, wetlands, and riparian areas, creating surface water salinity problems.

13.2.1   Mineral weathering sources of salts

The source of salts in soils is weathering of rocks and minerals that releases ions. In humid areas, soluble salts are carried through the soil profile by percolating rainwater and are ultimately transported to oceans or inland seas. In arid regions, salts accumulate because of the relative scarcity of rainfall, high evaporation, and plant transpiration rates, which is exacerbated when there is landlocked topography.

Without leaching, in-situ weathering of minerals eventually results in accumulation of soluble salts to levels detrimental for plant growth. In well-drained natural soil systems, excessive salt accumulation is rare because during weathering salts are slowly released, and even the occasional rains in arid regions are usually sufficient to leach most of the salts out of the rooting zone, and eventually to the sea, a land-locked lake, or a nearby saline seep.

13.2.2   Salinity from fossil salts

So-called fossil salts can introduce large amounts of salinity into soil and groundwater, as is occurring in the Murray-Darling Basin. A similar salinity problem

occurred in the 1960s in the Wellton-Mohawk irrigation project of Arizona, where saline groundwaters were discharged into the Gila River after irrigation raised the groundwater level in a valley underlain by saline deposits. The drainage water mixed with the Colorado River and significantly increased the river's salinity. Downstream farmers in Mexicali, Mexico, were understandably angered when the more saline water damaged their irrigated crops. The leaching of fossil salts from irrigation projects upstream of the Colorado River continues to contribute to the river's salinity.

Fossil salts can also be dissolved when water-storage or water-transmission structures (e.g., unlined canals) are placed over saline sediments. The Lake Mead reservoir behind Hoover Dam in southern Nevada overlies deposits of gypsiferous sediments. Dissolution of this gypsum substantially increases the salinity of the Colorado River during its passage through the reservoir.

13.2.3   Atmospheric salt sources

Appreciable salt can be deposited from the atmosphere. Rain droplets form around tiny condensation nuclei such as salt or dust particles. The total salt concentration of rainfall may be as high as 50 to 200 mg L^{-1} near the seacoast, but rapidly decreases to only a few mg L^{-1} in the continental interior. The exact pattern of the decrease depends on local topography and weather patterns.

The quantities of salt added from the atmosphere to arid and semiarid regions may amount to only a few kilograms per hectare per year, but over periods of tens to thousands of years, the amounts introduced can be substantial. The vegetation of such areas normally is adapted to the natural salt balance and incoming precipitation, so salts tend to accumulate below the surface at the average depth of soil wetting and plant rooting. During periodic high rainfall events, or when human activities change the annual water balance, accumulated atmospheric salts are flushed from the soil at relatively high concentrations.

13.2.4   Topographic influence on soil salt concentrations

In basin regions, soil salinity commonly results in formation of salt pans. Soils in low-lying areas, even in arid regions, may have high water tables. Water from groundwater tables within a meter or so of the surface

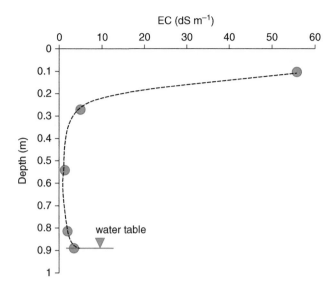

Figure 13.1 Typical salinity profile of a soil exposed to a high-water table. Source: Mohamed and Amer (1972).

can move by capillary fluxes to the soil surface to evaporate, leaving behind dissolved salts. **Figure 13.1** shows an example of salt distribution above a water table 90 cm below the soil surface. The soil salinity concentration in **Figure 13.1** is expressed as electrical conductivity, the common method of measurement (described below).

Large-scale examples of water collection and evaporation are the Great Salt Lake of Utah, a remnant of ancient landlocked Lake Bonneville that once covered much of the western United States; the Caspian Sea in Asia; Lake Chad in Africa; and Lake Ayre in Australia. More widespread examples of salt accumulation basins are the fringes of salt accumulation along arid-region rivers and drainage ways, and small playas. In some fields, low elevation areas cause *slick-spot* patches of sodic (sodium-rich) soil that are enriched in swelling clays and, when Na^+-saturated, limit drainage.

13.2.5   Human sources of soil salinity

Many salt-affected soils result from human activities. Evaporation and transpiration in irrigated soils increases the solution concentration of dissolved salts, which are leached to the extent of the irrigation water-wetting front. This causes salt concentration of the soil to increase with depth. **Figure 13.2** shows examples of salt distributions in soil profiles that were irrigated with different amounts of excess water (water not

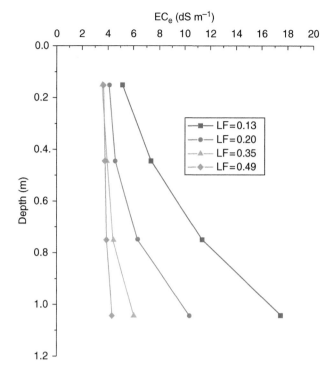

Figure 13.2 Electrical conductivity of soil profiles under different leaching fractions with irrigation water that is 4 dS m^{-1}. The leaching fraction is the fraction of irrigation water that flows through the root zone. Soil is a sandy loam planted with alfalfa. Adapted from Bower et al. (1969), with permission.

lost by evapotranspiration). The excess water is the **leaching fraction** because it leaches the salts from soil. As the leaching fraction increases, salt accumulation moves to deeper depths of the soil. Because water in arid regions is scarce and expensive, minimizing the amount of leach water required to avoid salinization of irrigated soils is important.

Buildup of salts in the soil should be prevented by leaching them out of the soil and transporting the soil leachate to rivers, the ocean, or drained into holding areas or evaporation ponds. However, in many cases, adequate disposal of drainage waters creates additional problems. As drainage waters or irrigation return flows evaporate, high concentrations of salts accumulate. An extreme example is the Salton Sea of southern California, which formed initially after a break in the dikes of the Colorado River during the early 1900s. The Salton Sea has become highly saline during subsequent accumulation and evaporation of irrigation return flows from the nearby Coachella and Imperial Valleys.

Oil-field development, waste-spreading operations, and fertilization can also add sizable quantities of soluble salts to soils. Development of tidal or formerly marine areas can lead to salinity from saltwater intrusion whenever freshwater is insufficient to keep out seawater. The seaward flow of freshwater is decreased by pumping from wells and diversion of streams for irrigation, exacerbating seawater intrusion. One management method used to prevent seawater intrusion is to pump treated municipal wastewater into the vadose zone. The tortuosity of the soil pores slows mixing of the freshwater with saltwater, so the freshwater acts as a *dam* against seawater movement inland.

13.3 Characterization of salinity in soil and water

Soil salinity includes soluble salts in soil water and salt solids in the soil. The soil salinity properties reported from salinity measurement depends on method used. In addition to measuring soil salinity, salinity of irrigation water is an important measure, and there are several methods for measuring or reporting the salinity of water. **Table 13.1** lists the various measurements used to determine soil and irrigation or drainage water salinity.

13.3.1 Total dissolved solids

Early appraisals of the salinity of irrigation waters were in terms of amounts of **total dissolved solids** (TDS). The total dissolved solids are determined by evaporating a known volume of water to dryness and the solids remaining are weighed. The presence of hygroscopic water in the resultant salt mixtures causes TDS values to strongly depend on the drying conditions

and the salt type. Concentrations of salts in most irrigation waters are less than 1000 mg L^{-1} TDS. Over half of the waters used for irrigation in the western United States have TDS values less than 500 mg L^{-1}; less than 10% have values greater than 1500 mg L^{-1}. Groundwater used for irrigation is usually higher in TDS than surface waters. Some groundwaters have been used successfully for irrigation despite TDS values approaching 5000 mg L^{-1}. For comparison, TDS of seawater is about 35 000 mg L^{-1}.

13.3.2 Electrical conductivity

Salinity of a solution is directly related to its **electrical conductivity** (**EC**). Advantages of determining salinity with EC are that ambiguities of TDS measurements are avoided, and EC measurement is quicker. To measure the EC of a solution, it is placed between two electrodes of constant geometry separated by a known distance (**Figure 13.3**). When an electrical potential is imparted to the electrodes, the electrical current varies directly with the total concentration of dissolved salts (ions in solution). The current is inversely proportional to the solution's resistance and can be measured with a resistance bridge. Conductance is the reciprocal of resistance and has units of reciprocal ohms (Ω) or **siemens** (formerly mhos). The EC of the **saturation extract** of the soil measures the soil's salinity, and the EC of irrigation water measures the water's salinity.

The absolute value of the conductance in a solution is a result of the solution's salt concentration and the geometry of the electrode cell. The effects of electrode geometry are embodied in the cell constant, which is related to the distance between electrodes and their

Table 13.1 Measurements of soil and water salinity.

Measurement	Abbreviation	Units	Description
Electrical conductivity	EC	dS m^{-1}	Measures the resistance between two electrodes, which is proportional to aqueous salt content.
Total dissolved solids	TDS	mg L^{-1}	Measures the mass of soluble salts in a solution by evaporation of the solution.
Osmotic potential	OP	kPa	The force of water diffusion across a semi-permeable membrane.
Sodium adsorption ratio	SAR	$mol^{1/2} L^{-1/2}$	Ratio of concentration of Na^+ to the square root of the concentration of Mg^{2+} and Ca^{2+} in solution (irrigation water or saturation extract).
Exchangeable sodium percent	ESP	%	Percent of Na^+ on cation exchange sites.
Exchangeable sodium ratio	ESR	–	Ratio of concentration Na^+ to concentration Mg^{2+} and Ca^{2+} on exchange sites.

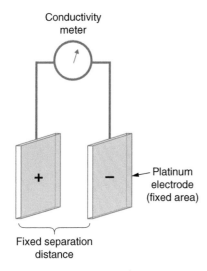

Conductivity
meter

+ − ← Platinum
electrode
(fixed area)

Fixed separation
distance

Figure 13.3 Illustration of electrical conductivity probe. The platinum electrodes need to be a fixed area and separated by a fixed distance so the cell constant can be accurately determined.

cross-sectional area. The cell constant is measured by calibration with KCl solutions of known concentration. The conductivity of KCl solutions is available in published tables. For example, calibration might yield a cell constant of 2.0 cm^{-1}; a test solution measuring 2000 Ω resistance (conductance of $1/2000\,\Omega^{-1}$ or 0.0005 siemens) has a conductivity of

$$EC = 0.0005\,\text{siemens} \times \frac{2.0}{\text{cm}} \times \frac{10\,\text{decisiemens}}{\text{siemen}} \times \frac{100\,\text{cm}}{\text{m}}$$
$$= 1.0\,\text{dS}\,\text{m}^{-1} \tag{13.1}$$

dS m^{-1} is the standard unit used for reporting EC of soil solutions and is equal to the former unit of conductivity mmho cm^{-1}.

EC measurement of soils is done by wetting a fixed mass of soil with solution. A saturated soil paste is common, but other wetting ratios are also used, including 1:1 and 1:5. In many methods, extraction of the solution is recommended. Equations to convert between the methods have been developed.

In-situ EC measurements of soil solution can be done using sensors embedded in porous ceramic, which maintains solution contact with the electrodes under varying soil water content. Groups of electrodes (commonly four) placed across the soil surface to measure the salinity of underlying soils are often used to account for heterogeneity. Mobile electromagnetic

wands carried across the landscape show some promise for more rapidly measuring soil salinity and allowing for mapping of spatial variability. However, the magnetic induction measures the conductivity of the soil solids, air, and liquid together, and soil solution conductivity needs to be extracted from the magnetic induction EC values. Methods for manipulation of the data from magnetic induction are not yet resolved.

Several empirical relationships exist for converting one type of water quality analysis to another. For solutions in the EC range from 0.1 to 5 dS m^{-1}, solution ion composition and TDS conversions are

Sum of cations or anions $\left(\text{mmol}\left(+\text{ or }-\right)\text{L}^{-1}\right)$
$$\approx EC\left(\text{dS}\,\text{m}^{-1}\right) \times 10 \tag{13.2}$$

$$TDS\left(\text{mg}\,\text{L}^{-1}\right) \approx EC\left(\text{dS}\,\text{m}^{-1}\right) \times 640 \tag{13.3}$$

For soil extracts in the EC range from 3 to 30 dS m^{-1}, the **osmotic potential** (OP) is

$$OP\left(\text{bars}\right) \approx EC\left(\text{dS}\,\text{m}^{-1}\right) \times -0.36 \tag{13.4}$$

The osmotic pressure or osmotic potential measures the tendency of water to diffuse across a membrane against a salinity gradient and indicates the effects of salinity on plant growth since plant roots are semipermeable membranes (they allow water, but not salt to enter the roots). Critical values of OP for plants are provided in reference tables.

Example: An irrigation water containing 3 mmol(+) L^{-1} Ca^{2+}, 2 mmol(+) L^{-1} Mg^{2+}, and 3 mmol(+) L^{-1} Na$^+$ has 8 mmol(+) L^{-1} total cations, and has a measured EC of approximately 0.8 dS m^{-1}. Because of the requirement for electrical neutrality, the anion concentration must equal 8 mmol(−) L^{-1}. The 0.8 dS m^{-1} EC corresponds to a TDS value of approximately 510 mg L^{-1}, and an OP of approximately −0.3 bars or −30 kPa.

Like EC, ionic strength (I) is a measure that is related to the sum of ion charges in solution (see Chapter 4). Thus, not surprisingly, there exists an empirical relationship between electrical conductivity and ionic strength:

$$I = 0.013 \times EC\left(\text{dS}\,\text{m}^{-1}\right) \tag{13.5}$$

The relationship is derived from correlation analysis of more than 100 river waters and soil extracts (the correlation coefficient was 0.996, indicating a very strong relationship) (Griffin and Jurinak, 1973). Calculating ionic strength from electrical conductivity is useful when a complete analysis of solution ion concentration is not available because it allows for estimation of activity corrections so that activities of ions in solution can be calculated (Chapter 4).

13.3.3 Sodium hazard

Another important measurement of water quality is its relative amount of sodium (**sodicity**). Irrigation waters with high sodium content tend to produce soils with high exchangeable sodium levels. In 1921, Scofield and Headley concluded from a series of alkali-soil reclamation experiments that

Hard water (Ca²⁺ and Mg²⁺ -rich) makes soft (friable) land, and soft water (Na⁺-rich) makes hard (cloddy, dense) land.

Sodic soils crust badly, swell, and disperse, decreasing a soil's hydraulic conductivity and water permeability. Clay swelling is a result of the increased diffuse double layer associated with Na^+ on the cation exchange sites as compared to divalent cations such as Ca^{2+} and Mg^{2+}.

Figure 13.4 shows the effect of sodium concentration of irrigation water (SAR) and salinity (EC) on infiltration rate of irrigation water. Clay particles disperse and block the soil-water flow channels as the SAR of the solutions increases. Increasing salt concentration counters, to some extent, the effect of increasing exchangeable Na^+, in agreement with Eq. 10.5. Thus, a very salty NaCl solution shrinks the diffuse double layer, causing flocculation, even though the monovalent Na^+ cation promotes an increased diffuse double layer as compared to Ca^{2+} or Mg^{2+}. Decreased permeability interferes with the drainage required for salinity control, and with the water supply and aeration required for plant growth.

Early estimates of sodicity risks from irrigation water were based on the percent sodium in the irrigation water. However, predictions and observations of sodicity were not accurate because percent Na^+ does not linearly correlate with the amount of Na^+ on the soil mineral exchange sites. This is because of the strong preference of most soil particles for divalent cations over monovalent cations. Thus, waters with appreciable concentrations of Ca^{2+} and Mg^{2+} and

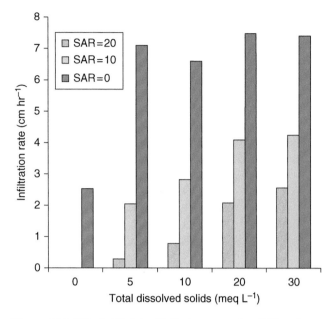

Figure 13.4 Effect of total salinity (expressed as EC) and sodium adsorption ratio (SAR) of irrigation water on infiltration rate. Adapted from Ayers and Westcot (1985).

relatively high Na^+ contents may still produce relatively low exchangeable Na^+ levels in soils.

Cation exchange selectivity equations are used to predict the distribution of Na^+, Ca^{2+}, and Mg^{2+} on the exchange sites. The Gapon exchange equation is the most commonly used cation exchange selectivity equation used to evaluate soil salinity because it is relatively simple and has been shown to be accurate for many salt-affected soils. Other exchange equations, such as the Vanselow, can also be used. As described in Chapter 10, the Gapon formulation of the Na^+–Ca^{2+} exchange reaction is

$$Na^+ + Ca_{1/2}X = NaX + \tfrac{1}{2}Ca^{2+} \tag{13.6}$$

The Gapon exchange constant for this reaction is

$$K_G = \frac{[NaX][Ca^{2+}]^{1/2}}{[Ca_{1/2}X][Na^+]} \tag{13.7}$$

Where brackets on aqueous species indicate concentrations (mol L⁻¹ or mmol L⁻¹), and exchanger phase concentrations are equivalent cation charge (mol(+) kg⁻¹ or mmol(+) kg⁻¹). For convenience, cation units are often listed as mmol(+) rather than mol(+). Rearranging

Eq. 13.7. to separate the aqueous and solid phase components yields

$$\frac{[NaX]}{[Ca_{1/2}X]} = \frac{[Na^+]}{[Ca^{2+}]^{1/2}} K_G \tag{13.8}$$

This shows that the ratio of Na^+ to Ca^{2+} on the exchange sites (left-hand side) is a function of the ratio of the cations in the aqueous solution multiplied by the exchange selectivity coefficient (K_G) (right-hand side). The range of K_G is typically 0.008 to 0.016 mmol$^{-1/2}$ L$^{1/2}$ for alkali soils; a value of 0.015 is commonly used for soils with ESR < 30. If K_G is known, the Gapon equation can be used to predict the distribution of Na^+ and Ca^{2+} between the solid and solution, and therefore predict the effects of irrigation water composition on the soil colloidal properties, i.e., the degree of dispersion or flocculation.

The presence of Mg^{2+} in most alkaline soil solutions and irrigation waters complicates prediction because the ternary Na^+-Ca^{2+}-Mg^{2+} exchange-reaction equilibrium state requires a more complex model than the Gapon equation. To simplify the calculation of Na^+, Ca^{2+}, and Mg^{2+} on the soil particles, the exchange behavior of Ca^{2+} and Mg^{2+} is assumed to be similar; thus, the sum of the two species can be substituted for the divalent aqueous and exchanger phase Ca^{2+} in the denominators of **Eq. 13.8** (i.e., $[Ca^{2+}]' = [Ca^{2+}] + [Mg^{2+}]$ and $[Ca_{1/2}X]' = [Ca_{1/2}X] + [Mg_{1/2}X]$, where the prime indicates the sum of divalent species to be used in the denominator of **Eq. 13.8**). This substitution yields:

$$\underbrace{\frac{[NaX]}{([Ca_{1/2}X] + [Mg_{1/2}X])}}_{\text{Exchanger phase}} = \underbrace{\frac{[Na^+]}{([Ca^{2+}] + [Mg^{2+}])^{1/2}}}_{\text{Solution phase}} \times K_G$$

$$\text{Exchanger phase} = \text{Solution phase} \times K_G \tag{13.9}$$

This equation allows prediction of the relative distribution of Na^+ on soil's exchange sites using the concentrations of Na^+, Mg^{2+}, and Ca^{2+} in solution and K_G as defined above. Recognizing the importance of the Ca^{2+} and Mg^{2+} concentrations of irrigation water on Na^+ adsorption, workers at the US Salinity Laboratory proposed using the solution phase part of **Eq. 13.9** as a relative assessment of the potential of irrigation water

to cause sodicity in a soil. The ratio of the solution composition in **Eq. 13.9** is called the **sodium adsorption ratio (SAR)**:

$$SAR = \frac{[Na^+]}{([Ca^{2+}] + [Mg^{2+}])^{1/2}} \tag{13.10}$$

When the units of charge equivalent per volume (mmol(+) L^{-1}) are used (as is commonly done) instead of molar concentrations, the Ca^{2+} and Mg^{2+} concentrations in **Eq. 13.10** are divided by two. The units of SAR are mmol(+)$^{1/2}$ L$^{-1/2}$, but in practice are not typically listed.

SAR provides a relative index of the *potential* of irrigation water to exchange Na^+ on soil clay exchange sites. High SAR values correspond to high Na^+ concentrations on soil exchange sites, and vice versa. Errors from using total cation concentrations, as opposed to free ion concentrations occur because of ion-pairing (aqueous complexation). SAR-type equations to account for ion pairing have been developed, but corrections are minimal for most aqueous solutions.

13.3.4 Exchangeable sodium percentage

SAR is a prediction of how sodicity of irrigation water will impact the behavior of soils irrigated with the water. The sodium concentration on the exchange sites can be directly measured, and is typically reported as the **exchangeable sodium percentage (ESP)**, which is based on the total Ca^{2+}, Mg^{2+}, and Na^+ cation charge occupation:

$$ESP = \frac{[NaX]}{([Ca_{1/2}X] + [Mg_{1/2}X] + [NaX])} \times 100 \tag{13.11}$$

where solid-phase concentrations are in units of charge per mass (mmol(+) kg^{-1}). In a system where Na^+, Mg^{2+}, and Ca^{2+} are the main cations, the denominator is equal to the cation exchange capacity, and **Eq. 13.11** becomes

$$ESP = \frac{[NaX]}{CEC} \times 100 \tag{13.12}$$

This relationship is useful because it relates the amount of Na^+ on the exchange site to CEC, a commonly reported soil parameter. ESP values below 25–30% are approximately numerically equivalent to the left-hand side of **Eq. 13.9**. Above 30%, the **exchangeable sodium ratio (ESR)**, the left-hand side of **Eq. 13.9**, can be used to predict ESP:

$$ESR = \frac{[NaX]}{([Ca_{1/2}X]+[Mg_{1/2}X])} \qquad (13.13)$$

As shown in **Eq. 13.9**, ESR can be predicted from solution composition (SAR) and an exchange coefficient (K_G). Substitution and simplification of **Eqs. 13.13** into **13.11** yields

$$ESP = \frac{ESR}{1+ESR} \times 100 \qquad (13.14)$$

This relationship allows ESP to be predicted from the SAR of irrigation water or saturated extract solution and an exchange coefficient (substituting ESR with SAR and K_G (**Eq. 13.9**)). For example, assuming a K_G of 0.01, ESP is calculated as follows:

$$ESP = \frac{\dfrac{[Na^+]}{([Ca^{2+}]+[Mg^{2+}])^{1/2}}}{100+\dfrac{[Na^+]}{([Ca^{2+}]+[Mg^{2+}])^{1/2}}} \times 100 \qquad (13.15)$$

Thus, exchangeable sodium percentage can be computed from cation composition of an irrigation water or saturated paste extract.

Empirical relationships between SAR and ESP have also been developed, as well as relationships between the ESP and saturated soil extract SAR values. **Figure 13.5** shows an empirical relationship that researchers from the US Salinity Laboratory developed between ESR and SAR of the saturated paste extract of the soil. The linear line indicates that the SAR value and ESR are directly correlated, and that ESP can be estimated from measurements of concentrations of cations in water (i.e., SAR). The slope of the line is a reasonable estimate of K_G in **Eq. 13.9**.

Management of soil salinity based on irrigation water SAR values is much easier than ESP because the exchangeable cations do not have to be measured. Additionally, salt compositions of irrigation water often vary through the irrigation season, and adjustments to the irrigation water can be made. For example, high SAR irrigation water can be diluted by mixing

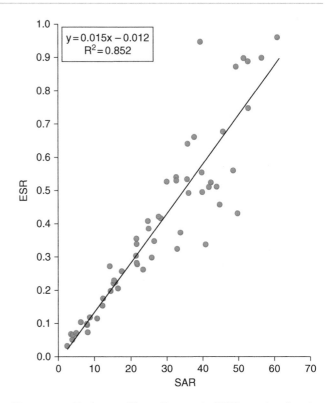

Figure 13.5 Exchangeable sodium ratio (ESR) as related to the soil extract sodium adsorption ratio (SAR). From Richards (1954), with permission.

with lower SAR water to eliminate problems associated with high ESP values. In practice, irrigation water SAR is used in conjunction with EC of irrigation water to manage water infiltration rate. Sodium toxicity problems are usually secondary to problems created from decreases in water infiltration rate. Farmers with soils susceptible to clogging from high ESP typically strive for irrigation water with ESP values less than 15%, and if there is a crop with high sodium sensitivity, ESP values less than 10% are recommended.

Example: An irrigation water will be used to irrigate field of citrus, which has high sensitivity for sodium damage. Thus, the farmer desires to maintain soil ESP as low as possible. The irrigation water contains 2.08 mmol L^{-1} of Ca^{2+}, 0.71 mmol L^{-1} of Mg^{2+}, 5.96 mmol L^{-1} of Na^+. Assuming a K_G of 0.015 and that the soil pore water will have the same composition as the irrigation water, the ESP can be calculated from the SAR and ESR:

$$SAR = \frac{[Na^+]}{([Ca^{2+}]+[Mg^{2+}])^{\frac{1}{2}}} = \frac{5.96}{(2.08+0.71)^{\frac{1}{2}}}$$
$$= 3.6\,mmol(+)^{1/2}\,L^{-1/2} \qquad (13.16)$$

$$ESR = SAR \times K_G = 3.6 \times 0.015 = 0.054 \qquad (13.17)$$

$$ESP = \frac{ESR}{1+ESR} \times 100 = \frac{0.054}{1+0.054} \times 100 = 5.1\% \qquad (13.18)$$

Thus, the ESP predicted from the irrigation water is relatively low and is suitable for irrigating the citrus crop. However, in most cases, the assumption that the soil pore water ion composition is the same as the irrigation water is not accurate. In the soil, evapotranspiration and addition and subtractions of ions to the soil solution by plants and mineral dissolution or precipitation will change its cation composition. If enough irrigation water is used, eventually the soil pore water may be close to the irrigation water's ion composition, but this would be an inefficient use of irrigation water. Empirical and mechanistic models to relate soil solution SAR, irrigation water SAR, and ESP have been developed that allow for more accurate prediction of ESP. Many of these are based on a saturated soil extract ion composition (e.g., **Figure 13.5**).

13.3.5 Bicarbonate hazard

Another property related to the sodium hazard of irrigation waters is the bicarbonate concentration. Bicarbonate toxicities associated with some waters generally arise from deficiencies of iron or other micronutrients caused by the high pH. Precipitation of calcium carbonate from alkaline waters removes bicarbonate:

$$Ca^{2+} + 2HCO_3^- = CaCO_3 + H_2O + CO_2 \qquad (13.19)$$

which lowers the concentration of dissolved Ca^{2+}, increases the SAR, and increases the exchangeable-sodium level (ESP) of the soil. $CaCO_3$ precipitation can be accounted for by using an *adjusted* SAR calculated from the bicarbonate concentration present in the irrigation water, and assuming the solution is saturated with respect to calcite.

Early workers used the residual sodium carbonate (RSC) to predict the tendency of calcium carbonate to precipitate from high-bicarbonate waters, and thus create a sodium hazard. The RSC was defined as

$$RSC = \left(\left[HCO_3^- \right] + \left[CO_3^{2-} \right] \right) - \left(\left[Ca^{2+} \right] + \left[Mg^{2+} \right] \right)$$
$$(13.20)$$

where all values have units of moles charge per liter. The concept of an adjusted SAR has found widespread applicability. Waters of RSC greater than 2.5 are considered hazardous under all conditions. RSC values between 1.25 and 2.50 are considered potentially hazardous, and waters with RSC values less than 1.25 are considered safe. These predictions work reasonably well.

A disadvantage of the RSC is that it treats all bicarbonate in the water as if it would precipitate. The amount of bicarbonate that actually precipitates depends on the degree to which salts are concentrated by evapotranspiration in the plant-root zone, and if the solution exceeds the saturation index of calcite. As an extreme example, if no water evaporated from the soil, and the solution remains at equilibrium or undersaturated with respect to calcite, bicarbonate will pass through the soil and will not precipitate. Conversely, if all the soil water evaporated through transpiration, all of the bicarbonate would precipitate, some as calcite and some as other bicarbonate salts. Hence, the quantity of bicarbonate precipitating depends on the proportion of water percolating through the soil, or the leaching fraction, and the aqueous ion composition. Modern models allow mineral and salt precipitation reactions to be included in the leaching fraction predictions (see, e.g., **Figure 13.6**), and thus more accurately predicts sodicity risks.

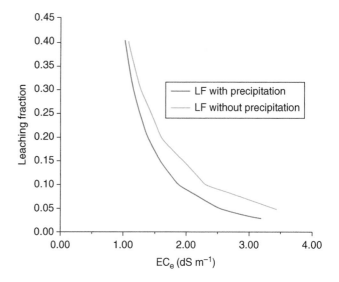

Figure 13.6 Simulation of required leaching fractions as a function of soil salinity using a leaching model that allows precipitation of salts compared to a model that does not allow salt precipitation. The salt precipitation model predicts a lower leaching fraction requirement at a given soil salinity. Source: Corwin et al. (2007). Reproduced with permission of Elsevier.

13.3.6 Other problematic solutes in irrigation water

In addition to sodium and bicarbonate, irrigation waters also contain the potentially toxic ions boron, lithium, and chloride. The boron concentrations of irrigation waters are particularly important because many crops are susceptible to even extremely low concentrations of this element. The differences between deficient and toxic boron concentrations are small. Sodium and chloride ions are hazardous to fruit and berry crops, and to woody plants. Their ranges of hazardous concentration are considerably higher than for boron. In addition to absorption through roots, toxic ions also can be absorbed through foliage. Sprinkling water high in sodium or chloride on the leaves of horticultural plants and vegetables, fruits, and berry crops causes plant damage as the water evaporates. Lithium ions (Li^+) have also been reported as a potential problem in some regions; however, the management that controls boron, sodium, and chloride also prevents lithium toxicity.

13.4 Describing salt-affected soils

The sodium status of soils is described by the soil's exchangeable sodium percentage (ESP, **Eq. 13.11**). Measuring ESP, however, is tedious and subject to error. The following are some of the steps and problems associated with measuring ESP:

1 To measure ESP, the exchanger phase ions must be exchanged using a high concentration of an index cation, such as NH_4^+.
2 The concentration of *soluble* sodium salts removed during the exchange-extraction must be subtracted from the total quantity of sodium extracted from the exchanger phase to obtain the exchangeable sodium.
3 The CEC is measured by removing the index cation (i.e., NH_4^+) from the initial exchange step using an additional exchange index cation (e.g., K^+).
4 Incomplete removal of the index salt solution during the wash step of CEC determinations can lead to high CEC values, and therefore low ESP estimates.
5 Hydrolysis during removal of the index salt solution, trapping of NH_4^+ from the index solution between soil mineral lattices, and calcium carbonate

or gypsum dissolution in the index or replacement solutions can all lead to low CEC values, and hence to high ESP estimates.

Because of added cost and potential errors in soil ESP determinations, and because of the generally good relationship between SAR of the soil solution and ESP of the soil, the SAR of the extract from a saturated soil is normally a satisfactory index to the exchangeable-sodium status of salt-affected soils. Since the saturation extract is already required to determine EC, measuring the SAR requires only a few additional chemical determinations be made on this extract.

The traditional classification of salt-affected soils uses soluble salt concentrations or electrical conductivities of extracted soil solutions, and the ESP of the soil. The EC dividing line for most plants between saline and nonsaline soils was established at 4 dS m^{-1} for water extracts from saturated soil pastes. Salt-sensitive plants, however, can be affected in soil with saturation extract ECs of 2 to 4 dS m^{-1}. The relationships of soil salinity parameters to EC, SAR, and ESP are shown in **Figure 13.7**.

13.4.1 Saline soils

Saline soils (white alkali) are those in which plant growth is reduced by excess soluble salts. Salinity in these soils can be by leaching the excess salts from the plant root zone. The pH of saline soils generally is less than 8.5, and they are normally well flocculated (i.e., as permeable as might be expected from soil texture alone). Plants growing on such soils may appear stunted and have thickened leaves and a dark green color. Substantial reductions in plant growth can occur without appreciable changes in plant appearance.

13.4.2 Saline-sodic soils

Soils containing both high soluble-salt and high exchangeable-sodium levels are called **saline-sodic soils**. Such soils also reduce plant growth because of their high soluble-salt content. Because the soluble salts prevent hydrolysis, the pH of saline-sodic soils is typically less than 8.5. The main hazard occurs when these soils are leached to remove salts. Leaching removes the salts faster than it removes exchangeable

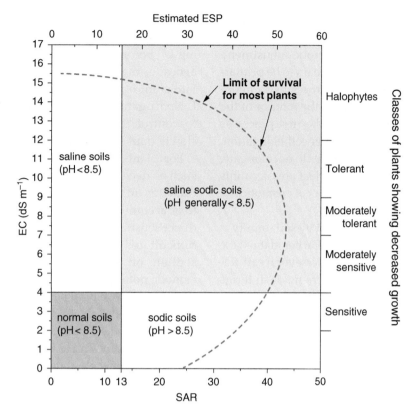

Figure 13.7 Classification of soil salinity and sodicity in relation to electrical conductivity, SAR, ESP, and pH. The curve shows the limit for survival of most plants. Range of EC for plants is also shown. The relation between soil extract SAR and ESP is derived from an empirical relationship, similar to the ESR-SAR relationship shown in **Figure 13.5** (see **Eq. 13.17**). Adapted from Brady and Weil (2002), with permission.

Na$^+$, causing conversion to sodic soils. This can severely reduce soil permeability or hydraulic conductivity and affect plant-water relations and the ability to leach for salinity control.

13.4.3 Sodic soils

Sodic soils are a particularly difficult management problem. The permeability of these soils to water is very slow. The pH of sodic soils is commonly greater than 8.5, and the clay and organic fractions are dispersed. Dispersed organic matter accumulates at the surface of poorly drained areas as water evaporates and imparts a black color to the surface, hence the historic identification as *black alkali soils*.

Sodic soils occur in many parts of the western United States. In some locations, they occur in small patches, *slick spots,* less than 0.5 ha in extent. Such patches occupy slight depressions, which become accentuated

as surface soil particles disperse and are blown away by wind erosion.

The percolation of insufficient water to satisfy plants and control salinity is the main problem associated with sodic soils. In addition, the relatively high sodium salt concentrations can result in direct sodium toxicities to sensitive plants.

13.5 Effects of salts on soils and plants

The main effect of soluble salts on plants is osmotic. Plants must expend large amounts of energy to absorb water from salty soil solution solutions – energy that would otherwise be used for plant growth and crop yield. Plant roots are semipermeable membranes permitting water to pass but rejecting most of the salt. Thus, water is osmotically more difficult to extract from increasingly saline solutions. Plants growing on saline media can increase their internal osmotic concentrations by producing organic acids or by absorbing salts. This process is called osmotic adjustment. The effect of salinity on the plant appears to be primarily energy diversion from growth processes to maintain the osmotic differential between the interior of the plant and the soil solution. One of the first processes from which this energy is diverted is cell elongation. Leaf tissue cells continue to divide but do not elongate. The occurrence of more cells per unit leaf area accounts for the typically dark green color of osmotically stressed plants.

The relative plant health as affected by soil salinity is referred to as plant salt tolerance. Earlier data were summarized by separating plants into several salt-tolerant groups. Subsequent listings were made in terms of relative plant growth at various salinity levels (EC) of the soil's saturation extract. Some recent listings of plant's salt tolerance list the EC value that initial yield decline is observed. Salt-tolerance values are also provided in terms of percent yield decrease per unit increase in salinity beyond this threshold. Most yield data were obtained from uniformly salinized field plots having nearly constant salinity with depth. Actual distributions under field conditions, however, more closely resemble those in **Figure 13.2**, and the plant may extract most of its water from the least-salinized portion of the profile.

Some plants are particularly sensitive to salinity during the germination or seedling stages when a restricted root zone makes the plant extremely vulnerable to osmotic stress. Alternate-furrow irrigation (where only one side of the crop row is irrigated at any one time) can be used to flush salts past the young seedling if single rows are used. If double-row beds are used, alternate-furrow irrigation can flush salts to the vicinity of the bed edge opposite the irrigated furrow and hence stress the seedlings near this edge. Drip irrigation generally flushes salts to the periphery of the wetted soil volume, but can also lead to serious salinity problems when high rates of fertilizer are added through the drip lines, upon replanting, or whenever rainfall flushes accumulated salts toward previously nonstressed plant roots.

In addition to the general osmotic effects, many plants are sensitive to specific ions in irrigation waters or soil solutions. Boron toxicity is probably the most common. For example, sugar beet, alfalfa, and lettuce are tolerant of relatively high levels of boron (2–4 mg L^{-1} boron in irrigation water). In contrast, orange, avocado, many stone fruits, and many nut trees are intolerant of boron concentrations in excess of 0.3 mg L^{-1} in irrigation water. Boron is generally more difficult to control than is salinity because it leaches more slowly than soluble salts.

For plants that are extremely sensitive to sodium, such as deciduous fruits, citrus, nuts, beans, and avocados, as little as 5% exchangeable sodium may lead to toxic accumulations of sodium in leaf tissues. However, direct sensitivity to exchangeable or soluble sodium is difficult to differentiate from the detrimental effects of sodium on soil properties (i.e., permeability and osmotic potential).

Chloride toxicity in some plants is also problematic in salt-affected soils. Excessive accumulations of chloride in tissues near plant tips, the end of the plant transpiration stream, lead to necrosis, death of leaf tips and margins, and eventual death of the plant. Some plants can screen ions such as chloride through their root membranes. In addition, different rootstocks may possess varying abilities to exclude sodium or chloride from above-ground parts. Some grape rootstocks exhibit up to 30-fold differences in their abilities to exclude chloride ions. Selection of a rootstock that is able to screen ions may prevent toxic accumulations in plant tops.

A third mechanism for salt injury to plants is nutritional imbalances. An example is the bicarbonate toxicities reported for some saline environments. The toxicity results primarily from reduced iron availability at the high pH common in high-bicarbonate soils, rather than from the bicarbonate ions themselves. The nutritional needs of plants may also vary with the types of salts present. For example, high sodium levels can lead to calcium and magnesium deficiencies. The high pH levels of sodic soils can accentuate deficiencies of many of the microelements. High soil pH levels also might lead to high concentrations of soluble aluminum, such as the aluminate $(Al(OH)_4^-)$ species (see Table 4.11). Salt tolerance also varies with soil fertility; especially when inadequate fertility limits yields. Nutritional effects of salinity on plants are poorly understood at present. Many of the supposed consequences are still largely speculative.

13.6   Salt balance and leaching requirement

Management of salt-affected soils once centered around maintaining the salt balance. This concept dictated that the quantity of salt leaving an area be equal to, or greater than, the quantity of salt entering the area. The concern was justified by the difficulty in maintaining long-term agriculture for many irrigated areas of the world, such as the Tigris and Euphrates valleys of Iraq, where farming has taken place for several millennia. Some irrigation projects, however, appear able to operate indefinitely at a negative salt balance (more salt entering than leaving) with few adverse effects on soils or plants. The key in such cases is the amount of salt precipitating in the soil that is inactivated with respect to plant availability. Normal plant growth can continue, provided that the quantities of salt precipitated do not lead to sodic soil conditions or to nutritional imbalances. If the salt can be kept below the rooting zone, detrimental salinity effects on plants can be avoided.

The most common approach to salinity management is to maintain a prescribed leaching fraction (LF), defined as

$$LF = \frac{\text{Volume of water leached below the root zone}}{\text{Volume of water applied}}$$
$$= \frac{D_{dw}}{D_{iw}} = \frac{EC_{iw}}{EC_{dw}} \tag{13.21}$$

where EC_{dw} and EC_{iw} are the electrical conductivities (salt concentrations) of the **drainage and irrigation waters**, and D_{iw} and D_{dw} are the amounts (volume) of irrigation and drainage water. The relationship is based on the assumptions that a salt balance exists (i.e., that $EC_{iw}D_{iw} = EC_{dw}D_{dw}$) and that the plant is a perfect semipermeable membrane removing only water from the soil solution and leaving all salts behind. The relationship is inaccurate when substantial salt precipitates in the plant-root zone, dissolves from soil minerals, or is taken up by the crop. Despite these constraints, leaching requirement calculations are sufficiently accurate for most crop management purposes.

In many arid regions, researchers and farmers have demonstrated that careful management of extremely low leaching requirement values can still maintain adequate salinity control. One source of error in estimating the leaching requirement is substituting EC of the saturation extract for EC_{dw} in the leaching requirement formula. Most salt-tolerance data refer to EC_{dw} values, while many measurements of soil salinity levels use EC of a saturated paste. Saturation paste solutions are commonly two to three times more dilute than in the water of a freely draining soil profile (field capacity). Hence, using saturation extract data in place of drainage water data gives an EC in the denominator of **Eq. 13.21** that is a half to a third too small, and a leaching requirement estimate that is two or three times too high.

Calculating the leaching requirement requires an estimate of the *desired* EC of the saturation extract. These values are obtained from existing salt-tolerance data tables. Soil texture, porosity, or some other soil parameter must then be used to convert the EC value to an estimated EC of the drainage water (EC_{dw}) for the soil-plant system. This value, and the EC of the irrigation water (EC_{iw}), can be used to estimate the fraction of leaching water that must pass through the plant root zone for salinity control. The excess water can then be compared to the soil infiltration rate, to plant tolerance at waterlogged conditions, and to drainage system capacity to see if salinity control is feasible under the chosen set of crop, soil, and water management conditions.

The high leaching requirements recommended in the past do not consider nonuniform water distribution of many irrigation systems, nor spatial variability of soil permeability. Excess water leaches through the more permeable parts of the field, while salts can remain in less permeable areas. Large quantities of water are therefore necessary to remove salts from the entire field. More uniform water application is possible with sprinkler irrigation or careful leveling of irrigated fields. More accurate estimates of the minimum leaching requirement can then be used. Modern salinity and leaching models include transient processes such as irrigation, transient water flow, plant growth, CO_2 (g) partial pressures, and chemical reactions to predict the leaching requirement. These models predict that leaching requirements to meet optimum plant productivity are about two-thirds the leaching requirements that are predicted based on steady-state (salt balance) leaching requirement predictions. The reduced leaching fraction is primarily a result of the salt precipitation reactions (**Figure 13.6**). In regions where water is scarce and drainage limited, reducing the leaching fraction is highly desirable.

13.7 Reclamation

The aims of reclamation are to make Ca^{2+} the major exchangeable ion and to reduce the salt concentration in the soil solution. The main requirement to reclaim salt-affected soils is that enough water pass through the plant-root zone to lower the salt concentration to acceptable values. Passing one meter of leaching water per meter of soil depth under ponded conditions normally removes approximately 80% of the soluble salt from soils (**Figure 13.8**). If leaching occurs under unsaturated conditions, such as with the use of intermittent ponding or sprinkler irrigation, this quantity of water may be reduced to as little as 350–200 mm of water because much of the water flow through large pores and cracks is avoided. Boron removal can require up to three times more water than removal of sodium and chloride salts because borate is retained to some extent by soils.

Several techniques have been developed for reclaiming salt-affected soils. Ponding is a traditional method involving the construction of a large dike around a field. A substantial depth of water (commonly 0.3 m or more) is then maintained inside the dike to leach salts from the soil. Such an approach requires drainage facilities capable of removing large quantities of drainage water. The reclamation process is relatively inefficient, because much of the water passes through large soil pores that have already been purged of salts, and

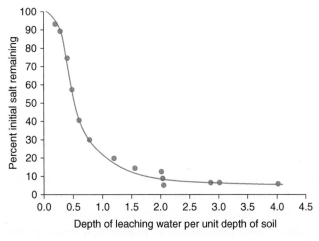

Figure 13.8 Depth of water per unit depth of soil required to leach a highly saline soil. Source: Reeve et al. (1955).

because the ponded water is susceptible to evaporation losses. Salt is removed only slowly from the fine pores of the adjacent soil mass.

A more efficient leaching technique is the basin-furrow method. The soil is nearly leveled, and irrigation water is allowed to meander back and forth across the field through adjacent sets of furrows. The water may take as long as a week to meander across the entire field under such conditions, but the quantities of water required are less than for ponded leaching. Another advantage is that this technique does not produce sterile strips corresponding to former dike positions, where large quantities of salt accumulate during ponded leaching.

Soluble divalent ions (generally Ca^{2+}) must be present during the reclamation of sodic soils. A common amendment for such purposes is gypsum ($CaSO_4 \cdot 2H_2O$), added at rates of up to several mega grams per hectare to provide Ca^{2+} as water percolates through the soil. Tests of exchangeable-soil sodium (ESP) levels should be made every two to three years to estimate the need for reapplication of gypsum.

Another Ca^{2+} source is the pedogenic $CaCO_3$ that occurs in many salt-affected soils. If a soil is only slightly sodic, and rather sandy in texture, tillage to bring subsoil lime to the surface before water application may be sufficient to maintain soil permeability during reclamation. Deep plowing to the 0.7- to 0.9-m depth has also proved helpful in opening the soil to maintain adequate water permeability during reclamation. In most instances, however, lime is not sufficiently soluble to serve as an amendment for sodic-soil reclamation. It can be used as a source of soluble calcium only if an acidifying amendment is applied to dissolve the lime before reclamation begins.

Common acidifying amendments for the reclamation of calcareous sodic soils are sulfuric acid and elemental sulfur. Sulfur must be oxidized to sulfuric acid by soil microorganisms before it becomes effective. A lead-time of several weeks or months is required for microbial oxidation before leaching begins. The reaction is

$$2S + 3O_2 + 2H_2O = 2H_2SO_4 \tag{13.22}$$

$$CaCO_3 + H_2SO_4 = Ca^{2+} + SO_4^{2-} + H_2O + CO_2 \tag{13.23}$$

$$2NaX + Ca^{2+} = CaX_2 + 2Na^+ \tag{13.24}$$

Gypsum produced by the acid reacting with soil lime may occur, but since gypsum is fairly soluble, it behaves like added Ca^{2+} and SO_4^{2-} in the reclamation process.

Another reclamation procedure for sodic soils is the high-salt water reclamation method. Saline-sodic soils will remain permeable as long as soil solution salt concentrations are high enough to flocculate the soil (see Figure 10.11). The soil is leached with successively more dilute water while trying to increase exchangeable Ca^{2+} by displacing exchangeable Na^+. Each increment of dilution is small enough to prevent the soil from swelling or dispersing. The initial step of adding high-salt water may increase either the soluble salt concentration or the exchangeable-sodium percentage of the soil. For example, treatments of soil with seawater, which has a salt concentration of 600 mmol(+) L^{-1} and an SAR of approximately 60, generally produces soils with exchangeable-sodium percentages of 40 to 50. A fourfold dilution of the seawater, by mixing one part with three parts of freshwater, produces a salt concentration of 150 mmol L^{-1} and an SAR of 30 (because the SAR changes as the square root upon dilution of salt concentration). A second fourfold dilution step produces a salt concentration of 37.5 mmol L^{-1}, an EC of 3.8 dS m^{-1}, and an SAR of 15. A third fourfold dilution produces an EC of 0.9 dS m^{-1} and an SAR of 7.5. This three-step leaching process would reclaim the soil if drainage facilities were adequate to remove the resultant large quantities of salty drainage water. The high-salt water reclamation method also depends on access to high-salt water supply. Saline ground and surface waters are common in salt-affected areas, and this is usually not a limitation. The ideal environment for high-salt water reclamation is near an ocean or saline lake, where both water access and disposal are available.

13.8 Summary of important concepts in soil salinity

Soil salinity is one of the major challenges to keeping soil productive so that adequate food can be produced to feed the growing population. Key factors to managing soil salinity are accurate measurement of soil salinity (EC, ESP, and ESR) and managing irrigation waters (SAR) to keep sodium and salinity concentrations in

soils in check. Understanding soil salinity and managing it requires accurate models. The SAR model uses the Gapon cation exchange model to estimate exchanger phase sodium percentage from solution ion composition. More advanced models that include active transport of soil salt solutions through the soil profile and chemical reactions, including cation exchange and precipitation and dissolution, can be developed. Such models may provide more accurate predictions and models to base soil salinity management, but require a high level of user input.

Adequate drainage is essential to managing irrigated agriculture sustainably, whether the concern is routine management (LF concept) or reclamation of soil salinity buildup. Failure to provide adequate drainage means salt will accumulate and soil salinity will be a problem. History is replete with examples where civilizations based on irrigated agriculture failed because adequate drainage was not provided.

Questions

1 An arid area receives 150 mm of rainfall annually with an average salt concentration of 10 mg L^{-1}. If the surface soil from this area contains an average of 30% water at saturation, how many years would be required for sufficient salt to be added from the atmosphere to increase the EC of the saturation extract by 1 dS m^{-1}?

2 An irrigation water contains 750 mg L^{-1} soluble salts. If used at an average leaching fraction of 0.15, what would be the average EC of the drainage water leaving the bottom of the crop root zone?

3 What is the EC (in dS m^{-1}) of a solution having an electrical resistance of 1500 in a conductivity cell with a cell constant of 5.0 cm^{-1}?

4 An irrigation water has an EC of 0.8 dS m^{-1} and a sodium concentration of 35 mg L^{-1}. Calculate:
   a Its osmotic potential.
   b Its SAR (*Hint:* Assume all cations in solution are sodium, calcium, and magnesium).
   c The equilibrium ESP for soils having a Gapon exchange constant of 0.015 (L mmol^{-1})$^{1/2}$.

5 An irrigation water contains 600 mg L^{-1} TDS and has an SAR of 6. If it is applied to a soil containing 40% water at saturation and 20% water at field capacity, what will be the EC and SAR of the saturation extract

of surface soil and the resultant salinity classification category after prolonged irrigation with this water? If the same water-holding characteristics are found at the bottom of the crop-root zone, what will be the EC and SAR of the saturation extract for soil from this portion of the profile after prolonged irrigation at a leaching fraction of 20%?

6 A 30-cm depth of surface soil contains 28% exchangeable sodium and has a CEC of 150 mmol(+) kg^{-1}. How many tonnes per hectare of sulfur will be required to lower its ESP to 5%?

7 Explain how overgrazing, conversion from shrubs to grasses, burning, and summer fallowing can lead to saline seeps.

8 Describe the use and limitations of EC for determining saline or sodicity status of soils.

9 A soil is equilibrated with a solution of SAR = 20. Using the Gapon equation, what would be its equilibrium exchangeable sodium percentage (ESP)? If the soil had instead been equilibrated with the same solution diluted fivefold with salt-free water, what would have been the corresponding SAR and ESP values?

10 What are the biological aspects of soil salinity that are detrimental for plants?

11 In natural ecosystems, soil salinity limitations for native plants are uncommon. Why?

12 In arid region soils, boron is often observed to be elevated and inhibits the soil's potential for crop growth. What are the soil biogeochemical processes that affect boron reactions in arid region soils?

Bibliography

Richards, L.A., ed. 1954. Diagnosis and Improvement of Saline and Alkali Soils. USDA, United States Salinity Laboratory, US Government Printing Office, Washington, DC.

Ayers, R.S., and D.W. Westcot. 1985. Water quality for agriculture: Irrigation and water quality paper. FAO, Rome 29 Rev. 1.

Bower, C.A., G. Ogata, and J.M. Tucker. 1969. Rootzone salt profiles and alfalfa growth as influenced by irrigation water salinity and leaching fraction. Agronomy Journal 61:783–785.

Brady, N.C., and R.R. Weil. 2002. The Nature and Properties of Soil. Prenctice Hall, Upper Saddle River, NJ.

Corwin, D.L., J.D. Rhoades, and J. Simunek. 2007. Leaching requirement for soil salinity control: Steady-state versus transient models. Agricultural Water Management 90:165–180.

Griffin, R.A., and J.J. Jurinak. 1973. Estimation of activity-coefficients from electrical conductivity of natural aquatic systems and soil extracts. Soil Science 116:26–30.

Mohamed, N.A., and F. Amer. 1972. Sodium carbonate formation in Ferhash area and possibility of biological dealkalization. Proc. Internat. Symp. on New Developments in the Field of Salt Affected Soils. 4–9 December 1972. Ministry of Agriculture, Cairo. p. 346.

Reeve, R., A. Pillsbury, and L. Wilcox. 1955. Reclamation of a saline and high boron soil in the Coachella Valley of California. Hilgardia 24:69–91.

Scofield, C.S., and F.B. Headley. 1921. Quality of Irrigation Water in Relation to Land Reclamation. US Government Printing Office, Washington, DC.

Westcot, D.W. 1976. Water Quality for Agriculture. Food and Agriculture Organization of the UN, Rome.

INDEX

Soil Chemistry, Fifth Edition. Daniel G. Strawn, Hinrich L. Bohn, and George A. O'Connor.
© 2020 John Wiley & Sons Ltd. Published 2020 by John Wiley & Sons Ltd.

Wiley The manufacturer's authorized representative according to the EU
General Product Safety Regulation is Wiley-VCH GmbH, Boschstr. 12,
69469 Weinheim, Germany, e-mail: Product_Safety@wiley.com.

Printed and bound by CPI Group (UK) Ltd, Croydon, CR0 4YY

21/03/2025

01835483-0003